Phase Transformation in Metals

Nestor Perez

Phase Transformation in Metals

Mathematics, Theory and Practice

 Springer

Nestor Perez
University of Puerto Rico
Mayaguez, Puerto Rico

ISBN 978-3-030-49170-3   ISBN 978-3-030-49168-0   (eBook)
https://doi.org/10.1007/978-3-030-49168-0

This Springer imprint is published by the registered company Springer Nature Switzerland AG
The registered company address is: Gewerbestrasse 11, 6330 Cham, Switzerland

My wife Neida
My daughters Jennifer and Roxie
My son Christopher

Foreword

It is indeed a privilege for me to introduce to engineering students, instructors, and practitioners Dr. Nestor Perez's third book: Mathematics of Phase Transformation. His first two books, Fracture Mechanics and Electrochemistry and Corrosion Science, published by Springer, were so successful that very quickly a second edition came up for each of the books, and a translation in Mandarin was also published in rapid succession. This time Dr. Perez wrote in the same line of thought as in the first two books, thinking about the academia as well as the industries. His mathematics, as the title of this book indicates, is very straightforward and to the point. The book is meant for the advanced undergraduate and the graduate level courses in all engineering disciplines that deal with both theory and practice of industrial materials. Dr. Perez has treated each chapter with careful caution such that its mathematics could help to understand and explain the physical phenomena, and would not "govern" them. There is no governing equation but simply mathematical analyses a priori in order to predict some experimental results or a posteriori for obtaining some empirical relationship out of the experimental results. *Apples would still fall even if Newton were not born!*

I should highly recommend this book not only for the students, instructors, and the engineering practitioners in metalworking industries, such as steel plants, materials forming as well as materials removal type manufacturing industries, but also for the consulting engineers in metallurgy and in the allied fields. Besides, it is a very good reference material in the libraries of the universities, technical colleges, and industrial firms. Furthermore, it is handy to keep a copy in the shelf of any retired engineer, for casual consulting and for hobby or pastime in order to keep in touch with lifelong learning.

Department of Mechanical Engineering   Jay (Jayanta) Banerjee
University of Puerto Rico at Mayaguez (UPRM)
Mayaguez, Puerto Rico
2020

Preface

The nature of this textbook is based on many topics such as crystallography, mass transport by diffusion, thermodynamics, heat transfer and related temperature gradients, thermal deformation, and even fracture mechanics in order to understand the physics of phase transformation and associated constraints from a metallurgical or materials science point of view, hence an interdisciplinary textbook for senior and graduate students and professionals in the metal-making and surface reconstruction industries.

The work presented in this textbook is mostly devoted to solidification and related analytical models based on heat transfer because the most fundamental physical event is the continuous evolution of latent heat of fusion for directional or nondirectional liquid-to-solid phase transformation at a specific interface with certain geometrical shape: planar or curved front.

The purpose of this book is to introduce mathematical and engineering approximation schemes for describing the phase transformation, mainly during solidification of pure metals and alloys. The text in each chapter is easy to follow by giving clear definitions and explanations of theoretical concepts and full detail of the derivation of formulae. Mathematics is kept simple so that the student does not have an obstacle for understanding the physical meaning of electrochemical processes, as related to the complex subject of phase transformation. Hence, understanding and learning the phase transformation behavior can be achieved when the principles or theoretical background is succinctly described with the aid of pictures, figures, graphs, and schematic models, followed by derivation of equations to quantify relevant parameters. Eventually, the reader's learning process may be enhanced by deriving mathematical models from principles of physical events followed by concrete examples containing clear concepts and ideas.

Example problems are included to illustrate the ease application of thermodynamics and heat transfer concepts for solving complex solidification problems in an easy and succinct manner. Thus, the reader is exposed to a compilation of mathematical models that are useful in finding solutions to solidification problems. In particular, solidification shrinkage and gas porosity are described as the main casting defects. Subsequently, cracks around a pore are analyzed using linear elastic fracture mechanics (LEFM).

The book has been written to suit the needs of Metallurgical Engineering, Materials Science and Mechanical Engineering senior/graduate students, and professional engineers for understanding phase transformations from the liquid and solid states. Phase transformation from the liquid state is the main focus of the book. However, the phase transformation from the solid state is related to heat treatment of casting components for homogenizing a microstructure or precipitating a secondary phase. Below is a brief description of each chapter included in this book.

Chapter 1: Characterization of Crystalline Lattice. This chapter describes the methodology for characterizing the crystal structure of solid surfaces using X-ray diffraction and vector algebra in orthogonal and non-orthogonal coordinate systems.

Chapter 2: Surface Reconstruction. This chapter deals with the deposition, mainly by adsorption, of monolayers on substrate surfaces and related crystallography of resultant crystal planes.

Chapter 3: Mass Transport by Diffusion. This chapter aims at the motion of a chemical species in a fluid or solid mixture caused by the random walk phenomenon. The range of application of diffusion is quite vast and it is substantially a part of the fundamentals of liquid-to-solid (solidification) and solid-state (heat treatment) phase transformation.

Chapter 4: Fundamentals of Solidification. This chapter includes fundamentals of solidification for producing metal castings and aims at the heat transfer related to solidification in green sand mold. Foundry technology is briefly discussed.

Chapter 5: Planar Metal Solidification. This chapter deals with the theoretical background on phase-change heat transfer related to melting and solidification of pure metals at a fixed melting or fusion temperature. Solidification is characterized using the concept of half-space in a semi-infinite region and it is described by a group of analytical solutions to heat equations for one-dimensional melting and solidification problems related to a planar solidification front.

Chapter 6: Contour Metal Solidification. This chapter uses the concept of half-space in a semi-infinite region, where there exists a contoured solidification front. It also treats a group of analytical solutions to heat equations for one-dimensional solidification problems in cylindrical coordinates with moving liquid–solid interfaces.

Chapter 7: Single-Phase Alloy Solidification. This chapter is devoted to solidification heat transfer in substitutional and interstitial binary alloys of solute composition C_o. It also characterizes the coupled heat and mass transport process at the solidified melt boundary using analytical solutions to one-dimensional heat equations, which admit similarity solutions.

Chapter 8: Two-Phase Alloy Solidification. This chapter describes the solidification of binary $A-B$ alloys of nominal solute composition C_o binary phase diagrams, where solidification occurs at freezing temperature range $T_s \leq T_f^* \leq T_l$. Also included in this chapter is the morphology of the solid phases, the mass fractions of the phases, and the degree of any undercooling ΔT.

Chapter 9: Solid-State Phase Transformation. This chapter introduces a compilation of analytical solutions to one-dimensional phase transformation problems in the solid state with grain boundary migration. Moreover, it also uses theoretical background on thermally induced and mechanically induced solid-state phase transformation.

Chapter 10: Solidification Defects. This chapter aims at the theory of solidification and cooling shrinkage problems during casting. It also includes a compilation of analytical solutions available in the literature for producing defect-free and dimensional fit metal castings.

Mayaguez, Puerto Rico   Nestor Perez
2020

Contents

Chapter 1
Crystallography

1.1 Introduction

This chapter describes crystallography as an experimental science and includes a methodology for characterizing the crystal structure of solid surfaces using X-ray diffraction and vector algebra in orthogonal ($\alpha = \beta = \gamma = 90°$), and non-orthogonal coordinate systems.

These topics are briefly introduced henceforth in order to show the procedure for revealing the initial conditions of solids or substrate surfaces prior to any experiment on solid-state phase transformation or the final structure due to solidification, which is a liquid-to-solid ($L \rightarrow S$) phase transformation. Hence, knowledge of X-ray diffraction and reciprocal lattice defined by a scattering vector g_{hkl} are essential to complement the analysis of crystallography and related atomic structure of a specimen surface.

In materials science, the unit cell is treated as the smallest particle that constitutes the repeating pattern to form the crystal lattice with symmetry. Thus, the edges (principal axes) of the unit cell and the corresponding angles between the axes defined the lattice parameters, which are needed for determining the size of the unit cell. Subsequently, these lattice parameters are used to characterize the type of crystal.

1.2 The Essence of Crystallography

Crystallography is the representation of the atomic arrangement of atoms in a solid phase, which is determined using a diffraction technique X-ray or electron diffraction data. Knowing the type of crystal structure is essential for characterizing solid materials subjected to mechanical deformation, phase transformation, and even to an electrochemical surface process. The latter has a variety of purposes,

© Springer Nature Switzerland AG 2020
N. Perez, *Phase Transformation in Metals*,
https://doi.org/10.1007/978-3-030-49168-0_1

including corrosion and corrosion protection, pure metal electrodeposition at atomic and industrial scales, production of electrical energy using batteries and fuel cells, surface reconstruction of solids by diffusion-induced and migration-induced adsorption, electron transfer and reaction mechanisms, kinetic of electrodes, and so forth. Hence, the field of electrochemical science and engineering has both fundamental and practical importance in a wide variety of electrochemistry-driven devices and relevant mechanisms [1].

Fundamentally, crystallography is part of an experimental surface science where atoms from a liquid phase can be attached on an solid surface as a closely packed layer of atoms due to diffusion from the liquid. Hence, solidification at a critical temperature or temperature range is a particular phase transformation process based on the reaction $L \rightarrow S$. On the other hand, if a solid structure (S_1) changes to another structure (S_2) due to cooling or heating, then the reversible solid reaction $S_1 \rightleftarrows S_2$ represents the solid-state phase transformation, subjected to a crystallographic analysis.

1.3   Crystal Defects

In general, surface defects affect material properties, atom adsorption, ion deposition, surface oxidation (corrosion) and, specifically, X-ray and electron diffraction patterns. For instance, X-ray diffraction is used to study crystalline and amorphous materials, even though diffraction peaks may be broadened and distorted due to the presence of surface defects and lack of periodicity. In essence, crystal structures and defects jointly controlled physical and mechanical properties, electrochemical behavior of materials, and so forth. This implies that solid surfaces are bond to have surface defects induced by processes such as solidification, electrochemical, corrosion, and mechanical deformation. For convenience, assumed that the crystal under study has an ideal crystal structure (perfect crystal) and it is free of features like

- Point defects: vacancies and impurity or foreign atoms (extrinsic defects)
- Linear defects: edge and screw dislocations, which are atomic mismatch
- Two-dimensional defects: grain boundaries and twin boundaries
- Volume defects: pores and precipitates.

Among crystal defects, voids and dislocations are the most common in crystalline solid materials. Therefore, crystal defects are potential (active) sites for coupled electrochemical reduction and oxidation reactions, which control the production of electrical energy or chemical energy within electrochemical cells or induce corrosion, which is a natural process due to the instability of refined (man-made) metals or alloys, specifically in the presence of water and oxygen.

1.4 X-Ray Crystallography

The determination of crystal structures (crystalline solid structures) is based on unit cells (three-dimensional atomic arrangements, also called Bravais unit cells) because it is an important crystallographic feature in any field of science. For instance, crystallography is the science of crystals and solidification is the science of crystal formation. On the other hand, X-ray crystallography (XRC) is essentially X-ray diffraction (XRD) of incident beams and it is a tool for characterizing the size of unit cells and corresponding sets of crystallographic $\{hkl\}$ planes and $\langle uvw \rangle$ directions. Individual planes and directions are denoted as (hkl) and $[uvw]$, respectively, and they represent integer numbers known as Miller indices.

Figure 1.1 illustrates the XRD model for parallel planes of atoms of a crystal [2, 3]. If waves 1 and 2 in the incident beam strike atoms "H" and "Q," then the waves 1' and 2' are diffracted or scattered, detected, and an X-ray pattern is generated. A crystallographic method requires the analysis of a diffraction pattern of a crystalline or amorphous solid sample targeted by a beam emitted by X-rays, electrons, or neutrons. The X-ray diffraction and low-energy electron diffraction (LEED) are the most common experimental techniques used for characterizing crystal structures.

The relationship between the interplanar spacing d_{hkl}, which may be denoted as the d-spacing between equally parallel (hkl) planes, and the diffraction angle θ derived by Sir W.H. Bragg and his son Sir W.L. Bragg in 1913 is known as the Bragg's equation or Bragg's law.

Thus, the diffraction from a set of parallel lattice (hkl) planes separated by $d = d_{hkl}$ obeys Bragg's law written as

$$n\lambda = 2d \sin \theta \tag{1.1}$$

where $n = 1, 2, \ldots$ is the order of diffraction. Letting $d_{hkl} = d/n$ for all reflections from a set of (hkl) planes simplifies Eq. (1.1) as [2]

$$\lambda = 2d_{hkl} \sin \theta \tag{1.2}$$

Fig. 1.1 Model for X-ray diffraction by planes of atoms. Adapted from [3]

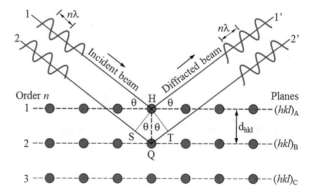

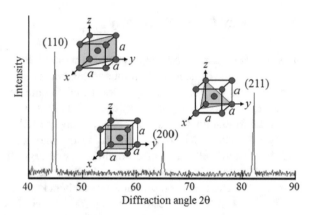

Fig. 1.2 X-ray diffraction pattern for polycrystalline α-iron (*BCC*). Adapted from Ref. [3]

This expression implies that observable diffracted waves are possible at a Bragg angle θ with respect to a set of parallel (hkl) planes separated by a common d_{hkl} distance, which is related to a scattered vector $\mathbf{g}_{hkl}$ of the diffracted beams. In fact, $\mathbf{g}_{hkl}$ is called the reciprocal lattice vector related to X-ray diffracted beams. See the Ewald sphere in Appendix 1B. Actually, many waves are commonly diffracted for a meaningful X-ray diffraction pattern, as partially shown in Fig. 1.2 by three peaks along with background noise [3]. There are three identical $\{hkl\}$ planes: $(hkl)_A$, $(hkl)_B$, and $(hkl)_C$, separated by a distance called interplanar spacing or d-spacing denoted as d_{hkl} [2]. Plane $(hkl)_A$ is on the surface of a sample. Waves 1' and 2' in the diffracted beam have the same wavelength λ as waves 1 and 2 in incident beam. The notation $n\lambda$, where $n = 1, 2, 3, \ldots$, represents the order of reflection with a specific wave length λ. Specifically, $n = 1$ is the first order on a sample surface (Fig. 1.1).

Actually, Fig. 1.2 shows an experimental X-ray diffraction pattern for pure iron (Fe) crystal identified as a body-centered cubic (BCC), known as $BCC\ \alpha\text{-}Fe$. This pattern is a plot of a function called intensity, $I = f(2\theta)$, due to the atom positions in the unit cell, where 2θ is twice the diffraction angle shown in Fig. 1.1. The intensity of the electromagnetic radiation is a measure of the rate of flow of energy per unit area *(joules/m^2 sec)* perpendicular to the direction of motion of the wave [2].

The incident wavelength λ (in *nm* units) for diffraction to occur as an electromagnetic radiation phenomenon at a Bragg angle θ is related to the energy (E_e) of a moving wave-like particle. Hence,

$$E_e = h\upsilon = \frac{hc}{\lambda} = \frac{1.2398}{\lambda} \quad (\text{in keV}) \tag{1.3}$$

where h is the Plank's constant, υ is the frequency, and c is the speed of light in the diffraction environment. Sharp peaks from any X-ray diffraction pattern correspond to a set of (hkl) planes, which are indexed to determine the type of unit cell (Fig. 1.2) [3].

Up to this point, knowing the surface (hkl) plane on an electrode crystal (substrate) helps determine the structure of a monolayer or overlayer in *adsorption* or *deposition* experiments. In fact, use of X-ray diffraction in adsorption experiments dates back to 1966 [4]. Actually, Fig. 1.2 shows a portion of an automatically recorded X-ray diffraction pattern. This is an X-ray diffraction chart representing the relative diffraction intensity proportional to counts per second vs. diffraction angle $2\theta > 0$ [3]. Actually, an XRD pattern is just a diffractogram with the intensity scale (I_d) on the y-axis. Thus, $I_d = f(2\theta)$ which has arbitrary units (au).

The interpretation of the X-ray diffraction using the well-known Bragg's equation is merely confined to the fact that a diffracted beam containing many X-rays is detectable if Bragg's law, Eq. (1.1), is obeyed, as indicated by the peaks at the corresponding scattering angle $2\theta > 0$. Notice that Bragg's law does not provide any information on the relative intensities of the scattered X-rays (reflections). This is accomplished by the Fourier transform (FT) method, which is included in Appendix 1A.

Note that each peak in Fig. 1.2 has its own diffraction intensity that relates to the relative strength of the diffracted beams. However, the peak height may be used as a qualitative measure of relative intensity. Most accurate measure of the peak intensity in an X-ray diffraction pattern is obtained by measuring the relative area under the peaks. These peaks are induced by the coherent scattering of the incident X-rays interacting with a target atom on a (hkl) plane. As a result, the incident and diffracted beams have the same energy and wavelength, while wave scattering is attributed to the interactions between X-ray particles and electrons.

Example 1.1 Consider an electromagnetic radiation to be diffracted as a first-order incident wave (X-ray diffraction) of a hypothetical crystalline solid at room temperature. (a) Derive Bragg's equation using the information given in Fig. 1.1. In addition, determine (b) the d-spacing for the first peak of the X-ray pattern in Fig. 1.2 and (c) the energy of the incident wave. Assume a first-order X-ray diffraction with Cu-K_α radiation with wavelength $\lambda = 0.1527$ nm.

Solution

(a) This example uses a simple trigonometric approach based on the model shown in Fig. 1.1. The resultant expression is the well-known Bragg's law. From Fig. 1.1, SQ and QT are defined by

$$SQ = QT = d_{hkl} \sin \theta$$

$$n\lambda = SQ + QT$$

$$n\lambda = 2d_{hkl} \sin \theta$$

(b) For the (110) crystallographic plane, the peak is at $2\theta = 44.2°$, $\theta = 22.1°$ (diffraction angle), and diffraction order $n = 1$,

$$d_{hkl} = \frac{\lambda}{2 \sin \theta} = \frac{0.1527 \, \text{nm}}{2 \sin (22.1°)} = 0.203 \, \text{nm}$$

(c) From Eq. (1.3),

$$E_e = \frac{hc}{\lambda} = \frac{\left(6.63 \times 10^{-34} \, \text{J s}\right) \left(3 \times 10^8 \, \text{m/s}\right)}{0.1527 \times 10^{-9} \, \text{m}} = 1.30 \times 10^{-15} \, \text{J}$$

$$E_e = \left(1.30 \times 10^{-15} \, \text{J}\right) / \left(1.62 \times 10^{-19} \, \text{eV/J}\right) = 8025 \, \text{eV} = 8.025 \, \text{keV}$$

Hence, this result is within an energy range, $100 \, \text{eV} < E_e < 15 \, \text{keV}$, of the incident electromagnetic wave suitable for studying crystal structures at $25 \, °\text{C} = 298 \, \text{K}$.

1.5   Crystal Lattice

The lattice is a regular, ordered, and symmetrical three-dimensional array of atoms, ions, or molecules and, in general, these can be treated as particles or points that have discrete translational symmetry described by a translation vector r_{hkl}. In crystallography, (hkl) are integers related to crystallographic coordinates a, b, c and corresponding α, β, and γ angles. These lead to the determination of the crystal lattice as an ordered array of points (atoms) that describes the unit cell, which is the smallest volume in the solid state.

The reciprocal lattice is the Fourier transform of the Bravais lattice, which also known as real lattice or direct lattice. The Bravais lattice is related to a periodic three-dimensional arrangement of atoms in real space and it is fundamentally important in the theory of diffraction for analytic studies of periodic crystal structures.

The initial physical condition of a substrate depends on the melting and solidification processes, machining and possibly heat treatment for producing desired specimens to be studied. Freshly prepared specimen surfaces are subjected to a natural process of thin film formation in air. Nonetheless, assume that the solid specimen has an ideal surface for determining its crystal structure using, say, the X-ray diffraction technique. This technique requires knowledge of reciprocal lattice for determining crystal dimensions and related geometry. Therefore, use of vector algebra for manipulation of vectors in two and three dimensions is essential in reciprocal lattice theory and surface reconstruction. In the end, determining the crystal lattice using X-ray diffraction is enormously helpful in material characterization schemes. This is the foundation for understanding the in-service performance of solid components.

1.6 Crystal Structures

This section includes relevant information on the crystal structure of solids, specifically metals composed of atoms that are modeled as hard-spheres forming a repeated unit cell (parallelepiped) with six (hkl) planes (parallelograms). And it is the crystal structure of solids along with some imperfections the main solid-state feature that controls most properties of engineering materials. Hence, crystalline or non-crystalline (amorphous) engineering materials arise. Therefore, the characterization of crystal structures is significantly important in order to understand the environmental effect on material behavior.

By definition, the crystal lattice is a three-dimensional network that defines the crystal geometry having lattice vectors **a**, **b**, **c** and translational symmetry (space periodicity). In crystallography, these seven types of crystal structures or crystal systems described by six lattice parameters: a, b, c, α, β, and γ. The combination of these parameters defines the smallest parallelepiped called unit cell with three crystallographic axes (edge lengths) a, b, c and three interaxial angles α, β, γ. Figure 1.3a illustrates repeated $(hkl) = (001)$ planes in space lattice and the general unit cell in a crystal lattice. Figure 1.3b shows the unit cell and the six lattice parameters in the a, b, c coordinate system. In this particular case, a, b, c are not orthogonal axes as drawn.

The usual features of unit cells include atoms being modeled as perfect spheres for determining the unit cell configurations based on the distance between the two nearest-neighbor atoms in the a, b, c axes, the distance between these atoms are edge lengths denoted as a, b, c, the plane angles $\alpha = \angle(a, c)$, $\beta = \angle(b, c)$ and $\gamma = \angle(a, b)$. These lattice parameters and the interplanar distance d_{hkl} between adjacent (hkl) planes are experimentally determined using, say, X-ray diffraction data. Hence, construction of conventional unit cells followed.

Table 1.1 lists the seven types of conventional unit cells and their respective lattice parameters that define crystalline materials, some of which have additional atoms being centered or face-centered in the cells. Hence, there are fourteen (14) Bravais lattices (BL) as indicated in Table 1.1 and they are known as direct or real lattices, defining infinite crystalline arrangements of discrete atoms in three-dimensional space having periodic structures. Hence, they form unit cells of unique sizes and geometries described by the vectors (**a**, **b**, **c**). In particular, the trigonal crystal system is also referred to as the rhombohedral lattice system.

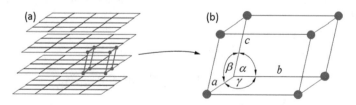

Fig. 1.3 (a) Crystal lattice and (b) conventional unit cell of a lattice structure

Table 1.1 Seven crystal structures and fourteen Bravais lattices (BL)

Unit cell	Cell edges	Cell angles	BL
Cubic	$a = b = c$	$\alpha = \beta = \gamma = 90°$	3
Tetragonal	$a = b \neq c$	$\alpha = \beta = \gamma = 90°$	2
Hexagonal	$a = b \neq c$	$\alpha = \beta = 90°$, $\gamma = 120°$	1
Orthorhombic	$a \neq b \neq c$	$\alpha = \beta = \gamma = 90°$	4
Trigonal	$a = b = c$	$\alpha = \beta = \gamma < 120°$, $\neq 90°$	1
Monoclinic	$a \neq b \neq c$	$\alpha \neq 90°$, $\beta = \gamma = 90°$	2
Triclinic	$a \neq b \neq c$	$\alpha \neq \beta \neq \gamma$	1

Notice that there are fourteen lattice types derived from seven crystal systems (Table 1.1). Nonetheless, the geometry for each Bravais lattice can be found in [3].

In principle, a single-crystal structure is composed of repeated unit cells and in general, grained metallic solids can be characterized in terms of incompressibility, rigidity, mechanical strength, and corrosion resistance. Thus, a particular crystalline material is essentially composed of many grains, which in turn, have similar crystal structure being oriented in a preferred manner. Moreover, a material may have more than one type of crystal structure (Table 1.1). When this is the case, one must deal with phases and subsequently, one must identify the crystal structure using, say, X-ray diffraction.

Specifically, only the cubic and hexagonal unit cells are considered henceforth for illustrating some important details on vector algebra used in crystallography studies.

1.7　Conventional Unit Cells and Lattice Planes

Figure 1.4 depicts the FCC, BCC, and HCP units cells and related (hkl) planes of atoms having dense atomic arrangements: the triangle FCC (111) plane (Fig. 1.4), a rectangular BCC (110) plane (Fig. 1.4c, d), and an HCP (0001) basal plane (Fig. 1.4e, f).

The 2D representation of these particular (hkl) planes shown in Fig. 1.4b, d, f gives additional details, such as small parallelograms defining oblique, square, and hexagonal lattice planes defined by an individual set of primitive vectors ($\mathbf{a}_1, \mathbf{a}_2$). Hence, the concept of primitive "p" lattice plane arises with crystal structure dimensions

$$\|\mathbf{a}_1\| = \|\mathbf{a}_2\| = \left(\sqrt{2}/2\right)a \quad \text{for } FCC \text{ (Fig. 1.4b)}$$

$$\|\mathbf{a}_1\| = \|\mathbf{a}_2\| = \left(\sqrt{3}/2\right)a \quad \text{for } BCC \text{ (Fig.1.4d)}$$

$$\|\mathbf{a}_1\| = \|\mathbf{a}_2\| = a \qquad\qquad \text{for } HCP \text{ (Fig. 1.4f)}$$

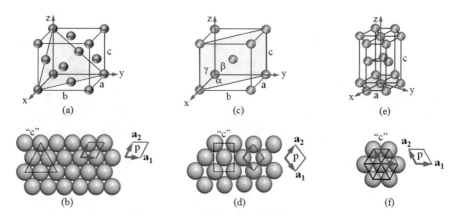

Fig. 1.4 Conventional unit cells and lattices for common crystal structures, primitive lattices "p" (smallest parallelogram with $\mathbf{a}_1$, $\mathbf{a}_2$ vectors), and conventional lattices "c". (**a**) FCC unit cell and (111) plane. (**b**) FCC (111) plane. (**c**) BCC unit cell and (110) plane. (**d**) BCC (110) plane. (**e**) HCP unit cell and (0001) plane. (**f**) HCP (0001) plane

Again, Fig. 1.4 illustrates, in general, the set of vectors denoted as conventional ($\mathbf{a}$, $\mathbf{b}$, $\mathbf{c}$) and primitive ($\mathbf{a}_1$, $\mathbf{a}_2$, $\mathbf{a}_3$).

In materials science, it is common to illustrate how to extract the conventional unit cell out of a three-dimensional array of atoms in space. By definition, a unit cell is a repetitive three-dimensional geometry representing the atomic arrangement of crystals at equilibrium due to atomic forces from the surrounding atoms. This is shown in Fig. 1.5.

A selected example (borrowed from Callister–Rethwisch book [3]) is depicted in Fig. 1.5a for an FCC crystal structure, which includes the surface plane shaded in yellow color. This surface plane is shown in Fig. 1.5b in order to compare the conventional "c" and primitive "p" lattice planes.

Figure 1.5c is a practical sketch of the unit cell shown in Fig. 1.5a since it shows the fraction of atoms per corner ($8 \times 1/8 = 1$) and face planes ($6 \times 1/2 = 3$). Thus, there are 4 atoms contained in the FCC unit cell with diagonal length $D = \sqrt{2}a = 4R$ so that $a = 4R/\sqrt{2}$. Recall that "R" and "a" denote the atomic radius of a metallic element and the lattice parameter of the conventional unit cell, respectively.

This information is now used, as specific example, to determine the atomic packing factor as

$$APF_{FCC} = \frac{4V_{sphere}}{V_{cell}} = \frac{4\,(4/3)\,\pi\,R^3}{a^3} = \frac{4\,(4/3)\,\pi\,R^3}{\left(4R/\sqrt{2}\right)^3} \tag{1.4}$$

$$APF_{FCC} = 0.74 \text{ or } 74\% \tag{1.5}$$

This result indicates that the unit cell is 74% packed with atoms and it contains a total of 26% empty space, which is essentially divided among packed atoms. Each

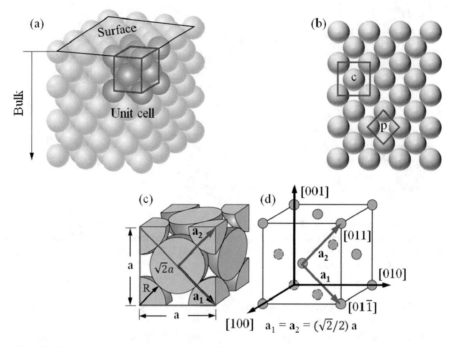

Fig. 1.5 Face-centered cubic (FCC). (**a**) Hard-sphere model of crystal structure, (**b**) dense lattice planes (conventional lattice "c" and primitive lattice "p"), (**c**) unit cell with lattice parameter "a" and radius "R," and (**d**) unit cell model (adapted from [3])

individual empty space is known as a potential lattice site for interstitial atoms to accommodate themselves. For instance, carbon "C" atoms into iron Fe-FCC interstitial lattice sites at high temperatures.

Also shown in Fig. 1.5c, d is the chosen origin of the primitive lattice vectors ($\mathbf{a}_1$, $\mathbf{a}_2$) on the (100) plane. The common magnitude of these vectors (Fig. 1.5d) is determined using simple trigonometry. Hence,

$$\|\mathbf{a}_1\| = \|\mathbf{a}_2\| = \left(\frac{\sqrt{2}}{2}\right) a \qquad (1.6)$$

Figure 1.5d, on the other hand, is a model for the unit cell in Fig. 1.5a showing the selected lattice vectors ($\mathbf{a}_1$, $\mathbf{a}_2$) along with their corresponding crystallographic directions $[01\bar{1}]$ and [011]. Additional details on crystallography can be found in [3].

1.8   Vector Algebra in Crystallography

In crystallography, orthogonal and non-orthogonal coordinate systems are regularly used to described a particular crystal structure in terms of unit cells. Hence, vector transformation from conventional to reciprocal lattices is a common operation in characterizing crystal morphology, specifically the geometry and size of the unit cell. Techniques such as X-ray and neutron diffraction are commonly used to determine the atomic structure of crystals. Moreover, the crystal structure is defined as a three-dimensional repetition of unit cells.

Consider an element with orthogonal x, y, z axes depicted in Fig. 1.6 illustrating a set of two position vectors, $\mathbf{p}$ and $\mathbf{q}$, along the following crystallographic directions:

$$[u_1 v_2 w_3] = [210] \text{ and } [u_2 v_2 w_2] = [111] \quad \text{(Fig. 1.6a)} \tag{1.7}$$

$$[u_1 v_2 w_3] = [102] \text{ and } [u_2 v_2 w_2] = [111] \quad \text{(Fig. 1.6b)} \tag{1.8}$$

Recall that a vector has magnitude and direction, such as force. On the other hand, a scalar has magnitude.

Let $\mathbf{p}$ and $\mathbf{q}$ be defined by

$$\mathbf{p} = u_1\mathbf{a} + v_1\mathbf{b} + w_1\mathbf{c} \tag{1.9a}$$

$$\mathbf{q} = u_2\mathbf{a} + v_2\mathbf{b} + w_2\mathbf{c} \tag{1.9b}$$

The goal now is to implement the use of the dot product as a scalar product (also referred to as the inner product) between these non-zero vectors in order to determine the interplanar angle θ and some properties related to crystallography.

Subsequently, the cross product $\mathbf{p} \times \mathbf{q}$, on the other hand, defines the area of a parallelogram, namely a crystallographic (hkl) plane. Both vector algebra tools are essential for determining the volume of a unit cell in real and reciprocal lattices.

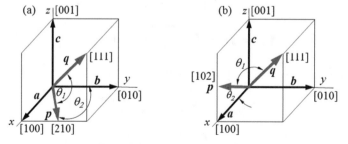

Fig. 1.6 Translational vectors in crystallography (a, b, c) coordinates and $[uvw]$ directions in an orthogonal unit cells

Other vector characteristics (magnitude and a direction) follow as these tools are implemented.

Dot Product The dot product is fundamentally defined by

$$\mathbf{p} \cdot \mathbf{q} = \|\mathbf{p}\| \, \|\mathbf{q}\| \cos \theta \tag{1.10}$$

The general mathematical form of the vector product $\mathbf{p} \cdot \mathbf{q}$ using Eq. (1.9) is based on simple vector operations and it is given here in detail to show the proper use of some dot product properties. Hence,

$$\mathbf{p} \cdot \mathbf{q} = (u_1 \mathbf{a} + v_1 \mathbf{b} + w_1 \mathbf{c}) \cdot (u_2 \mathbf{a} + v_2 \mathbf{b} + w_2 \mathbf{c}) \tag{1.11a}$$

$$\mathbf{p} \cdot \mathbf{q} = u_1 u_2 \, (\mathbf{a} \cdot \mathbf{a}) + u_1 v_2 \, (\mathbf{a} \cdot \mathbf{b}) + u_1 w_2 \, (\mathbf{a} \cdot \mathbf{c}) \tag{1.11b}$$

$$+ \, v_1 u_2 \, (\mathbf{b} \cdot \mathbf{a}) + v_1 v_2 \, (\mathbf{b} \cdot \mathbf{b}) + v_1 w_2 \, (\mathbf{b} \cdot \mathbf{c})$$

$$+ \, w_1 u_2 \, (\mathbf{c} \cdot \mathbf{a}) + w_1 v_2 \, (\mathbf{c} \cdot \mathbf{b}) + w_1 w_2 \, (\mathbf{c} \cdot \mathbf{c})$$

Orthogonal Properties
- Two vectors, $\mathbf{p}$ and $\mathbf{q}$, are orthogonal (perpendicular: $\mathbf{p} \perp \mathbf{q}$ at $90°$) if $\cos \theta = 1$ and $\mathbf{p} \cdot \mathbf{q} = 0$
- Two vectors are parallel (co-directional: $\mathbf{p} \, \| \, \mathbf{q}$ at $0°$) if $\cos \theta = 0$ and $\mathbf{p} \cdot \mathbf{q} \neq 0$
- Two vectors are non-orthogonal if $0 < \cos \theta < 1$ and $\mathbf{p} \cdot \mathbf{q} \neq 0$
- Two unit vectors, $\mathbf{a}$ and $\mathbf{b}$, give $\mathbf{a} \cdot \mathbf{a} = \mathbf{b} \cdot \mathbf{b} = \mathbf{c} \cdot \mathbf{c} = 1$
- Two unit vectors, $\mathbf{a}$ and $\mathbf{b}$, give, $\mathbf{a} \cdot \mathbf{b} = \mathbf{b} \cdot \mathbf{a} = 0$, $\mathbf{a} \cdot \mathbf{c} = \mathbf{c} \cdot \mathbf{a} = 0$ and $\mathbf{b} \cdot \mathbf{c} = \mathbf{c} \cdot \mathbf{b} = 0$
- Absolute entities: $\|\mathbf{a}\| = a$, $\|\mathbf{b}\| = b$ and $\|\mathbf{c}\| = c$.

Now, substituting the proper properties into Eq. (1.14b) yields

$$\mathbf{p} \cdot \mathbf{q} = (u_1 u_2) \, a^2 + (v_1 v_2) \, b^2 + (w_1 w_2) \, c^2 \tag{1.12}$$

Moreover, the magnitude of these vectors is clearly defined by

$$\|\mathbf{p}\| = \sqrt{u_1^2 \, \|\mathbf{a}\|^2 + v_1^2 \, \|\mathbf{b}\|^2 + w_1^2 \, \|\mathbf{c}\|^2} \tag{1.13a}$$

$$\|\mathbf{q}\| = \sqrt{u_2^2 \, \|\mathbf{a}\|^2 + v_2^2 \, \|\mathbf{b}\|^2 + w_2^2 \, \|\mathbf{c}\|^2} \tag{1.13b}$$

where $\|\mathbf{a}\|^2 = a^2$ and $\|\mathbf{a}\|^2 = \|\mathbf{b}\|^2 = \|\mathbf{c}\|^2$. Then,

$$\|\mathbf{p}\| = \sqrt{u_1^2 a^2 + v_1^2 b^2 + w_1^2 c^2} \tag{1.14a}$$

$$\|\mathbf{q}\| = \sqrt{u_2^2 a^2 + v_2^2 b^2 + w_2^2 c^2} \tag{1.14b}$$

The interplanar angles are also defined by the inner products as a generalization of the dot products $\mathbf{b} \cdot \mathbf{c}$, $\mathbf{c} \cdot \mathbf{a}$, and $\mathbf{a} \cdot \mathbf{b}$. Thus,

$$\cos \alpha = \frac{\mathbf{b} \cdot \mathbf{c}}{\|\mathbf{b}\| \, \|\mathbf{c}\|}; \quad \cos \beta = \frac{\mathbf{c} \cdot \mathbf{a}}{\|\mathbf{c}\| \, \|\mathbf{a}\|}; \quad \cos \gamma = \frac{\mathbf{a} \cdot \mathbf{b}}{\|\mathbf{a}\| \, \|\mathbf{b}\|} \tag{1.15}$$

Finally, the dot product angle θ becomes

$$\cos \theta = \frac{\mathbf{p} \cdot \mathbf{q}}{\|\mathbf{p}\| \, \|\mathbf{q}\|} \quad \text{for} \quad -\pi \le \theta \le \pi \tag{1.16a}$$

$$\cos \theta = \frac{(u_1 \mathbf{a} + v_1 \mathbf{b} + w_1 \mathbf{c}) \cdot (u_2 \mathbf{a} + v_2 \mathbf{b} + w_2 \mathbf{c})}{\sqrt{u_1^2 a^2 + v_1^2 b^2 + w_1^2 c^2} \, \sqrt{u_2^2 a^2 + v_2^2 b^2 + w_2^2 c^2}} \tag{1.16b}$$

$$\cos \theta = \frac{(u_1 u_2) \, a^2 + (v_1 v_2) \, b^2 + (w_1 w_2) \, c^2}{\sqrt{u_1^2 a^2 + v_1^2 b^2 + w_1^2 c^2} \, \sqrt{u_2^2 a^2 + v_2^2 b^2 + w_2^2 c^2}} \tag{1.16c}$$

For crystal structures having non-orthogonal x, y, z axes as shown in Fig. 1.3b, combine Eqs. (1.11b) and (1.15) to give

$$\mathbf{p} \cdot \mathbf{q} = (u_1 u_2) \, a^2 + (v_1 v_2) \, b^2 + (w_1 w_2) \, c^2 + (u_1 v_2 + v_1 u_2) \, ba \cos \gamma$$
$$+ (v_1 w_2 + v_1 w_2) \, bc \cos \alpha + (u_1 w_2 + w_1 u_2) \, ca \cos \beta \tag{1.17}$$

For cubic, tetragonal, and orthorhombic crystal structures, $\alpha = \beta = \gamma = 90°$ and Eq. (1.17) reduces to (1.12).

Cross Product   The cross product $\mathbf{p} \times \mathbf{q}$ is defined explicitly by

$$\mathbf{p} \times \mathbf{q} = \|\mathbf{p}\| \, \|\mathbf{q}\| \sin \theta \tag{1.18}$$

From Eq. (1.9),

$$\mathbf{p} \times \mathbf{q} = (u_1 \mathbf{a} + v_1 \mathbf{b} + w_1 \mathbf{c}) \times (u_2 \mathbf{a} + v_2 \mathbf{b} + w_2 \mathbf{c}) \tag{1.19a}$$

$$\mathbf{p} \times \mathbf{q} = u_1 u_2 \, (\mathbf{a} \times \mathbf{a}) + u_1 v_2 \, (\mathbf{a} \times \mathbf{b}) + u_1 w_2 \, (\mathbf{a} \times \mathbf{c}) \tag{1.19b}$$
$$+ v_1 u_2 \, (\mathbf{b} \times \mathbf{a}) + v_1 v_2 \, (\mathbf{b} \times \mathbf{b}) + v_1 w_2 \, (\mathbf{b} \times \mathbf{c})$$
$$+ w_1 u_2 \, (\mathbf{c} \times \mathbf{a}) + w_1 v_2 \, (\mathbf{c} \times \mathbf{b}) + w_1 w_2 \, (\mathbf{c} \times \mathbf{c})$$

Useful Vector Properties
- The cross product of two vectors, $\mathbf{p}$ and $\mathbf{q}$, is equal to the area of the parallelogram: $A = \|\mathbf{p} \times \mathbf{q}\|$
- Two vectors, $\mathbf{p}$ and $\mathbf{q}$, are orthogonal (perpendicular: $\mathbf{p} \perp \mathbf{q}$ at 90°) if $\sin \theta \ne 0$ and $\|\mathbf{p} \times \mathbf{q}\| > 0$
- Two vectors, $\mathbf{p}$ and $\mathbf{q}$, are non-orthogonal if $\sin \theta = 0$ and $\mathbf{p} \times \mathbf{q} = 0$ (also $\mathbf{p} \cdot \mathbf{q} \ne 0$)
- Entities:

$$\mathbf{a} \times \mathbf{b} = -(\mathbf{b} \times \mathbf{a}) = \mathbf{c} \qquad (1.20a)$$

$$\mathbf{b} \times \mathbf{c} = -(\mathbf{c} \times \mathbf{b}) = \mathbf{a} \qquad (1.20b)$$

$$\mathbf{c} \times \mathbf{a} = -(\mathbf{a} \times \mathbf{c}) = \mathbf{b} \qquad (1.20c)$$

$$\mathbf{a} \times \mathbf{a} = \mathbf{b} \times \mathbf{b} = \mathbf{c} \times \mathbf{c} = 0 \qquad (1.20d)$$

Notice that the dot product and the cross product are simple mathematical procedures covered in a traditional vector calculus course. Indeed, the former measures θ and the latter measures the area spanned by two vectors.

Applying Eq. (1.20) to (1.19b) gives

$$\mathbf{p} \times \mathbf{q} = (v_1 w_2 - w_1 v_2)\,\mathbf{a} - (u_1 w_2 - w_1 u_2)\,\mathbf{b} + (u_1 v_2 - v_1 u_2)\,\mathbf{c} \qquad (1.21)$$

In symmetric matrix form,

$$\mathbf{p} \times \mathbf{q} = \begin{bmatrix} \mathbf{a} & \mathbf{b} & \mathbf{c} \\ u_1 & v_1 & w_1 \\ u_2 & v_2 & w_2 \end{bmatrix} = \det \begin{bmatrix} \mathbf{a} & \mathbf{b} & \mathbf{c} \\ u_1 & v_1 & w_1 \\ u_2 & v_2 & w_2 \end{bmatrix} \qquad (1.22a)$$

$$\mathbf{p} \times \mathbf{q} = \mathbf{a} \begin{bmatrix} v_1 & w_1 \\ v_2 & w_2 \end{bmatrix} - \mathbf{b} \begin{bmatrix} u_1 & w_1 \\ u_2 & w_2 \end{bmatrix} + \mathbf{c} \begin{bmatrix} u_1 & v_1 \\ u_2 & v_2 \end{bmatrix} \qquad (1.22b)$$

which defines Eq. (1.21a). The cross product is now used to find the angle θ between two vectors

$$\sin\theta = \frac{\mathbf{p} \times \mathbf{q}}{\|\mathbf{p}\|\,\|\mathbf{q}\|} \quad \text{for} \ \ 0 \le \theta \le \pi \qquad (1.23)$$

Law of Cosines At times, the law of cosines is useful for finding the length of vector $\mathbf{c}_1$ that connects $(\mathbf{a_1}, \mathbf{a_2})$. Hence,

$$c_1^2 = a_1^2 + a_2^2 - 2a_1 a_2 \cos\theta \qquad (1.24)$$

Basically, the law of cosines is used for solving triangles. Furthermore, below are some useful expressions related to dot-product and cross-product notations

$$(\mathbf{a_1} + \mathbf{a_2}) \cdot (\mathbf{a_3} - \mathbf{a_4}) = (\mathbf{a_1} \cdot \mathbf{a_3}) - (\mathbf{a_1} \cdot \mathbf{a_4}) + (\mathbf{a_2} \cdot \mathbf{a_3}) - (\mathbf{a_2} \cdot \mathbf{a_4}) \qquad (1.25a)$$

$$\mathbf{a_1} \cdot (\mathbf{a_2} \times \mathbf{a_3}) = \mathbf{a_2} \cdot (\mathbf{a_3} \times \mathbf{a_1}) = \mathbf{a_3} \cdot (\mathbf{a_1} \times \mathbf{a_2}) \qquad (1.25b)$$

$$\mathbf{a_1} \times (\mathbf{a_2} \times \mathbf{a_3}) = \mathbf{a_2}(\mathbf{a_1} \cdot \mathbf{a_3}) - \mathbf{a_3}(\mathbf{a_1} \cdot \mathbf{a_2}) \qquad (1.25c)$$

$$(\mathbf{a_1} \times \mathbf{a_2}) = -(\mathbf{a_2} \times \mathbf{a_1}) \qquad (1.25d)$$

$$\mathbf{a_1} \times (\mathbf{a_2} + \mathbf{a_3}) = (\mathbf{a_1} \times \mathbf{a_2}) + (\mathbf{a_1} \times \mathbf{a_3}) \qquad (1.25e)$$

Properties of the cross product for crystallographic directions follow

$$u \times v = -(v \times u) = w \tag{1.26a}$$

$$v \times w = -(w \times v) = u \tag{1.26b}$$

$$w \times u = -(u \times w) = v \tag{1.26c}$$

$$u \times u = v \times v = w \times w = 0 \tag{1.26d}$$

In essence, the preceding vector treatment has its place in crystallography. For instance, the laws of cosines (cosine rule) and sines (sine rule) can be used to find the angles of right, obtuse, and acute triangles when all sides are known.

Example 1.2 Determine if the vectors $\mathbf{a_1} = 2i - 2j + 3k$ and $\mathbf{b_1} = 4i - j + 3k$ are perpendicular, parallel, or neither perpendicular nor parallel based on their dot product.

Solution Dot product between $\mathbf{a_1} = \langle 2, -2, 3 \rangle$ and $\mathbf{b_1} = \langle 4, -1, 3 \rangle$ gives

$$\mathbf{a_1} \cdot \mathbf{b_1} = (2)(4) + (-2)(-1) + (3)(3) = 19$$

$$\|\mathbf{a_1}\| = \sqrt{(2)^2 + (-2)^2 + (3)^2} = \sqrt{17}$$

$$\|\mathbf{b_1}\| = \sqrt{(4)^2 + (-1)^2 + (3)^2} = \sqrt{26}$$

and

$$\cos \theta = \frac{\mathbf{a_1} \cdot \mathbf{b_1}}{\|\mathbf{a_1}\| \|\mathbf{b_1}\|} = \frac{19}{\sqrt{17}\sqrt{26}} = 0.90374$$

$$\theta = 0.44237 \; rad = 25.35°$$

These vectors are neither perpendicular nor parallel because $0 < \theta < 90°$.

Example 1.3 Consider a cubic crystal and the crystallographic $[uvw]$ directions depicted in Fig. 1.6 to calculate the angles (a) θ_1 using the dot product and (b) θ_2 using both the dot product and the cross product. This example shows that these mathematical tools are very powerful not only in vector algebra, but also in crystallography. Recall that a cubic crystal has orthogonal x, y, z axes.

Solution

(a) Dot product. For a cubic crystal, use $a = b = c$, $[uvw]_1 = [210]_1$ and $[uvw]_2 = [111]_2$. Using Eq. (1.16c) gives the value of the angle θ_1 between vectors $\mathbf{p}$ and $\mathbf{q}$. Hence,

$$\cos \theta_1 = \frac{a^2 [210]_1 \cdot [111]_2}{a^2 \sqrt{u_1^2 + v_1^2 + w_1^2} \sqrt{u_2^2 + v_2^2 + w_2^2}}$$

$$= \frac{(2)\,(1) + (1)\,(1) + (0)\,(1)}{\sqrt{2^2 + 1^2 + 0^2}\sqrt{1^2 + 1^2 + 1^2}}$$

$$\cos\theta_1 = \frac{3}{\sqrt{5}\sqrt{3}} \quad \Longrightarrow \quad \theta_1 = 39.23°$$

(b) Cross product: Let, $[uvw]_1 = [210]_1$, $[uvw]_2 = [010]_2$, and $a = b = c$. In order to avoid confusion, let $\mathbf{m}$ be the vector along the y-axis. Thus,

$$\mathbf{p} = (u_1\mathbf{a} + v_1\mathbf{b} + w_1\mathbf{c}) = (u_1\mathbf{a} + v_1\mathbf{b})$$

$$\mathbf{m} = (u_2\mathbf{a} + v_2\mathbf{b} + w_2\mathbf{c}) = v_2\mathbf{b}$$

and

$$\mathbf{p} \times \mathbf{m} = (u_1\mathbf{a} + v_1\mathbf{b}) \times (v_2\mathbf{b}) = u_1v_2\,(\mathbf{a} \times \mathbf{b}) + v_1v_2\,(\mathbf{b} \times \mathbf{b})$$

$$= u_1v_2\mathbf{c} + v_1v_2\,(0) = u_1v_2\mathbf{c}$$

$$\|\mathbf{p} \times \mathbf{m}\| = (2)\,(1)\,a^2 = 2a^2$$

with

$$\|\mathbf{p}\| = \sqrt{(2)^2\,a^2 + (1)^2\,a^2 + (0)^2\,a^2} = a\sqrt{5}$$

$$\|\mathbf{m}\| = \sqrt{(0)^2\,a^2 + (1)^2\,a^2 + (0)^2\,a^2} = a$$

Then from Eq. (1.18), the angle θ_2 between vectors $\mathbf{p}$ and $\mathbf{m}$ is

$$\sin\theta_2 = \frac{\|\mathbf{p} \times \mathbf{m}\|}{\|\mathbf{p}\|\,\|\mathbf{m}\|} = \frac{2}{\sqrt{5}\sqrt{1}} = \frac{2}{\sqrt{5}}$$

$$\theta_2 = 63.44°$$

From Eq. (1.16c), the dot product gives

$$\mathbf{p} \cdot \mathbf{m} = (u_1\mathbf{a} + v_1\mathbf{b} + w_1\mathbf{c}) \cdot (u_2\mathbf{a} + v_2\mathbf{b} + w_2\mathbf{c}) = (u_1\mathbf{a} + v_1\mathbf{b}) \cdot (v_2\mathbf{b})$$

$$\cos\theta_2 = \frac{\mathbf{p} \cdot \mathbf{m}}{\|\mathbf{p}\|\,\|\mathbf{m}\|} = \frac{(u_1\mathbf{a} + v\mathbf{b}) \cdot (v_2\mathbf{b})}{a\sqrt{5}a\sqrt{1}} = \frac{v_1v_2b^2}{a^2\sqrt{5}} = \frac{(1)\,(1)\,a^2}{a^2\sqrt{5}} = \frac{1}{\sqrt{5}}$$

$$\theta_2 = 63.44°$$

Therefore, both cross-product and dot-product methods, as expected, yield the same result for the angle θ_2 between the given vectors.

1.9 Rotation Matrix

Assume that a parallelogram is described by the general vectors $(\mathbf{a}_1, \mathbf{a}_2)$ and that $(\mathbf{b}_1, \mathbf{b}_2)$ are the resultant vectors after counterclockwise rotation at an angle θ about the origin of the vector system. In this case, the individual 2D matrices are

$$M_a = \begin{bmatrix} a_{11} & a_{12} \\ a_{21} & a_{22} \end{bmatrix} \tag{1.27a}$$

$$M_b = \begin{bmatrix} b_{11} & b_{12} \\ b_{21} & b_{22} \end{bmatrix} \tag{1.27b}$$

Now, the general rotation matrix for $(\mathbf{a}_1, \mathbf{a}_2) \rightarrow (\mathbf{b}_1, \mathbf{b}_2)$ at a rotation angle θ can be described by

$$M_R = \begin{bmatrix} \cos\theta & -\sin\theta \\ \sin\theta & \cos\theta \end{bmatrix} \tag{1.28}$$

For (a_1, b_1) vertices, M_R, $\det M_R$, and $\cos\theta$ are, respectively [5]

$$M_R = \begin{bmatrix} b_1 & a_1 \\ -a_1 & b_1 - a_1 \end{bmatrix} \tag{1.29a}$$

$$\det M_R = a_1^2 - a_1 b_1 + b_1^2 \tag{1.29b}$$

$$\cos\theta = \frac{b_1 - a_1/2}{\sqrt{\det M_R}} \tag{1.29c}$$

where $\det M_R$ and $\sqrt{\det M_R}$ are the area and edge length of a parallelogram, respectively.

In crystallography, the size of the parallelogram can be defined using the notation $\left(\sqrt{\det M_R} \times \sqrt{\det M_R}\right)$. This, then, identifies the crystallographic condition for a particular material X having a surface defined by the (hkl) plane. Here, (hkl) are integer numbers called Miller indices of a crystallographic plane.

The recognized definition of a perfect monolayer covering a substrate surface is a theoretical concept of fundamental importance for the subsequent analysis of surface properties. In reality, it is inevitable to obtain a perfect monolayer without crystal defects, such as voids.

The main goal now is to characterize the crystallographic orientation angle between the monolayer and substrate, as mathematically indicated by Eq. (1.29c). Once a substrate (specimen) undergoes a surface reconstruction by at least one monolayer, the specimen surface crystallography is of significant importance for determining the surface properties.

All these signify that a simple matrix analysis, as cited above by Eqs. (1.28) and (1.29), is significantly necessary for characterizing the monolayer as a collection of atoms on a (hkl) plane. Thus, the concept of rotation matrix becomes important in the crystallography field.

It follows from Eq. (1.29c) that the rotational angle θ is strictly dependent on the determinant of the rotational matrix: $\det M_R$. In fact, for an effective rotational matrix, the rotational angle must be $0 < \theta < \pi/2$ or $0 < \theta < 90°$. However, it is recognized that the rotation angle θ is restricted by the compatibility of rotational and translational symmetry in two-dimensional lattices and that the $\theta = 360°/n$, where $n = 2, 3, 4, 6$ ([5, p. 35]).

Example 1.4 Consider the (001) plane at the bottom and top of the simple cubic (SC) unit cell shown below.

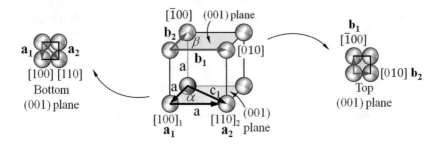

Calculate (a) the angle α and β between vectors $(\mathbf{a}_1, \mathbf{a}_2)$ and $(\mathbf{a}_3, \mathbf{a}_4)$, Respectively, and (b) prove that $c_1 = a$, where a is the lattice parameter. (c) Determine which set of vectors defines a primitive lattice, which is a significant key in surface reconstruction of a substrate surface during adsorption or deposition, and (d) the matrix for each set of vectors and their vertices.

Solution By inspection, $\|a_1\| = \|b_1\| = \|b_2\| = a$ and $\|a_2\| = \sqrt{2}a$, and $\|c_1\| = a$ for the (001) plane. The angles are $\alpha = 45°$ and $\beta = 90°$. Let us prove that this is correct using the dot product and the law of cosines.

(a) For the bottom (001) plane, the dot product gives

$$\mathbf{a}_1 \cdot \mathbf{a}_2 = [uvw]_{a_1}\, a \cdot [uvw]_{a_2}\, a = [100]_{a_1} \cdot [110]_{a_2}\, a^2$$

$$\mathbf{a}_1 \cdot \mathbf{a}_2 = [1 \times 1 + 0 \times 1 + 0 \times 0]\, a^2 = a^2$$

The length of these vectors is

$$\|\mathbf{a}_1\| = a_1 = \sqrt{u_1^2 + v_1^2 + w_1^2}\, a = \sqrt{1^2 + 0^2 + 0^2}\, a = a$$

$$\|\mathbf{a}_2\| = a_2 = \sqrt{u_2^2 + v_2^2 + w_2^2}\, a = \sqrt{1^2 + 1^2 + 0^2}\, a = \sqrt{2}a$$

The angle between these vectors is

$$\cos\alpha = \frac{\mathbf{a}_1 \cdot \mathbf{a}_2}{\|\mathbf{a}_1\| \, \|\mathbf{a}_2\|} = \frac{a^2}{(a)\left(\sqrt{2}a\right)} = \frac{1}{\sqrt{2}}$$

$$\alpha = \arccos\left(\frac{1}{\sqrt{2}}\right) = 45°$$

For the top (001) plane,

$$\mathbf{b}_1 \cdot \mathbf{b}_2 = [uvw]_{b_1}\, a \cdot [uvw]_{b_2}\, a = \left[\overline{1}00\right]_{b_1} \cdot [010]_{b_2}\, a^2$$

$$\mathbf{b}_1 \cdot \mathbf{b}_2 = [(-1) \times (0) + (0) \times (1) + (0) \times (0)]\, a^2 = 0$$

The length of these vectors is

$$\|\mathbf{b}_1\| = b_1 = \sqrt{u_3^2 + v_3^2 + w_3^2}\, a = \sqrt{(-1)^2 + (0)^2 + (0)^2}\, a = a$$

$$\|\mathbf{b}_2\| = b_2 = |\mathbf{a}_2| = \sqrt{u_4^2 + v_4^2 + w_4^2}\, a = \sqrt{(0)^2 + (1)^2 + (0)^2}\, a = a$$

As expected, $\|\mathbf{b}_1\| = \|\mathbf{b}_2\| = a$ and the angle is

$$\cos\beta = \frac{\mathbf{b}_1 \cdot \mathbf{b}_2}{\|\mathbf{b}_1\| \, \|\mathbf{b}_2\|} = \frac{0}{(a)\,(a)} = 0$$

$$\beta = \arccos(0) = 90°$$

Therefore, $\alpha = 45°$ and $\beta = 90°$ are correct. The magnitude of these angles can be determined by pure inspection of the unit cell.

(b) Using the law of cosines yields

$$c_1^2 = a_1^2 + a_2^2 - 2a_1 a_2 \cos\alpha$$

$$c_1^2 = a^2 + \left(\sqrt{2}a\right)^2 - 2a\left(\sqrt{2}a\right)\left(\frac{1}{\sqrt{2}}\right) = a^2 + \left(\sqrt{2}a\right)^2 - 2a^2$$

$$c_1^2 = a^2 + 2a^2 - 2a^2 = a^2$$

$$c_1 = a$$

Therefore, $c_1 = a$ is the lattice parameter of the SC unit cell.

(c) Only the set of vectors $(\mathbf{b}_1, \mathbf{b}_2)$ defines a primitive lattice due to

$$\|\mathbf{b}_1\| = \|\mathbf{b}_2\| = a$$

which is the shortest distance on the (001) plane. On the other hand, the conventional vectors $(\mathbf{a}_1, \mathbf{a}_2)$ gives

$$\|\mathbf{a}_2\| = \sqrt{2}\,\|\mathbf{a}_1\| = \sqrt{2}a$$

Hence, $(a_1, a_2) = \left(1, \sqrt{2}\right) a$ and $(b_1, b_2) = (1, 1)\, a$.

(d) In matrix notation, $(a_1 \times a_2) = \left(1 \times \sqrt{2}\right)$, $(b_1 \times b_2) = (1 \times 1)$ (rows) and so that their matrices and determinants become

$$M_a = \begin{pmatrix} 1 & 0 \\ 0 & \sqrt{2} \end{pmatrix}$$

$$\det M_a = \sqrt{2}$$

and

$$M_b = \begin{pmatrix} 1 & 0 \\ 0 & 1 \end{pmatrix}$$

$$\det M_b = 1$$

The physical meaning of $\det M_a$ and $\det M_b$ is that they describe the scaling factors of the linear transformation of their corresponding square the matrices. Hence, $n \times n = 2 \times 2$ is a square matrix as defined in matrix calculus. Additionally, a determinant can be treated as a matrix-dependent function or as a mathematical object that outputs a real number from its matrix. The goal here is to show that matrix calculus can be applied to crystallography is a simple and understandable basic approach.

1.10   Reciprocal Lattice in Crystallography

This section includes the theoretical background on reciprocal lattice of a perfect and symmetric arrangement of atoms called Bravais lattice in a three-dimensional space. The reciprocal lattice is based on Fourier transform of a primitive lattice and it is a convenient mathematical technique to describe crystal structures of solids having atomic periodicity. Hence, real and reciprocal lattices are connected via Fourier transform. This is achieved by a vector algebra procedure for characterizing the geometry of unit cells based on crystallographic (hkl) planes and $[uvw]$ directions.

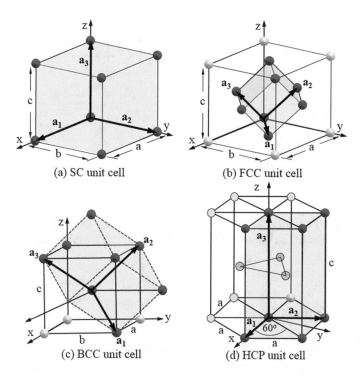

Fig. 1.7 Conventional and primitive crystal structures (**a**) *SC*, (**b**) *FCC*, (**c**) *BCC*, and (**d**) *HCP* unit cells

1.10.1   Primitive Lattice Vectors

Figure 1.7 depicts the most common conventional unit cells as large parallelepipeds with a set of translational lattice vectors **a**, **b**, **c** and a set of edge length a, b, c.

If the crystallographic coordinates are orthogonal, then $(\mathbf{a}, \mathbf{b}, \mathbf{c}) = (\mathbf{x}, \mathbf{y}, \mathbf{z})$ as in Fig. 1.7 depicts common primitive unit cells as the small shared parallelepipeds, which represent the principal domain for translation symmetry of the lattice with primitive translation vectors $(\mathbf{a}_1, \mathbf{a}_2, \mathbf{a}_3)$ as the coordinate axes. There is also another type of primitive unit cell known as Wigner–Seitz cell with a polyhedral shape in momentum space and its reciprocal lattice is called the Brillouin zone. Characteristics and related detail features of the Wigner–Seitz cell can be found in [5, 7, 10].

The simple cubic *SC* (Fig. 1.7a) has its own primitive lattice, but the face-centered cubic *FCC* (Fig. 1.7b) and body-centered cubic *BCC* (Fig. 1.7c) have primitive cells as rotated parallelepipeds. The hexagonal-closed pack *HCP* (Fig. 1.7d) has an hexagon as its primitive cell. Nonetheless, these unit cells represent three-dimensional symmetric arrangement of atoms described by their respective set of vectors. Fundamentally, a primitive unit cell contains only eight

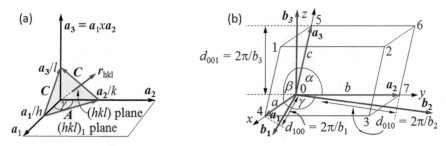

Fig. 1.8 Illustration of vector directions. (a) Cross product. (b) Triclinic conventional and reciprocal lattice vectors

corner points and only one lattice point to choose as the origin of $(\mathbf{a}_1, \mathbf{a}_2, \mathbf{a}_3)$ vectors. This particular characteristic of the primitive unit cell is clearly shown in Fig. 1.7.

1.10.2   Reciprocal Lattice Vector

In general, in order to describe a unit cell in the field of crystallography science, materials science, or solid-state physics, use of the reciprocal lattice scattering vector $\mathbf{g}_{hkl}$ related to a crystallographic (hkl) plane can be described by a set of vectors written as $\mathbf{b}_i = (\mathbf{b}_1, \mathbf{b}_2, \mathbf{b}_3)$ [2]. Thus, any point in a lattice (three-dimensional parallelepiped) can be described by a translational vector (position vector) $\mathbf{r}_{uvw}$ depicted in Fig. 1.8a and defined by Cullity [2]

$$\mathbf{r}_{uvw} = u\mathbf{a}_1 + v\mathbf{a}_2 + w\mathbf{a}_3 \tag{1.30}$$

where u, v, w are integers called Miller indices of the crystal structure and $(\mathbf{a}_1, \mathbf{a}_2, \mathbf{a}_3)$ are vector directions.

These vectors span the crystal smallest parallelepiped along the conventional $\mathbf{a}_1, \mathbf{a}_2, \mathbf{a}_3$ vector coordinates shown in Fig. 1.8a as orthogonal coordinates and illustrated in Fig. 1.8b as non-orthogonal coordinates for a triclinic unit cell.

The three-dimensional space defining the crystal structure is called the reciprocal lattice (direct space). The corresponding reciprocal lattice scattering vector, $\mathbf{g}_{hkl}$, is related to d_{hkl} ([5, p. 68]) and $\omega = 1$ prevails in the crystallography literature ([2, p. 482] and [6, p. 12]).

Furthermore, this vector algebra approach is useful in characterizing an X-ray diffraction pattern of a crystal (Fig. 1.2) since $d_{hkl} \propto 1/\|\mathbf{g}_{hkl}\|$, $\mathbf{g}_{(hkl)} \perp (hkl)$ and $d_{hkl} = f[\|\mathbf{a}_i\|, (hkl)]$. These are unique characteristic features of the reciprocal lattice approach related to vector algebra.

Mathematically, the components of the reciprocal lattice vector, $(\mathbf{b}_1, \mathbf{b}_2, \mathbf{b}_3)$, are defined in terms of the primitive lattice vectors $(\mathbf{a}_1, \mathbf{a}_2, \mathbf{a}_3)$ as defined by the group of equations given below

$$\mathbf{b}_1 = \frac{\omega\,(\mathbf{a}_2 \times \mathbf{a}_3)}{\mathbf{a}_1 \cdot (\mathbf{a}_2 \times \mathbf{a}_3)} = \frac{\omega\,(\mathbf{a}_2 \times \mathbf{a}_3)}{V} \tag{1.31a}$$

$$\mathbf{b}_2 = \frac{\omega\,(\mathbf{a}_3 \times \mathbf{a}_1)}{\mathbf{a}_1 \cdot (\mathbf{a}_2 \times \mathbf{a}_3)} = \frac{\omega\,(\mathbf{a}_3 \times \mathbf{a}_1)}{V} \tag{1.31b}$$

$$\mathbf{b}_3 = \frac{\omega\,(\mathbf{a}_1 \times \mathbf{a}_2)}{\mathbf{a}_1 \cdot (\mathbf{a}_2 \times \mathbf{a}_3)} = \frac{\omega\,(\mathbf{a}_1 \times \mathbf{a}_2)}{V} \tag{1.31c}$$

Here, $V = \mathbf{a}_1 \cdot (\mathbf{a}_2 \times \mathbf{a}_3)$ is the volume of the original unit cell, i, j are dummy indices, and the vectors $\mathbf{b}_j \perp \mathbf{a}_i$ for $i \neq j$, $\mathbf{b}_j$ are perpendicular to each other, and most importantly they have dimensions of reciprocal length (inverse length) in crystallography. Apparently, the value of ω in Eq. (1.31) is a matter of choice.

It can be stated that $(\mathbf{a}_1, \mathbf{a}_2, \mathbf{a}_3)$ and $(\mathbf{b}_1, \mathbf{b}_2, \mathbf{b}_3)$ are equally reciprocal, provided that they satisfy the dot product $\mathbf{a}_i \cdot \mathbf{b}_j$ with the following Laue conditions:

$$\mathbf{a}_i \cdot \mathbf{b}_j = \omega\delta_{ij} \text{ if } i = j \text{ and } \delta_{ij} = 1 \tag{1.32a}$$

$$\mathbf{a}_i \cdot \mathbf{b}_j = 0 \text{ if } i \neq j \tag{1.32b}$$

where the Kronecker unit matrix is

$$\delta_{ij} = \begin{bmatrix} 1 & 0 & 0 \\ 0 & 1 & 0 \\ 0 & 0 & 1 \end{bmatrix} \tag{1.33}$$

In addition, the interaxial angles, $\alpha^* = \angle\,(\mathbf{b}_2, \mathbf{b}_3)$, $\beta^* = \angle\,(\mathbf{b}_3, \mathbf{b}_1)$, and $\gamma^* = \angle\,(\mathbf{b}_1, \mathbf{b}_2)$ in the reciprocal lattice angles are defined by

$$\cos\alpha^* = \frac{\mathbf{b}_2 \cdot \mathbf{b}_3}{\|\mathbf{b}_2\|\,\|\mathbf{b}_3\|}; \;\; \cos\beta^* = \frac{\mathbf{b}_3 \cdot \mathbf{b}_1}{\|\mathbf{b}_3\|\,\|\mathbf{b}_1\|}; \;\; \cos\gamma^* = \frac{\mathbf{b}_1 \cdot \mathbf{b}_2}{\|\mathbf{b}_1\|\,\|\mathbf{b}_2\|} \tag{1.34}$$

Accordingly, the Bravais primitive lattice vectors along with the volume of the reciprocal lattice $V^* = \mathbf{b}_1 \cdot (\mathbf{b}_2 \times \mathbf{b}_3) = \omega^3/V$ can be determined as indicated by the following group of equations ([2] and [7, p. 170]):

$$\mathbf{a}_1 = \frac{\omega\,(\mathbf{b}_2 \times \mathbf{b}_3)}{\mathbf{b}_1 \cdot (\mathbf{b}_2 \times \mathbf{b}_3)} = \frac{\omega\,(\mathbf{b}_2 \times \mathbf{b}_3)}{V^*} \tag{1.35a}$$

$$\mathbf{a}_2 = \frac{\omega\,(\mathbf{b}_3 \times \mathbf{b}_1)}{\mathbf{a}_1 \cdot (\mathbf{b}_2 \times \mathbf{b}_3)} = \frac{\omega\,(\mathbf{a}_3 \times \mathbf{a}_1)}{V^*} \tag{1.35b}$$

$$\mathbf{a}_3 = \frac{\omega\,(\mathbf{b}_1 \times \mathbf{b}_2)}{\mathbf{a}_1 \cdot (\mathbf{b}_2 \times \mathbf{a}_3)} = \frac{\omega\,(\mathbf{b}_1 \times \mathbf{b}_2)}{V^*} \tag{1.35c}$$

1.10.3 Interplanar Spacing

In addition, Fig. 1.8a illustrates the cross product $\mathbf{a}_3 = \mathbf{a}_1 \times \mathbf{a}_2$, where $\mathbf{a}_1 \times \mathbf{a}_2$ is the area of the $(hkl)_1$ plane (parallelogram), $\mathbf{a}_3 \perp (hkl)_1$ (perpendicular), and $\mathbf{a}_1 \times \mathbf{a}_2$ gives the direction of the line vector $\mathbf{a}_3$. Accordingly, $\mathbf{r}_{(hkl)} \perp (hkl)$ with $\mathbf{a}_1/h$, $\mathbf{a}_2/k$ and $\mathbf{a}_3/l$ intercepts.

For the sake of clarity, the unit cell is simply an arbitrary parallelepiped with edges a, b, c and angles α, β, γ. By proper choice of these edges, a primitive cell can be constructed (Fig. 1.7) and be characterized using Eq. (1.35). Moreover, the primitive cell has the lowest volume and that it has one lattice point, which corresponds to one atom per unit cell.

As illustrated in Figs. 1.1 and 1.8b, the interplanar spacing d_{hkl} between parallel (hkl) planes can be defined by Kelly and Knowles [6]

$$d_{hkl} = \frac{\mathbf{a}_1}{h} \cdot \frac{\mathbf{g}_{hkl}}{\|\mathbf{g}_{hkl}\|} = \frac{\mathbf{a}_1 \cdot h\mathbf{b}_1}{h\,\|\mathbf{g}_{hkl}\|} = \frac{h\mathbf{a}_1 \cdot \mathbf{b}_1}{h\,\|\mathbf{g}_{hkl}\|} \tag{1.36a}$$

$$d_{hkl} = \frac{2\pi}{\|\mathbf{g}_{hkl}\|} = \frac{1}{\sqrt{h^2\,\|\mathbf{b}_1\|^2 + k^2\,\|\mathbf{b}_2\|^2 + l^2\,\|\mathbf{b}_3\|^2}} \tag{1.36b}$$

Fundamentally, Eq. (1.36b) is applicable to cubic crystals. Notice that the greater the interplanar spacing d_{hkl}, the shorter the reciprocal lattice vector. This means that d_{hkl} is inversely proportional to g_{hkl}.

Moreover, from Eq. (1.36b), the dot product $\mathbf{g}_{hkl} \cdot \mathbf{g}_{hkl} = g_{hkl}^2 = \|\mathbf{g}_{hkl}\|^2$ yields ([6, Appendix 3])

$$g_{hkl}^2 = (h\mathbf{b}_1 + k\mathbf{b}_2 + l\mathbf{b}_3) \cdot (h\mathbf{b}_1 + k\mathbf{b}_2 + l\mathbf{b}_3) \tag{1.37a}$$

$$g_{hkl}^2 = h^2\,(\mathbf{b}_1)^2 + k^2\,(\mathbf{b}_2)^2 + l^2\,(\mathbf{b}_3)^2 \tag{1.37b}$$
$$+ 2kl\,(\mathbf{b}_2 \cdot \mathbf{b}_3) + 2lh\,(\mathbf{b}_3 \cdot \mathbf{b}_1) + 2hk\,(\mathbf{b}_1 \cdot \mathbf{b}_2)$$

Solving the dot product terms in Eq. (1.36) and rearranging the resultant expression give

$$g_{hkl}^2/(2\pi)^2 = h^2\,(b_1)^2 + k^2\,(b_2)^2 + l^2\,(b_3)^2 \tag{1.38}$$
$$+ 2klb_2b_3\cos\alpha^* + 2lhb_3b_1\cos\beta^* + 2hkb_1b_2\cos\gamma^*$$

Combining Eq. (1.33) and (1.34) yields the interaxial angles $\alpha^* = \angle\,(\mathbf{b}_2, \mathbf{b}_3)$, $\beta^* = \angle\,(\mathbf{b}_3, \mathbf{b}_1)$ and $\gamma^* = \angle\,(\mathbf{b}_1, \mathbf{b}_2)$ in terms of α, β, γ [6]

$$\cos\alpha^* = \frac{\cos\beta\cos\gamma - \cos\alpha}{\sin\beta\sin\gamma} \tag{1.39a}$$

$$\cos\beta^* = \frac{\cos\alpha\cos\gamma - \cos\beta}{\sin\alpha\sin\gamma} \tag{1.39b}$$

$$\cos \gamma^* = \frac{\cos \alpha \cos \beta - \cos \gamma}{\sin \alpha \sin \beta} \tag{1.39c}$$

These angles are important in the field of crystallography for determining the shape of the reciprocal lattice.

1.10.4 Practical Aspects of the Reciprocal Lattice

For practical purposes, the resultant mathematical definitions of the conventional or direct $\mathbf{a}_1, \mathbf{a}_2, \mathbf{a}_3$ and reciprocal primitive lattice $\mathbf{b}_1, \mathbf{b}_2, \mathbf{b}_3$ vectors for BCC, FCC, and HCP crystal structures, as shown in Fig. 1.7, are conveniently written as

FCC Crystal:

$$\mathbf{a}_1 = \frac{a}{2}(\mathbf{x} + \mathbf{y}) \qquad \mathbf{b}_1 = \frac{2\pi}{a}(-\mathbf{x} + \mathbf{y} + \mathbf{z}) \tag{1.40a}$$

$$\mathbf{a}_2 = \frac{a}{2}(\mathbf{y} + \mathbf{z}) \qquad \mathbf{b}_2 = \frac{2\pi}{a}(\mathbf{x} - \mathbf{y} + \mathbf{z}) \tag{1.40b}$$

$$\mathbf{a}_3 = \frac{a}{2}(\mathbf{x} + \mathbf{z}) \qquad \mathbf{b}_3 = \frac{2\pi}{a}(\mathbf{x} + \mathbf{y} - \mathbf{z}) \tag{1.40c}$$

BCC crystal:

$$\mathbf{a}_1 = \frac{a}{2}(\mathbf{x} + \mathbf{y} - \mathbf{z}) \qquad \mathbf{b}_1 = \frac{2\pi}{a}(\mathbf{x} + \mathbf{y}) \tag{1.41a}$$

$$\mathbf{a}_2 = \frac{a}{2}(-\mathbf{x} + \mathbf{y} + \mathbf{z}) \qquad \mathbf{b}_2 = \frac{2\pi}{a}(\mathbf{y} + \mathbf{z}) \tag{1.41b}$$

$$\mathbf{a}_3 = \frac{a}{2}(\mathbf{x} - \mathbf{y} + \mathbf{z}) \qquad \mathbf{b}_3 = \frac{2\pi}{a}(\mathbf{x} + \mathbf{z}) \tag{1.41c}$$

These are very common crystal structures found in some commercial alloys, such as α-brass and carbon steels at room temperature, respectively.

Also common is the hexagonal-closed pack (HCP) crystal structure found in Ti, Mg, and Zr metals. Thus,

HCP crystal:

$$\mathbf{a}_1 = a\mathbf{x} \qquad \mathbf{b}_1 = \frac{2\pi}{a}\left(\mathbf{x} + \frac{1}{\sqrt{3}}\mathbf{y}\right) \tag{1.42a}$$

$$\mathbf{a}_2 = \frac{a}{2}\left(-\mathbf{x} + \sqrt{3}\mathbf{y}\right) \qquad \mathbf{b}_2 = \frac{2\pi}{a}\left(\frac{2}{\sqrt{3}}\mathbf{y}\right) \tag{1.42b}$$

$$\mathbf{a}_3 = c\mathbf{z} \qquad \mathbf{b}_3 = \frac{2\pi}{a}(\mathbf{z}) \tag{1.42c}$$

where

$$\mathbf{a}_2 = a\left[-\sin\left(30°\right)\mathbf{x} + \sin\left(30°\right)\mathbf{y}\right] = (a/2)\left(-\mathbf{x}+\sqrt{3}\mathbf{y}\right) \tag{1.43a}$$

$$\mathbf{a}_2 = (a/2)\left(-\mathbf{x}+\sqrt{3}\mathbf{y}\right) \tag{1.43b}$$

Notice that $(\mathbf{a}_1, \mathbf{a}_2, \mathbf{a}_3)_{FCC} = (\mathbf{b}_1, \mathbf{b}_2, \mathbf{b}_3)_{BCC}$ and vice versa. This is also shown in a later example using different definitions for the BCC lattice vectors. Moreover, the $(\mathbf{b}_1, \mathbf{b}_2, \mathbf{b}_3)$ for the HCP crystal has been determined for comparison purposes.

Recall that parallel (hkl) planes are separated by a common d_{hkl} distance known as the interplanar spacing, which in turn, is related to a scattered vector $\mathbf{g}_{hkl}$ of the diffracted X-ray beams, $d_{hkl} \propto 1/\|\mathbf{g}_{hkl}\|$ or $\|\mathbf{g}_{hkl}\| = 1/d_{hkl}$ being reciprocal to the interplanar spacing for an arbitrary (hkl) plane, where h, k, l are called Miller indices.

Additionally, a particular (hkl) plane repeats itself within a continuous crystal and subsequently, the interplanar spacing d_{hkl} is the same between these successive planes.

Furthermore, using Eq. (1.36b) for the three (hkl) face planes depicted in Fig. 1.8b gives the corresponding interplanar spacing d_{hkl} (cited as d-spacings in the literature) as

$$d_{100} = \frac{2\pi}{\|\mathbf{g}_{100}\|} = \frac{2\pi}{b_1} \text{ for } (hkl) = (100) \tag{1.44a}$$

$$d_{010} = \frac{2\pi}{\|\mathbf{g}_{010}\|} = \frac{2\pi}{b_2} \text{ for } (hkl) = (010) \tag{1.44b}$$

$$d_{001} = \frac{2\pi}{\|\mathbf{g}_{001}\|} = \frac{2\pi}{b_3} \text{ for } (hkl) = (001) \tag{1.44c}$$

For convenience,

$$(hkl) = (100)_{1234} \| (100)_{5670} \ \& \ d_{100} = \frac{2\pi}{\|\mathbf{b}_1\|} \tag{1.45a}$$

$$(hkl) = (010)_{4150} \| (010)_{3267} \ \& \ d_{010} = \frac{2\pi}{\|\mathbf{b}_2\|} \tag{1.45b}$$

$$(hkl) = (00\bar{1})_{4073} \| (001)_{1562} \ \& \ d_{001} = \frac{2\pi}{\|\mathbf{b}_3\|} \tag{1.45c}$$

which are important definitions of the interplanar spacing (Fig. 1.8b).

For FCC-(111) and BCC-(110) planes drawn in Fig. 1.4a, c, respectively, the corresponding reciprocal lattice vectors being perpendicular to these planes are denoted as $\mathbf{g}_{(111)} \perp (111)$ and $\mathbf{g}_{(110)} \perp (110)$.

According to Eq. (1.36b), the d-spacings d_{hkl} related to these vectors are defined by

$$d_{111} = \frac{2\pi}{\|\mathbf{g}_{(111)}\|} = \frac{2\pi}{\|\mathbf{b}_1 + \mathbf{b}_2 + \mathbf{b}_3\|} \tag{1.46a}$$

$$d_{110} = \frac{2\pi}{\|\mathbf{g}_{(110)}\|} = \frac{2\pi}{\|\mathbf{b}_1 + \mathbf{b}_2\|} \tag{1.46b}$$

This simple analysis makes the reciprocal lattice approach a convenient and important method for determining d_{hkl} between identical and parallel crystallographic (hkl) planes.

Combine Eqs. (1.36a) and (1.37b) for the crystal systems in Fig. 1.7 to get d_{hkl} for a cubic crystal

$$d_{hkl}^{cubic} = \sqrt{\frac{1}{\left(h^2 + k^2 + l^2\right) \|\mathbf{b}_1\|^2}} = \frac{a}{\sqrt{\left(h^2 + k^2 + l^2\right)}} \tag{1.47a}$$

with

$$\|\mathbf{b}_1\| = \|\mathbf{b}_2\| = \|\mathbf{b}_3\| = \frac{2\pi}{a} \quad \text{(cubic)} \tag{1.47b}$$

$$\alpha^* = \beta^* = \gamma^* = 90° \tag{1.47c}$$

For a hexagonal (hex) crystal, d_{hkl} becomes

$$d_{hkl}^{(hex)} = \sqrt{\frac{1}{\left(h^2 + k^2 + hk\right) \|\mathbf{b}_1\|^2 + l^2 \|\mathbf{b}_3\|^2}} \tag{1.48a}$$

$$d_{hkl}^{(hex)} = \sqrt{\frac{3 \|\mathbf{b}_1\|^2 \|\mathbf{b}_3\|^2}{\left(4h^2 + k^2 + hk\right) \|\mathbf{b}_3\|^2 + 3l^2 \|\mathbf{b}_1\|^2}} \tag{1.48b}$$

where

$$\|\mathbf{b}_1\| = \|\mathbf{b}_2\| = 2\pi \left[2/\left(a\sqrt{3}\right)\right] \quad \text{(hexagonal)} \tag{1.49a}$$

$$\|\mathbf{b}_3\| = 2\pi/c \tag{1.49b}$$

$$\alpha^* = \beta^* = 90° \text{ and } \gamma^* = 60° \tag{1.49c}$$

The d_{hkl} equations for other crystal systems can be found in [2, 6]. Moreover, the given theoretical background for defining d_{hkl} is of significant importance in X-ray diffraction and crystallography for characterizing unit cell features and identifying the crystal structure of a solid crystal structure [8–11].

Additionally, careful analysis of an X-ray diffraction pattern (diffractogram) and the subsequent indexing procedure required for this data is a conventional method for determining the morphology of a unit cell in crystal structure. In other words, this leads to the determination of the atomic arrangement of a crystal structure having a specific geometry and size described by the absolute values of $\|\mathbf{a}_i\|$ and $\|\mathbf{b}_i\|$ lattice vectors.

Example 1.5 This example illustrates a simple procedure (method) for indexing diffraction peaks. Determining Miller indices for the (hkl) planes depicted in Fig. 1.2 and the lattice parameter for BCC α-Fe is called ferrite in the steel field. The method requires use of Eqs. (1.2) and (1.48) along with the diffraction angles 2θ for each diffraction peak and the wavelength $\lambda = 0.1527$ nm used to obtain an X-ray diffraction pattern.

Solution Let $K = \lambda/(2d_{hkl})$ in Eq. (1.2) and tabulate the computations by (1) assigning a real number to K based on trial and error so that $K \sin^2 \theta$ is a small integer number, which must be even number for a BCC structure, (2) assigning this integer to $(h^2 + k^2 + l^2)$ and (3) finding a suitable combination of the Miller indices for the (hkl) planes. Hence,

$$\sin^2 \theta = \frac{\lambda^2}{4a^2} \left(h^2 + k^2 + l^2 \right)$$

The partial results are

2θ	θ	$\sin^2 \theta$	K	$K \sin^2 \theta$	$(h^2 + k^2 + l^2)$	(hkl)
44.20	22.10	0.14154	14.130	2	2	110
64.40	32.20	0.28758	14.087	4	4	200
81.32	40.66	0.44097	14.133	6	6	211

Notice that the trial and error calculations for obtaining nearly constant K values are based on three decimal places. Afterward, calculate *the d-spacing* d_{hkl} from $K = \lambda/(2d_{hkl})$ values and the lattice parameter a using Eq. (1.47). These results are subsequently used for further calculations. The new tabulated results are shown below

θ	(hkl)	d_{hkl} (nm)	a (nm)
22.10	110	0.20294	0.28700
32.20	200	0.14328	0.28656
40.66	211	0.11718	0.28703

Take the average lattice parameter a so that

$$a = 0.28686 \, \text{nm} \simeq 0.287 \, \text{nm}$$

$$a = b = c = 0.287 \, \text{nm}$$

$$\alpha = \beta = \gamma = 90°$$

and the volume of the α-BCC unit cell is

$$V = abc = a^3 = 0.02364 \, \text{nm}^3$$

Notwithstanding the fact that the X-ray diffraction pattern showing the relative positions of the diffraction peaks is not included in this example, only three crystallographic planes are sufficient to determine the size and shape of the BCC unit cell. Suffice it to say that additional crystallographic planes lead to more accurate results. Obviously, indexing crystal structures using the above procedure is nowadays carried out using commercial software.

Example 1.6 The primitive lattice vectors $(\mathbf{a_1, a_2, a_3})$ for (a) SC, (b) BCC, and (c) FCC crystals are drawn below in orthogonal x, y, z axes. Determine the reciprocal lattices $(\mathbf{a_1, b_2, b_3})$ for each crystal. Explain.

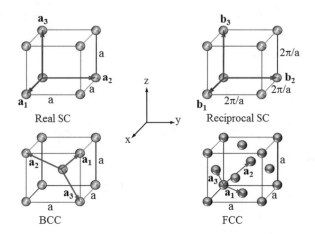

Solution

(a) The lattice vectors for the conventional SC unit cell are

$$\mathbf{a_1} = a\mathbf{x}$$

$$\mathbf{a_2} = a\mathbf{y}$$

$$\mathbf{a_3} = a\mathbf{z}$$

since $b = a$ and $c = a$. Therefore, the conventional SC unit cell geometry is itself the reciprocal unit cell and the lattice. From Eq. (1.31) along with the volume of the primitive unit cell $V = \mathbf{a_1} \cdot (\mathbf{a_2} \times \mathbf{a_3})$,

$$b_1 = \frac{\omega\,(a_2 \times a_3)}{a_1 \cdot (a_2 \times a_3)} = \frac{\omega\,(ay) \times (az)}{ax \cdot [(ay) \times (az)]} = \frac{\omega}{a}x$$

$$b_2 = \frac{\omega\,(a_3 \times a_1)}{a_1 \cdot (a_2 \times a_3)} = \frac{\omega\,(az) \times (ax)}{a^3 x \cdot [y \times z]} = \frac{\omega}{a}y$$

$$b_3 = \frac{\omega\,(a_1 \times a_2)}{a_1 \cdot (a_2 \times a_3)} = \frac{\omega\,(ax) \times (ay)}{a^3 x \cdot [y \times z]} = \frac{\omega}{a}z$$

From Eqs. (1.31) and (1.36b),

$$g_{hkl} = \frac{\omega}{a}(h x + k y + l z)$$

$$d_{hkl} = \frac{\omega}{\|g_{hkl}\|} = \frac{a}{\sqrt{h^2 + k^2 + l^2}}$$

which is commonly cited in the literature as a standalone equation.

(b) For the BCC crystal,

$$a_1 = \frac{a}{2}\,(x + y + z)\,, \quad a_2 = \frac{a}{2}\,(x - y + z)\,, \quad a_3 = \frac{a}{2}\,(x + y - z)$$

$$V = a_1 \cdot (a_2 \times a_3) = \frac{a^3}{2}$$

From Eq. (1.34), the reciprocal lattice vectors are

$$b_1 = \frac{\omega\,(a_2 \times a_3)}{a_1 \cdot (a_2 \times a_3)} = \frac{\omega}{V}\,(a_2 \times a_3) = \omega\left(\frac{a^2/4}{a^3/2}\right)(x - y + z) \times (x + y - z)$$

$$= \frac{\omega}{a}\,(z + y - z + x + y + x) = \frac{\omega}{a}\,(y + z)$$

$$b_2 = \frac{\omega\,(a_3 \times a_1)}{a_1 \cdot (a_2 \times a_3)} = \frac{\omega}{a}\,(x + z)$$

$$b_3 = \frac{\omega\,(a_1 \times a_2)}{a_1 \cdot (a_2 \times a_3)} = \frac{\omega}{a}\,(x + y)$$

From Eq. (1.31) and (1.39b),

$$g_{hkl} = \frac{\omega}{a}\,[h\,(y + z) + k\,(x + z) + l\,(x + y)]$$

$$d_{hkl} = \frac{\omega}{\|g_{hkl}\|} = \frac{a}{\sqrt{h^2 + k^2 + l^2}}$$

(c) For a conventional FCC crystal,

$$\mathbf{a}_1 = \frac{a}{2}\,(\mathbf{x} + \mathbf{y})\,,\ \ \mathbf{a}_2 = \frac{a}{2}\,(\mathbf{y} + \mathbf{z})\,,\ \ \mathbf{a}_3 = \frac{a}{2}\,(\mathbf{x} + \mathbf{z})$$

Then, the reciprocal lattice vectors become

$$\mathbf{b}_1 = \frac{\omega\,(\mathbf{a}_2 \times \mathbf{a}_3)}{\mathbf{a}_1 \cdot (\mathbf{a}_2 \times \mathbf{a}_3)} = \frac{\omega}{a}\,(-\mathbf{x} + \mathbf{y} + \mathbf{z})$$

$$\mathbf{b}_2 = \frac{\omega\,(\mathbf{a}_3 \times \mathbf{a}_1)}{\mathbf{a}_1 \cdot (\mathbf{a}_2 \times \mathbf{a}_3)} = \frac{\omega}{a}\,(\mathbf{x} - \mathbf{y} + \mathbf{z})$$

$$\mathbf{b}_3 = \frac{\omega\,(\mathbf{a}_1 \times \mathbf{a}_2)}{\mathbf{a}_1 \cdot (\mathbf{a}_2 \times \mathbf{a}_3)} = \frac{\omega}{a}\,(\mathbf{x} + \mathbf{y} - \mathbf{z})$$

From Eq. (1.31) and (1.39b),

$$\mathbf{g}_{hkl} = \frac{\omega}{a}\,[h\,(-\mathbf{x} + \mathbf{y} + \mathbf{z}) + k\,(\mathbf{x} - \mathbf{y} + \mathbf{z}) + l\,(\mathbf{x} + \mathbf{y} - \mathbf{z})]$$

$$d_{hkl} = \frac{\omega}{\|\mathbf{g}_{hkl}\|} = \frac{a}{\sqrt{h^2 + k^2 + l^2}}$$

Therefore, $(\mathbf{b}_1, \mathbf{b}_2, \mathbf{b}_3)_{BCC} = \omega\,(\mathbf{a}_1, \mathbf{a}_2, \mathbf{a}_3)_{FCC}$ where $\omega = 2\pi$ in solid-state physics and $\omega = 1$ in crystallography. In fact, the 2π factor is just a matter of mathematical convenience related to Fourier transform. For BCC α-Fe with $a = 0.287$ nm and $(h11)$ planes, the above d_{hkl} equation yields

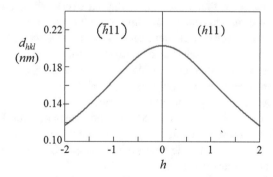

1.11 Single-Crystal Structure

In essence, the main goals are to find relations between the $X\,(hkl)$ and $(hkl)_{red}$ surface structures and associated surface properties using single-crystal electrodes. On the other hand, electrochemical reactions on polycrystalline surface electrodes are more difficult to characterize due to, at least, grain boundary effect and different (hkl) surface planes in each grain.

Despite that the making of single crystals is out of the scope of this chapter, the characterization of them follows the cited theory of X-ray diffraction and the Science of Crystallography since the atoms are symmetrically arranged in a three-dimensional periodic lattice. Hence, a simple translational vector, $\mathbf{r}_{uvw}$, defined by Eq. (1.30) and a reciprocal lattice scattering vector, $\mathbf{g}_{hkl}$, described by Eq. (1.37) are the relevant entities in crystallographic analysis of crystalline structures.

This section includes aspects of electrochemical science related to surface protection or reconstruction and explains the level of importance of a solid crystallographic surface based on the type of (hkl) planes. Conversely, a polycrystalline solid surface has different (hkl) planes with preferred orientations and is separated by grain boundaries, which in turn, are considered as the natural form of surface defect due to atomic mismatch. The gain boundaries are potential sites for undesirable and unlikable electrochemical events, such as corrosion and unstable layer deposition. Other types of surface defects are inevitably present and may affect the surface properties of the solid.

Ideally, a single-crystal substrate without surface defects and with a (hkl) surface plane is commonly referred to as $X\,(hkl)$, where X is a metallic element. For example, the notation $Cu\,(111)$ implies that the substrate is $X = Cu$ and $(hkl) = (111)$ is the lattice surface plane. Thus, knowing the (hkl) plane is essential in surface science.

In general, the preparation of a single-crystal or a polycrystal electrode requires some sensitive steps related to cleanliness. Firstly, the electrode in contact with the electrolyte has to be electrically connected to the electrochemical cell under potential control. This is obviously an important step in order to avoid natural surface reconstruction in the form of an adherent overlayer or ion deposition in the form of a thin film. This can be accomplished by selecting the proper electric potential E to maintain the electrochemical cell in a passive mode so that Faraday's current does not flow prior to the start of an experiment.

In principle, this particular electric potential may correspond to an electrical double layer on the electrode surface. The theory of electrical double layer, which is also known as the double layer, includes the electrostatic properties and potential energy for interactions between charged electrode surface. This leads to charge distribution of ions in a suitable electrolyte (electrically conductive solution) [12]. Interestingly, if two parallel double layers are in contact, then they get aligned with opposite charge at the their interface.

Amazingly, the principles related to electrical double layer are reasonably understood in the electrochemical field. This is a particular topic the reader ought to explore, at least, from a theoretical point of view to gain some insights on this subject matter. A literature survey would reveal several models explaining the ionic characteristics related to the ionic behavior within electrochemical systems. All models have particular physical characteristics based on assumptions.

1.12 Polycrystalline Crystal Structure

As previously mentioned, grain boundaries and matrix–particle interface are two-dimensional defects that have the appearance of irregular lines under the microscope. Figure 1.7a shows the microstructure of an AISI 304 stainless steel and Fig. 1.7b depicts a partially amorphous microstructure of $Ni_{53}Mo_{35}B_9Fe_2$ alloy. The former is a typical FCC microstructure and the latter contains boride particles embedded in the $Ni_{53}Mo_{35}B_9Fe_2$ amorphous phase (no grain boundaries), as revealed by X-ray diffraction [13].

The grain boundaries in crystalline solids represent high-energy areas due to the atomic mismatch and therefore, they are considered microstructural defects, which corrode more rapidly than the grain surfaces. Figure 1.7 illustrates two microstructures of $AISI$ 304 stainless steel being annealed at 1000 °C for 0.5 and 24 h at 1100 °C for a rapidly solidified and consolidated $Ni_{53}Mo_{35}B_9Fe_2$ (Devitrium 7025) alloy. Figure 1.7a shows the grain boundaries as dark lines because of the severe chemical attack using an Aqua Regia etching solution (80%HCl + 20%HNO_3).

In general, grain and interface boundaries affect the electrical and thermal conductivity of the material, but they are the preferred sites for the onset of corrosion and for the precipitation of new phases from the solid state.

Figure 1.7b shows crystalline particles embedded in an Ni-Mo matrix after being etched with Marble's reagent. Denote that the RSA alloy does not have visible grain boundaries, but it is clear that the severe chemical attack occurred along the matrix–particle interfaces due to localized galvanic cells [13]. These interfaces appear as bright areas due to optical effects.

It suffices to say that one has to examine the surface microstructure and its topology for a suitable microstructural interpretation. Thus, the fundamental understanding of corrosion mechanism on crystalline and amorphous alloys is enhanced when the microstructural interpretation is satisfactorily accomplished.

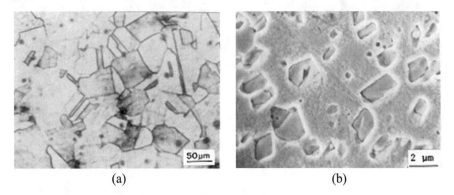

(a) (b)

Fig. 1.7 (**a**) Crystalline microstructure of AISI 304 stainless steel and (**b**) partially amorphous microstructure $Ni_{53}Mo_{35}B_9Fe_2$ alloy containing embedded boride particles [13]

Corrosion mechanisms, for example, must be analyzed, preferably at all scales possible using microscopic techniques by separating the relative importance of microstructural effects on critical events such as stress corrosion crack (SCC) initiation, subsequent growth, and consequent unstable fracture. In fact, stress corrosion cracking is a very complex topic many researchers have been studying for decades.

Figure 1.8 depicts the dislocation networks for the 304 stainless steel and $Ni_{53}Mo_{35}B_9Fe_2$ alloys. These dislocations are defects generated during plastic deformation in tension mode.

Another metallurgical aspect to consider is the dislocation network encounter in plastically deformed alloys. In general, dislocations are linear defect, which can act as high-energy lines and consequently, they are susceptible to corrode as rapidly as grain boundaries in a corrosive medium. Figure 1.8 illustrates dislocation networks in an $AISI$ 304 stainless steel and in RSA $Ni_{53}Mo_{35}B_9Fe_2$. The relevant pretreatment conditions can be found in [13].

With respect to Fig. 1.8b, there is a clear sub-grain boundary shown as a dark horizontal line across the upper part of the TEM photomicrograph. The small white areas surrounded by dislocations are called sub-grains, which are crystal having an FCC structure for both alloys.

Despite the few microstructural features shown in this chapter, it is clear that morphology of defects on crystal surfaces may be detrimental to a solid component design life. This is an important issue that must be considered in order to keep a structural integrity as prolong as possible. Thus, the assessment of crystal structures

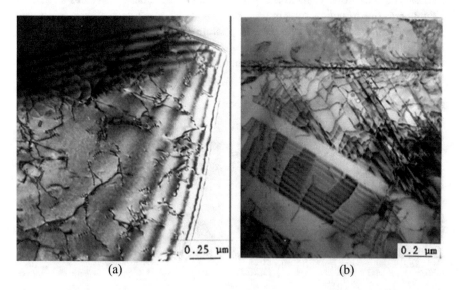

(a)   (b)

Fig. 1.8 Bright field TEM photomicrographs showing dislocation networks for (**a**) AISI 304 stainless steel and (**b**) $Ni_{53}Mo_{35}B_9Fe_2$ alloy [13]

and defects is part of material's characterization using available techniques, such as X-ray diffraction and related crystallography.

Interestingly, the current solar panel technology calls for either monocrystalline or polycrystalline materials. Apparently, the efficiency of solar panels based on climatic data measurements is a decisive factor for selecting the proper material. In this particular case, the efficiency is simply related to the conversion of solar radiation to useful energy.

Furthermore, Devitrium alloy 7025, denoted as an $Ni_{53}Mo_{35}B_9Fe_2$ rapidly solidified alloy [13, 14], is an extruded Ni-based superalloy with its properties being influenced by the extrusion process. Of particular interest is the alloy X-ray diffraction of heat-treated microstructures containing stable precipitated boride Mo_2NiB_2 (intermetallic phase) which precipitated at high temperatures and subsequent stress corrosion cracking studies [14].

Lastly, rapidly solidified alloys (RSAs) can be produced using techniques like melt spinning (MS) and planar flow casting (PFC) by forcing a molten metal pass through a nozzle and immediately makes contact with a water-cooled copper (Cu) or copper beryllium (Cu-Be) rotating wheel (moving substrate). As a result, $100\,\mu$m thick ribbons or flakes with ultrafine grained or amorphous microstructure are obtained by a rapid solidification (RS).

1.13 Industrial Applications

In the aerospace industry, directionally solidified nickel-based alloys are used to manufacture turbine helical blades and vanes for jet engines. The ultimate goal is to avoid defects and impurities in the crystal structure. One particular reason for using single crystals in high temperature applications is to avoid grain boundary sliding due to a deformation mechanism called creep, which is a common cause of failure in polycrystalline turbine blades.

In the semiconductor industry, the Bridgman and Czochralski techniques are nowadays used to produce single crystals in induction heating environments. The latter technique, described in a later chapter, is a common time-consuming and expensive method for producing large silicon crystals, known as monocrystalline crystals, which are subsequently sliced into wafers for further treatment as semiconductors.

In the electronic industry, the use of single-crystal components is an important choice since polycrystalline materials containing grain boundaries decrease the electric conductivity.

In electrochemical science, a defect-free single crystal represented by X (hkl) is the ideal electronic conductor for characterizing reduction or oxidation reactions at pressure P and temperature T. Normally, reduction reactions prevail in most cases for revealing the $(hkl)_{red}$ plane that forms on the X (hkl) surface and evaluating the adsorption of organic molecules with a voltage sweep method or the deposition of a metallic ions on an X (hkl) electrode surface.

Corrosion, in general, occurs when unstable refined materials tend to reach a stable state by forming metal oxides (Fe_2O_3, MgO, Al_2O_3, etc.), hydroxides (iron hydroxide $Fe(OH)_2$, aluminum hydroxide, $Al(OH)_3$, etc.) or sulfides (chalcocite Cu_2S, Pyrite FeS_2, etc.).

Recall from chemistry that adsorption is an adhesion process of atoms, ions, or molecules that form overlayers on substrate surfaces. Hence, crystallography and electrochemistry are coupled in order to make atomic adjustments on surface (hkl) planes through adsorption and deposition of atoms from a gas or liquid phase (state of matter with uniform chemical composition and physical state).

Furthermore, electrodeposition is a plating process that can be treated as an electrochemically induced liquid–solid phase transformation at the electrode surface containing crystal defects. In particular, the solid phase is a state of matter with uniform chemical composition and physical crystal state with respect to the atomic arrangement that forms the smallest particle-like called unit cell. Thus, a large group of unit cells form a crystal lattice and many crystals with different orientations in a solid material form grains, separated by the atomic mismatch arrangement known as grain boundaries.

In crystalline solids, the solid phase is simply the random arrangement of crystals and in amorphous solid materials, the solid phase does not have unit cells but a non-uniform arrangement of spherical atoms. Therefore, an amorphous structure lacks a long-range order (translational periodicity).

For the sake of clarity, a long-range order is fundamentally a translational periodicity over large numbers of atomic diameters. This concept is mostly applicable to metallic solids as a quantitative measurement of the atomic distance within the crystal lattice, whereas liquids have short-range order.

1.14　Summary

A description of crystallography as a surface science implies that electrode crystal structure must be determined in order to have known crystallographic surface initial conditions. This has been shown to be essential using X-ray diffraction technique and related reciprocal lattice vectors for revealing the unit cells of crystals. On the other hand, the mechanical behavior and mechanical properties of solids being subjected to external loads strongly are affected by solid surface defects and the aggressiveness of the environment.

X-ray diffraction is briefly described in order to elucidate how this technique provides means to reveal the atomic structure of crystalline solids. Hence, crystalline solids subjected to static and dynamic forces, surface reactions, or a combination of these respond in a particular manner based on the type and orientation of unit cells. Therefore, it is important to determine the type of crystal structure of a material prior to an application within a suitable environment. However, a crystalline material may be modeled as a defect-free structure, but this is would be unrealistic because defects have a high degree of control on properties.

Since the main topic in this chapter is crystallography, the general description of crystal structures plays a significant role on film or overlayer formation, corrosion, and related industrial applications. Specifically, the significant aspects of crystallography have been introduced as conventional and primitive unit cells and the corresponding crystallographic planes and directions. This leads to study vector algebra associated with reciprocal lattice, rotation matrix, and the well-recognized dot and cross products. Moreover, the interplanar d_{hkl}-spacing being mathematically defined as the inverse of the magnitude of the reciprocal lattice vector $\mathbf{g}_{hkl}$ is an essential crystallographic variable related to X-ray diffraction and crystal size.

Lastly, single and polycrystalline structures are briefly introduced as potential solid structures that have important industrial applications. Special interest is given to crystalline structures and related linear crystal defects called dislocations.

1.15   Appendix 1A Fourier Transform

This appendix is devoted to a brief description of the Fourier transform (FT) related to X-ray diffraction intensity (I) in a crystal or unit cell. Essentially, X-rays in the form of a high-energy beam travel in space parallel to a vector $\overrightarrow{s}$, causing scattering events known as diffraction at a particular Bragg's angle (θ) with respect to a vector $\overrightarrow{r}$. Figure 1A.1a shows an X-ray diffraction model indicating the trajectories of the incident and diffracted beams at a particular target. On the other hand, Fig. 1A.1b illustrates the X-ray pattern ($I = f(2\theta)$ of a polycrystalline tungsten (W) sample ([3, p. 103]). Moreover, a polycrystalline material shows more X-ray peaks than its single crystal form due to the randomly oriented grains.

Mathematically, the structure factor $F\left(\overrightarrow{s}\right)$ and the scattering factor $f\left(\overrightarrow{r}\right)$ are included in a general FT equation of the form ([15, p. 22])

$$F\left(\overrightarrow{s}\right) = \int \int \int f\left(\overrightarrow{r}\right) \exp\left[+2\pi i \left(\overrightarrow{s} \cdot \overrightarrow{r}\right)\right] dV \qquad (1A)$$

with the inverse FT defined as

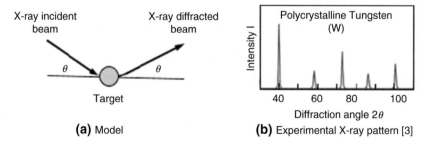

(a) Model   **(b)** Experimental X-ray pattern [3]

Fig. 1A.1 (a) X-ray diffraction model and (b) polycrystalline X-ray pattern

$$f\left(\vec{r}\right) = \int \int \int F\left(\vec{s}\right) \exp\left[-2\pi i \left(\vec{s} \cdot \vec{r}\right)\right] dV \qquad (2A)$$

Here, $dV = dxdydz$ denotes the volume of a cell, $\vec{r}$ denotes a particular atomic position in the unit cell, $i = \sqrt{-1}$ denotes a complex number, and $\vec{s} \cdot \vec{r}$ denotes the dot product defined as

$$\vec{s} \cdot \vec{r} = \left(h\vec{a}^* + k\vec{b}^* + l\vec{c}^*\right) \cdot \left(x\vec{a} + y\vec{b} + z\vec{c}\right) \qquad (3A.1)$$

$$\vec{s} \cdot \vec{r} = hx + ky + lz \qquad (3A.2)$$

where (a^*, b^*, c^*) denote the reciprocal lattice and (a, b, c) denote the conventional lattice parameters of a unit cell, (h, k, l) denote the Miller indices of a plane, and (x, y, z) denote the Cartesian coordinates.

Substituting Eq. (3A.2) into (1A) and (2A), and letting $F\left(\vec{s}\right) = F_{hkl}$ and $f\left(\vec{r}\right) = f_{hkl}$ yield

$$F_{hkl} = \int \int \int f_{hkl} \exp\left[+2\pi i \left(hx + ky + lz\right)\right] dV \qquad (4A.1)$$

$$f_{hkl} = \int \int \int F_{hkl} \exp\left[-2\pi i \left(hx + ky + lz\right)\right] dV \qquad (4A.2)$$

The sought X-ray diffraction intensity (I) is defined by the square absolute value of the structure factor F_{hkl}, which represents the electromagnetic wave amplitude. Thus,

$$I = |F_{hkl}|^2 \qquad (5A)$$

In addition, the reader is recommended to evaluate Chapter 4 and sec. 10.9 in Cullity's book [3] for details on the structure factor F_{hkl}.

1.15.1   Energy Field

According to Prof. Paul Heiney (University of Pennsylvania) video found at https://repository.upenn.edu/xray_scattering_math/1/, X-rays travel in space with an electric field energy $\vec{E}\left(r, t\right)$ dependent on position (r) and time (t). For an electromagnetic wave traveling along the x-axis with speed of light c, wavelength λ, and frequency f. The electric field vector $\vec{E}\left(r, t\right)$ can be written as

$$\vec{E}\left(r, t\right) = \vec{E}_o \cos\left[\frac{2\pi \left(x - ct\right)}{\lambda}\right] = \vec{E}_o \cos\left(kz - \omega t\right) \qquad (6A.1)$$

$$\vec{E}\,(r,t) = \mathrm{Re}\left(\vec{E}_o \exp\left[i\,(kz - \omega t)\right]\right) \qquad (6A.2)$$

where $\vec{E}_o$ is a vector pointing in some direction, $\cos(kz - \omega t)$ describes the traveling electromagnetic wave, and

$$k = \frac{2\pi}{\lambda} \qquad (7A.1)$$

$$\omega = 2\pi f = \frac{2\pi c}{\lambda} = ck \qquad (7A.2)$$

Denote that an equivalent Euler's exponential formulation, $\exp(\pm i\phi) = \cos\phi \pm i\sin\phi$ with $\phi = kz - \omega t$, can be used in Eq. (6A.2), where "Re" is the real part of the complex notation.

The X-ray diffraction intensity (I) is now defined as the square absolute value of $\vec{E}\,(r,t)$. Thus,

$$I = \left|\vec{E}\,(r,t)\right|^2 \qquad (8A)$$

which is a form of energy amplitude.

1.16   Appendix 1B Ewald Sphere

The Ewald sphere (ES) is also known as the sphere of reflection with radius $1/\lambda$ passing through the origin O of the reciprocal lattice, where λ is the wavelength of the X-ray incident beam. For convenience, Fig. 1B.1 shows the Ewald circle of radius $1/\lambda$ and the reciprocal lattice vector $\mathbf{CP} = \mathbf{g}_{hkl}$.

This circle is constructed such that the crystal plane is located at the center of the circle. The proper construction of the circle is as follows:

- Draw a circle of radius $OB = 1/\lambda$
- The circumference must intercept a lattice point "P"
- Draw the BP line to define 2θ
- Draw the OP line to define θ
- Draw the CP line to define $\|g_{hkl}\| = 1/d_{hkl}$.

Thus, any point (atom) on the Ewald's circle is diffracted and subsequently, detected and recorded for further X-ray or electron analysis. The latter is an attractive technique since the wavelength of the moving electrons is smaller than the spacing of atomic (hkl) planes and as a result, a strong diffraction intensity is achieved. This brief introduction of the Ewald's circle model allows one to conclude that the corresponding geometric construction of the lattice of points of

Fig. 1B.1 (a) Lattice points
(P's) and (b) Ewald's circle
for defining the reciprocal
lattice vector g_{hkl}

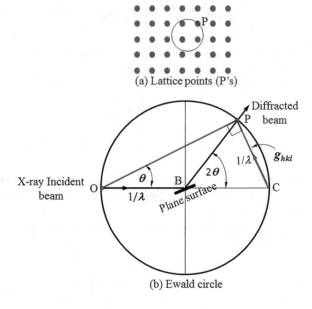

(a) Lattice points (P's)

(b) Ewald circle

equal symmetry is based on the relationship between the wavelength of the incident
and diffracted X-ray beams.

1.17   Problems

1.1 (a) Derive Eq. (1.28) for the rotated coordinate system shown below to derive
the two-dimensional rotation matrix and (b) determine the reducible matrix for
twofold $(2\pi/n = 2\pi/2 = \pi)$ and threefold $(2\pi/n = 2\pi/3)$ rotations, and the
eigenvalues λ.

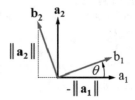

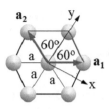

1.2 (a) Find the lattice vector equations for the translation of the hexagonal space lattice in orthogonal Cartesian $(\mathbf{x}, \mathbf{y}, \mathbf{z})$ unit vectors. The primitive lattice is the small oblique parallelogram (red) shown above. (b) Calculate the area of the parallelogram with a lattice parameter of graphite $a = 25$ nm.

1.3 (a) Determine if $\mathbf{a}_1 = 2i - j + 3k$ and $\mathbf{b}_1 = 4i - j + 3k$ are perpendicular, parallel or neither. (b) Find $\mathbf{a}_1 \times \mathbf{b}_1$.

1.4 Using the idea of the dot product on the general unit cell shown in the sketch below, (a) derive a 3×3 matrix for the general tensor r_{ij} defined by

$$r_{ij} = \mathbf{a}_i \cdot \mathbf{a}_j = \|\mathbf{a}_i\| \|\mathbf{a}_j\| \cos\theta$$

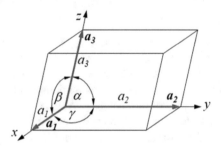

(b) Find $\det r_{ij}$ for a BCC crystal if $a_1 = a_2 = a_3 = a = 1/3$ nm.

1.5 Consider an hexagonal crystal structure described by the lattice vectors

$$\mathbf{a}_1 = a\mathbf{x} = a\,(1, 0, 0)$$

$$\mathbf{a}_2 = \frac{a}{2}\mathbf{x} + \frac{\sqrt{3}a}{2}\mathbf{y} = a\left(\frac{1}{2}, \frac{\sqrt{3}}{2}, 0\right)$$

$$\mathbf{a}_3 = c\mathbf{z} = c\,(0, 0, 1)$$

Determine (a) $\mathbf{b}_1, \mathbf{b}_2, \mathbf{b}_3$ and (b) θ between $\mathbf{a}_1$ and $\mathbf{b}_1$.

1.6 (a) Calculate the angle θ between [110] and [111] directions (vectors) in a cubic crystal using the dot product. (b) Determine if these directions are orthogonal, non-orthogonal, or parallel to each other. Recall that a cubic crystal has orthogonal x, y, z axes.

1.7 Combine Eqs. (1.4) and (1.35b) to derive the d_{hkl} expression for a simple cubic (*SC*).

1.8 Calculate length of $\|\mathbf{b}_i\|$, $\|\mathbf{g}_{hkl}\|$,and d_{hkl} associated with the (200) plane in a cubic structure. Assume a lattice constant of $a = 0.287$ nm.

1.9 It is well known that diffraction of X-rays is described by the Bragg equation and that the reciprocal lattice scattering vector $\mathbf{g}_{hkl}$ is related to the inverse of the interplanar spacing d_{hkl}. Using this information, calculate the diffraction angle θ for (110), (200), and (211) planes shown in Fig. 1.2. Determine the order of reflection taking into account if the Miller indices (*hkl*) are reducible or not. Let the lattice parameter and the electromagnetic radiation wavelength be $a = 0.287$ nm and $\lambda = 0.1527$ nm, respectively.

1.10 Consider an orthogonal coordinate system containing the (*hkl*) plane described by the vectors $\mathbf{a}_1, \mathbf{a}_2$ given below. (a) Find the coefficients p_i and q_i for the primitive reciprocal lattice vectors b_1, b_2 using Eqs. (1.35a) and (1.35b) and (b) the angle $\theta = \measuredangle (\mathbf{a}_1, \mathbf{b}_1)$.

$$\mathbf{a}_1 = a\mathbf{x} = a\,(1, 0)\,; \qquad\qquad \mathbf{b}_1 = p_1\mathbf{x} + q_1\mathbf{y}$$

$$\mathbf{a}_2 = \frac{a}{2}\mathbf{x} + \frac{\sqrt{3}a}{2}\mathbf{y} = a\left(\frac{1}{2}, \frac{\sqrt{3}}{2}\right); \quad \mathbf{b}_2 = p_2\mathbf{x} + q_2\mathbf{y}$$

1.11 (a) Show that g_{hkl} is perpendicular, $g_{hkl} \perp (hkl)$, to nonparallel vectors $\mathbf{A}, \mathbf{B}, \mathbf{C}$ vectors as shown in Fig. 1.8a. This implies that g_{hkl} and (*hkl*) are orthogonal so that the angle between them is $90° = \pi/2$ radiants. This work requires that vectors $\mathbf{A}, \mathbf{B}, \mathbf{C}$ on the (*hkl*) plane edges be defined in terms of the intercept points. Subsequently, take the dot product $g_{hkl} \cdot \mathbf{A}$, $g_{hkl} \cdot \mathbf{B}$ and $g_{hkl} \cdot \mathbf{C}$. (b) Derive a general expression for d_{hkl} for a cubic crystal and (c) calculate values of $\|\mathbf{b}_i\|$, $\|\mathbf{g}_{hkl}\|$ and d_{hkl} for the three (*hkl*) planes shown in Fig. 1.2 if the lattice constant is $a = 0.287$ nm.

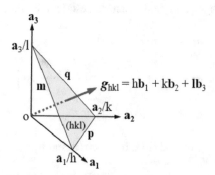

1.12 (a) Combine the Bragg equation and the reciprocal lattice scattering vector $\mathbf{g}_{hkl}$ to calculate the diffraction angles for the (hkl) planes shown in Fig. 1.2. (b) Compare the results with those given in Example 1.5. Is $\mathbf{g}_{hkl}$ comparable with Bragg equation? (c) Deduce the order of diffraction n for these planes by reducing the Miller indices to their smallest integer values.

1.13 For a simple cubic, the reciprocal lattice vectors are defined by

$$\mathbf{b}_1 = \frac{2\pi}{a}\, \vec{x}, \ \ \mathbf{b}_2 = \frac{2\pi}{a}\, \vec{y}, \ \ \mathbf{b}_3 = \frac{2\pi}{a}\, \vec{z}$$

1.14 Prove that $e^{i\mathbf{g}\cdot\mathbf{r}} = 1$ and n is an integer number in Eq. (1.33).
(a) Derive an equation for the interplanar spacing d_{hkl} and (b) show that

$$\mathbf{b}_1 \cdot (\mathbf{b}_2 \times \mathbf{b}_3) = \frac{(2\pi)^3}{\mathbf{a}_1 \cdot (\mathbf{a}_2 \times \mathbf{a}_3)}$$

1.15 Use the triangular (hkl) plane to show that $\mathbf{g}_{hkl} \perp (hkl)$.

References

1. R.C. Alkire, R.D. Braatz, Electrochemical engineering in an age of discovery and innovation. AIChE J. **50**(9), 2000–2007 (2004)
2. B.D. Cullity, *Elements of X-Ray Diffraction*, 2nd edn. (Addison-Wesley, New York, 1978). ISBN 0-201-01174-3
3. W.D. Callister, D.G. Rethwisch, *Materials Science and Engineering: An Introduction* (Wiley, New York, 2014). ISBN:978-1-118-32457-8
4. D. Senich, X-ray diffraction and adsorption isotherm studies of the calcium montmorillonite-H_2O System, Ph.D. Dissertation, Digital Repository, Iowa State University, Ames, 66-10 438 (1966)
5. K. Hermann, *Crystallography and Surface Structure: An Introduction for Surface Scientists and Nanoscientists* (Wiley, Berlin, 2011)
6. A. Kelly, K.M. Knowles, *Crystallography and Crystal Defects*, 2nd edn. (Wiley, West Sussex, 2012)
7. Y. Waseda, E. Matsubara, K. Shinoda, *X-Ray Diffraction Crystallography: Introduction, Examples and Solved Problems* (Springer, Berlin, 2011). ISBN 978-3-642-16634-1
8. U. Shmueli, Reciprocal space in crystallography, in *International Tables for Crystallography*, ed. by U. Shmueli. International Union of Crystallography 2001, vol. B, 2nd edn., Chapter 1.1 (Kluwer Academic Publishers, Amsterdam, 2006), pp. 2–9. ISBN 0-7923-6592-5
9. G. Bricogne, Fourier transforms in crystallography: theory, algorithms and applications, in *International Tables for Crystallography*, ed. U. Shmueli. International Union of Crystallography 2001, vol. B, 2nd edn., Chapter 1.3 (Kluwer Academic Publishers, Amsterdam, 2006), pp. 25–98. ISBN 0-7923-6592-5
10. C. Kittel, *Introduction to Solid State Physics*, 8th edn. (Wiley, New York, 2005). ISBN 0-471-41526-X
11. G.-C. Wang, T.-M. Lu, *RHEED Transmission Mode and Pole Figures: Thin Film and Nanostructure Texture Analysis* (Springer, New York, 2014). ISBN 978-1-4614-9286-3

12. L.A. Kibler, *Preparation and Characterization of Noble Metal Single Crystal Electrode Surfaces*. Department of Electrochemistry, University of Ulm (2003), pp. 1–55. Copyright (C) International Society of Electrochemistry
13. N.L. Perez, Evaluation of thermally degraded rapidly solidified Fe-base and Ni-base alloys. Dissertation, University of Idaho, Moscow (1988)
14. N.L. Perez, T.A. Place, X-ray diffraction of a heat-treated rapidly solidified Ni53Mo35Fe9B2 Alloy. J. Mater. Sci. Lett. **9**(8), 940–942 (1990)
15. R. Guinebretiere, *X-Ray Diffraction by Polycrystalline Materials* (ISTE USA, Newport Beach, 2007)

Chapter 2
Surface Reconstruction

2.1 Introduction

Surface reconstruction (SR) may be thermally induced as in solid–gas environments at relatively high temperatures, electrochemically induced as in solid–liquid half-cells subjected to an electric potential range or chemically induced due to chemical surface reactions at relatively low temperatures. Nonetheless, the initial stage of deposition during surface reconstruction dictates the pattern for the adsorption layer.

For the sake of clarity, surface reconstruction refers to a remedy or modification of a surface topography due to deviations or alterations of the atomic periodicity, leading to surface defects, which in turn, affect the properties of a material. On the other hand, surface relaxation refers to retained atomic plane periodicity parallel to the surface of a substrate.

In addition, adsorption or deposition is the adhesion of particles or species (atoms, ions, and molecules from a medium) to a solid surface and diffusion is the mechanism of particle motion. As a whole, electrochemical deposition is an electrolysis process for tightly adhered planes of atoms (known as a film or coating) onto the surface of an electronic conductor (substrate or electrode).

The analysis of surface reconstruction included henceforth is based on ideal cases of atomically clean surface. However, real crystal structures do have atomically imperfect surfaces and the surface reconstruction is most elaborated, complicated than the ideal case.

2.2 Overlayer Lattice Planes

In electrochemically induced surface reconstruction, it is assumed that the fundamental electric double layer is disrupted by a type of adsorbate or ion since the adsorbate adheres or attaches and an ion bonds to the electrode (substrate) surface

© Springer Nature Switzerland AG 2020
N. Perez, *Phase Transformation in Metals*,
https://doi.org/10.1007/978-3-030-49168-0_2

having a particular crystallographic plane. However, adsorbates may act as barriers
for smooth electrodeposition of a metal ion.

Hypothetically, it can be assumed that any electrochemical process at the
electrode–electrolyte interface begins with the initial adsorption of species, provided
that the cell electric potential (E) is not within the limits for inducing electrodeposition. Hence, this process depends on the electrochemical system, the concentration
and type of ions in solution, temperature, and pressure.

For the moment, it is assumed henceforth that the relevant electric potential is
within a suitable range for reconstruction to be possible. The methodology involved
in crystallography for surface reconstruction details steps in an elementary manner.

The surface lattice, in general, has numerous defects, such as terraces, voids,
adatoms, kinks, valleys, steps and to an extent, grain boundaries and particle–grain
interfaces. These defects control surface properties, such as surface energy (γ_s) and
surface texture like surface roughness. Figure 2.1 shows a sketch for two common
surface defects (adatom and vacancy) and a partial monolayer.

This monolayer is assumed to grow along the substrate surface as the reconstruction process progresses. A surface reconstruction with more than one monolayer
defines an overlayer (multilayer).

Fundamentally, adsorption and electrodeposition on electrodes may occur in
sequential manner. The former is the initial and spontaneous process when the
electrode is in contact with the electrolyte. Electrodeposition follows after a short
interval in time elapses at the expense of a slight potential drop [1].

Nowadays it is common to conduct in-situ electrochemical experiments to reveal
images of substrate crystallographic surfaces. Thus, the traditional current-potential
diagrams at a macroscale can be interpreted with a higher precision with the aid of
STM and AFM, which are imaging tools for structural characterization of surface
electrodes. Therefore, knowledge of crystallography is essential to understanding
adsorption by adherence and electrodeposition by bonding in a suitable environment
at pressure P and temperature T. However, the electrochemical cell potential for
adsorption and electrodeposition is a controlling factor during surface reconstruction. Details on this matter are addressed in subsequent chapters.

In this chapter, a monolayer is an insoluble single closely packed layer of atoms.
It can also be a single layer of molecules known in the literature as the Langmuir
monolayer. According to the model in Fig. 2.1, only a single type of atom is selected
as an illustration of the formation of an atom-thick homogeneous layer. However,

Fig. 2.1 Schematic surface
defects on an irregular
substrate surface

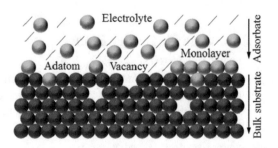

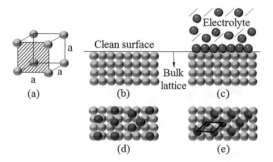

Fig. 2.2 Illustration of surface reconstruction of the (100) plane. (**a**) SC unit cell showing the (100) plane, (**b**) cross-sectional view of repetitive (100) plane, (**c**) formation of a monolayer on the surface of the (100) plane, (**d**) top view of an amorphous monolayer, and (**e**) top view of rotated crystalline monolayer

a monolayer can contain more than one type of atoms. Hence, a mixed monolayer due to adsorption (adhesion of atoms, ions, or molecules) or electrodeposition from an ionic solution.

First of all, Fig. 2.2a depicts the simple cubic (SC) unit cell and the (100) plane. Figure 2.2b shows the cross-sectional view of the (100) plane for clarity. Now, assume that this plane is on the surface of a substrate subjected to be covered by a monolayer of atoms from solution. This process can be adsorption-induced or deposition-induced process as indicated in Fig. 2.2c.

Once the monolayer is in place, it may acquire a particular structure such as a partial amorphous monolayer as illustrated in Fig. 2.2d. In this particular case, the monolayer does not have a crystalline structure to reveal an adsorbate lattice. On the other hand, Fig. 2.2e exhibits a (2×2) primitive matrix (read 2 by 2) monolayer at 45° with respect to the substrate (100) plane orientation and consequently, the substrate with the surface (100) crystallographic plane is being reconstructed by a 45° rotation of the monolayer, which has a crystalline structure with missing rows of atoms. Again, this is a partial monolayer with a specific atomic arrangement purposely drawn at 45° and without point defects (missing atoms).

Additionally, surface reconstruction is assumed to be faster at relative high temperature than at ambient temperature. This, then, allows atoms to migration or to diffuse towards a substrate (specimen) surface defects. Fundamentally, surface defect formation is inevitable, but controllable. Consequently, the time-dependence of diffusion can induce an undesirable surface roughness.

Furthermore, surface defects are inevitably present in most cases and are potential sites for the initial adsorption or electrodeposition process. They can be eliminated or covered significantly by adsorption of different gas atoms or ions from the bulk lattice. This is a surface reconstruction that can be achieved by adsorbate-induced or ion-induced monolayer.

Assume that adsorption of at least one monolayer takes place on an ideally uniform and defect-free (clean) substrate (hkl) plane. Thus, an adsorbate or deposited (hkl) plane is henceforward treated as a flat monolayer of atoms that forms a two-dimensional structure referred to as a lattice.

If a monolayers shows semiconductor properties, then a metal-based semiconductor device can be designed and manufactured using a surface reconstruction method. Thus, one can obtain a reconstructed surface with a heterogeneous two-dimensional monolayer made out of a pure metal or even a metal oxide. However, a sufficiently thick layer composed of many monolayers is a material that may have a dimensional-dependent property characteristics.

2.3   Two-Dimensional Bravais Lattices

A Bravais lattice is a three-dimensional (3D) array of discrete points, such as atoms or ions, that forms a unique geometry. Thus, the Bravais lattice concept is used to formally define a crystalline arrangement and subsequently, classify crystalline solids such as body-centered cubic (BCC), face-centered cubic (FCC), hexagonal-closed pack (HCP), tetragonal, and monoclinic. Moreover, Fig. 2.3 shows the two-dimensional (2D) Bravais lattice planes.

In order to characterize the surface reconstruction of a substrate, one needs to consider, at least, one of the five symmetrical planes shown in Fig. 2.3.

This can be achieved by an adsorption-induced method as the lattice controlling the behavior and surface properties of a material related to these crystallographic planes having repetitive primitive-lattice morphology. Denote that $\mathbf{a}_1$ and $\mathbf{a}_2$ are discrete primitive vectors defining the shortest vectors of primitive lattices.

In two-dimensional (2D) space, the above 5 Bravais planar lattices in Fig. 2.3 identify a set of 2D crystallographic planes, which are denoted, in general, as specific $\{hkl\}$ Miller indices. Their corresponding shapes are

Fig. 2.3 The five symmetrical Bravais lattices with (hkl) planes

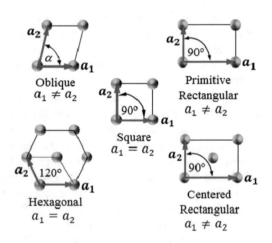

- *Oblique*: Monoclinic and triclinic {100} planes, $(1 \times x)$
- *Square*: SC, FCC, and BCC {100} planes
- *Rectangular*: Tetragonal and orthorhombic {100} planes
- *Centered rectangular*: BCC and body-centered orthorhombic {110} planes
- *Hexagonal*: Hexagonal-closed pack (HCP) (0001) basal plane.

Considering the complex nature of the surface reconstruction mechanism, one can assume for the moment that the initial surface reconstruction process may be due to diffusion-controlled adsorption, coadsorption, deposition, or codeposition. The reader should be aware of the specific features of these topics, which require extensive studies for the proper understanding of the mechanisms.

Furthermore, the symmetry principle can be applied to the two-dimensional Bravais lattices depicted in Fig. 2.3. For instance, the square and rectangular Bravais lattices are considered to have high symmetry since the a,b,c crystallographic coordinates are at right angles (90°). On the other hand, the least symmetric Bravais lattice in two dimensions is the oblique lattice because the coordinate angles are not at right angles. All this means that symmetry represents the consistency of the repetition of the unit cells in a single crystal.

2.4 Matrix and Wood Notations

This section treats the linear translation of vectors in two dimensions for studying SC, FCC, and BCC lattices. Most often the SC lattice characterizes FCC and BCC lattices due to its simple geometry, which defines a primitive lattice having the shortest edge length between neighboring atoms. A mathematical justification for this approach is available in [2, 3].

This simple case of surface reconstruction provides sufficient information on how to modify electrode surfaces for any particular application. There are complicated surface reconstruction processes that depend on the type of substrate lattice or unit mesh, and the electrolyte species in a regime of temperature and pressure conditions.

Surface reconstruction can be induced or affected by the adsorption or bonding of different atoms onto the lattice surface due to changes in interatomic forces. Consequently, surface properties may be affected by this process. This atomic process is highly dependent on (1) the initial physical or chemical state of the substrate (lattice) and adsorbate (foreign atoms or ions) and (2) the environmental conditions. As a result, the surface energy (γ_s) may change beneficially or detrimentally.

Initially, surface reconstruction is ideally achieved by a monolayer or single plane of atoms. In such a case, a rotation angle (θ) describes the crystallographic differences between the monolayer and the bulk lattice. However, surface reconstruction by multilayer adhesion or deposition having a different surface lattice is normally called a superlattice, supercell, or superstructure, which may be treated as a thin film.

The two-dimensional *adsorption lattice* orientation with respect to the *substrate lattice* can be characterized by the conventional Wood notation using the surface vectors. Apparently, Wood notation is applicable only when the adsorbate and substrate have the same lattice geometry regardless of the lattice sizes. The latter is chosen to be a primitive lattice $p(1 \times 1)$ matrix and the former normally has a large size.

In order to be consistent in terminology, the phrase adsorbate lattice will be used henceforward to imply and assume that the surface reconstruction is achieved by a uniform plane of atoms being adhered or deposited on the substrate surface. This is appropriate for explaining and using simple mathematical equations based on elementary vector algebra.

Actually, surface reconstruction can be characterized by *Wood notation* [4] or *Matrix notation* [5]. Both notations describe the relationship between the absorbed and bulk lattices. Generally, the lattice with translational vectors describing the relation between the adsorbate $(\mathbf{b}_1, \mathbf{b}_2)$ and the substrate $(\mathbf{a}_1, \mathbf{a}_2)$ primitive lattice vectors is an essential concept for defining the general Matrix notation as proposed by Park and Madden [5]. Thus,

$$\mathbf{b}_1 = m_{11}\mathbf{a}_1 + m_{12}\mathbf{a}_2 \tag{2.1a}$$

$$\mathbf{b}_2 = m_{21}\mathbf{a}_1 + m_{22}\mathbf{a}_2 \tag{2.1b}$$

Here, $(\mathbf{a}_1, \mathbf{a}_2)$ are the translational unit vectors of the substrate surface described by the (hkl) plane, $(\mathbf{b}_1, \mathbf{b}_2)$ are the translational vectors of the reconstructed surface, namely the adsorbate overlayer, and m_{ij} are coefficients.

The rotation matrix or *Matrix notation* (M_R) and its determinant $(\det M_R)$ take the simple form

$$M_R = \begin{pmatrix} m_{11} \ m_{12} \\ m_{21} \ m_{22} \end{pmatrix} \tag{2.2a}$$

$$\det M_R = m_{11}m_{22} - m_{12}m_{21} \neq 0 \tag{2.2b}$$

Recall from Linear Algebra that $\det M_R = 0$ means that M_R is not invertible and the two vectors, $(\mathbf{b}_1, \mathbf{b}_2)$, are collinear since points or atoms lie on the same straight line. Hence, $\det M_R$ is commonly used whenever convenient to describe the relationship between the crystallography of the reconstructed substrate and the overlayer.

On the other hand, *Wood notation* (W_N) [4] considers the physical contact of the adsorbate and substrate layers that may or may not have the same periodicity or symmetry and nonetheless, it describes the type of adsorbate lattice on the surface of a substrate X with a (hkl) plane being reconstructed by foreign atoms. The general short hand equations for Wood notation are

$$W_N = X\,(hkl) \left(\frac{\|\mathbf{b}_1\|}{\|\mathbf{a}_1\|} \times \frac{\|\mathbf{b}_2\|}{\|\mathbf{a}_2\|} \right) R\theta \tag{2.3a}$$

$$W_N = X\,(hkl)\left(\frac{b_1}{a_1} \times \frac{b_2}{a_2}\right) R\theta = X\,(hkl)\,(m \times n)\,R\theta \qquad (2.3b)$$

where $\|\mathbf{a}_1\| = a_1$ and $\|\mathbf{b}_1\| = b_1$, $X\,(hkl)\,(m \times n)\,R\theta$ is an $(m \times n)$ lattice rotated by θ from the substrate primitive-lattice direction.

Apparently, the Wood notation applies to commensurate (coherent) adsorption when the adsorbate and substrate are linked by lattice symmetry. Conversely, in incommensurate (incoherent) adsorption, the Matrix notation seems appropriate due to the lack of lattice symmetry or atomic mismatch.

According to Hermann [2, 3] mathematical treatment (without proof henceforth) on Wood notation, the adsorbate rotation matrix (M_R), and the rotation angle θ, defined by the set of vectors $(\mathbf{b}_1, \mathbf{a}_1)$, indirectly depend on the angle $\alpha = \measuredangle\,(\mathbf{b}_1, \mathbf{b}_2)$ included in Fig. 2.3. Thus,

$$M_R = \begin{pmatrix} b_1 & a_1 \\ -a_1 & b_1 - a_1 \end{pmatrix} \quad \text{for } \alpha > 90° \qquad (2.4a)$$

$$\cos\theta = \frac{b_1 - a_1/2}{\sqrt{\det M_R}} = \frac{b_1 - a_1/2}{\sqrt{b_1^2 - a_1 b_1 + a_1^2}} \quad \text{for } \alpha > 90° \qquad (2.4b)$$

and the cosine function is written as

$$M_R = \begin{pmatrix} b_1 & a_1 \\ -a_1 & b_1 + a_1 \end{pmatrix} \quad \text{for } \alpha < 90° \qquad (2.5a)$$

$$\cos\theta = \frac{b_1 + a_1/2}{\sqrt{\det M_R}} = \frac{b_1 + a_1/2}{\sqrt{b_1^2 + a_1 b_1 + a_1^2}} \quad \text{for } \alpha \leq 90° \qquad (2.5b)$$

These equations are used henceforward, whenever appropriate, for determining the Wood notation of a material X being (hkl) surface reconstructed.

For instance, Fig. 2.4a shows a model for possible coadsorption or codeposition of foreign atoms and Fig. 2.4b, c, d exhibits three randomly selected STM images from the literature based on the type of lattice morphology [6–8]. This is shown in Fig. 2.4 for selected images.

Once more, Fig. 2.4a depicts a model indicating a general two-dimensional arrangement of foreign atoms (gray, blue, and red colors) referred to as adsorbate lattices on a substrate surface (green) having a particular (hkl) plane. These lattices can be adsorption-induced monolayers and overlayers (layers of adatoms adsorbed onto a surface), which in turn, are subsequently characterized with respect to crystallographic features, such as $[uvw]$ directions, $\{hkl\}$ planes, and type of lattices. Henceforth, the morphology of adsorption-induced overlayers, clusters, and defects (if any) is studied in detail as per experimental or simulation conditions based on the available instrumentation.

Fig. 2.4 Two-dimensional lattices. (**a**) Model containing foreign adsorbates on the substrate surface, (**b**) Carbon C_{60} molecular monolayer with $< 112 >$ directions on Au(111) surface plane [6], (**c**) halide-substituted benzoic acid lattice and principal $[uvw]$ directions [7], and (**d**) Sn monolayer (lighter atoms) on Ag(111) surface plane [8]

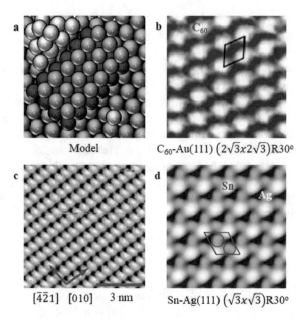

Model

C_{60}-Au(111) $(2\sqrt{3}x2\sqrt{3})R30°$

$[\overline{4}\overline{2}1]$ [010] 3 nm

Sn-Ag(111) $(\sqrt{3}x\sqrt{3})R30°$

Moreover, Fig. 2.4b shows an organic compound 3, 5-diiodosalicylic acid (DISA) (halide-substituted benzoic) monolayer on a calcium carbonate ($CaCO_3$), known as calcite, lattice. The surface crystallographic directions, [010] and $[\overline{4}21]$, of this organic compound are also shown as an additional crystallographic feature [7].

Figure 2.4c illustrates the Sn lattice on the $Ag(111)$ surface plane at 30° rotation. The red oblique parallelogram outlines the size of the monolayer lattice. Denote the nomenclature used for identifying this type of surface plane. The oblique shape on this figure is the $Ag(111)$ primitive lattice and the two circles identify the two Ag atoms within this lattice [8]. On the other hand, Fig. 2.4d elucidates the fullerene monomer (C_{60}) monolayer lattice having $< 112 >$ directions on the $Au(111)$ plane with $< 110 >$ directions. Here, the black oblique parallelogram outlines the size of the monolayer lattice. Denote the notation below this image, C_{60}-Au (111) $\left(2\sqrt{3} \times 2\sqrt{3}\right) R30°$ [6]. The C_{60} structure contains several hexagons and pentagons forming spherical-molecule clusters in a crystallographic plane on the Au (111) substrate. As a result, the C_{60} $\left(2\sqrt{3} \times 2\sqrt{3}\right) R30°$ monolayer reconstructs the $Au(111)$ surface in a rather unique symmetrical plane. This specific notation for the carbon C_{60} structure has drawn attention in the scientific community.

Additionally, an oxygen-rich electrolyte containing chrome cations (Cr^{+6}) is fundamentally favorable for the formation of the Cr_2O_3-$Au(111)$ $\left(\sqrt{3} \times \sqrt{3}\right) R30°$ lattice under diffusion-controlled adsorption mechanism. As a matter of fact, one can highlight the surface reconstruction from the STM images depicted in Figs. 2.4b–d as a complex reconstruction process, where the Wood notation is rather complicated. In such a case, the Matrix notation is an option for describing surface reconstruction.

Accordingly, the surface reconstruction can be labeled as

- Organic-controlled adsorption as in C_{60}-Au (111) $\left(2\sqrt{3} \times 2\sqrt{3}\right)$ $R30°$ (Fig. 2.4b).
- Metal-controlled adsorption as in Sn-$Ag(111)$ $\left(\sqrt{3} \times \sqrt{3}\right)$ $R30°$ (Fig. 2.4d)
- Ceramic-controlled adsorption as in Cr_2O_3-$Cr(110)$ $\left(\sqrt{3} \times \sqrt{3}\right)$ $R30°$ (No figure is included)

This general classification of monolayers elucidates the characteristics of the reconstructed substrate. Thus, a combination of monolayers on a metal or metal oxide substrate surface is possible (Fig. 2.4a). Nonetheless, the formation of an overlayer onto $X(hkl)$ surfaces is a mass transport mechanism induced by a diffusion-controlled or migration-controlled deposition process. Mass transport theory will be included in a later chapter.

To this point, the two-dimensional surface reconstruction process has been assumed to occur individually and uniformly by steady diffusion and adsorption mechanisms. This suggests that any surface topology is conveniently excluded from the present analysis. This is simply a convenient approach for succinctly explaining the principles related to surface reconstruction of a particular crystallographic (hkl) surface plane. Thus, one conveys clear and accurate information on the current matter.

Fundamentally, there are seven crystal systems and five plane lattices with common base centered, body centered, and face centered. Conveniently, a substrate surface denoted as $X(hkl)$ can experience surface reconstruction under natural or artificial closed-control experimental conditions. Firstly, if the substrate-surface unit cell is crystallographically not distorted, then it is assumed to be under a relaxation condition; otherwise, it is significantly reconstructed. Secondly, a surface reconstruction implies that the substrate (hkl) lattice plane periodicity is definitely altered, leading to new surface characteristics.

Example 2.1 Show that det (M_R) can be defined as an area ratio.

Solution From Eq. (2.1),

$$\begin{pmatrix} b_1 \\ b_2 \end{pmatrix} = \begin{pmatrix} m_{11} & m_{12} \\ m_{21} & m_{22} \end{pmatrix} \begin{pmatrix} a_1 \\ a_2 \end{pmatrix} = M_R \begin{pmatrix} a_1 \\ a_2 \end{pmatrix}$$

$$A_b = b_1 \times b_2 = (m_{11}a_{11} + m_{12}a_2) \times (m_{21}a_{11} + m_{22}a_2)$$

$$A_b = (m_{11}m_{22} - m_{12}m_{21})(a_1 \times a_2)$$

$$A_b = \det(M_R) A_a$$

Thus ([2, p. 171]),

$$\det (M_R) = \frac{A_b}{A_a}$$

This is the proof.

Example 2.2 (Carbon (C) Atoms Onto a Gold (Au) Substrate)

(a) Determine the Wood notation for the scanning tunneling microscopy (STM) gold (Au) image in Fig. 2.4d using the dot product approach. This notation was determined by Yoshimoto et al. [6] during in-situ STM experiments. Assume a singular crystallographic direction from the set of directions $< 110 >$ for the gold $Au(111)$ plane and vector $\mathbf{a}_1$, and one from $< 112 >$ for carbon (fullerene monomer) C_{60} molecular overlayer and vector $\mathbf{b}_1$.

(b) Calculate the C_{60} overlayer lattice parameter $a_{C_{60}}$, if $a_{Au} = 0.289$ nm. Assume that the given information is sufficiently adequate to determine the fullerene-substrate, C_{60}-$Au(111)$, relation using the Wood notation.

Solution

(a) The proper selection of the crystallographic directions is important in this particular example. For $\left[1\bar{1}2\right]_{C_{60}}$ and $\left[0\bar{1}1\right]_{Au}$,

$$\|\mathbf{a}_1\| = a_1 = \left(\sqrt{u_1^2 + v_1^2 + w_1^2}\right) a = \left(\sqrt{(0)^2 + (-1)^2 + (1)^2}\right) a = \sqrt{2}a$$

$$\|\mathbf{a}_2\| = a_2 = \sqrt{u_2^2 + v_2^2 + w_2^2}\, a = \sqrt{(1)^2 + \left(\bar{1}\right)^2 + (0)^2}a = \sqrt{2}a$$

Then, the size of the primitive Au (111) lattice plane is

$$Au\ (111)\left(\sqrt{2} \times \sqrt{2}\right) a$$

where $a = a_{Au}$ (lattice parameter). The rotation angle between vectors a_1 and b_1 is

$$\cos\theta = \frac{\mathbf{a}_1 \cdot \mathbf{b}_1}{\|\mathbf{a}_1\|\,\|\mathbf{b}_1\|} = \frac{\left[0\bar{1}1\right]_{Au} \cdot \left[1\bar{1}2\right]_{C_{60}}}{\left(\sqrt{u_1^2 + v_1^2 + w_1^2}\right)_{Au}\left(\sqrt{u_1^2 + v_1^2 + w_1^2}\right)_{C_{60}}}$$

$$\cos\theta = \frac{(0)\,(1) + (-1)\,(-1) + (1)\,(2)}{\sqrt{(0)^2 + (-1)^2 + (1)^2}\sqrt{(1)^2 + (-1)^2 + (2)^2}} = \frac{3}{\sqrt{2}\sqrt{6}}$$

$$\cos\theta = \frac{3}{2\sqrt{3}}$$

$$\theta = \arccos\left(\frac{3}{2\sqrt{3}}\right) = 30°$$

Now, the Wood notation becomes exactly the same as previously reported by Yoshimoto et al. [6]. Thus,

$$C_{60}\text{-}Au\ (111)\left(2\sqrt{3}\times 2\sqrt{3}\right)R30°$$

(b) The C_{60} overlayer lattice parameter of the basal plane is approximately

$$a_{C_{60}}=\left(2\sqrt{3}\right)a_{Au}=\left(2\sqrt{3}\right)(0.289\,\text{nm})\simeq 1\,\text{nm}$$

Notice that this lattice parameter $a_{C_{60}}\simeq 1\,\text{nm}$ for the fullerene crystal C_{60} is quite large as compared with $a_{Au}=0.289\,\text{nm}$. That is, $a_{C_{60}}=3.46a_{Au}$. Nonetheless, the C_{60} crystal is assumed to have a well-defined morphology at a certain temperature T.

2.5 Lattice Planes

2.5.1 Square Primitive Lattice

First of all, a crystalline substrate is basically an array of atoms packed together forming a pattern with a unique arrangement called unit cell, which in turn, is the smallest solid particle that repeats through space without rotation. Thus, a crystal can be characterized as a *parallelogram* in 2D and as a *parallelepiped* in 3D. The former prevails in this section in order to illustrate the shape and size of the primitive and monolayer atomic arrangements.

It is now appropriate to define a *lattice* as set of points in space with translational symmetry since one point is surrounded by identical point. Hence, a primitive unit parallelogram contains one lattice point.

Figure 2.5a depicts a square substrate plane along with the primitive ("p") square lattice, oblique adsorbate lattices A and B, and a large square adsorbate lattice C. In fact, all adsorbates shown in Fig. 2.5a are primitive lattices. In addition, Fig. 2.2b shows the simple linear vector translations of adsorbate A and B lattices exhibiting rotation relative to "p" with principal vectors $\mathbf{a}_1$ and $\mathbf{a}_2$. For a simple cubic (SC) crystal structure, the vectors $\mathbf{a}_1$ and $\mathbf{a}_2$ have magnitudes equal to the lattice parameter a. This means that $a_1=\|\mathbf{a}_1\|=a$ and $a_2=\|\mathbf{a}_2\|=a$. However, complications arise for assessing monoclinic and triclinic structures.

In general, the lattice points (atoms) for vectors $(\mathbf{a}_1,\mathbf{a}_2)$ and $(\mathbf{b}_1,\mathbf{b}_2)$ define the primitive lattices for the substrate and adsorbates, respectively. Denote that the area of the primitive substrate lattice is smaller than that of the A, B, and C adsorbates shown in Fig. 2.5a. Also shown in Fig. 2.5 is the corresponding vector notation in two dimensions.

A simple vector matrix approach follows by considering that the adsorbate vectors are connected by linear transformations and that the primitive substrate

Fig. 2.5 Substrate square lattice for a single cubic (SC) crystal showing the primitive "p" lattice and three overlayer lattice planes (A, B, and C) and (**b**) set of vector for A, B, and C overlayer lattices

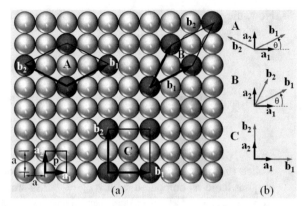

(a)   (b)

lattice is an orthogonal matrix (square matrix). The Matrix notation and Wood notation for the lattices shown in Fig. 2.5a are determined using Park and Madden [5] and Hermann [2, 3] equations.

Essentially, these equations can be used to determine the adsorbate or deposited overlayer lattice on the substrate surface without identifying the substrate material X and the Miller indices for the corresponding (hkl) plane, $[uvw]_{a_1}$ and $[uvw]_{b_1}$ directions, which are essential for defining the $\mathbf{a}_1$ and $\mathbf{b}_1$ vectors, respectively. This, then, makes the incoming procedure simple, to an extent, and easy to use for determining a monolayer notation. Hence, surface science deals with the characterization of well-defined crystal surface structures subjected to a surface reconstruction procedure under special conditions.

Next, the drawn lattice planes are individually characterized below. The main goal is to determine the size of the corresponding adsorbates assuming undisturbed surface reconstruction process, which in turn, is assumed to occur under steady-state diffusion.

Lattice Plane A Fig. 2.5a at $\alpha > 90°$. This is a parallelogram denoted "Lattice plane A" just for convenience. The monolayer matrix notation (M_R) with $(a_1, a_2) = (1, 1)$ and $(b_1, b_2) = (2, 2)$ vertices is

$$\mathbf{b}_1 = 2\mathbf{a}_1 + \mathbf{a}_2 \tag{2.6a}$$

$$\mathbf{b}_2 = -2\mathbf{a}_1 + \mathbf{a}_2 \tag{2.6b}$$

$$M_R = \begin{pmatrix} 2 & 1 \\ -2 & 1 \end{pmatrix} \quad \& \quad \det M_R = 4 \tag{2.6c}$$

Note that this is a simple matrix that yields the magnitude of the determinant, from which the size of the adsorbate (D_R) lattice is defined as a square matrix of the form

$$D_R = \left(\sqrt{\det M_R} \times \sqrt{\det M_R} \right) = (2 \times 2) \tag{2.7}$$

Accordingly, the Wood notation, Eq. (2.3a), for the adsorbate size becomes

$$\left(\frac{\|b_1\|}{\|a_1\|} \times \frac{\|b_2\|}{\|a_2\|} \right) = \left(\frac{2}{1} \times \frac{2}{1} \right) = (2 \times 2) \tag{2.8}$$

Now, the adsorbate rotation matrix (M_R) with $(b_1, a_1) = (2, 1)$ vertices and the $\cos \theta$ when $\alpha > 90°$ are determined using Eqs. (2.4a) and (2.4b), respectively. Thus, the angle of rotation is

$$M_R = \begin{pmatrix} b_1 & a_1 \\ -a_1 & b_1 - a_1 \end{pmatrix} = \begin{pmatrix} 2 & 1 \\ -1 & 1 \end{pmatrix} \tag{2.9a}$$

$$\det M_R = b_1^2 - a_1 b_1 + a_1^2 = 3 \tag{2.9b}$$

$$\cos \theta = \frac{b_1 - a_1/2}{\sqrt{\det M_R}} = \frac{2 - 1/2}{\sqrt{3}} = \frac{3}{2\sqrt{3}} \tag{2.9c}$$

$$\theta = \arccos \left(\frac{3}{2\sqrt{3}} \right) = 30° \tag{2.9d}$$

Consequently, the edge length (D_R) of the adsorbate relative to the substrate lattice and the Wood notation (W_N) become

$$D_R = \left(\sqrt{\det M_R} \times \sqrt{\det M_R} \right) = \left(\sqrt{3} \times \sqrt{3} \right) \tag{2.10a}$$

$$W_N = X \, (hkl) \, (D_R) \, R\theta \tag{2.10b}$$

$$W_N = X \, (hkl) \left(\sqrt{3} \times \sqrt{3} \right) R30° \tag{2.10c}$$

Some substrate overlayers can be assigned a notation defined by

$$W_N = X \, (hkl) \left(2\sqrt{3} \times 2\sqrt{3} \right) R30° \tag{2.11}$$

where the factor $2\sqrt{3}$ in the $\cos \theta$ function is a diagonal length of the parallelogram denoted as Lattice A. Inspection of the drawn arrangement of the atoms in the monolayer laying on the substrate surface shows that the monolayer parallelogram (Lattice A) is an unpacked array of atoms such that $b_1 = b_2$ and the angle between these vectors is less than $90°$. Hence, an oblique parallelogram with vectors $b_1 = b_2$.

Note that the monolayer atoms are not in contact with each other. This is the reason for defining this hypothetical parallelogram as an unpacked crystal plane without impurities, such as oxygen and nitrogen. This can be treated as a special case.

Lattice Plane B Fig. 2.5a at $\alpha < 90°$. In this case, M_R and D_R with $(a_1, a_2) = (1, 1)$ and $(b_1, b_2) = (2, 2)$ vertices are

$$\mathbf{b}_1 = 2\mathbf{a}_1 + \mathbf{a}_2 \tag{2.12a}$$

$$\mathbf{b}_2 = -2\mathbf{a}_1 + \mathbf{a}_2 \tag{2.12b}$$

$$M_R = \begin{pmatrix} 2 & 1 \\ -2 & 1 \end{pmatrix} \quad \& \quad \det M_R = 4 \tag{2.12c}$$

$$D_R = \left(\sqrt{\det M_R} \times \sqrt{\det M_R} \right) = (2 \times 2) \tag{2.12d}$$

Again, the Wood notation, Eq. (2.3a), gives

$$\left(\frac{\|b_1\|}{\|a_1\|} \times \frac{\|b_2\|}{\|a_2\|} \right) = \left(\frac{2}{1} \times \frac{2}{1} \right) = (2 \times 2) \tag{2.13}$$

From Eqs. (2.5a) and (2.5b) along with $(b_1, a_1) = (2, 1)$,

$$M_R = \begin{pmatrix} b_1 & a_1 \\ -a_1 & b_1 + a_1 \end{pmatrix} = \begin{pmatrix} 2 & 1 \\ -1 & 3 \end{pmatrix} \tag{2.14a}$$

$$\det M_R = b_1^2 + a_1 b_1 + a_1^2 = 7 \tag{2.14b}$$

$$\cos \theta = \frac{b_1 + a_1/2}{\sqrt{\det M_R}} = \frac{2 + 1/2}{\sqrt{3}} = \frac{5}{2\sqrt{7}} \tag{2.14c}$$

$$\theta = \arccos \left(\frac{5}{2\sqrt{7}} \right) = 19.1° \tag{2.14d}$$

Thus,

$$D_R = \left(\sqrt{\det M} \times \sqrt{\det M} \right) = \left(\sqrt{7} \times \sqrt{7} \right) \tag{2.15a}$$

$$W_N = X \, (hkl) \left(\sqrt{7} \times \sqrt{7} \right) R19.1° \tag{2.15b}$$

Again, $2\sqrt{7}$ may be used as part of the pertinent substrate-overlayer notation

$$W_N = X \, (hkl) \left(2\sqrt{7} \times 2\sqrt{7} \right) R19.1° \tag{2.16}$$

Recall that X represents a particular substrate material. Moreover, if the substrate [111] and overlayer [112] directions define the rotation matrix, then the dot product is written as

$$\cos \theta = \frac{\mathbf{a}_1 \cdot \mathbf{b}_1}{\|\mathbf{a}_1\| \, \|\mathbf{b}_1\|} = \frac{[111]_{Au} \cdot [112]_{C_{60}}}{\left(\sqrt{u_1^2 + v_1^2 + w_1^2} \right)_{Au} \left(\sqrt{u_1^2 + v_1^2 + w_1^2} \right)_{C_{60}}} \tag{2.17a}$$

so that the rotation angle becomes

$$\cos\theta = \frac{(1)\,(1) + (1)\,(1) + (1)\,(2)}{\sqrt{(1)^2 + (1)^2 + (1)^2}\sqrt{(1)^2 + (1)^2 + (2)^2}}$$

$$= \frac{4}{\sqrt{3}\sqrt{6}} = \frac{4}{3\sqrt{2}} \tag{2.17b}$$

$$\theta = \arccos\left(\frac{4}{3\sqrt{2}}\right) = 19.5° \tag{2.17c}$$

Accordingly, the Wood notation would be

$$C_{60}\text{-}Au\ (111)\ \left(3\sqrt{2} \times 3\sqrt{2}\right) R19.5° \tag{2.18}$$

Lattice Plane C Fig. 2.5a at $\alpha = 90°$. Similarly, M_R and D_R with $(a_1, a_2) = (1, 1)$ and $(b_1, b_2) = (2, 2)$ vertices become

$$\mathbf{b}_1 = 2\mathbf{a}_1 + 0\mathbf{a}_2 \tag{2.19a}$$

$$\mathbf{b}_2 = 0\mathbf{a}_1 + 2\mathbf{a}_2 \tag{2.19b}$$

$$M_b = \begin{pmatrix} 2 & 0 \\ 0 & 2 \end{pmatrix} \quad \&\quad \det M_b = 4 \tag{2.19c}$$

$$D_b = \left(\sqrt{\det M_b} \times \sqrt{\det M_b}\right) = (2 \times 2) \tag{2.19d}$$

Actually, the (2×2) matrix is denoted as $c\,(2 \times 2)$ because of the centered atom in the adsorbate lattice. From Wood notation, the adsorbate matrix is determined or identified as

$$c\left(\frac{\|b_1\|}{\|a_1\|} \times \frac{\|b_2\|}{\|a_2\|}\right) = c\left(\frac{2}{1} \times \frac{2}{1}\right) = c\,(2 \times 2) \tag{2.20}$$

Fort $\theta = 0°$,

$$D_b = c\left(\sqrt{\det M_b} \times \sqrt{\det M_b}\right) = c\left(\sqrt{4} \times \sqrt{4}\right) = c\,(2 \times 2) \tag{2.21a}$$

$$W_N = X\,(hkl)\,c\,(2 \times 2) \tag{2.21b}$$

Note that Eqs. (2.20) and (2.21a) yield the same result in a concise and clear form. So far the above examples are considered simple for determining the Wood notation W_N of an unknown substrate material X.

2.5.2 Overlayer Coherency

In principle, an overlayer coherency can be classified in relation to the topmost crystallographic (hkl) plane of the substrate. Thus, the surface crystallography of an overlayer is described by the set of vectors $(\mathbf{b}_1, \mathbf{b}_2)$ and those for the substrate are $(\mathbf{a}_1, \mathbf{a}_2)$. The overlayer can be characterized as

- ordered and coherent with the substrate
- ordered and semi-coherent with the substrate
- disordered and incoherent with the substrate.

The general matrix notation using these vectors is defined by Eq. (2.1). This means that the reconstructed topmost substrate $(hkl)_{sub}$ plane is partial or fully covered with an adsorbate $(hkl)_{ad}$ plane called overlayer, which can have one of the above types of coherency condition.

2.5.3 Hexagonal Primitive Lattice

The primitive lattices shown in Fig. 2.6a have non-orthogonal axes since the angles are $\alpha = \measuredangle (\mathbf{a}_1, \mathbf{a}_2) > 90°$ and $\beta = \measuredangle (\mathbf{b}_1, \mathbf{b}_2) > 90°$. Nonetheless, the planar substrate has its primitive lattice defined as a $p\,(a_1 \times a_2) = p(1 \times 1)$ matrix since $\mathbf{a}_1$ and $\mathbf{a}_2$ are the shortest vectors when one atom is in contact with the nearest neighbor in the shown vector directions.

Figure 2.6a also shows the conventional lattices (triangle and hexagonal) and Fig. 2.6b depicts the set of vectors for substrate and adsorbate lattices. Fundamentally, $\alpha = \measuredangle (\mathbf{a}_1, \mathbf{a}_2) > 90°$ is the angle of the smallest parallelogram with vertices $(a_1, a_2) = \left(\sqrt{2}/2\right) a$ for face-centered cubic (FCC) and $(a_1, a_2) = (a, a)$ for

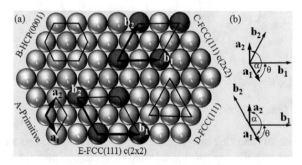

E-FCC(111) c(2x2)

Fig. 2.6 Crystal lattice model denoted as A, B, C, D, and E. **(a)** Non-orthogonal primitive lattice planes: A-substrate p(1×1) lattice, B-conventional HCP, C-overlayer with $FCC(111)$ c(2×2), D-conventional $FCC(111)$, and E-overlayer with $FCC(111)$ c(2×2), and **(b)** vector directions for lattices

hexagonal lattices. The Matrix notation and Wood notation for the C and E overlayer lattices are characterized in a similar manner as carried out for the cases shown in Fig. 2.5. Moreover, complications in determining the Wood notation may arise if the temperature T is included in the procedure. Fortunately, this is not required for characterizing monolayers and overlayers.

Lattice Plane C Figure 2.6a, $(b_1, a_1) = (2, 1)$ vertices at an angle θ. For $\alpha < 90°$, the monolayer or overlayer matrix notation (M_R) with $(a_1, a_2) = (1, 1)$ and $(b_1, b_2) = (2, 2)$ vertices is

$$\mathbf{b}_1 = 2\mathbf{a}_1 + 0\mathbf{a}_2 \tag{2.22a}$$

$$\mathbf{b}_2 = 0\mathbf{a}_1 + 2\mathbf{a}_2 \tag{2.22b}$$

$$M_b = \begin{pmatrix} 2 & 0 \\ 0 & 2 \end{pmatrix} \tag{2.22c}$$

which is a simple matrix. By inspection, the Wood notation for this monolayer containing a center atom can be represented as a simple $c(2 \times 2)$ matrix. For $(b_1, a_1) = (2, 1)$ vertices,

$$M_R = \begin{pmatrix} b_1 & a_1 \\ -a_1 & b_1 - a_1 \end{pmatrix} = \begin{pmatrix} 2 & 1 \\ -1 & 1 \end{pmatrix} \tag{2.23a}$$

$$\det M_R = b_1^2 - a_1 b_1 + a_1^2 = 3 \tag{2.23b}$$

$$\cos \theta = \frac{b_1 - a_1/2}{\sqrt{\det M_R}} = \frac{2 - 1/2}{\sqrt{3}} = \frac{3}{2\sqrt{3}} \tag{2.23c}$$

$$\theta = \arccos \left(\frac{3}{2\sqrt{3}} \right) = 30° \tag{2.23d}$$

From Artaud et al. [9],

$$\tan \theta = \frac{\sqrt{3} b_2}{2 b_1 - b_2} = \frac{\left(\sqrt{3} \right) (1)}{(2)(2) - 1} = \frac{\sqrt{3}}{3} \tag{2.24}$$

$$\theta = 30°$$

The general Wood notation for both C and E overlayers is the same. Thus,

$$W_N = X(hkl) \left(\sqrt{3} \times \sqrt{3} \right) R30° \quad \text{(Common model)} \tag{2.25}$$

It has been reported [1, 10] that the surface reconstruction of $X(hkl)$ during Ag and Cu deposition took place according to one of the adsorbate models without a centered atom shown in Fig. 2.6a. Hence, the specific Wood notation in these cases is [10, 11]

$$Ag \text{ on } Pt(111) \left(\sqrt{3} \times \sqrt{3}\right) R30° \tag{2.26a}$$

$$Cu \text{ on } Au(111) \left(\sqrt{3} \times \sqrt{3}\right) R30° \tag{2.26b}$$

$$Cu \text{ on } Pt(110) \left(\sqrt{3} \times \sqrt{3}\right) R30° \tag{2.26c}$$

In addition, if $2\sqrt{3}$ in Eq. (2.9a) is considered as a correction factor to $\left(\sqrt{3} \times \sqrt{3}\right)$, then W_N is described by

$$W_N = X(hkl) \left(2\sqrt{3} \times 2\sqrt{3}\right) R30° \quad \text{(Corrected model)} \tag{2.27}$$

which denotes a monolayer as an unpacked and clean parallelogram (crystal plane) rotated 30°.

For $(b_1, a_1) = (3, 1)$ vertices, the matrix and determinant are

$$M_R = \begin{pmatrix} b_1 & a_1 \\ -a_1 & b_1 - a_1 \end{pmatrix} = \begin{pmatrix} 3 & 1 \\ -1 & 2 \end{pmatrix} \tag{2.28a}$$

$$\det M_R = b_1(b_1 - a_1) - (-a_1)(a_1) = a_1^2 - a_1 b_1 + b_1^2 = 7 \tag{2.28b}$$

and the rotation angle is

$$\cos \theta = \frac{b_1 - a_1/2}{\sqrt{\det M_R}} = \frac{3 - 1/2}{\sqrt{7}} = \frac{5}{2\sqrt{7}} \tag{2.29a}$$

$$\theta = 19.1° \tag{2.29b}$$

Thus,

$$W_N = X(hkl) \left(\sqrt{7} \times \sqrt{7}\right) R19.1° \quad \text{(Common model)} \tag{2.30}$$

Examples of surface reconstruction of $X(hkl)$ substrates that follow this notation Eq. (2.30) are given below [12–14], respectively

$$Nb \text{ on } SrTiO_3(111) \left(\sqrt{7} \times \sqrt{7}\right) R19.1° \tag{2.31a}$$

$$C_4H_4N_2 \text{ on } Pt(111) \left(\sqrt{7} \times \sqrt{7}\right) R19.1° \tag{2.31b}$$

$$CH_3S \text{ on } Ag(111) \left(\sqrt{7} \times \sqrt{7}\right) R19.1° \tag{2.31c}$$

In addition, $SrTiO_3$ is an n-type single crystal used as an insulator, while $C_4H_4N_2$ (pyrazine-PZ) and CH_3S (methanethiolate) are organic compounds. If

$2\sqrt{7}$ is used instead of $\sqrt{7}$ to determine the Wood notation, then it may be taken as a correction factor for this type of surface reconstruction. Hence,

$$W_N = X(hkl)\left(2\sqrt{7} \times 2\sqrt{7}\right) R19.1° \quad \text{(Corrected model)} \qquad (2.32)$$

where the crystallographic (hkl) plane must be determined experimentally and the angle of rotation shown in Eq. (2.32) can be calculated using the dot product. This is accomplished in the example given below.

Example 2.3 Consider a thin gold (Au) plate being heat treated in a suitable furnace containing a carbon-rich environment (carbon atoms) at a relatively high temperature. Assume that carbon C_{60} (fullerene) forms a monolayer on the Au (111) crystallographic plane and that the lattice vectors a_1 and b_1 for Au and C_{60} unit cells are aligned along the $[111]_{Au}$ and $[211]_{C_{60}}$ directions, respectively. This suggests that there must be a rotation matrix and a rotation angle $\theta > 0$ between the Au and C_{60} interface. (a) Calculate the rotation angle using the dot product and (b) the marginal percentage error with respect to the rotation angle in Eq. (2.32). Explain the results.

Solution

(a) For $[111]_{Au}$ and $[211]_{C_{60}}$ directions, the dot product yields

$$\cos\theta = \frac{\mathbf{a}_1 \cdot \mathbf{b}_1}{\|\mathbf{a}_1\|\,\|\mathbf{b}_1\|} = \frac{[111]_{Au} \cdot [211]_{C_{60}}}{\left(\sqrt{u_1^2 + v_1^2 + w_1^2}\right)_{Au}\left(\sqrt{u_1^2 + v_1^2 + w_1^2}\right)_{C_{60}}}$$

$$\cos\theta = \frac{(1)(2) + (1)(1) + (1)(1)}{\sqrt{(1)^2 + (1)^2 + (1)^2}\sqrt{(2)^2 + (1)^2 + (1)^2}} = \frac{4}{\sqrt{3}\sqrt{6}} = \frac{4}{3\sqrt{2}}$$

$$\theta = \arccos\left(\frac{4}{3\sqrt{2}}\right) = 0.94281 \text{ rad.} = 19.5°$$

(b) The percentage error is calculated as

$$\text{error} = \frac{19.5 - 19.1}{19.1} \times 100 = 2.09\%$$

This means that the group of positive $\langle 211 \rangle_{C_{60}}$ directions are at $\theta = 19.5°$ with respect to the $[111]_{Au}$ crystallographic direction. Despite that the calculated rotation angle, $\theta = 19.5°$, has not been documented in the literature, the use of the dot product in this example serves as a theoretical approach to predict the rotation angle. Comparing $\theta = 19.5°$ and $\theta = 19.1°$ yields a remarkable small 2.09% error. Furthermore, in the end, a powerful experimental microscopic work using scanning tunneling microscopy (STM) along with X-ray photoelectron spectroscopy (XPS), low-energy electron diffraction (LEED), and so forth can provide reliable evidence on the characteristics of an overlayer morphology, atomic structure, and the corresponding rotation angle θ. Nonetheless, these methods have their fundamental importance in conducting surface science research.

2.6 Honeycomb and Hexagonal Lattices

Graphite is an allotrope of crystalline carbon and it is fundamentally a stack of hexagonal arrangement of carbon atoms, while graphene is basically one single layer of carbon forming a honeycomb type structure. Essentially, graphite is a stacking of graphene sheets. Figure 2.7 depicts an STM image exhibiting the honeycomb structure of the graphene lattice on an HCP graphite surface [15].

The STM image exhibits a flat and symmetric honeycomb structure of a single-layer graphene crystal [15]. A similar STM honeycomb graphene layer on a silicon dioxide (SiO_2) surface has been reported [16] after electrode deposition in an ultra-high vacuum (UHV) environment. Applications of graphene include electronic devices such as graphene-based touch screens and solar cells.

Interestingly, the known allotropes of carbon crystals are defined as diamond, graphite, graphene, nanotube, and fullerene (buckyball). According to Katsnelson [17], the atomic models for these structures are shown in Fig. 2.8 [18].

For instance,

Fig. 2.7 STM image of honeycomb graphene on HCP graphite [15]

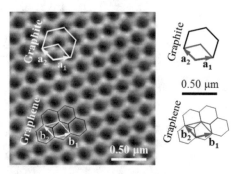

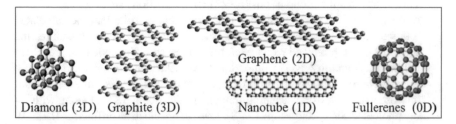

Fig. 2.8 Carbon structures. Adapted from Ref. [18]

- Diamond and graphite are three-dimensional structures (3D). The former has useful industrial applications and high social value, while the latter is used in the metallurgy field for high temperature, automotive industry for electric motor brushes, and electrochemistry field for battery applications.
- Graphene and fullerene are two-dimensional structures in the form of sheets (2D)
- Nanotube is a one-dimensional structure (1D)
- Fullerene is a zero-dimensional structure (0D).

In addition, the traditional usage of carbon to produce ferrous alloys (steels and cast irons) has excelled due to the diverse structures of the allotropes of carbon. The reader is encouraged to seek additional information on these structures in the literature. Nonetheless, these allotropes have different properties.

2.7 Basics for Electrodeposition

Electrodeposition, in general, is a temperature-dependent electrochemical process that must be induced by sufficient thermal and electrical energies. As a result, the flux of reactants towards the substrate has to overcome the initial interfacial barrier imposed by an electric double layer or adsorbed atomic monolayer.

The optimum electrodeposition at a stationary or rotating electrode depends on the flux of reactants induced by the cell electric potential and it is influenced by the temperature at the interfacial region (substrate–electrolyte interface). This process is generally performed at a moderate temperature range to avoid thermal expansion of the electrode and codeposition (secondary reactions), which may be inevitable in certain cases. Of course, codeposition leads to electrodeposition of an alloy. This may be desirable or not.

Apparently, metal deposition begins at surface imperfections and is normally represented by a reversible reaction defined by

$$M^{z+} + ze^- \rightleftarrows M \tag{2.33}$$

Here, the arrow $\rightarrow$ represents a forward reaction (to the right direction) known as the reduction reaction because the cation M^{z+} (positively charged ion) accepts ze^- electrons to become a neutral metal M. Here, z is the oxidation state or valence and e^- is the electron with negative charge. On the other hand, the arrow $\leftarrow$ represents an oxidation reaction (to the left direction).

2.7.1 Electrochemistry-Induced Surface Reconstruction

When a solid surface undergoes oxidation, it changes its surface properties leading to corrosion due to the interactions between the solid surface and the environment.

Therefore, corrosion is an environment-induced degradation of solids; it is a field of research in materials science and engineering to a great extent.

The natural occurrence of corrosion is inevitable since the oxidized solid, normally metals, reverts to its natural chemical form at very slow or fast rate, normally the former predominates in industrial structures. However, a minor corroded metal surface on an electronic chip conveys to a significant problem ion current flow between components. Nonetheless, corrosion is a detrimental atomic process that destroys solid surfaces, which in turn, can be reconstructed by electrochemical deposition. Hence, electrochemistry-induced surface reconstruction.

Theoretically, the mechanism of electrodeposition of a metal M or a polymer is essentially due to reduction reactions coupled with anodic reactions supplying the electrons under appropriate cell control. Normally, it can be studied by cyclic voltammetry and potential-step experiments.

Electrodeposition is driven by the Gibbs energy change in relation to a natural electric potential of an electrochemical cell or by an applied potential to force reduction reactions to proceed on electrode surfaces, whereby atom bonding must occur due to the attractive or Coulombic forces from the atoms constituting the surface (hkl) plane of a single crystal or planes of polycrystals.

In principle, electrodeposition starts at surface defects and continues under the loosely definition of nucleation and growth concepts [19]. Moreover, adding organic agents to the aqueous electrolyte is an industrial practice for brightening and grain refinement of the finish surface deposition. Subsequently, a lattice structure must develop on the cathode area, provided that metal ions acquire a significant mobility along with the flow current to initially overcome the barrier at the cathode–electrolyte interface. This barrier can be the electric double layer or an adsorbed foreign species.

Electrodeposition can be a natural or forced process. The latter has a significant technological application in the field of electrometallurgy, which focuses on a controlled electrochemical process for mass production of a pure metal M, such as copper, or an alloy due to a simultaneous codeposition of a secondary metal or artificially induced by mechanical entrapment [20]. Either the pure metal or alloy can be deposited on a substrate called starter sheet. Industrial production of a pure metal is commonly achieved on approximately $1 \, m^2$ thin sheets partially submerged in an aqueous solution called electrolyte.

2.7.2 Electrochemical Cells

From a practical point of view, electrodeposition is basically a controlled mass-transfer process that requires the determination of the rate of electrodeposition and the potential-current correlations under a well-defined current distribution for a high cell efficiency.

In general, Fig. 2.9 schematically depicts three electrochemical cells called (a) galvanic cell, (b) electrolytic cell, and (c) concentration cell.

Fig. 2.9 Electrochemical cells

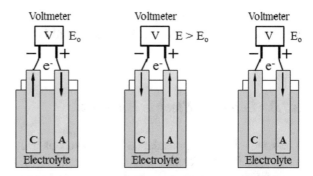

These type of cells can be classified as a one-electrolyte and two-electrode system. In particular, the concentration cell is an electrolytic cell with a high initial concentration of an ion at the anodic electrode. Eventually, this high concentration is diluted as reactions at the cathode surface proceed.

Electrochemical cells have their half-cells separated by either a porous diaphragm or a salt bridge. They are energy producing devices (galvanic cells) or energy consumption devices (electrolytic cell). The application of these devices as technological tools lead to electrochemical engineering, which deals with electroplating, electrowinning, electrorefining, batteries, fuel cells, sensors, and so forth.

Conveniently, electrochemical cells are then classified as (1) galvanic cells which convert chemical energy into electrical energy while electron and current flows exist. The electrochemical reactions are spontaneous. (2) Electrolytic cells for converting electrical energy into chemical energy. The electrochemical reactions are not spontaneous. (3) Concentration cells when the two half-cells are equal, but the concentration of a species j is higher in the anode half-cell than in the cathodic counterpart.

In the electrochemical field, the half-cell components of a galvanic cell system containing a suitable electrolyte are M_1 acting as the anode and metal M_2 as the cathode. The common irreversible reactions metals for M_1 and M_2, and for the redox reaction are

$$M_1 \rightarrow M_1^{z+} + ze^- \quad \text{(Anode)} \tag{2.34a}$$

$$M_2^{z+} + ze^- \rightarrow M_2 \quad \text{(Cathode)} \tag{2.34b}$$

$$M_1 + M_2^{z+} \rightarrow M_1^{z+} + M_2 \quad \text{(Redox)} \tag{2.34c}$$

The metallic cation M_1^{z+} produced at the anodic α-phase (negative terminal) goes into ϵ-phase solution and the number of electrons ze^- move along the wires to arrive at the cathodic electrode–electrolyte interface (positive terminal). As a result, the cation M_2^{z+} gains ze^- electrons and reduces on the cathode β-phase. Thus, M_2^{z+} cations become atoms that deposit or electroplate on the surface of the cathodic

β-phase, specifically at surface defects. The overall reaction, known as the redox reaction, is the sum of Eqs. (2.34a) and (2.34b) in which the electrons ze^- cancel out.

2.7.3 Standard Electric Potential

The standard electric potential ($E°$) of a reversible metal reaction is normally measured against the reference hydrogen electrode (SHE) under standard conditions. Commonly, $E°$ is known as the electromotive force (emf) under thermodynamic equilibrium conditions, such as unit activity, 25 °C, and 1 atm (101 kPa) pressure. Table 2.1 lists $E°$ for some reducing metallic cations (M^{z+}).

The above standard states (Table 2.1) conform to the convention adopted by the International Union of Pure and Applied Chemistry (IUPAC), which requires that the tabulated electromotive force (emf) or standard electrode potential values ($E°$) be so with respect to the standard hydrogen electrode (SHE) for reduction reactions of pure metals, $M^{z+} + ze^- \rightarrow M$. The spontaneous redox reaction in Fig. 2.9a is

$$H_2 = 2H^+ + 2e^- \qquad \text{(anode)} \qquad (2.35a)$$

$$M^{z+} + 2e^- = M \qquad \text{(reduction)} \qquad (2.35b)$$

$$M^{z+} + H_2 = M + 2H^+ \qquad \text{(redox)} \qquad (2.35c)$$

The usefulness of the galvanic series (Table 2.1) is illustrated by the assessment of Fe-based couplings to form galvanic cells. Iron Fe is located between copper (Cu) and zinc (Zn). Now, suppose that one steel plate is coated with Zn and another with Cu. Consequently, Fe is the anode for Cu and the cathode for Zn couplings.

In the latter case, Zn becomes a sacrificial anode, which is the principle of coupling for galvanized steel sheets and pipes. On the other hand, if Cu coating breaks down, steel is then exposed to an electrolyte and becomes the anode, and therefore, it oxidizes. The cell potential of each coupling cases is

$$Fe\text{-}Zn: \ E°_{cell} = E°_{Fe} + E°_{Zn} = -0.44\,\text{V} +0.763\,\text{V} = 0.323\,\text{V} \qquad (2.36a)$$

$$Fe\text{-}Cu: \ E°_{cell} = E°_{Fe} + E°_{Cu} = +0.44\,\text{V} +0.337\,\text{V} = 0.777\,\text{V} \qquad (2.36b)$$

Note that the sign of the Fe reduction potential in Eq. (2.36b) is changed to a positive value.

Fundamentally, surface reconstruction of a particular $X(hkl)$ crystallographic plane exposed to a suitable electrolyte can be achieved at an electrode potential more negative than the open circuit potential (OCP).

The general aspect of surface reconstruction of a particular $X(hkl)$ electrode surface using an electrochemical cell depends on the type of ions in solution, temperature, and type of cell. Commonly, the cells operate at atmospheric pressure.

Table 2.1 Standard potential for metal reduction

Type	Reduction reaction	$E°(V_{SHE})$
Noble	$Au^{3+}+3e^-=Au$	+1.498
↑	$O_2+4H^++4e^-=2H_2O$	+1.229
	$Pt^{2+}+2e^-=Pt$	+1.200
	$Pd^{2+}+2e^-=Pd$	+0.987
	$Ag^++e^-=Ag$	+0.799
	$Fe^{3+}+e^-=Fe^{2+}$	+0.770
	$Cu^{2+}+2e^-=Cu$	+0.337
↕	$2H^++2e^-=H_2$	0.000
	$Fe^{3+}+3e^-=Fe$	−0.036
	$Pb^{2+}+2e^-=Pb$	−0.126
	$Ni^{2+}+2e^-=Ni$	−0.250
	$Co^{2+}+2e^-=Co$	−0.277
	$Cd^{2+}+2e^-=Cd$	−0.403
	$Fe^{2+}+2e^-=Fe$	−0.440
	$Cr^{3+}+3e^-=Cr$	−0.744
	$Zn^{2+}+2e^-=Zn$	−0.763
	$Ti^{2+}+2e^-=Ti$	−1.630
	$Al^{3+}+3e^-=Al$	−1.662
	$Mg^{2+}+2e^-=Mg$	−2.363
	$Na^++e^-=Na$	−2.714
↓	$K^++e^-=K$	−2.925
Active	$Li^++e^-=Li$	−3.045

However, the electrochemical cell may be an important experimental variable during surface reconstruction.

2.7.4 Epitaxy

In electrochemistry, *epitaxy* refers to the deposition of a single crystalline lattice plane or film on a crystalline substrate. Thus, *epitaxial growth* occurs if the epitaxial film or epitaxial layer adopt the same substrate crystal lattice; otherwise, film is by *crystal growth*.

Fundamentally, *epitaxial growth* is an extension of the substrate crystal lattice due to cohesive forces between metal atoms [21]. For equal or fairly similar film-substrate metal epitaxy, the deposition process is known as *homoepitaxy*; otherwise, *heteroepitaxy*. In general, epitaxial growth is related to the crystallographic relationship between the substrate and deposited layer.

Furthermore, epitaxy growth of thin films is very important in nanotechnology and semiconductor fabrication procedures. For instance, heteroepitaxy of gallium nitride GaN on $Si(111)$ provides good thermal conductivity in semiconductor applications, such as GaN-based violet laser diodes [22].

2.8   Summary

Surface reconstruction, as described in this chapter, is a two-dimensional process, where structural stabilization results from the atomic positions on a substrate surface. For instance, epitaxy is briefly described in general terms for film formation and related industrial applications. However, deposition is actually a process that refers to reduction reactions on electrode surfaces. Hence, reduction reactions are considered the main driving force along with the Gibbs energy change (ΔG) needed for deposition of atoms that leads to film or overlayer formation and film growth.

Matrix and Wood notations are introduced in order to study common a crystal structure, such a single crystal (SC), face-centered cubic (FCC), body-centered cubic (BCC), hexagonal-closed pack (HCP), or a honeycomb shaped carbon structure. Nonetheless, surface reconstruction is induced by adsorption of atoms onto the substrate surface. As a result, at least a monolayer is formed with specific crystallographic characteristics, which can be analyzed using either the matrix or Wood notation. This leads to the concept of conventional and primitive lattice planes, and related vector notation.

2.9   Problems

2.1 Starting from an overlayer having $(b_1, a_1) = (4, 1)$ vertices, determine the Matrix and Wood notations (a) for a hypothetical material $X(111)$ subjected to surface reconstruction on its surface crystallographic (111) plane and (b) for Silicene $[R_2SiO]_n$ on $Ag(111)$. Here, $[R_2SiO]_n$ has a honeycomb structure similar to that of graphene and R is an organic group such as methyl, ethyl, or phenyl.

2.2 Assume that a (100) surface plane of a metal X is surface reconstructed by a foreign atom as shown in the figure given below.

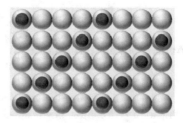

(a) Draw the primitive lattices for the substrate and overlayer (red dots) and their respective set of vectors (a_1, a_2) and (b_1, b_2), respectively. (b) Determine the set of linear equations based on the linear transformation of the substrate vectors (a_1, a_2) into the overlayer vectors (b_1, b_2), the rotation matrix (M_R) and the Matrix notation, and (c) the Wood notation.

2.3 Determine (a) the Matrix notation and (b) the Wood notation for the (b_1, b_2) lattice on an $FCC(100)$ substrate (Fig. P2.3).

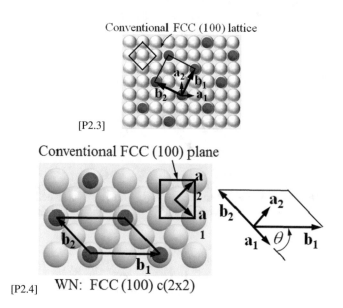

Conventional FCC (100) lattice

[P2.3]

Conventional FCC (100) plane

[P2.4]   WN: FCC (100) c(2x2)

2.4 Determine the Matrix notation and Wood notation for surface reconstruction of an FCC (100) plane using the parallelogram and the set of vectors given below. Denote that the primitive substrate lattice is a $p(1 \times 1)$ matrix and that of the adsorbate is a centered $c(2 \times 2)$ matrix. See Fig. P2.4.

2.5 Determine the Matrix notation and the Wood notation for surface reconstruction of an FCC (111) plane using the parallelogram and the vector directions given below.

2.6 Compute (a) the angle θ between $[uvw]_a$ and $[uvw]_b$ directions in an FCC unit cell. Deduce a theoretical Wood notation based on these directions.

(b) Derive the area A of the lattice (parallelogram) as a function the lattice parameter "a," and (c) calculate the parallelogram area A if the lattice parameter of the unit cell is $a = 0.289$ nm. Has the resultant Wood notation been reported in the literature?

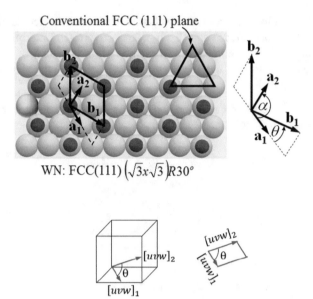

Conventional FCC (111) plane

WN: FCC(111)$\left(\sqrt{3}x\sqrt{3}\right)R30°$

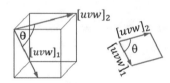

2.7 Calculate (a) the angle θ between $[uvw]_a$ and $[uvw]_b$ directions in an FCC unit cell. Deduce a theoretical Wood notation based on these directions.

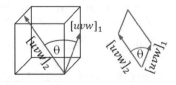

(b) Derive the area A of the lattice (parallelogram) as a function the lattice parameter "a" and (c) calculate the parallelogram area A if the lattice parameter of the unit cell is $a = 0.289$ nm. [Solution: (a) $\theta = 26.6°$ and (c) $A = 3a^2$]. Has the resultant Wood notation been reported in the literature?

2.8 Compute (a) the angle θ between $[uvw]_a$ and $[uvw]_b$ directions in an FCC unit cell. Deduce a theoretical Wood notation based on these directions.

(b) Derive the area A of the lattice (parallelogram) as a function the lattice parameter "a" and (c) calculate the parallelogram area A if the lattice parameter of the unit cell is $a = 0.289$ nm. Has the resultant Wood notation been reported in the literature?

2.9 Compute (a) the angle θ between $[\bar{3}01]$ and $[\bar{1}22]$ directions in an FCC unit cell. Deduce a theoretical Wood notation based on these directions. (b) Derive the area A of the lattice (parallelogram) as a function the lattice parameter "a" and (c) calculate the parallelogram area A if the lattice parameter of the unit cell is $a = 0.289$ nm.

2.10 Calculate (a) the angle θ between $[\bar{4}01]$ and $[\bar{1}22]$ directions in an FCC unit cell. Deduce a theoretical Wood notation based on these directions. (b) Derive the area A of the lattice (parallelogram) as a function the lattice parameter "a" and (c) calculate the parallelogram area A if the lattice parameter of the unit cell is $a = 0.289$ nm.

References

1. J. Isidorsson, T. Lindstrom, C.G. Granqvist, M. Herranen, Adsorption and electrodeposition on SnO_2 and WO_3 electrodes in 1 M $LiClO_4/PC$: in situ light scattering and in situ atomic force microscopy studies. J. Electrochem. Soc. **147**(7), 2784–2795 (2000)
2. K. Hermann, *Crystallography and Surface Structure: An Introduction for Surface Scientists and Nanoscientists* (Wiley, Weinheim, 2011)
3. K. Hermann, Periodic overlayers and moire patterns: theoretical studies of geometric properties. J. Phys. Condens. Matter **24**(31), 314210 (13pp.) (2012)
4. E.A. Wood, Vocabulary of surface crystallography. J. Appl. Phys. **35**(4), 1306–1312 (1964)
5. R.L. Park, H.H. Madden, Annealing changes on the (100) surface of palladium and their effect on CO adsorption. Surface Sci. **11**(2), 188–202 (1968)
6. S. Yoshimoto, R. Narita, E. Tsutsumi, M. Matsumoto, K. Itaya, O. Ito, K. Fujiwara, Y. Murata, K. Komatsu, Adlayers of fullerene monomer and [2 + 2]-type dimer on Au(111) in aqueous solution studied by in situ scanning tunneling microscopy. Langmuir **18**, 8518–8522 (2002)
7. M. Kittelmann, P. Rahe, M. Nimmrich, C.M. Hauke, A. Gourdon, A. Kuhnle, On-surface covalent linking of organic building blocks on a bulk insulator. J. ACS Nano **5**(10), 8420–8425 (2011)
8. J.R. Osiecki, R.I.G. Uhrberg, Alloying of Sn in the surface layer of Ag(111). Phys. Rev. B **87**, 075441-1–075441-5 (2013)
9. A. Artaud, L. Magaud, T. Le Quang, V. Guisset, P. David, C. Chapelier, J. Coraux, Universal classification of twisted, strained and sheared graphene moire superlattices, Scientific reports, Nature Publishing Group, vol. 6 (2016), pp. 1–14
10. E.D. Mishina, N. Ohta, Q.-K. Yu, S. Nakabayashi, Dynamics of surface reconstruction and electrodeposition studied in situ by second harmonic generation. Surface Sci. **494**(1), L748–L754 (2001)
11. M.S. Zei, G. Ertl, Electrodeposition of Cu onto Reconstructed Pt(100) and Pt(110) Surfaces, Zeitschrift für Physikalische Chemie, Berlin, Germany, Bd. 202, S. 5–19 (1997)
12. B.C. Russell, M.R. Castell, $\left(\sqrt{13}x\sqrt{13}\right)$ $R13.9°$ and $\left(\sqrt{7}x\sqrt{7}\right)$ $R19.1°$ reconstruction of the polar $SrTiO_3$ (111) surface. Phys. Rev. B **75**, 155433-1–155433-7 (2007)
13. Y.G. Kim, S.L. Yau, K. Itaya, In situ scanning tunneling microscopy of highly ordered adlayers of aromatic molecules on well-defined Pt(111) electrodes in solution: benzoic acid, terephthalic acid, and pyrazine. Langmuir **15**, 7810–7815 (1999)

14. G.S. Parkinson, A. Hentz, P.D. Quinn, A.J. Window, D.P. Woodruff, P. Bailey, T.C.Q. Noakes, Methylthiolate-induced reconstruction of Ag (111): a medium energy ion scattering study. Surface Sci. **601**(1), 50–57 (2007)
15. A. Luican, G. Li, E.Y. Andrei, Scanning tunneling microscopy and spectroscopy of graphene layers on graphite. Solid State Commun. **149**(27–28), 1151–1156 (2009)
16. E. Stolyarova, K.T. Rim, S. Ryu, J. Maultzsch, P. Kim, L.E. Brus, T.F. Heinz, M.S. Hybertsen, G.W. Flynn, High-resolution scanning tunneling microscopy imaging of mesoscopic graphene sheets on an insulating surface. Proc. Natl. Acad. Sci. **104**(22), 9209–9212 (2007)
17. J.W. Hill, R.H. Petrucci, *General Chemistry*, 3rd edn. (Prentice Hall, New Jersey, 2002)
18. M. Katsnelson, *Graphene: Carbon in Two Dimensions*. Materials Today. Institute for Molecules and Materials, vol. 10, no. 1–2 (Radboud University Nijmegen, Nijmegen, 2007)
19. D.M. Kolb, The initial stages of metal deposition as viewed by scanning tunneling microscopy, in *Advances in Electrochemical Science and Engineering*, ed. by R.C. Alkire, D. M. Kolb, vol. 7 (Wiley, Weinheim, 2002), pp. 107–150
20. P.W. Martin, R.V. Williams, *Proceedings of Interfinish 64*, British Iron and Steel Research Association, London (1964), pp. 182–188
21. J.L. Stickney, Electrochemical atomic layer epitaxy (EC-ALE): nanoscale control in the electrodeposition of compound semiconductors, in *Advances in Electrochemical Science and Engineering*, ed. by R.C. Alkire, D.M. Kolb, vol. 7 (Wiley, Weinheim, 2002), pp. 1–106
22. A. Dadgar, F. Schulze, M. Wienecke, A. Gadanecz, J. Bläsing, P. Veit, T. Hempel, A. Diez, J. Christen, A. Krost, Epitaxy of GaN on silicon-impact of symmetry and surface reconstruction. New J. Phys. **9**(10), 389–398 (2007)

Chapter 3
Mass Transport by Diffusion

3.1 Introduction

It is a well-known fact that physical and mechanical properties of crystalline solid materials depend on the type of atomic structure and related grain morphology. Of special interest are some steels (Fe), silicon (Si) and germanium (Ge) since they have been studied extensively. Steels are important in the engineering field for constructing large structures, while silicon is used to manufacture small wafers used in the electronic industry.

The crystal structure of some crystalline materials, such as steel and silicon, is sensitive to heat treatment processes in specific gaseous environments since the atomic structure of thin surface layers can be altered by diffusion of interstitial atoms during heat treatment at relatively high temperatures and for time $t > 0$. Diffusion is a mass transport mechanism due to the motion of atoms, molecules, or ions into a medium. This is an atomic process related to the random walk theory, which is a statistical physics model used for understanding the kinetics of particle (atom, molecular, or ion) diffusion. The random walk model gives insight of the random collisions of species while traveling to a destination within a solid, liquid, and gas media, and a composite medium such as the human body. Nonetheless, the diffusion of a species j into a substrate is an interdiffusion process due to interstitial or substitutional atoms undergoing atomic motion at a temperature T.

3.2 Random Walks

In physics and materials science, the mass transport by diffusion is a fundamentally random thermally activated motion of atoms in a solid crystal structure (lattice). The source of the diffusing atoms can be from a solid in contact with the solid host or from a gas containing the diffusing solute. Thus, solid–solid or gas–solid

© Springer Nature Switzerland AG 2020
N. Perez, *Phase Transformation in Metals*,
https://doi.org/10.1007/978-3-030-49168-0_3

Fig. 3.1 Random walk
model for atom jump

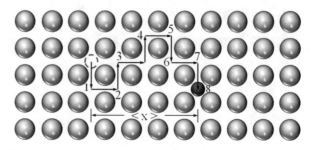

diffusion are the most common terminologies in the fields of science. Nonetheless,
the mechanism for mass transport of impurity atoms into a solid is generally referred
to as interdiffusion or impurity diffusion, which is classified as either interstitial
diffusion if the impurity atoms are smaller than the host solid or substitutional
diffusion if the impurity and host atoms have the same or similar in size.

The main objective of this chapter is to introduce mathematical models to
describe the mass transport phenomenon by diffusion under steady-state and
transient conditions as defined by the Fick's laws of diffusion, which is physically
based on the random walk theory. The fundamental of random walk of particles
(atoms and molecules) by succession of random steps is based on probability and
statistics to predict system behavior.

Consider the hypothetical particle motion within a limited domain depicted in
Fig. 3.1 and assume a particle motion in a random manner in the x-direction at a
small time interval τ. This can be treated as a system of one particle that undergoes
a series of relatively small steps at random. Hence, the particle randomly changes
position within a limited domain. The reader is encouraged to be acquainted with
the variety of the random walks.

For a total number of steps k, the average distance $\langle x \rangle$ for equal steps to the right
and left can be deduced as

$$\langle x \rangle = \frac{\left[\sum (x_k)_{right} + \sum (-x_k)_{left}\right]}{k} = 0 \tag{3.1}$$

Nevertheless, the *mean-square distance* is non-zero variable and it is mathematically
defined by

$$\left\langle x^2 \right\rangle = \frac{\left[\sum (x_k)^2_{right} + \sum (-x_k)^2_{left}\right]}{k} = \lambda^2 \neq 0 \tag{3.2}$$

If τ is the time step at $+x$ and $-x$ directions, the *mean-square distance* $\left\langle x^2 \right\rangle$ for
a total random walk time t and the average statistical travel distance $\langle x \rangle$ are time-
dependent variables.
Thus,

$$\langle x^2 \rangle = \left(\frac{t}{\tau}\right) \lambda^2 = Dt \quad \text{(random)} \tag{3.3}$$

$$\langle x \rangle = \sqrt{Dt} \quad \text{(random)} \tag{3.4}$$

$$\langle x \rangle = \left(\frac{\lambda}{\tau}\right) t = vt \quad \text{(average)} \tag{3.5}$$

where D and v are the diffusion coefficient and particle speed, respectively. From Eqs. (3.3) and (3.4), D and the diffusion step length L are

$$D = \frac{\lambda^2}{\tau} \tag{3.6}$$

$$L = \sqrt{Dt} \tag{3.7}$$

Consider the independently and identically distributed two-dimensional (2D) random steps of an interstitial atom (treated as a particle) shown in Fig. 3.1. Thus, the mathematical definitions of the particle speed for the random and the average walks are compared as indicated below

$$v_{ran} = \frac{L}{t} = \frac{L}{L^2/D} = \frac{D}{L} \quad \text{(Random)} \tag{3.8}$$

$$v_{ave} = \frac{\lambda}{\tau} \quad \text{(Average)} \tag{3.9}$$

Combining Eqs. (3.3) and (3.9) yields the average diffusion coefficient

$$D = \lambda v_{ave} \tag{3.10}$$

In general, the particle displacement by random walk due to diffusion is $x = 2\langle x \rangle$ and from Eq. (3.4), x is cast as the characteristic distance or diffusion length. Thus,

$$x = 2\sqrt{Dt} \tag{3.11}$$

from which a common dimensionless variable used in statistical physics, transport phenomena, and atomic diffusion is defined by

$$n = \frac{x}{2\sqrt{Dt}} \tag{3.12}$$

Mathematically, n is the argument in the error function erf(n) and the exponential function $\exp\left(-n^2\right)$ that occur in solutions to partial differential equations. Nonetheless, the above mathematical approach is a general description of the relation between random walk and atomic diffusion [1–3].

Interestingly, the random walks of atoms in solids is based on random jumps. This implies that individual particle velocity may be predicted by Einstein's theory for the average kinetic energy of particles influenced by the absolute temperature T of the system [4]. Hence, Einstein's random walk theory is based on Brownian motion (random motion of particles suspended in a fluid).

For one-dimensional analysis, the average kinetic energy and the thermal energy are assumed to be equal. Under such an assumption in the solid state, the energy balance and the particle (electron) or drift velocity at an absolute temperature T are defined by

$$\frac{mv_d^2}{2} = \frac{k_B T}{2} \tag{3.13}$$

$$v_d = \sqrt{\frac{k_B T}{m}} \tag{3.14}$$

where m = Mass of the particle (atom or molecule)

v_d = Average velocity of a particle in the x-direction

$m = A_w/N_o$ = (Atomic weight)/(Avogadro's number)

$m = M_w/N_o$ = (Molecular weight)/(Avogadro's number)

$k_B = R/N_o = 1.38 \times 10^{-23}$ J/(K) = Plank's constant

$k_B = 1.38 \times 10^{-20}$ g·m^2/ $\left(\text{s}^2\text{K}\right)$ = 8.62×10^{-5} eV/K

$R = 8.3145$ J/ (mol · K) = Universal gas constant

$N_o = 6.022 \times 10^{23}$ = Avogadro's number

$k_B T = 4.14 \times 10^{-14}$ g · cm^2/s^2 = 0.026 eV at 300 K (25 °C)

Notice that v_d depends on T and subsequently, it can be characterized as the thermal velocity, which in turn must be different from that influenced by an electric field. Also, the random motion of a particle can occur with alternating velocities in certain science fields. For the moment, Eq. (3.14) remains as an average drift velocity. Hence, $v_d = f(T)$ for $T < T_f$, where T_f denotes the fusion or melting temperature of an entity. According to Eq. (3.14), the particle has a finite velocity v_d during its random motion and as a result, the random walk can be treated as real physical event.

Remarkably, random walk in the solid state is a quite complicated theory, but the above simplified approach is considered adequate for the reader to understand this phenomenon as clear as possible. Hence, the atomic diffusion theory is based on the random walk concept.

Continuing the two-dimensional (2D) analysis of particle motion by diffusion (Fig. 3.1), the comprehensive analysis of one particle (interstitial atom) random walk on a lattice plane describes a time-dependent path, which consists of a succession of random steps. Thus, a random motion can be modeled as a random walk of subsequent diffusion steps in the solid-state domain.

The simultaneous motion of many particles is a complicated process related to probability and discrete time for a sequence of random variables. Thus, the particle motion not only depends on time, but also on the interactions among moving particles on a lattice domain, where an individual particle has a limited area to randomly walk before coming to a complete stop.

For instance, the motion of an interstitial atom along the one-dimensional lattice shown in Fig. 3.1 is completely a random walk because there is a lack of bias. Consequently, there is an equal probability that the atom moves in random directions.

Conceivably, putting the ideas of random walk to work through an example can help the reader understand the current theory. This implies, again, that the reader is encouraged to be totally acquainted with the theories describing atomic diffusion and related random walk.

Example 3.1 Use of the random walk theory in Metallurgy and Biology. Calculate the drift velocity v_d, the diffusion coefficient D, and the time step τ at $T = 300\,\text{K}$ and $\lambda = 5 \times 10^{-10}\,\text{cm}$ for (a) carbon atoms in steel and (b) Albumin protein (a family of globular proteins). The atomic weight of carbon is $12\,\text{g/mol}$ and that of Albumin is $68,000\,\text{g/mol}$.

Solution

(a) For carbon, $A_w = 12\,\text{g/mol}$ at $T = 300\,\text{K}$, $\lambda = 5 \times 10^{-10}\,\text{cm}$ and

$$m = \frac{A_w}{N_o} = \frac{12\,\text{g/mol atom}}{6.022 \times 10^{23}\,\text{atom/mol}} = 1.9927 \times 10^{-23}\,\text{g}$$

$$v_d = \sqrt{\frac{k_B T}{m}} = \sqrt{\frac{4.4 \times 10^{-14}\,\text{g} \cdot \text{cm}^2/\text{s}^2}{1.9927 \times 10^{-23}\,\text{g}}} \simeq 46,990\,\text{cm/s}$$

and from Eq. (3.10) with $v_d = v_{ave}$

$$D = \lambda v_d = \left(5 \times 10^{-10}\,\text{cm}\right)(47,000\,\text{cm/s}) = 2.35 \times 10^{-5}\,\text{cm}^2/\text{s}$$

From Eq. (3.6),

$$\tau = \frac{\lambda^2}{D} = \frac{\left(5 \times 10^{-10}\,\text{cm}\right)^2}{2.35 \times 10^{-5}\,\text{cm}^2/\text{s}} = 1.06 \times 10^{-15}\,\text{s} \simeq 0.00106\,\text{ps}$$

(b) For an Albumin protein structure with $M_w = 68,000\,\text{g/mol}$ at $T = 300\,\text{K}$, $\lambda = 5 \times 10^{-10}\,\text{cm}$ and $\tau = 1\,\text{ms}$,

$$m = \frac{M_w}{N_o} = \frac{68,000\,\text{g/mol atom}}{6.022 \times 10^{23}\,\text{atom/mol}} = 1.1292 \times 10^{-19}\,\text{g}$$

$$v_d = \sqrt{\frac{k_B T}{m}} = \sqrt{\frac{4.14 \times 10^{-14}\, g \cdot cm^2/s^2}{1.1292 \times 10^{-19}\, g}} \simeq 605.50\, cm/s$$

and

$$D = \lambda v_d = \left(5 \times 10^{-10}\, cm\right)(605.50\, cm/s) = 3.03 \times 10^{-7}\, cm^2/s$$

Subsequently, Eq. (3.6) gives

$$\tau = \frac{\lambda^2}{D} = \frac{\left(5 \times 10^{-10}\, cm\right)^2}{3.03 \times 10^{-7}\, cm^2/s} \simeq 8.25 \times 10^{-13}\, s = 0.825\, ps$$

Percent error in the drift velocities:

$$error = \frac{46{,}990 - 605.50}{(46{,}990 + 605.50)\,/2} \times 100 = 195\%$$

Therefore, these calculations are meant to show the reader that the random diffusion of particles vary from types of particles in diverse media. Denote that carbon diffuses faster than Albumin due to the fact that the former is a small single particle composed of an atom and the latter is a large particle (group of atoms) defined by the chemical formula $C_{123}H_{193}N_{35}O_{37}$. Additionally, there is 195% difference (average percentage error) in v_d.

3.3   Solid-State Diffusion

Solid-state diffusion is a mass transport method for introducing interstitial atoms in ferrous materials or dopant atoms into semiconductors. Figure 3.2 schematically shows the main diffusion mechanisms known as **vacancy or substitutional diffusion**, which is the atomic motion of an atom from one lattice site in Fig. 3.2a to an adjacent vacancy in Fig. 3.2b.

Usually substitutional atoms undergo this type of mechanism. On the other hand, **interstitial diffusion** is the atomic motion of interstitial (small) atoms from one interstitial site (Fig. 3.2c) to another interstitial site (Fig. 3.2d). Common interstitial atoms are hydrogen, carbon, boron, and nitrogen. This type of diffusion is said to be faster than vacancy diffusion.

In general, steady-state diffusion and transient diffusion are the two conditions one has to consider in solving diffusion problems. For instance, interstitial diffusion of carbon or nitrogen atoms is a process used to change the chemical composition of the substrate to a certain depth. The main objective is to harden the steel surface for improving wear resistance.

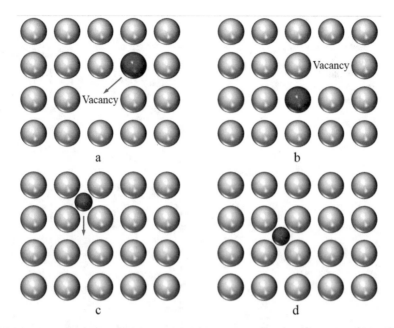

Fig. 3.2 Mass transport by diffusion. (**a**) Vacancy and substitutional atom, (**b**) substitutional diffusion, (**c**) interstitial atom, and (**d**) interstitial diffusion

For ferrous conductors, Fe-C-X alloy, as an example, Fe-C is the basis for plain carbon steels and X is for alloy steels. Both C and X may be subjected to simultaneous interstitial and substitutional diffusion mechanisms, respectively. The latter occurs, provided that vacancies are available in the Fe matrix.

For semiconductors, substitutional diffusion is carried out for adjusting electrical properties of silicon wafers. This is achievable by doping a species j, say boron (B), arsenic (As), or phosphorous (P), to a certain junction depth x_j. A common practice in the semiconductor industry is to use ion implantation followed by diffusion controlling the amount of dopant species into silicon wafers.

3.3.1 Octahedral and Tetrahedral Sites

The preferred interstitial sites for small impurity atoms (solute) are the lattice empty spaces called voids or interstices within the substrate (solvent or host). As a result, the original chemical composition of the substrate is altered by interstitial diffusion of impurity atoms that form a small diffusion layer from the substrate surface to a diffusion depth x, which in turn sets the limit for interstitial solid solution with decreasing concentration [5].

Fig. 3.3 Octahedral sites in
(**a**) FCC and (**b**) *BCC* unit
cells. Tetrahedral sites in (**c**)
FCC and (**d**) *BCC* crystal
structures. The interstitial
sites are represented by the
blue and red dots

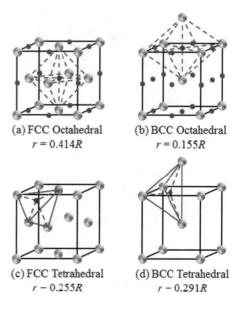

(a) FCC Octahedral   (b) BCC Octahedral
$r = 0.414R$ $r = 0.155R$

(c) FCC Tetrahedral   (d) BCC Tetrahedral
$r - 0.255R$ $r - 0.291R$

For comparison, Fig. 3.3 shows the conventional FCC and BCC unit cells (host crystal structures) along with deduced octahedral and tetrahedral geometries, which in turn, exhibit interstitial sites or voids (holes) that can be occupied by other type of atoms or ions.

These sites are empty spaces located in the center of these structures and are treated as interstitial voids for the obvious locations of interstitial atoms. This means that six (6) atoms packed together form the octahedral and four (4) atoms form the tetrahedral arrangements. Nevertheless, these sites are distinguished by the number of nearest-neighbor atoms called the coordination number (CN): $CN = 4$ for a tetrahedral and $CN = 6$ for octahedral sites.

The tetrahedral and octahedral geometries can also be drawn at different locations or be rotated. Interestingly, impurity atoms reside in the interstitial sites occupying either the tetrahedral or octahedral voids. This implies that the size of the impurity atoms is a determining factor in interstitially alloying metals. As an example, properties of carbon-iron alloy systems are enhanced since the carbon atoms at these sites induce a slight distortion of the iron (Fe) lattice.

Another important information on the current topic is the size of the interstitial atoms that can accommodate in the substrate lattice causing little or no distortion of the lattice known as lattice strain. Using the perfect sphere model, one can calculate the interstitial largest sphere that can be placed in a tetrahedral or octahedral void without causing distortion.

For instance, X-ray diffraction (XRD) technique can be used to determine the distribution of carbon among octahedral and tetrahedral interstitial sites (also known as interstices) in $Fe\text{-}FCC$ iron lattice. If the Fe is cooled from an austenitic temperature, then it transforms to BCC and the carbon atoms remain in these interstitial sites.

3.4 Fick's Laws of Diffusion

Diffusion is a well-known mechanism of mass transport during heat treatment of a substrate in an impurity-rich environment. For steady-state conditions, the mass transfer of interstitial carbon, nitrogen, or hydrogen atoms in gas–solid is called interstitial diffusion and solid–solid couplings is a substitutional diffusion method known as interdiffusion.

3.4.1 First Law: Steady-State Diffusion

In one-dimensional analysis, the *steady-state diffusion* implies that the concentration rate must be $\partial C/\partial t = 0$ as a primary condition at the substrate–environment interface. Thus, the molar flux J_x of impurity atoms j can be predicted by

$$J_x = C v_{ran} \quad \& \quad J_x = \frac{m}{A_s t} \tag{3.15a}$$

$$C = \frac{N}{V} \tag{3.15b}$$

where $C = C(x, t)$ is the concentration of the impurity atom $(atoms/cm^3)$, m is the mass of the diffusing impurity into a substrate, A_s is the substrate exposed surface area, t is the diffusion time, N is the number of particles, and V is the volume.

For diffusion of an impurity atom j in a specific homogeneous medium, Fick's first law of diffusion defines the molar flux as

$$J_x = -D \frac{\partial C}{\partial x} \tag{3.16}$$

Here, $\partial C/\partial x$ is the concentration gradient that acts as the driving force for $J_x > 0$ and D is assumed to be constant within a small temperature range ΔT. Figure 3.4a illustrates a two-dimensional model for interstitial planes (red dots) which represent the concentration $C = C(x, t)$. Fig. 3.4b shows the schematic concentration profile as a function of diffusion depth for time $t > 0$.

This is an ideal case for interstitial diffusion due to a concentration gradient $\partial C/\partial x$. However, it is important to determine the type of lattice voids these impurity atoms reside in the interstitial sites occupying either the tetrahedral or octahedral void as shown in Fig. 3.3 [5]. Theoretically, atomic diffusion of a species j can occur in large crystalline and amorphous substrates to a limited depth x or through

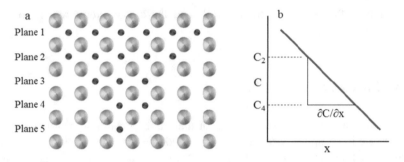

Fig. 3.4 (a) Model for interstitial diffusion within a cubic crystal plane and (b) concentration profile and its concentration gradient

a thin sheet. Nonetheless, interstitial diffusion is faster than substitutional diffusion due to the impurity-substrate atom size effects.

The physical interpretation of Eq. (3.16) dictates that the molar or mass flux J_x follows a similar behavior as D does because steady-state diffusion depends on the nature of the species j and the concentration gradient $\partial C/\partial x < 0$ in a time period. In particular, D is mathematically a multiplying factor for J_x.

Recall that metallic materials have relatively high atomic packing factors, the interstitial positions (empty spaces or voids between packed atoms in the lattice) are relatively small and consequently, the atomic diameter of an interstitial impurity atom must be substantially smaller than that of the substrate atoms. This can be investigated further by using the atom radius ratio $r/R = c_x$ or $r = c_x R$, where c_x is a constant (cited in Fig. 3.3) and by assuming diffusion of carbon atoms, for example, into FCC iron (austenite phase) at $T > 912\,°C$. For instance, octahedral and tetrahedral voids in iron (Fe) have their corresponding relative radii as

$$r_{Fe,oct} = 0.414\,(0.124\,\text{nm}) = 0.051\,\text{nm}$$

$$r_{Fe,tet} = 0.255\,(0.124\,\text{nm}) = 0.032\,\text{nm}$$

Using the carbon radius $r_C = 0.071\,\text{nm} > r_{Fe,oct} > r_{Fe,tet}$ one can deduce that carbon atoms will occupy the octahedral sites introducing some lattice distortion. The tetrahedral size is relatively too small for carbon to occupy these sites. Therefore, iron (Fe) is hardened by adding carbon to its octahedral lattice sites. Graphical representation of carbon concentration profile is schematically depicted in Fig. 3.4b.

Furthermore, combining Eqs. (3.15a) and (3.16) yields the concentration gradient (Fig. 3.4b) related to the random walk speed

$$\frac{\partial C}{\partial x} = -\left(\frac{C}{D}\right) v_{ran} \tag{3.17}$$

Diffusion Coefficient Now assume that the diffusion coefficient D (also known as intrinsic diffusivity) obeys the Arrhenius-type rate equation so that

$$D = D_o \exp\left(-\frac{Q_d}{RT}\right) = D_o \exp\left(-\frac{Q_d}{k_B T}\right) \qquad (3.18)$$

where D_o is a proportionality constant and Q_d is the activation energy for diffusion. The product RT or $k_B T$ in Eq. (3.18) is a matter of choice based on units. Anyway, $Q_d > k_B T$ so that an impurity atom j acquires a high mobility at high temperatures. Actually, the diffusion coefficient, as defined by Eq. (3.18), is a temperature-dependent variable, $D = f(T)$.

Convenient graphical representations of $D = f(T)$, as defined by Eq. (3.18), and $D = f(1/T)$ are shown in Fig. 3.5a, b, respectively. Denote that D is strongly dependent on a temperature range $0 < T < T_m$, where T_m is the melting temperature of the substrate. In solid-state diffusion, if $T \to 0$, then $D \to 0$ since $\exp(-\infty) = 0$ and consequently, diffusion of a species j requires an incubation time in a particular environment. On the other hand, if mathematically $T \to T_m \to \infty$, then $D \to D_o$ due to $\exp(-0) = 1$.

Furthermore, Fig. 3.5b exhibits a linear plot suggesting that it is a practical approach for analyzing diffusion data and determining (1) the thermal activation energy Q_d through the slope S_R and (2) the proportionality constant D_o. This plot is obtained by taking the natural logarithm of Eq. (3.18) and then converted to common (base-10) logarithm.

Actually, Fig. 3.5b is a graphical representation of the linearized Arrhenius empirical expression, Eq. (3.18), used in science fields. A small data set is sufficient for determining the slope S_R, from which the activation energy Q_d is calculated.

Example 3.2 Linearize Eq. (3.18) and explain the meaning of the slope in terms of the physical constants R and k_B individually.

Solution Take the natural logarithm on both sides of Eq. (3.18) to get

$$\ln D = \ln D_o - \frac{Q_d}{R}\frac{1}{T} \qquad (a)$$

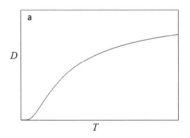
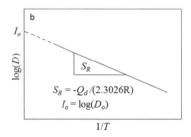

Fig. 3.5 (a) Non-linear and (b) linear Arrhenius plots

$$\log D = \log D_o - \left(\frac{Q_d}{2.3026R}\right)\frac{1}{T} = I_o - S_R\left(\frac{1}{T}\right) \tag{b}$$

where $S_R = Q_d/(2.3026R)$ has units of $(J/mol)/(J/mol\ K) = K$ and it is the slope in Fig. 3.5b along with the intercept $I_o = \log D_o$.

Taking common logarithm on both sides of Eq. (3.18) yields a similar expression written as

$$\log D = \log D_o - \left(\frac{Q_d}{2.3026\,k_B}\right)\frac{1}{T} = I_o - S_B\left(\frac{1}{T}\right) \tag{c}$$

Then,

$$S_B = \frac{Q_d}{2}.3026\,k_B \tag{d}$$

Similarly, the slope S_B has units of $(eV/atom)/(eV/atom\ K) = K$. Note that both S_R and S_B have the same meaning as well as the same absolute temperature unit, degrees Kelvin.

Despite of the extreme simplicity in laying down the mathematical procedure in this example, the slope of the straight line is an important variable for determining the activation energy Q_d.

Example 3.3 Consider thin sheet of metal 2 mm thick being exposed to atomic hydrogen on both sides as shown in the sketch below. First of all, adsorption and dissociation of gas A occur on the sheet surfaces at relatively high temperatures. Assume that the hydrogen concentration (in mol/m^3) and diffusion coefficient are hypothetically defined by

$$C_H = 73\sqrt{P}\exp\left(-\frac{12,000\,J/mol}{RT}\right)$$

$$D_H = (5\,m^2)\exp\left(-\frac{4000\,J/mol}{RT}\right)$$

where pressure P is in MPa. If pressures $P_1 = 4\,MPa$ and $P_2 = 0.1\,MPa$ $T = 400\,°C$ and $\Delta x = 1\,mm$, then calculate(a) the hydrogen diffusion flux and (b) deduce the diffusional area for $m = 1$ mol and $t = 10$ s.

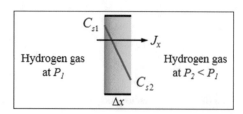

Solution

$$C_{H,P_1} = 73\sqrt{4}\exp\left(-\frac{12,000\,\text{J/mol}}{(8.3145\,\text{J/mol K})\,(400 + 273)}\right) = 17.10\,\text{mol/m}^3$$

$$C_{H,P_2} = 73\sqrt{0.1}\exp\left(-\frac{12,000\,\text{J/mol}}{(8.3145\,\text{J/mol K})\,(400 + 273)}\right) = 2.70\,\text{mol/m}^3$$

$$D_H = (5\,\text{m}^2)\exp\left(-\frac{4,000\,\text{J/mol}}{(8.3145\,\text{J/mol K})\,(400 + 273)}\right)$$

$$D_H = 2.45\,\text{m}^2/\text{s}$$

(a) Thus, the hydrogen diffusion flux is

$$J_x = -D_H\frac{C_{H,P_1} - C_{H,P_2}}{x_{P_1} - x_{P_2}}$$

$$J_x = -(2.45\,\text{m}^2)\left[\frac{(17.10 - 2.70)\,\text{mol/m}^3}{0 - 1\times 10^{-3}\,\text{m}}\right]$$

$$J_x = 35{,}280\,\text{mol/}\left(\text{m}^2\,\text{s}\right)$$

which is relatively a small molar flux.

(b) The diffusional area is

$$J_x = \frac{m}{A_d t}$$

$$A_d = \frac{m}{J_x t} = \frac{1\,\text{mol}}{(35{,}280\,\text{mol/m}^2\,\text{s})\,(10\,\text{s})} = 2.83\times 10^{-6}\,\text{m}^2$$

$$A_d = 2.83\times 10^6\,\mu\text{m}^2 = 2.83\,\text{nm}$$

3.4.2   Second Law: Transient Diffusion

Rectangular Element Consider the central plane as the reference point in the volume element (Fig. 3.6). The diffusing plane at position 2 moves along the x-direction at a distance $x\text{-}dx$ from position 1 and $x + dx$ to position 3 [3].

Thus, the rate of diffusion that enters and leaves the volume element is

$$R_x = dydz\left[J_x - \frac{\partial J_x}{\partial x}dx\right] - dydz\left[J_x + \frac{\partial J_x}{\partial x}dx\right] \tag{3.19a}$$

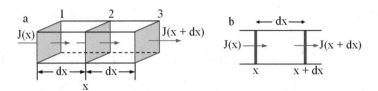

Fig. 3.6 Diffusion in an isotropic and homogeneous medium. (**a**) Rectangular element and (**b**) rectangular planes for one-dimensional diffusion flux J_x in the opposite direction of the concentration gradient dC/dx

$$R_x = -2dxdydz\frac{\partial J_x}{\partial x} = -2dV\frac{\partial J_x}{\partial x} \tag{3.19b}$$

Similarly,

$$R_y = -2dxdydz\frac{\partial J_y}{\partial y} = -2dV\frac{\partial J_y}{\partial y} \tag{3.20a}$$

$$R_z = -2dxdydz\frac{\partial J_z}{\partial z} = -2dV\frac{\partial J_z}{\partial z} \tag{3.20b}$$

For continuity of conservation of mass (transport phenomenon), the concentration rate is written as [3]

$$\frac{\partial C}{dt} = \frac{1}{2dV}\sum R_i = -\frac{\partial J_x}{\partial x} - \frac{\partial J_y}{\partial y} - \frac{\partial J_z}{\partial z} \tag{3.21}$$

The general three-dimensional (3D) steady-state diffusion can be described by Fick's first law. Thus, the diffusion flux equation is defined by

$$J = -D\left(\frac{\partial C}{\partial x} + \frac{\partial C}{\partial y} + \frac{\partial C}{\partial z}\right) \tag{3.22}$$

Combining Eqs. (3.21) and (3.22) yields the governing partial differential equation (PDF) for *transient diffusion* fundamentally described by Fick's second law of diffusion (Fick's second law equations) for a single component

$$\frac{\partial C}{dt} = D\left[\frac{\partial^2 C}{\partial x^2} + \frac{\partial^2 C}{\partial y^2} + \frac{\partial^2 C}{\partial z^2}\right] = D\nabla^2 C \tag{3.23a}$$

$$\frac{\partial C}{dt} = D\frac{\partial^2 C}{\partial x^2} \quad \text{(One-dimension)} \tag{3.23b}$$

where ∇^2 denotes the Laplace operator and Eq. (3.23b) is derived as Eq. (3A.9) in Appendix 3A.

Rectangular Plane According to the model in Fig. 3.6b, the concentration rate can be defined by

$$\frac{\partial C}{dt} = \frac{J\,(x) - J\,(x + dx)}{dx} \tag{3.24}$$

Expand $J\,(x + dx)$ using Taylor series

$$J\,(x + dx) = J\,(x) + \frac{\partial J\,(x)}{\partial x}dx + \frac{\partial^2 J\,(x)}{\partial x^2}\frac{dx^2}{2} + \dots . \tag{3.25}$$

Substituting the first two terms into Eq. (3.24) yields

$$\frac{\partial C}{dt} = \frac{J\,(x) - J\,(x) - [\partial J\,(x)/\partial x]\,dx}{dx} = -\frac{\partial J\,(x)}{\partial x} \tag{3.26}$$

Combining Eqs. (3.23b) and (3.26) gives

$$\frac{\partial J_x}{\partial x} = -D\frac{\partial^2 C}{\partial x^2} \tag{3.27}$$

For an isotropic medium of infinite length and unit cross-sectional area, the total amount particles N (atoms, molecules, or ions) is defined as

$$N = \int_{-\infty}^{\infty} C\,(x, t)\,dx \tag{3.28a}$$

$$C\,(x, t) = \frac{N}{V_m} = \frac{1}{V_m}\int_{-\infty}^{\infty} C\,(x, t)\,dx \tag{3.28b}$$

where V_m denotes molar volume. This integral, Eq. (3.28), can be solved once $C = C\,(x, t)$ is derived based on initial and boundary conditions. This is treated in the next section.

It is clear now that the theoretical background on the analytical derivation of Fick's laws of diffusion is of fundamental importance. Specifically, the governing partial differential equation (PDE) for transient diffusion, Eq. (3.23), requires a special treatment based on the type of diffusing species j in a suitable substrate.

Despite that the fundamental importance of transient diffusion in many engineering and science fields has been documented throughout decades, the solution to a transient diffusion problem is mathematically derived in accord with recognizable initial and boundary conditions related to the complexity of particular situations. Thus, one deals with linear and non-linear diffusion problems.

Numerical methods may be required for solving complicated diffusion problems, but the analytical approach will be implemented henceforth for solving simple diffusion problems, mostly related to carbon/hydrogen in steels and boron/phosphorus in silicon single crystals. Although steady-state diffusion and corresponding solutions are important, specially in characterizing quasi-equilibrium systems, transient diffusion prevails as the most realistic condition. Interestingly, an equilibrium state may exist momentarily, but it is altered by dynamic atomic forces to become a dynamic state. Hence, transient diffusion.

Actually, transient diffusion is also referred to the relative unsteady motion of species in a solid solution or simply in liquid mixture due to concentration gradients near a target surface. In special cases, both heat transfer and transient diffusion occur simultaneously.

3.5   Solution to Fick's Second Law Equation

3.5.1   Method of Separation of Variables

The solution to one-dimensional Fick's second law of diffusion, mathematically defined by Eq. (3.27) is based on the transformation of a partial differential equation (PDE) to ordinary differential equations (ODE). This can be achieved if the concentration $C(x, t)$ of a species j is initially defined as a $C(x, t) = f(x, t)$ function or as $C(x, t) = f(x) g(t)$ for the well-known separation of variables method [6].

The transformation of PDE to ODE is achieved by equating the function that depends on time t on one side and the one depending on x on the other side. Subsequently, the solution will contain some constants that must be evaluated according to the chosen IC and BC.

For non-linear diffusion problems, Eq. (3.27) can be approximated using unidimensional analysis and its solution can be derived by converting PDE to ODE, provided that $C = C(x, t) = f(n)$ with n being defined by Eq. (3.12).

From Eq. (3.12), dn/dx and dn/dt are

$$n = \frac{x}{2\sqrt{Dt}} \tag{3.29a}$$

$$\frac{dn}{dx} = \frac{1}{2\sqrt{Dt}} \tag{3.29b}$$

$$\frac{dn}{dt} = -\frac{x}{4\sqrt{Dt^3}} = -\frac{x}{4t\sqrt{Dt}} \tag{3.29c}$$

Using the chain rule of differentiation yields

$$\frac{\partial C}{\partial t} = \frac{dC}{dn}\frac{\partial n}{\partial t} = -\frac{x}{4t\sqrt{Dt}}\frac{dC}{dn} \tag{3.30a}$$

$$\frac{\partial C}{\partial x} = \frac{dC}{dn}\frac{\partial n}{\partial x} = \frac{1}{2\sqrt{Dt}}\frac{dC}{dn} \tag{3.30b}$$

$$\frac{\partial^2 C}{\partial x^2} = \frac{d^2 C}{dn^2} \left(\frac{\partial n}{\partial x}\right)^2 + \frac{dC}{dn} \frac{\partial^2 n}{\partial x^2} = \left(\frac{1}{2\sqrt{Dt}}\right)^2 \frac{d^2 C}{dn^2} \tag{3.30c}$$

$$\frac{\partial^2 C}{\partial x^2} = \frac{1}{4Dt} \frac{d^2 C}{dn^2} \tag{3.30d}$$

where $\partial^2 n/\partial x^2 = 0$. Substitute Eqs. (3.30a) and (3.30d) into (3.27) to get the sought ODE

$$-\frac{x}{4t\sqrt{Dt}} \frac{dC}{dn} = \frac{D}{4Dt} \frac{d^2 C}{dn^2} \tag{3.31a}$$

$$-2n\frac{dC}{dn} = \frac{d^2 C}{dn^2} \tag{3.31b}$$

Note that Eq. (3.31b) is an ordinary differential equation (ODE) containing one independent variable, namely n. Accordingly, the transformation from PDE to ODE has been accomplished. In actual fact, this is a partial analytical solution to the transient diffusion governing equation since the ODE has be solved for practical purposes.

Letting $f = dC/dn$ and $df = d^2C/dn^2$ in Eq. (3.31b) yields

$$-2nf = df \tag{3.32a}$$

$$\frac{df}{f} = -2n \tag{3.32b}$$

$$\int \frac{df}{f} = -2\int n\,dn \tag{3.32c}$$

$$\ln(f) = -n^2 + h \tag{3.32d}$$

where h is a constant of integration. Solving for f yields

$$f = \exp\left(-n^2 + h\right) = A\exp\left(-n^2\right) \tag{3.33a}$$

$$\frac{dC}{dn} = A\exp\left(-n^2\right) \tag{3.33b}$$

$$dC = A\exp\left(-n^2\right)dn \tag{3.33c}$$

In order to integrate Eq. (3.33c) and solve for the constant A, one needs to describe the diffusion problem and, set the initial condition (IC) and boundary conditions (BC) as the limits of integration.

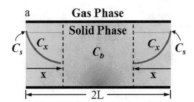
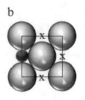

Fig. 3.7 Interstitial diffusion into a metallic plane. (**a**) Concentrations at different locations due to a diffusion process and (**b**) FCC plane showing positions of an impurity atom

3.5.1.1 Case 1: Bulk Concentration $C_b > 0$

Figure 3.7 schematically represents a specimen, such as steel, pure iron, or silicon plate in a semi-infinite region, and assume that it is exposed to a type of diffusing interstitial atom (carbon, nitrogen, hydrogen, or boron) at a relatively high temperature for a prolong time t. Thus, one-dimensional diffusion suffices the preceding analysis.

Carburizing This is a case-hardening process for hardening the surface of relatively low-carbon steels to a certain depth. Some examples are gears, firearm parts, engine camshafts, drilling screws, and the like. For the sake of clarity, assume that Fig. 3.7a represents a model for a low-carbon steel sample placed in a suitable furnace containing a carbon-rich gas (say, methane), which dissociates, providing a constant concentration of carbon atoms on the steel specimen surface at constant temperature T and pressure P. Thus, carbon atoms (Fig. 3.7b) diffuse into the FCC austenitic phase (γ-Fe) of the steel containing an initial carbon concentration termed bulk concentration or background concentration $C(x, t) = C_b$ and as a result, the carbon concentration is increased to a certain depth from the surface, and the steel hardens.

Dopant Diffusion In the field of metallurgy, the Czochralski process is used to produce silicon (Si) single crystals in the form of ingots, which in turn, are sliced in substrates called wafers in the order of $6 \times 6 \times 0.4\,\text{mm}$ [3]. These wafers are subsequently prepared to have smooth surfaces and subsequently are exposed to impurity-rich environments (AsH_3, PH_3, and B_2H_6) at relatively high temperatures for adjusting electrical properties by diffusion.

The wafers are subjected to a dopant diffusion process in two steps commonly known as predeposition or ion-implantation and drive-in diffusion. Diffusion of As, P, and Sb (electron donors) in silicon wafers produces n-type semiconductors, while B, In, and Al (hole donors) in Si produce p-type semiconductors.

Ion Implantation This is a technique used to introduce specific amounts of impurity ions (B^+, H^+, As^+, P^+) into a semiconductor substrate at an accelerated rate (high ion kinetic energy) at relatively low temperatures. Obviously, the ions penetrate the substrate (silicon wafer) by electronic collisions with electrons in the substrate lattice, creating dissipation of heat. This is a reason for having a controlled environment during ion implantation to produce either *n*-type or *p*-type semiconductors.

Nowadays, semiconductor processors are commonly known as integrated circuits (ICs) or microchips and can be exposed to an ion implantation process for doping impurity ions into silicon wafers. In fact, microchips are essential electronic devices in cell phones, cars, and computers and consequently, these devices are an important part of modern life.

Mathematical Analysis In order to analyze the diffusion process for hardening ferrous and for adjusting electrical properties in non-ferrous materials, Fig. 3.5a illustrates a general plane containing an initial concentration of an impurity species *j* (carbon, boron, phosphorous, and the like) that can be used to define meaningful initial (IC) and boundary (BC) conditions before and after transient atomic diffusion starts.

For an isotropic material as a semi-infinite $(0 < x < \infty)$ medium,

$$\text{IC} \rightarrow C = C_b \quad \text{for } n = \infty \text{ at } x > 0, \quad t = 0 \tag{3.34a}$$

$$\text{BC} \rightarrow C = C_s \quad \text{for } n = 0 \text{ at } x = 0, \quad t > 0 \tag{3.34b}$$

$$\text{BC} \rightarrow C = C_b \quad \text{for } n = \infty \text{ at } x = \infty, t > 0 \tag{3.34c}$$

Integrate Eq. (3.33c) using the BC given by Eq. (3.34c) so that

$$\int_{C_b}^{C_s} dC = A \int_0^\infty \exp\left(-n^2\right) dn = A\frac{\sqrt{\pi}}{2} \tag{3.35a}$$

$$A = \frac{2\,(C_s - C_b)}{\sqrt{\pi}} \tag{3.35b}$$

Integrate Eq. (3.33c) again with IC defined by (3.34a) to get

$$\int_{C_b}^{C} dC = A \int_0^n \exp\left(-n^2\right) dn \tag{3.36a}$$

$$C - C_b = (C_s - C_b)\frac{2}{\sqrt{\pi}} \int_n^\infty \exp\left(-n^2\right) dn \tag{3.36b}$$

from which

$$C - C_b = (C_s - C_b)\,\text{erfc}\,(n) \tag{3.37a}$$

$$C - C_b = (C_s - C_b)\,[1 - \text{erf}\,(n)] \tag{3.37b}$$

$$\text{erf}\,(n) = 1 - \text{erfc}\,(n) \tag{3.37c}$$

where erf(n) is the Gaussian error function of sigmoidal shape and erfc(n) is the complementary error function. See Appendix 3A for values of erf(n).

Mathematically,

$$\text{erf}\,(n) = \frac{2}{\sqrt{\pi}} \int_0^n \exp\left(-n^2\right) dn \tag{3.38a}$$

$$= \frac{2}{\sqrt{\pi}} \int_o^n \sum_{k=0}^{\infty} \frac{(-1)^k\, n^{2k}}{k!} dn \tag{3.38b}$$

$$\text{erf}\,(n) = \frac{2}{\sqrt{\pi}} \sum_{k=0}^{\infty} \frac{(-1)^k\, n^{2k+1}}{k!\,(2k+1)} \qquad \text{(Maclaurin's series)} \tag{3.38c}$$

$$\text{erf}\,(n) = \frac{2}{\sqrt{\pi}} \left(n - \frac{n^3}{3} + \frac{n^5}{10} - \frac{n^7}{42} + \ldots\right) \tag{3.38d}$$

Rearranging Eq. (3.37a) along with (3.29a) gives the common solution to Fick's second law of diffusion, Eq. (3.27), in terms of concentration ratio (see Appendix 3A for analytical details)

$$\frac{C - C_b}{C_s - C_b} = \text{erfc}\left(\frac{x}{2\sqrt{Dt}}\right) = 1 - \text{erf}\left(\frac{x}{2\sqrt{Dt}}\right) \quad \text{(inwards)} \tag{3.39a}$$

$$C = C_b + (C_s - C_b)\,\text{erfc}\left(\frac{x}{2\sqrt{Dt}}\right) \tag{3.39b}$$

From Eq. (3.39b), the concentration gradient is defined by

$$\frac{\partial C}{\partial x} = (C_s - C_b)\left(\frac{1}{\sqrt{\pi Dt}}\right) \exp\left(-\frac{x^2}{4Dt}\right) \tag{3.40}$$

It is important to note that if $\partial C / \partial x = 0$, then there is no diffusion flux in and out of the source.

Assuming that $\partial C / \partial x \neq 0$ and inserting Eq. (3.40) into (3.27) yields the diffusion flux as

$$J_x = -D\,(C_s - C_b)\left(\frac{1}{\sqrt{\pi Dt}}\right) \exp\left(-\frac{x^2}{4Dt}\right) \tag{3.41}$$

From Eq. (3.40),

$$\frac{\partial^2 C}{\partial x^2} = -(C_s - C_b)\left(\frac{x}{2Dt\sqrt{\pi Dt}}\right) \exp\left(-\frac{x^2}{4Dt}\right) \tag{3.42}$$

Thus, Eq. (3.27) along with (3.42) becomes

$$\frac{\partial C}{\partial t} = -\frac{(C_s - C_b)\,x}{2t\sqrt{\pi\,Dt}}\exp\left(-\frac{x^2}{4Dt}\right) \tag{3.43}$$

3.5.1.2 Case 2: Bulk Concentration $C_b = 0$

For pure iron (Fe) specimen under the same carburizing process shown in Fig. 3.5 with $C_b = 0$ and Eq. (3.39a) or (3.39b) simplifies to (see Appendices A and B for details)

$$\frac{C}{C_s} = 1 - \mathrm{erf}\left(\frac{x}{2\sqrt{Dt}}\right) = \mathrm{erfc}\left(\frac{x}{2\sqrt{Dt}}\right) \quad \text{(inwards)} \tag{3.44a}$$

$$C = C(x,t) = C_s\left[1 - \mathrm{erf}\left(\frac{x}{2\sqrt{Dt}}\right)\right] = C_s\,\mathrm{erfc}\left(\frac{x}{2\sqrt{Dt}}\right) \tag{3.44b}$$

The first and second derivatives of Eq. (3.44b) are

$$\frac{\partial C}{\partial x} = -\frac{C_s}{\sqrt{4\pi\,Dt}}\exp\left(-\frac{x^2}{4Dt}\right) \tag{3.45a}$$

$$\frac{\partial^2 C}{\partial x^2} = \frac{xC_s}{4Dt\sqrt{\pi\,Dt}}\exp\left(-\frac{x^2}{4Dt}\right) \tag{3.45b}$$

Hence, Eq. (3.27) becomes

$$\frac{\partial C}{\partial t} = \frac{xC_s}{4t\sqrt{\pi\,Dt}}\exp\left(-\frac{x^2}{4Dt}\right) \tag{3.46}$$

For evaporation of the interstitial atoms, the surface concentration is $C_s = 0$ in Eq. (3.39a) or (3.39b) and

$$\frac{C}{C_b} = \mathrm{erf}\,(n) \quad \text{(outwards)} \tag{3.47}$$

from which

$$C = C_b\,\mathrm{erf}\left(\frac{x}{2\sqrt{Dt}}\right) \tag{3.48}$$

where C_b is the preexisting bulk or background concentration of an impurity X atom due to a previous heat treatment at a relatively high temperature or simply due to the nominal chemical composition of a substrate.

In fact, the source of C_b is not a relevant issue in this particular theoretical treatment of solid-state atomic diffusion. It is initially a constant in Eq. (3.48) and treated as such as diffusion progresses.

3.5.2   Amount of Diffusing Solute

Substituting Eq. (3.44b) into (3.28) yields the equation for the total amount of the diffusing species j (dose) per unit area, normally used in dopant diffusion and ion implantation followed by diffusion in silicon wafers. Thus,

$$N = \int_0^\infty C dx = C_s \int_0^\infty \text{erfc}\left(\frac{x}{2\sqrt{Dt}}\right) dx \tag{3.49a}$$

$$N = \frac{C_s}{\sqrt{\pi}}\left(2\sqrt{Dt}\right) = 2C_s\sqrt{\frac{Dt}{\pi}} \tag{3.49b}$$

According to Crank [3], the following equation is a solution to the transient diffusion problem, mathematically defined by Eq. (3.27),

$$C = \frac{B}{\sqrt{t}}\exp\left(-\frac{x^2}{4Dt}\right) = \frac{B}{\sqrt{t}}\exp\left(-n^2\right) \tag{3.50}$$

where B is an unknown constant. From Eq. (3.29a),

$$x = (2\sqrt{Dt})n \tag{3.51a}$$

$$dx = (2\sqrt{Dt})dn \tag{3.51b}$$

Combining Eqs. (3.49) and (3.51b), and integrating yields equations needed for further analysis

$$N = \int_0^\infty C dx = \int_0^\infty \frac{2B\sqrt{Dt}}{\sqrt{t}}\exp\left(-n^2\right) dn \tag{3.52a}$$

$$N = \left(2A\sqrt{D}\right)\int_0^\infty \exp\left(-n^2\right) dn = \left(2B\sqrt{D}\right)\frac{\sqrt{\pi}}{2} \tag{3.52b}$$

from which

$$B = \frac{N}{\sqrt{\pi D}} \tag{3.53}$$

Inserting Eq. (3.53) and (3.29a) into (3.50) yields

$$C = C(x, t) = \frac{N}{\sqrt{\pi Dt}}\exp\left(-\frac{x^2}{4Dt}\right) \tag{3.54}$$

With regard to known values of N and D, Eq. (3.54) can be solved for the diffusion depth or penetration x from the surface of a specimen at a predetermined concentration $C = C(x, t) > 0$. That is,

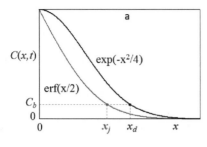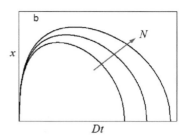

Fig. 3.8 Schematic plots for (**a**) normalized concentration distribution using $\mathrm{erf}\, c(x/2)$ and $exp(-x^2/4)$ for interstitial diffusion in a semi-infinite medium, and (**b**) interstitial diffusion and junction depth as function of $\sqrt{Dt}$ for increasing N

$$x = \left[(4Dt) \ln \left(\frac{N}{C\sqrt{\pi Dt}} \right) \right]^{1/2} \tag{3.55}$$

which is a convenient non-linear equation for predicting the junction depth x in a transient diffusion problem.

Plotting Eqs. (3.48) and (3.54) yields comparable $\mathrm{erf}\,(x/2)$ and $\exp\left(-x^2/4\right)$ non-linear profiles. This is schematically depicted in Fig. 3.8a with $Dt = 1$.

Figure 3.8b illustrates the diffusion depth profile defined by $x = f(Dt)$, Eq. (3.54), with increasing N. The interpretation of these profiles suggests that diffusion ceases when x reaches a maximum value corresponding to the diffusion depth. This can be a particular design condition in dopant diffusion technology.

Clearly, the curves in Fig. 3.8a exhibit rapid and similar decay as x increases, indicating that diffusion in semi-infinite substrates is limited to a certain diffusion length x.

According to Fig. 3.8a, it is possible to have $C(x, t) < C_b$ as diffusion progresses for a prolong time ($t \to \infty$). For a substrate containing $C_b > 0$, $\mathrm{erf}\,(-x/2)$ in Eq. (3.48) suggests that $C(x, t) = C_b$ at $x = x_j$; the junction depth used silicon semiconductors.

On the other hand, evaluating $\exp\left(-x^2/4\right)$ as per Eq. (3.54) at $C(x, t) = C_b$ gives the diffusion depth $x_d > x_j$ (Fig. 3.8a). This deduction is, at least, probable from a theoretical point of view and one can speak of error function diffusion and exponential diffusion due to $\mathrm{erf}\,(-x/2)$ and $\exp\left(-x^2/4\right)$ functions, respectively. Both functions give a similar trend with respect to the depletion of the concentration of impurity species j from the surface as the diffusion continues. This is attributed to atomic collisions within the diffusion domain. This is an expected result since diffusion is limited when the substrate is sufficiently thick.

Furthermore, the diffusion impurity penetrates the surface and subsequently stops at a diffusion length or depth x as schematically depicted in Fig. 3.8a due to collisions. It is also expected that the diffusion depth is in the order of a micrometer (μm).

Theoretically, this diffusion process can occur at low temperatures, but it may take an infinite time and therefore, it would be an impractical approach. Instead, this process is readily accomplished at relatively high temperatures in order to thermally activate the substrate lattice and the diffusing impurity for an effective atomic mechanism.

The above mathematical framework for characterizing atomic diffusion can be defined as a macroscopic diffusion approach related to relevant boundary conditions. Unfortunately, all atomic mechanism and related atomic interactions are not fully known. This, then, relates the transient diffusion problem, to an extent, to uncertainties in a dynamic state. However, the preceding analytical approach is of fundamental importance.

Example 3.4 Consider carburizing low-carbon steel (AISI 1025) and pure iron in a furnace containing a carbon-rich gas environment. The latter at $900\,^{\circ}C$ and the former at $950\,^{\circ}C$ for 10 hours each. Assume that the semi-infinite ($0 < x < \infty$) specimens can be modeled as shown in Fig. 3.4, the carbon concentration, in weight percent, on the specimen surfaces is kept constant at $C_s = 1\%$, and that the diffusion coefficient is $D = 6 \times 10^{-6}\,mm^2/s$. Let $C_b = 0.25\%$ for steel and $C_b = 0\%$ for pure iron. Assume that carburizing of the AISI 1025 is carried out in a gas-controlled furnace in order to avoid decarburization. For your information, the controlled reactions for mass transport by carburizing are written as

$$2CO \rightarrow C_{\gamma\text{-}Fe} + CO_2$$

$$CH_4 \rightarrow C_{\gamma\text{-}Fe} + 2H_2$$

$$CO + H_2 \rightarrow C_{\gamma\text{-}Fe} + H_2O$$

where $C_{\gamma\text{-}Fe}$ is carbon concentration in the FCC austenite phase ($\gamma\text{-}Fe$). Once carbon has diffused into the ferrous lattice, the diffusion surface layer is hardened preferably by quenching in water or oil from the carburizing temperature. It can also be slowly cooled to room temperature, reaustenitize and quench. (a) Plot $C = f(x)$ at $0 \leq x \leq 2\,mm$ for low-carbon steel (AISI 1025) and pure iron. (b) Calculate the concentration C at $x = 2\,mm$ for steel and for iron. Assume that both steel and pure iron are quenched after diffusion is completed. (c) If the mass density of iron or steel and the atomic weight of carbon are $7.87\,g/cm^3$ and $12\,g/mol$, respectively, convert weight percent to concentration of carbon in units of mol/cm^3. (d) Calculate the amount of carbon on the surface of the specimens.

Solution

(a) For a steel specimen containing an initial bulk concentration C_b, Eq. (3.39a) along with the given data provides the concentration equation in weight %

$$\frac{C - C_b}{C_s - C_b} = 1 - erf\left(\frac{x}{2\sqrt{Dt}}\right) = erfc\left(\frac{x}{2\sqrt{Dt}}\right)$$

$$C = C_b + (C_s - C_b)\, \mathrm{erfc}\left(\frac{x}{2\sqrt{Dt}}\right)$$

$$C = 0.25 + (1 - 0.25)$$

$$\times\, \mathrm{erfc}\left(\frac{x}{2\sqrt{(6 \times 10^{-6}\,\mathrm{mm^2/s})\,(10 \times 60 \times 60\,\mathrm{s})}}\right)$$

$$C = 0.25 + 0.75\,\mathrm{erfc}\,(1.0758x) \quad \text{(in percentage \%)}$$

which is a useful and simple non-linear equation for determining the concentration of carbon atoms at a depth x. Similarly, for pure iron, Eq. (3.44a) yields the concentration equation in weight percentage (%) defined as

$$\frac{C}{C_s} = 1 - \mathrm{erf}\left(\frac{x}{2\sqrt{Dt}}\right) = \mathrm{erfc}\left(\frac{x}{2\sqrt{Dt}}\right)$$

$$C = C_s\left[1 - \mathrm{erf}\left(\frac{x}{2\sqrt{Dt}}\right)\right] = C_s\,\mathrm{erfc}\left(\frac{x}{2\sqrt{Dt}}\right)$$

$$C = (1)\,[1 - \mathrm{erf}\,(1.0758x)] = \mathrm{erfc}\,(1.0758x) \quad \text{(in \%)}$$

The plot is given below.

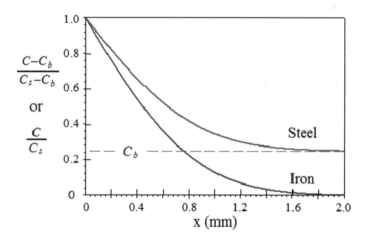

(b) The concentrations as per Eqs. (3.39a) and (3.44a) or from part (a) are

$$C_{steel} = 0.25 + 0.75\,\mathrm{erfc}\,[(1.0758)\,(2)] \simeq 0.25$$

$$C_{iron} = (1)\,\mathrm{erfc}\,[(1.0758)\,(2)] = 2.34 \times 10^{-3} \simeq 0$$

(c) The concentration of carbon on the surface of the specimens is

$$C_s = \left(\frac{7.87\,\text{g/cm}^3}{12\,\text{g/mol}}\right)\left(\frac{1\%}{100\%}\right) = 6.5583 \times 10^{-3}\,\text{mol/cm}^3$$

$$C_s = \left(6.5583 \times 10^{-3}\,\text{mol/cm}^3\right)\left(6.022 \times 10^{23}\,\text{atoms/mol}\right)$$

$$C_s \simeq 4 \times 10^{21}\,\text{atoms/cm}^3$$

where the constant 6.022×10^{23} atoms/mol is the Avogadro's number. The background or the nominal composition of carbon (concentration) is

$$C_b = \left(\frac{7.87\,\text{g/cm}^3}{12\,\text{g/mol}}\right)\left(\frac{0.25\%}{100\%}\right) = 1.6396 \times 10^{-3}\,\text{mol/cm}^3$$

$$C_b = 1.6396 \times 10^3\,\text{mol/m}^3 = 1639.60\,\text{mol/m}^3$$

$$C_b = \left(1.6396 \times 10^{-3}\,\text{mol/cm}^3\right)\left(6.022 \times 10^{23}\,\text{atoms/mol}\right)$$

$$C_b \simeq 10^{21}\,\text{atoms/cm}^3$$

In effect, the concentration in mol/cm^3 or mol/m^3 are the most common units in the academic and scientific communities.

(d) From Eq. (3.49b),

$$N = 2C_s\sqrt{\frac{Dt}{\pi}}$$

$$N = (2)\left(6.5583 \times 10^{-3}\,\text{mol/cm}^3\right)\sqrt{\left(6 \times 10^{-8}\,\text{cm}^2/\text{s}\right)(10 \times 60 \times 60\,\text{s})/\pi}$$

$$N = 3.4393 \times 10^{-4}\,\text{mol/cm}^2$$

$$N = \left(3.4393 \times 10^{-4}\,\text{mol/cm}^2\right)\left(6.022 \times 10^{23}\,\text{atoms/mol}\right)$$

$$N \simeq 2 \times 10^{20}\,\text{atoms/cm}^2$$

Convert moles to grams so that

$$N = \left(3.4393 \times 10^{-4}\,\text{mol/cm}^2\right)(12\,\text{g/mol})$$

$$N = 4.1272 \times 10^{-3}\,\text{g/cm}^2 = 0.65\,\text{g/mm}^2$$

Additionally,

$$\frac{dN}{dt} = C_s \sqrt{\frac{D}{\pi t}} = \left(6.5583 \times 10^{-3} \, \text{mol/cm}^3\right) \sqrt{\frac{\left(6 \times 10^{-8} \, \text{cm}^2/\text{s}\right)}{\pi \left(10 \times 60 \times 60 \, \text{s}\right)}}$$

$$\frac{dN}{dt} = 4.78 \times 10^{-9} \, \frac{\text{mol}}{\text{cm}^2 \, \text{s}} = 5.74 \times 10^{-8} \, \frac{\text{g}}{\text{cm}^2 \, \text{s}}$$

$$\frac{dN}{dt} \simeq 3 \times 10^{15} \, \frac{\text{atoms}}{\text{cm}^2 \, \text{s}}$$

This is the rate of carbon atoms supplied on the specimen surfaces.

Example 3.5 Consider a silicon (Si) n-type semiconductor wafer being doped with constant boron (B) at 1050 °C for 1 h (3600 s). If the amount of boron is 2.4×10^{20} atoms/cm^2, then calculate the diffusion depth x for a concentration of 6×10^{20} atoms/cm^3. Data: the activation energy and the proportionality constant for the Arrhenius equation are 3.45 eV and 1.20 cm,^{2}/s respectively.

Solution The boron diffusion coefficient is

$$D = D_o \exp\left(-\frac{Q_d}{k_B T}\right)$$

$$D = \left(1.20 \text{cm}^2/\text{s}\right) \exp\left[-\frac{3.45 \, \text{eV}}{\left(8.6173 \times 10^{-5} \, \text{eV/K}\right)\left(1050 + 273\right) \, \text{K}}\right]$$

$$D = 8.65 \times 10^{-14} \, \text{cm}^2/\text{s}$$

$$Dt = \left(8.65 \times 10^{-14} \, \text{cm}^2/\text{s}\right)\left(3600 \, \text{s}\right) = 3.114 \times 10^{-10} \, \text{cm}^2$$

From Eq. (3.49b), the surface concentration of boron is

$$C_s = \frac{N}{2\sqrt{Dt/\pi}} = \frac{2.4 \times 10^{20} \, \text{atoms/cm}^2}{2\sqrt{\left(3.114 \times 10^{-10} \, \text{cm}^2\right)/\pi}}$$

$$C_s = 1.2053 \times 10^{25} \, \text{atoms/cm}^3$$

Using Eq. (3.44a) yields the diffusion depth as

$$\frac{C}{C_s} = 1 - \text{erf}\left(\frac{x}{2\sqrt{Dt}}\right)$$

$$\text{erf}\left(\frac{x}{2\sqrt{Dt}}\right) = \left(1 - \frac{C}{C_s}\right) = 1 - \frac{6 \times 10^{20}}{1.2053 \times 10^{25}} = 0.99995$$

From an error function table, $x = (2.86)\left(2\sqrt{Dt}\right)$ so that

$$x = 2\sqrt{3.114 \times 10^{-10}\,\text{cm}^2}\,(2.86) = 1.0094 \times 10^{-4}\,\text{cm}$$

$$x \simeq 1.01\,\mu\text{m}$$

Using Eq. (3.55) and $Dt = 3.114 \times 10^{-10}\,\text{cm}^2$ yields

$$x = \left[(4Dt)\ln\left(\frac{N}{C\sqrt{\pi Dt}}\right)\right]^{1/2}$$

$$A = \ln\left[\frac{2.4 \times 10^{20}\,\text{atoms/cm}^2}{\left(6 \times 10^{20}\,\text{atoms/cm}^3\right)\sqrt{\pi\left(3.114 \times 10^{-10}\,\text{cm}^2\right)}}\right]$$

$$x = \sqrt{\left(4 \times 3.114 \times 10^{-10}\,\text{cm}^2\right)(A)} = 1.0853 \times 10^{-4}\,\text{cm} \simeq 1.09\,\mu\text{m}$$

Therefore, there is approximately $(1.09 - 1.01)/((1.09 + 1.01)/2) = 0.0762$ or 7.62% error in calculating the diffusion depth x from Eqs. (3.44a) and (3.55). This percentage error may be considered adequate.

3.6   Kirkendall Effect

The Kirkendall effect is fundamentally based on substitutional diffusion by vacancy mechanism in binary alloy systems, where an A-B diffusion couple is in physical contact at a common interface known as the Kirkendall reference plane. In this case, the diffusion fluxes are $J_A = -J_B$, $J_A > J_B$, or $J_B > J_A$. It all depends on the diffusion coefficient D of the element with the lowest melting temperature (T_m) so that $D_A > D_B$ or $D_B > C_A$ and the corresponding concentration gradient. Thus, the substitutional diffusion between A and B is referred to as interdiffusion since it is an exchange process where Fick's first law is assumed to obey the vacancy mechanisms

$$J_A = -D_A\frac{\partial C_A}{\partial x} \tag{3.57a}$$

$$J_B = -D_B\frac{\partial C_B}{\partial x} \tag{3.57b}$$

In fact, this is an interdiffusion process where the resultant metallic product during a heat treatment called annealing at $T_A < T_{m,A}$ or $T_B < T_{m,B}$ is a binary alloy, such as Ni-Ti, Cu-Ni, and Al-Si. . Details on this type of diffusion can be found in the literature under Darken's analysis of binary systems [7] and Kirkendall effect [6, 8].

Furthermore, these transient diffusion equations indicate that each species (A and B) has its characteristics and it is more likely that the diffusion coefficients be different in value. Obviously, in a two-species system diffusing into a single substrate or matrix, one species diffuses faster than the other, provided that $D_A > D_B$ or $D_B > D_A$.

3.7 Effects of Defects on Diffusion

Diffusion is a structure-sensitive property and lattice irregularity or lattice features, such as vacancies, grain and interface boundaries and dislocations, are attributed to enhance diffusion by random walk, which in turn is enhanced by a more open or less packed atomic structure (lower atomic packing density). Therefore, diffusion in plastically deformed (cold-worked) metals is faster than in the annealed state due to these lattice defects and distorted crystallographic planes. If the diffusion coefficient D obeys the Arrhenius equations, Eq. (3.16), then the temperature-induced diffusivity increases with increasing number of lattice defects.

For self-diffusion and interdiffusion to occur, there must exist vacancies so that the direction of diffusing atoms within a substrate lattice is towards the adjacent vacancy sites. Conversely, interstitial diffusion needs the empty interstitial sites for this process to occur at a relative high temperature $T < T_m$. Interstitial diffusion is not always a beneficial process as in carburizing. For instance, hydrogen diffusion in steel pipes is a recognized phenomenon that causes embrittlement. Most metallic substrates contain dislocations, grain boundaries, and lattice-inclusion interfaces. These defects are considered high-energy sites for hydrogen diffusion. From an engineering point of view, hydrogen diffusion can be the main cause for low steel toughness and yield strength. These properties are essential in designing codes for assuring structural integrity.

3.8 Measurements of Diffusion Coefficient

In general, the measured diffusion coefficient D depends on the structure of the sample and the environment, and the technique being used. For instance, radioactive tracers are the most common techniques for such a purpose. Chronologically, the work of Wasik and McCulloh [9], Crank [3], Murch and Nowick [10], and Thrippleton et al. [11] for experimental and theoretical details on how to measure the diffusion coefficient D.

It should be mentioned that it is common to determine D at $T < T_m$. However, D measurements at relatively low temperatures, $T << T_m$, are important in phase precipitation during "aging" of hardenable non-ferrous alloys (Al-alloys), metal oxidation of metallic alloys (steels, brass, etc.), radiation damage, and the like.

3.9 Summary

The theory of diffusion is still an ongoing research topic due to either new discoveries in the vast field of materials science or solid-state physics and associated mathematical approaches to describe a particular atomic process. Hence, mass transport by diffusion.

Atomic diffusion is a mass transport mechanism due to mainly high temperature-induced particle motion within a particular medium. During this process, particle motion is related to the concept of random walk, which gives insights of the random collisions of species. Particularly, diffusion of a species j into a substrate can be treated as a solid-state interdiffusion process due to interstitial or substitutional atoms undergoing motion, preferably at relatively high temperatures. Moreover, mass transport is normally characterized by the Fick's laws of diffusion under steady-state or transient conditions. Nonetheless, the diffusion coefficient of a species j is empirically defined as an Arrhenius type equation.

In addition, the phrases "dislocation diffusion" and "grain boundary diffusion" imply diffusion of impurity atoms towards these defects. The motion of dislocations is attributed to a mechanical action during deformation and the motion of grain boundaries is due to grain growth during heat treatment at a temperature $T < T_m$. With respect to heat treatment, the carburization process is commonly used to harden steel surfaces by carbon diffusion and the doping process is typically used to induce boron or phosphorus into single-crystal silicon. This brief summary implies that atomic diffusion has a significant place in industrial applications.

3.10 Appendix 3A Second Law of Diffusion

3.10.1 Diffusion in a Rectangular Element

Fick's second law of diffusion for a non-steady-state or transient conditions defines the concentration rate of a species j as $dC_j/dt \neq 0$. Using Crank's model ([3, p. 3]) for the rectangular element shown in Fig. 3A.1.

Consider the central plane as the reference point in the rectangular volume element and assume that the diffusing plane at position 2 moves along the x-direction at a distance $x\text{-}dx$ from position 1 and $x + dx$ to position 3. Thus, the

Fig. 3A.1 Rectangular element for one-dimensional diffusion model

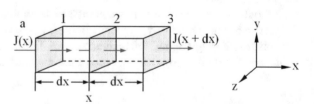

rate of diffusion in the x,y,z directions that enters the volume element at position 1 and leaves at position 3 can be deduced as written

$$R_x = dydz \left[J_x - \frac{\partial J_x}{\partial x} dx \right] - dydz \left[J_x + \frac{\partial J_x}{\partial x} dx \right] \tag{3A.1}$$

$$R_x = -2dxdydz \frac{\partial J_x}{\partial x} = -2dV \frac{\partial J_x}{\partial x} \tag{3A.2}$$

$$R_y = -2dxdydz \frac{\partial J_y}{\partial y} = -2dV \frac{\partial J_y}{\partial y} \tag{3A.3}$$

$$R_z = -2dxdydz \frac{\partial J_z}{\partial z} = -2dV \frac{\partial J_z}{\partial z} \tag{3A.4}$$

Using the principle of continuity for conservation of mass yields the concentration rate written as

$$\frac{\partial C}{dt} = \frac{1}{2dV} \sum R_i \tag{3A.5}$$

$$\frac{\partial C}{dt} = -\frac{\partial J_x}{\partial x} - \frac{\partial J_y}{\partial y} - \frac{\partial J_z}{\partial z} \tag{3A.6}$$

For steady-state diffusion $\partial C / \partial t = 0$ and $\partial J_k / \partial k > 0$, Fick's first law of diffusion describes the molar flux in Cartesian coordinates as

$$\frac{\partial J_k}{\partial k} = -\sum D_k \frac{\partial C_k}{\partial k} = -D_x \frac{\partial C_x}{\partial x} - D_y \frac{\partial C_y}{\partial y} - D_z \frac{\partial C_z}{\partial z} \tag{3A.7}$$

Combining Eqs. (3A.6) and (3A.7), with constant diffusion coefficient D under transient diffusion conditions $\partial C / \partial t > 0$, yields Fick's second law equation

$$\frac{\partial C}{dt} = D \left[\frac{\partial^2 C}{\partial x^2} + \frac{\partial^2 C}{\partial y^2} + \frac{\partial^2 C}{\partial z^2} \right] \tag{3A.8}$$

For unidirectional (one-dimensional) mass transport by diffusion, the concentration rate in Eq. (3A.8) becomes the most familiar Fick's second law equation of diffusion [see Eq. (3.23b)]

$$\frac{\partial C}{dt} = D \frac{\partial^2 C}{\partial x^2} \tag{3A.9}$$

If D is not constant, then Eq. (3A.9) yields the concentration rate in the x-direction as

$$\frac{\partial C}{dt} = \frac{\partial}{dx}\left[D\frac{\partial C}{\partial x}\right] \tag{3A.10}$$

$$\frac{\partial C}{dt} = \frac{\partial D}{\partial x}\frac{\partial C}{\partial x} + D\frac{\partial^2 C}{\partial x^2} \tag{3A.11}$$

In fact, Eq. (3A.9) is the most used form of Fick's second law and it is the general diffusion expression for one-dimensional analysis under non-steady-state or transient condition.

3.10.2 Diffusion Through a Rectangular Plane

Consider a simplified version of the one-dimensional diffusion model in Fig. 3A.1. In this case, Fick's second law equation can also be derived using Fig. 3A.2 and Taylor series on $J(x + dx)$ ([12, p. 84]).

Define the rate concentration as [2]

$$\frac{\partial C}{dt} = \frac{J(x) - J(x + dx)}{dx} \tag{3A.12}$$

Expanding $J(x + dx)$ so that

$$J(x + dx) = J(x) + \frac{\partial J(x)}{\partial x}dx + \frac{\partial^2 J(x)}{\partial x^2}\frac{dx^2}{2} + \dots. \tag{3A.13}$$

Substituting the first two terms of Eq. (3A.13) into (3A.12) yields

$$\frac{\partial C}{dt} = \frac{J(x) - J(x) - \frac{\partial J(x)}{\partial x}dx}{dx} \tag{3A.14}$$

$$\frac{\partial C}{dt} = -\frac{(\partial J(x)/\partial x)dx}{dx} = -\frac{\partial J(x)}{\partial x} = -\frac{\partial J_x}{\partial x} \tag{3A.15}$$

which is exactly the same as Eq. (3.19) for transient diffusion in the x-direction.

Fig. 3A.2 Rectangular planes for one-dimensional diffusion in the opposite direction of the concentration gradient (dC/dx)

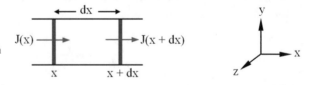

Table 3.1 Values of the error functionerf (n)

n	erf (n)	n	erf (n)	n	erf (n)
0	0	0.55	0.5633	1.30	0.9340
0.025	0.0282	0.60	0.6039	1.40	0.9523
0.05	0.0564	0.65	0.6420	1.50	0.9661
0.10	0.1125	0.70	0.6778	1.60	0.9763
0.15	0.1680	0.75	0.7112	1.70	0.9838
0.20	0.2227	0.80	0.7421	1.80	0.9891
0.25	0.2763	0.85	0.7707	1.90	0.9928
0.30	0.3286	0.90	0.7970	2.00	0.9953
0.35	0.3794	0.95	0.8209	2.10	0.9970
0.40	0.4284	1.00	0.8427	2.20	0.9981
0.45	0.4755	1.10	0.8802	2.30	0.9989
0.50	0.5205	1.20	0.9103	2.40	0.9993

3.10.2.1   Error Function Table

This appendix includes a convenient table for selected values of the error function erf (n) and some mathematical details (Table 3.1).

$$\text{erf}\,(n) = \frac{2}{\sqrt{\pi}} \int_0^n \exp\left(-n^2\right) dn \tag{3A.1a}$$

$$\text{erfc}\,(n) = \frac{2}{\sqrt{\pi}} \int_n^\infty \exp\left(-n^2\right) dn \tag{3A.1b}$$

$$n = \frac{x}{2\sqrt{Dt}} \tag{3A.1c}$$

$$\frac{\partial\,\text{erf}\,(n)}{\partial n} = \frac{2}{\sqrt{\pi}} \exp\left(-n^2\right) \tag{3A.1d}$$

$$\frac{\partial^2\,\text{erf}\,(n)}{\partial n^2} = \frac{4n}{\sqrt{\pi}} \exp\left(-n^2\right) \tag{3A.1e}$$

Properties and selected values of error functions:

$$\text{erf}\,(n) + \text{erfc}\,(n) = 1 \tag{3A.2a}$$

$$\text{erf}\,(-n) = -\,\text{erf}\,(n) \tag{3A.2b}$$

$$\text{erfc}\,(-n) = 2 - \text{erf}\,(n) \tag{3A.2c}$$

$$\text{erf}\,(0) = 0 \quad \& \quad \text{erfc}\,(0) = 1 \tag{3A.2d}$$

$$\text{erf}\,(\infty) = 1 \quad \& \quad \text{erfc}\,(\infty) = 0 \tag{3A.2e}$$

In fact, the error function erf(n) and the complementary error function erfc(n) appear in the solutions of atomic diffusion, heat transfer by conduction, physics, and other fields of study. Nonetheless, the error function is mathematically known as the Gauss error function.

In addition, diffusion is a mass transport phenomenon associated with the random thermal motion of atoms caused by the presence of a concentration gradient of a species j. Therefore, the error function, as described by Eq. (3A.1a), is essentially the representation of the significance of the error function as the integral of the standard normal distribution.

3.11 Appendix 3B Laplace Transformation

3.11.1 Introduction

Laplace transform (LT) is an integral transform method to solve partial differential equations (PDE) by changing a function of time t into ordinary differential equations (ODE) of a function of frequency s. This method is useful in solving engineering and mathematical problems by changing a time domain into a frequency domain function. This can be represented by $f(t)$ into $F(s)$ or $f(x, t)$ into $F(x, s)$, where s is a constant.

In general, multiply $f(x, t)$ by the function e^{-st} and integrate to get the Laplace notation $L\{f(x, t)\} = F(x, s)$ as

$$L\{f(x, t)\} = \int_0^\infty f(x, t) e^{-st} dt = F(x, s) \tag{3B.1}$$

Once $F(x, s)$ is derived, the function $f(x, t)$ is recovered by taking the inverse transformation of $F(x, s)$. Hence,

$$f(x, t) = L^{-1}\{F(x, s)\} = \frac{1}{2\pi i} \int_{-\infty}^{+\infty} F(x, s) e^{-st} ds \tag{3B.2}$$

where $i = \sqrt{-1}$, and $f(x, t)$ and $f(x, s)$ represent the same physical problem of diffusion by replacing time t with a constant s.

Laplace transform for a derivative of order n is defined by

$$L\left\{f^{(n)}(x, s)\right\} = s^n L\{f(x, t)\} - s^{n-1} f(x, 0) - s^{n-2} f'(x, 0) \tag{3B.3}$$

$$- s^{n-3} f'(x, 0) - \dots f^{(n-1)}(x, 0)$$

For $n = 0$, no derivative

$$L\left\{f^{(0)}(x,s)\right\} = L\{1\} = \frac{1}{s} \qquad (3B.4)$$

For $n = 1$,

$$L\left\{f^{(1)}(x,s)\right\} = L\left\{f'(x,s)\right\} \qquad (3B.5)$$

$$= sL\{f(x,t)\} - f(x,0)$$

$$= sF(x,s) - f(x,0)$$

Furthermore, if $f_t(x,t) = \partial f(x,t)/\partial t$, then LT yields

$$L\left\{f_t^{(1)}(x,s)\right\} = L\{f_t(x,t)\} = \int_0^\infty f_t(x,t)e^{-st}dt \qquad (3B.6)$$

$$= f_t(x,t)e^{-st}\,\big|_0^\infty + s\int_0^\infty f_t(x,t)e^{-st}dt$$

$$= f_t(x,t)e^{-\infty} - f_t(x,t)e^\circ + sF(x,s)$$

$$= f_t(x,t)(0) - f_t(x,t)(1) + sF(x,s)$$

$$= sF(x,s) - f(x,0)$$

For $n = 2$,

$$L\left\{f^{(2)}(x,s)\right\} = L\left\{f''(x,s)\right\} \qquad (3B.7)$$

$$= s^2 L\{f(x,t)\} - sf(x,0) - f'(x,0)$$

$$= s^2 F(x,s) - sf(x,0) - f'(x,0$$

Similarly, if $f_x(x,t) = \partial f(x,t)/\partial x$ and $f_{xx}(x,t) = \partial^2 f(x,t)/\partial x^2$, then

$$L\{f_x(x,t)\} = \int_0^\infty f_x(x,t)e^{-st}dt = F_x(x,s) \qquad (3B.8)$$

$$L\{f_{xx}(x,t)\} = \int_0^\infty f_{xx}(x,t)e^{-st}dt = F_{xx}(x,s) \qquad (3B.9)$$

where $F_x(x,s) = dF(x,s)/dx$ and $F_{xx}(x,s) = d^2F(x,s)/dx^2$.

3.11.2   Fick's Second Law of Diffusion and Solution

The solutions to Fick's second law of diffusion for a homogeneous linear sample can be derived by using appropriate initial (IC) and boundary (BC).

For diffusion of a species j into a host metal M, the concentration rate $\partial C(x,t)/\partial t = f_t(x,t) \rightarrow f(x,s)$, the Laplace transform (LT) is a tool used to convert PDE to ODE. Hence,

$$\frac{\partial C(x,t)}{\partial t} = D\frac{\partial^2 C(x,t)}{\partial x^2} \tag{3B.10}$$

$$L\left\{\frac{\partial C(x,t)}{\partial t}\right\} = DL\left\{\frac{\partial^2 C(x,t)}{\partial x^2}\right\} \tag{3B.11}$$

In order to transform the time t domain into a frequency s domain, it is necessary to use the corresponding LT series for $\partial C(x,t)/\partial t$ as per Eqs. (3B.6) and (3B.9). Thus,

$$L\{C_t(x,t)\} = L\left\{\frac{\partial C(x,t)}{\partial t}\right\} = sC(x,s) - C(x,0) \tag{3B.12}$$

$$L\{C_{xx}(x,t)\} = D\frac{d^2 C(x,s)}{dx^2} \tag{3B.13}$$

Substituting Eqs. (3B.12) and (3B.13) into Eq. (3B.11) yields the homogeneous linear ODE with constant coefficients s, D, and $C(x,0)$

$$sC(x,s) - C(x,0) = D\frac{d^2 C(x,s)}{dx^2} \tag{3B.14}$$

$$D\frac{d^2 C(x,s)}{dx^2} - sC(x,s) = -C(x,0) \tag{3B.15}$$

Denote that a LT series for $\partial^2 C(x,t)/\partial x^2$ is not necessary for developing the ODE. Instead, the partial differential notation $\partial^2 C(x,t)/\partial x^2$ is transformed into ordinary differential notation $d^2 C(x,t)/dx^2$.

Let the initial and boundary conditions be

$$\text{IC: } C(x,0) = C_o \quad \text{for } x \geq 0, t = 0 \tag{3B.16a}$$

$$\text{BC: } C(0,t) = C_s \quad \text{for } x = 0, t > 0 \tag{3B.16b}$$

$$\text{BC: } C(x,t) = C_o \quad \text{for } x = \infty, t > 0 \tag{3B.16c}$$

Hence, Eq. (3B.15) with the IC given by (3B.16a) becomes

$$\frac{d^2C\,(x,s)}{dx^2} - \frac{s}{D}C\,(x,s) = -\frac{C_o}{D} \qquad (3B.17)$$

which has a general solution of the form

$$C\,(x,s) = A_1\,(s)\exp\,(r_1 x) + A_2\,(s)\exp\,(r_2 x) \qquad (3B.18)$$

Defining the characteristic or auxiliary expression for Eq. (3B.18) with $r^2 = d^2C\,(x,s)\,/dx^2$ and $b = -s/D$ yields the corresponding roots, r_1 and r_2, for a homogeneous differential equation. From Eq. (3B.17),

$$\frac{d^2C\,(x,s)}{dx^2} - \frac{s}{D}C\,(x,s) = 0 \qquad (3B.19)$$

$$r^2 - b = 0 \qquad (3B.20)$$

$$r_{1,2} = \pm\sqrt{b} = \pm\sqrt{\frac{s}{D}} \qquad (3B.21)$$

Substituting Eq. (3B.21) into (3B.18) gives

$$C\,(x,s) = A_1\,(s)\exp\left(\sqrt{\frac{s}{D}}x\right) + A_2\,(s)\exp\left(-\sqrt{\frac{s}{D}}x\right) \qquad (3B.22)$$

Using BC $C\,(x,s) = C_o$ when $x = \infty$, $s > 0$ yields

$$C_o = A_1\,(s)\exp\,(\infty) + A_2\,(s)\exp\,(-\infty) \qquad (a)$$

$$C_o = A_1\,(s)\,(\infty) + A_2\,(s)\,(0) = A_1\,(s)\,(\infty) \qquad (b)$$

$$A_1\,(s) = \frac{C_o}{\infty} = 0 \qquad (3B.23)$$

and Eq. (3B.22) becomes

$$C\,(x,s) = A_2\,(s)\exp\left(-x\sqrt{\frac{s}{D}}\right) \qquad (3B.24)$$

Use BC $C\,(x,s) = C\,(0,s) = C_s/s$ when $x = 0$, $s > 0$ so that Eq. (3B.24) along with $\exp\,(0) = 1$ yields $A_2\,(s) = C_s/s$. Then, Eq. (3B.24) changes to

$$C\,(x,s) = \frac{C_s}{s}\exp\left(-x\sqrt{\frac{s}{D}}\right) \qquad (3B.25)$$

From Crank's book [3, p. 377,Table 2.2, item 8], the inverse Laplace transform of Eq. (3B.25) is defined as $\mathcal{L}^{-1}\left\{(1/s)\exp\left(-x\sqrt{s/D}\right)\right\}$. Subsequently, the complementary error function takes the form $\mathrm{erfc}(x/\sqrt{4Dt}) = 1 - \mathrm{erf}\left(x/\sqrt{4Dt}\right)$ and

$$C(x,t) = C_s \mathrm{erfc}\left(\frac{x}{2\sqrt{Dt}}\right) \tag{3B.26a}$$

$$\frac{C(x,t)}{C_s} = 1 - \mathrm{erf}\left(\frac{x}{2\sqrt{Dt}}\right) \tag{3B.26b}$$

Notice that Eq. (3B.26) is equal to (3.44).

3.12   Problems

3.1 Consider a particle in the x-direction and calculate v_d, D_x, and τ at $T = 300$ K and $\lambda = 5 \times 10^{-10}$ cm for (a) nitrogen and (b) hydrogen atoms in steel. Which element or particle diffuses faster? Why?

3.2 Consider a wafer made of silicon single-crystal exposed to a boron-rich environment in a furnace at relatively high temperatures for a two-step dopant diffusion process: predeposition diffusion and drive-in diffusion. The purpose of diffusing boron into silicon is to adjust the electrical properties of the wafer in an n-conductive integrated circuit (IC). Let the surface concentration and the background concentration be 7×10^{25} atoms/m^3 and 10^{20} atoms/m^3 boron (B) atoms, respectively. The diffusion coefficient is defined by

$$D = \left(2.55 \times 10^{-3}\,\mathrm{m^2/s}\right) \exp\left[-\frac{4\,\mathrm{eV/atom}}{\left(8.62 \times 10^{-5}\,\mathrm{eV/atom\,K}\right)T}\right]$$

According to the terminology used in the semiconductor industry, calculate (a) the amount of boron atoms on the surface of the silicon wafer during the predeposition process at 900 °C for 40 min and (b) the junction depth x_j in the drive-in diffusion process at 1100 °C for 2 h.

3.3 (a) Derive Eq. (3.49b) using (3.44b) and Fick's first law of diffusion. (b) Calculate the diffusion depth if $Dt = 3 \times 10^{-10}$ cm^2 and $C_s/C = 2 \times 10^4$.

3.4 For interstitial diffusion, the concentration gradient is given by Eq. (3.54). If $C_s \gg C_b$, then show that

$$\frac{C}{C_s} = \frac{1}{\sqrt{4\pi Dt}}\left(\frac{x^3}{12Dt} - x\right)$$

3.5 If interstitial diffusion of an impurity atom is to occur, then (a) plot the concentration rate $\partial C/\partial t$ in the x-direction at 1 mm from a substrate surface at interval $[0, 600\,\text{s}]$. (b) Plot $\partial C/dt = f(t)$ at $[0, 30\,\text{s}]$ interval. Is transient diffusion an instantaneous process? (c) Plot $\partial C/dt = f(t)$ at $[0, 600\,\text{s}]$ interval. How long does it take $\partial C/dt$ to reach a maximum? Given data: $D = 10^{-5}\,\text{cm}^2/\text{s}$, $C_s = 10^{-4}\,\text{mol/cm}^3$. [Solution: $(\partial C/dt)_{\text{max}} = 462.55 \times 10^{-10}\,\text{mol}/\left(\text{cm}^3\,\text{s}\right)$ at $166.67\,\text{s}$].

3.6 Consider an AISI 1020 steel gear and a pure iron gear exposed to a carburizing gas in a suitable furnace at $900\,^\circ\text{C}$. Assume that the diffusion coefficient and the carbon surface concentration are $6 \times 10^{-6}\,\text{mm}^2/\text{s}$ and 1%, respectively. For comparison purposes, calculate the heat treatment time (a) for steel (t_{steel}) and (b) for pure iron (t_{Fe}) when the carbon concentration is 0.4% at a diffusion depth of $0.5\,\text{mm}$ below the gear surface. Explain. [Solution: $t_{Fe} \simeq 1.87 t_{steel}$].

3.7 Consider an AISI 1025 steel gear and a pure iron gear exposed to a carburizing gas in a suitable furnace at $900\,^\circ\text{C}$. Assume that the diffusion coefficient and the carbon surface concentration are $5.80 \times 10^{-6}\,\text{mm}^2/\text{s}$ and 1%, respectively. For comparison purposes, calculate the heat treatment time (a) for steel (t_{steel}) and (b) for pure iron (t_{Fe}) when the carbon concentration is 0.4% at a diffusion depth of $0.5\,\text{mm}$ below the gear surface. Explain. [Solution: $t_{Fe} \simeq 2.8 t_{steel}$].

3.8 Consider a 0.25% C-steel gear exposed to a carburizing gas in a suitable furnace at $900\,^\circ\text{C}$. Assume that the diffusion coefficient and the carbon surface concentration are $5.80 \times 10^{-6}\,\text{mm}^2/\text{s}$ and 1%, respectively. Calculate the heat treatment time when the carbon concentration is 0.43% at a diffusion depth of $0.5\,\text{mm}$ below the gear surface. [$t = 5.18\,\text{s}$].

3.9 Assume that the Arrhenius equation defines the diffusion coefficient D of carbon in carbon steels (γ-*phase*) at an austenitic temperature range $1173\,\text{K} \leq T \leq 1373\,\text{K}$

$$D = \left(2 \times 10^{-5}\,\text{m}^2/\text{s}\right) \exp\left(-\frac{17{,}401\,\text{K}}{T}\right)$$

For a carburized steel sheet at $T = 1173\,\text{K}$ for $3.70\,\text{h}$, determine the nominal carbon composition (bulk composition C_b) in a steel plate when the carbon content 0.40% (by weight) at a depth of $0.5\,\text{mm}$. Assume a carbon surface concentration 1%. [Solution: $C_b = 0.20\%$].

3.10 For a carburized steel sheet at $T = 1200\,\text{K}$ for $3\,\text{h}$, determine the nominal carbon composition (bulk composition C_b) in a steel plate when the carbon content is 0.43% (by weight) at a depth of $0.5\,\text{mm}$. Use the diffusion coefficient equation given in problem P3.9 and assume a carbon surface concentration 1%.

3.11 Assume that a wafer made of silicon single-crystal is exposed to a boron-rich environment in a furnace at relatively high temperatures for a two-step dopant diffusion process: predeposition diffusion and drive-in diffusion. Let the surface concentration and the background concentration be 7.5×10^{25} atoms/m^3 and 1.5×10^{20} atoms/m^3 boron (B) atoms, respectively. The diffusion coefficient is defined by

$$D = \left(2.55 \times 10^{-3}\,\text{m}^2/\text{s}\right) \exp\left[-\frac{46,404\,\text{K}}{T}\right]$$

Calculate (a) the amount of boron atoms on the surface of the silicon wafer during the predeposition process at 900 °C for 40 min and (b) the junction depth x_j in the drive-in diffusion process at 1100 °C for 2 h.

3.12 Assume that $D = 10^{-5}\,\text{cm}^2/\text{s}$ and $C_s = 10^{-4}\,\text{mol}/\text{cm}^3$ for interstitial diffusion of an impurity atom at 1 h treatment. Based on this information and a diffusion depth of 1 mm, calculate (a) the concentration gradient $\partial C/\partial x$ and the diffusion flux J_x as per Fick's first law of diffusion, and (b) $\partial^2 C/dx^2$ and the concentration rate $\partial C/\partial t$ as per Fick's second law of diffusion. Solution: [(a) $J_x = 7.42 \times 10^{-10}\,\text{mol}/\left(\text{cm}^2\,\text{s}\right)$].

3.13 Consider $D = 2 \times 10^{-5}\,\text{cm}^2/\text{s}$ and $C_s = 2 \times 10^{-4}\,\text{mol}/\text{cm}^3$ for interstitial diffusion of an impurity atom at 2 h treatment. Based on this information and a diffusion depth of 2 mm, calculate (a) the concentration gradient $\partial C/\partial x$ and the diffusion flux J_x as per Fick's first law of diffusion, and (b) $\partial^2 C/dx^2$ and the concentration rate $\partial C/\partial t$ as per Fick's second law of diffusion. Solution: [(a) $J_x = 2.77 \times 10^{-9}\,\text{mol}/\left(\text{cm}^2\,\text{s}\right)$].

3.14 Assume that a particle in the x-direction. Calculate v_x, D_x, and τ at $T = 400\,\text{K}$ and $\lambda = 5 \times 10^{-10}\,\text{cm}$ for (a) nitrogen and (b) hydrogen atoms in steel. Which element or particle diffuses faster? Why?

3.15 Calculate v_x, D_x, and τ at $T = 600\,\text{K}$ and $\lambda = 5 \times 10^{-10}\,\text{cm}$ for (a) nitrogen and (b) hydrogen atoms in steel. Which element or particle diffuses faster? Why?

3.16 Assume that an AISI 1020 steel gear and a pure iron gear are exposed to a carburizing gas in a suitable furnace at 910 °C. Assume that the diffusion coefficient and the carbon surface concentration are $6.2 \times 10^{-6}\,\text{mm}^2/\text{s}$ and 1%, respectively. For comparison purposes, calculate the heat treatment time (a) for steel (t_{steel}) and (b) for pure iron (t_{Fe}) when the carbon concentration is 0.4% at a diffusion depth of 0.5 mm below the gear surface. Explain. [Solution: $t_{Fe} \simeq 1.863 t_{steel}$].

3.17 For a carburized steel sheet at $T = 1200\,\text{K}$ for 4 h, determine the nominal carbon composition (bulk composition C_b) in a steel plate if $C_x = 0.43\%$ (carbon content by weight) at a depth of 0.5 mm. Use the diffusion coefficient equation given in problem P3.9 and assume a carbon surface concentration 1%. [Solution: $C_b = 0.12\%$].

3.18 Assume that $D = 2 \times 10^{-5}\,\text{cm}^2/\text{s}$ and $C_s = 2 \times 10^{-4}\,\text{mol}/\text{cm}^3$ for interstitial diffusion of an impurity atom at 1 h treatment. Based on this information and a diffusion depth of 1 mm, calculate (a) the concentration gradient $\partial C/\partial x$ and the diffusion flux J_x as per Fick's first law of diffusion, and (b) $\partial^2 C/dx^2$ and the concentration rate $\partial C/dt$ as per Fick's second law of diffusion. Solution: [(a) $J_x = 2.97 \times 10^{-9} \, \text{mol}/(\text{cm}^2 \, \text{s})$].

3.19 For a carburized steel sheet at $T = 1200 \, \text{K}$ for 2 h, determine the nominal carbon composition (bulk composition C_b) in a steel plate if $C_x = 0.43\%$ (carbon content by weight) at a depth of 0.5 mm. Use the diffusion coefficient equation given in problem P3.9 and assume a carbon surface concentration 1%. [Solution: $C_b = 0.30\%$].

3.20 Consider a carburized steel sheet at $T = 1200 \, \text{K}$ for 2 h. Determine the nominal carbon composition (bulk composition C_b) in a steel plate if $C_x = 0.43\%$ (carbon content by weight) at a depth of 0.5 mm. Use the diffusion coefficient equation given in problem P3.9 and a carbon surface concentration 1.10%. [Solution: $C_b = 0.27\%$].

References

1. W.M. Saltzman, *Engineering Principles for Drug Therapy* (Oxford University Press, New York,, 2001)
2. H.C. Berg, *Random Walks in Biology* (Princeton University Press, New York, 1993)
3. J. Crank, *The Mathematics of Diffusion* (Oxford University Press, New York, 1979)
4. A. Einstein, A new determination of molecular dimensions. Ann. Phys. **19**, 289–306 (1906)
5. D.A. Porter, K.E. Easterling, M.Y. Sherif, *Phase Transformations in Metals and Alloys*, 3rd edn. (CRC Press, Boca Raton, 2009)
6. D.R. Gaskell, *An Introduction to Transport Phenomena in Materials Engineering*, 2nd edn. (Momentum Press LLC, New Jersey, 2013)
7. L.S. Darken, Diffusion, mobility and their interrelation through free energy in binary metallic systems. Trans. Met. Soc. AIME **175**, 184–201 (1948)
8. A. Paul, The Kirkendall effect in solid state diffusion. Ph.D. Dissertation, Eindhoven University of Technology, NUR 813 (2004). ISBN 90-386-2646-0
9. S.P. Wasik, K.E. McCulloh, Measurements of gaseous diffusion coefficients by a gas chromatographic technique. J. Res. Notional Bureau Standards A. Phys. Chem. **73A**(2), 207–211 (1969)
10. G.E. Murch, A.S. Nowick (eds.), *Diffusion in Crystalline Solids* (Academic, New York, 1984). ISBN 0-12-522662-4
11. M.J. Thrippleton, N.M. Loening, J. Keeler, A fast method for the measurement of diffusion coefficients: one-dimensional DOSY. Magn. Reson. Chem. **41**, 441–447 (2003)
12. C.M.A. Brett, A.M.O. Brett, *Electrochemistry Principles, Methods and Applications* (Oxford University Press, New York, 1994)

Chapter 4
Solidification

4.1 Introduction

Producing metal castings for engineering applications involves foundry technology and heat transfer during melting and solidification. The most relevant aspect of foundry technology in this chapter is the pouring of a melt (liquid metal) into a sand or metal mold containing a hollow cavity and the subsequent liquid-to-solid (L-S) phase transformation, known as solidification, due to outward thermal energy flow, which in turn depends on the type of thermal resistance encounter within the solidification domain Γ. Therefore, thermophysical properties of the mold and the solidifying melt are essential for making high quality castings by nucleation and growth of the solid phase. In the end, surface quality, dimensional tolerances, surface and machining allowances are essential features that must be controlled for high quality castings.

On the one hand, sand molds are poor thermal conductors with certain degree of porosity for high temperature (T) gases to escape while quasi-static solidification takes place, mainly at the moving mold–liquid interface. In fact, sand molds are commonly used due to cost- effective foundry materials and high degree of recyclability. Metallic molds, on the other hand, are good thermal conductors that induce faster solidification than sand molds. Accordingly, this chapter describes the analytical procedures related to the thermal energy transfer process during solidification of pure metals and binary A-B alloys.

4.2 Foundry Sand

The common foundry sand is a mixture of sieved sand particles, clay, and water. Thus, its name green sand (foundry sand) because it is wet. The foundry sand can also be referred to as molding sand or casting sand, and it is usually silica or

© Springer Nature Switzerland AG 2020
N. Perez, *Phase Transformation in Metals*,
https://doi.org/10.1007/978-3-030-49168-0_4

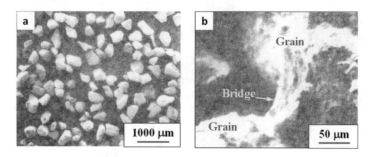

Fig. 4.1 SEM photomicrographs of (**a**) washed sub-angular α-quartz sand grains and (**b**) clay-grain bridge [1]

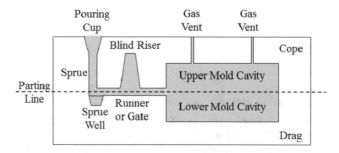

Fig. 4.2 Schematic sand mold and related terminology

α-quartz sand (SiO_2), chromite sand ($FeCr_2O_4$), zircon sand ($ZrSiO_4$), or even a combination of these sands. The fundamental properties of the sand molds are strength related to plasticity and permeability at relatively high temperatures.

Figure 4.1a depicts natural sub-angular α-quartz sand grains and Fig. 4.1b exhibits a typical clay-sand bridge between sand grain surfaces. This bridge is fundamentally the bonding between sand grains and the water-clay mixture needed for producing foundry sand molds with strength and plasticity suitable for sand castings. In fact, the grain shape affects porosity and permeability of the sand mold and the bridging between grains [1].

Figure 4.2 illustrates a schematic green-sand mold and related terminology. Denote the parting line that divides the mold in halves; cope and drag. Typically, green-sand molds (foundry sands) are supported by flasks; otherwise, the molds would collapse during pouring liquid metals.

Figure 4.3 exhibits regression surfaces for green compressive strength σ_c (Fig. 4.3a) and green permeability K_g (Fig. 4.3b) as functions of percentage moisture x and added Na-bentonite clay y. Moreover, dry strength is also a sand property used to prevent erosion by liquid metal during pouring since the inner mold surfaces dry out at $T \geq T_f$ (melting point).

Additionally, the foundry sand should be plastic so that sand grains and clay particles and water are held together by electrostatic attraction forces. Thus, the

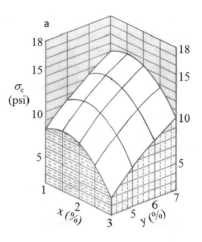
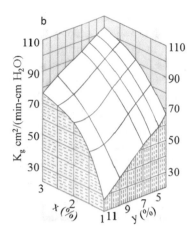

Fig. 4.3 (a) Green compressive strength $\sigma_c = f(x, y)$ and (b) green permeability $K_g = g(x, y)$ of a foundry sand. Here, x and y are the percentage by weight of moisture (H_2O) and Na-bentonite, respectively. The regression surfaces are $\sigma_c = 4.4357 - 0.0038x + 1.4643y - 0.8134x^2 - 0.0681y^2 + 0.3347xy$ and $K_g = 50.0046 + 41.7005x - 5.2717y - 6.7403x^2 - 0.0289y^2 + 0.8124xy$ [1]

Fig. 4.4 Actual foundry practice for steel making castings. (a) Pouring liquid steel from an arc furnace into a ladle and (b) unfinished steel valve

basic mixture of these ingredients is a cost-effective approach in the foundry engineering field.

Green permeability K_g measures the amount of air and hot gases that flow through the mold walls. When $K_g \to \infty$, mold voids are accessibly large and the liquid metal easily penetrates the mold walls. Conversely, if $K_g \to 0$, then air and gases may not escape the mold and consequently, gas entrapment may occur during metal solidification.

Figure 4.4a shows a ladle being filled with liquid steel (melt) to be poured into sand molds and Fig. 4.4b exhibits an actual semi-finished steel valve that needs further processing to mechanically remove the excess solidified steel.

The making of a steel casting is a long and complicated process because it involves sand mold technology, melting steel scrap (recyclable steel) with relevant chemical composition, and quality control assurance prior to pouring the liquid steel into molds. Among foundry issues, contaminated or wet steel scrap is the main source for hydrogen entrapment in the melt. Consequently, hydrogen gas forms bubbles during the solidification and as a result, gas porosity becomes a significant problem in producing high quality castings. Similar problems occur in the production of other metallic casting.

In addition, shrinkage porosity is also a casting defect caused by unsuitable mold design, mainly the gating system and riser size, which is the principal source for liquid feeding the desired casting geometry. All this is of fundamental importance for obtaining defect-free castings.

Example 4.1 Consider the image and the sketch of a 2-m diameter ladle containing 70 kg of liquid steel. Calculate the time to empty the steel through a 3-cm diameter hole at the bottom of the ladle.

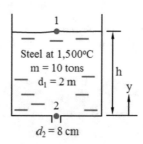

Solution Using the fundamental mass flow rate and the flow velocity equations one can solve the pouring time problem, which is the time for emptying the ladle of known diameter and height. Thus,

$$\frac{dm}{dt} = \frac{dm_{in}}{dt} - \frac{dm_{out}}{dt} = (\rho \upsilon A)_1 - (\rho \upsilon A)_2$$

$$\upsilon = \frac{dy}{dt} = \sqrt{2gy}$$

where ρ denotes the steel density, υ denotes the liquid steel velocity, and A denotes the cross-sectional area. Once the ladle is filled with liquid steel, $dm_1/dt = dm_{in}/dt = 0$ and

$$\frac{dm}{dt} = -(\rho \upsilon A)_{out} = -\frac{\pi}{4}d_2^2\rho\sqrt{2gy} \tag{a}$$

$$\frac{dm}{dt} = \frac{\pi}{4}d_1^2\rho\frac{dy}{dt} \tag{b}$$

Equating Eqs. (a) and (b) yields

$$dt = -\left(\frac{d_1}{d_2}\right)^2 \frac{1}{\sqrt{2g}} \frac{dy}{\sqrt{y}} = -\left(\frac{d_1}{d_2}\right)^2 \frac{1}{\sqrt{2g}} \int_{h_1}^{h_2} \frac{dy}{\sqrt{y}} \tag{c}$$

$$t = \left(\frac{d_1}{d_2}\right)^2 \frac{2}{\sqrt{2g}} \left(\sqrt{h_1} - \sqrt{h_2}\right) \tag{d}$$

For $d_1 = 2\,\text{m}$, $d_2 = 8\,\text{cm}$, $\rho = 7.80\,\text{g/cm}^3$, $m = 10\,\text{ton} = 10{,}000\,\text{kg}$, the steel column height is determined from the fundamental mass density equation $\rho = m/V = m/A_1 h_1$. Solving for h_1 and upon substitution of the available data yields the following result

$$h_1 = \frac{m}{\rho A_1} = \frac{10 \times 10^6\,\text{g}}{\left(7.80\,\text{g/cm}^3\right)\left(\pi/4\right)\left(2 \times 10^2\,\text{cm}\right)^2} = 4\,\text{m}$$

From Eq. (d) with $h_2 = 0$, it takes $t = 564.40\,\text{s} \simeq 9.41\,\text{min}$ to empty the ladle steadily. It has been assumed a steady laminar flow to fill the hypothetical casting; otherwise, a turbulent flow would cause problems, such as gas porosity.

4.3 Thermodynamics of Phase Transformation

Thermodynamics of phase transformation is the study of behavior of systems, such as solidification or melting. The driving force for liquid-to-solid ($L \to S$) phase transformation (solidification) is known as the Gibbs energy change (ΔG).

Consider the solidification of a pure metal. Figure 4.5 schematically shows the assumed linear behavior of $H = H(T)$ and $G = G(T)$. The slopes of the plots are the specific heats c_l, c_s and the entropies S_l, S_s for the liquid and solid, respectively.

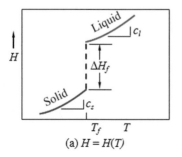

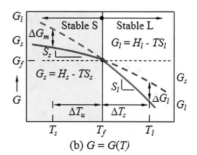

Fig. 4.5 Schematic energy trends for liquid and solid phases. (**a**) Enthalpy and (**b**) Gibbs energy

In addition, phase stability is graphically illustrated in Fig. 4.5b, where $G_f = f(T_f)$ is the equilibrium Gibbs energy at the fusion or melting temperature T_f. Notice that solidification occurs when $\Delta G = \Delta G_m = (G_s - G_l) < 0$ at $T = T_s$, induced by an undercooling $\Delta T_u = (T_f - T_s) > 0$. Here, ΔG_m denotes the Gibbs energy of mixing. For melting, $\Delta G = (G_l - G_s) < 0$ at $T = T_l$, where $\Delta T_s = (T_l - T_f) > 0$ denotes the degree of superheating the liquid.

Mathematically, the Gibbs energy at constant temperature T is defined by

$$G = H - TS \tag{4.1a}$$

$$dG = dH - TdS - SdT = dH - TdS \tag{4.1b}$$

Combining the first and second laws of thermodynamics yields

$$Q - W = \Delta H + \Delta KE + \Delta PE \tag{4.2a}$$

$$T\Delta S - W = \Delta H + \Delta KE + \Delta PE \tag{4.2b}$$

For steady-state solidification ($\Delta KE = 0$, $\Delta PE = 0$),

$$Q - W = \Delta H \tag{4.3a}$$

$$\Delta G = -W = \Delta H - Q \tag{4.3b}$$

Here, W is the work done, Q is the amount of heat transfer, and ΔH is the enthalpy change of the thermodynamic system.

During liquid-to-solid (L-S) phase transformation, $\Delta H_f = \Delta H = H_l - H_s$ is the latent heat of fusion of a pure metal at the freezing temperature T_f and for comparison, ΔH_f^* is the latent heat of fusion of an alloy at the freezing temperature range $T_s < T_f^* < T_l$ corresponding to a nominal composition C_o, and related to the liquidus temperature T_l and solidus temperature T_s. Moreover, these variables (C_o, T_l, T_s) are directly related to a particular phase diagram.

From the second law of thermodynamics during an isothermal process, the amount of heat transfer Q and the entropy change ΔS are related in the following mathematical form

$$Q = T\Delta S = T(S_l - S_s) \tag{4.4}$$

Combining Eqs. (4.3b) and (4.4) yields the Gibbs energy change

$$\Delta G = \Delta H - T\Delta S \tag{4.5}$$

For small undercooling ΔT, $\Delta G \to 0$ at $T \to T_f$, $\Delta H_f(T) = \Delta H_f(T_f)$ and $\Delta S_f(T) = \Delta S_f(T_f)$. Thus, Eq. (4.5) yields the entropy of fusion

$$\Delta S_f = \frac{\Delta H_f}{T} \tag{4.6}$$

Combining Eqs. (4.5) and (4.6) gives

$$\Delta G = \Delta H_f - T_f \frac{\Delta H_f}{T} = \Delta H_f \left(\frac{T - T_f}{T} \right) \tag{4.7a}$$

$$\Delta G = - \left(\frac{\Delta H_f}{T} \right) \Delta T = - \Delta S_f \Delta T \tag{4.7b}$$

$$\Delta T = -\Delta G / \Delta S_f \quad \& \quad \Delta T = (T_f - T) > 0 \tag{4.7c}$$

where ΔT denotes the degree of undercooling. Moreover, the entropy of fusion ΔS_f, in general, is a measure of disorganized energy, while the freezing temperature T_f is a measure of the atomic bond strength.

In addition, both product $\Delta S_f T_f$ and $\Delta S_f (T - T_f)$ terms represent two different conditions for heat transfer during solidification. On the one hand, the product $\Delta S_f T_f$ is for solidification at $T = T_f$. On the other hand, the term $\Delta S_f (T - T_f)$ indicates that solidification starts at $T < T_f$ since the melt is undercooled by a small degree of undercooling denoted by $\Delta T = T - T_f$. This implies that the evolution of latent of fusion ΔH_f for a pure metal occurs at a temperature written as $T < T_f$.

Essentially, the rate of solidification is strongly dependent on the rate of latent heat of fusion. Hence, slow solidification and rapid solidification. The former prevails in this book as the method for producing conventional casting and single crystals. The latter is characterized in a later chapter.

It is clear now that the thermodynamics of solidification is a useful tool for determining the fundamental variables, such as the Gibbs energy change and related undercooling. Eventually, complete analysis of solidification includes the solidification time, solidification velocity, temperature gradient (liquid and solid), heat transfer through the mold walls, X-ray diffraction and of course, all necessary microscopic work.

Moreover, equilibrium phase diagrams are also determined using thermodynamics since one can predict an alloy solidification path, calculate the partition coefficients, deduce the slopes of the liquidus m_l and the solidus m_s phase boundaries. These slopes are very useful in determining the alloy undercooling ΔT. Actually, ΔT for an alloy is derived in Chap. 8.

The main goal now is to analyze the heat transfer phenomena related to predictions of the solidification front position. The latter is briefly described in a later chapter.

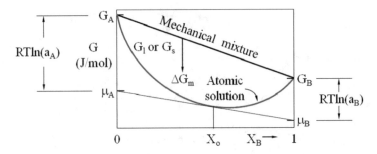

Fig. 4.6 Gibbs free energy diagram for a function $G = f(X_B)$. (**a**) Before mixing and (**b**) after mixing at temperature T and fixed pressure P

4.3.1   Single Phase Solutions

Figure 4.6 schematically shows the Gibbs energy diagram for an ideal solution of components (elements) A and B atoms at temperature T and pressure P. Here, μ_A, μ_B are chemical potentials.

The Gibbs energy functions $G = f(X_B)$ and $\Delta G = f(T)$ for ideally mixing A and B components are defined by

$$\Delta G = X_A G_A + X_B G_B = (1 - X_B) G_A + X_B G_B = f(X_B) \tag{4.8a}$$

$$\Delta G = \Delta H - T \Delta S = f(T) \tag{4.8b}$$

where X_A, X_B denotes the mole fractions in the A and B mix, G_A, G_B denote the Gibbs energy of pure components (elements) A and B, ΔH denotes the enthalpy change due to atomic bonding between neighboring atoms, ΔS denotes the entropy change due to atomic disorder in the mix and $G_A, G_B = \mu_A^o, \mu_B^o$, where μ_A^o, μ_B^o denote the standard or reference chemical potentials.

4.3.2   Thermodynamic Criteria for Reactions

On the whole, the Gibbs energy ΔG defined by Eq. (4.8b) serves as a criterion for determining the direction and condition of a reaction as written. For instance, consider the following reactions as written

$$
\begin{array}{lll}
Cu \rightarrow Cu^{+2} + 2e^- & \text{Electrochemical reaction} & \text{(oxidation)} \\
2NO + O_2 \rightarrow 2NO_2 & \text{Chemical reaction} & \text{(reduction)} \\
L \rightarrow \alpha + \beta & \text{Physical reaction} & \text{(phase change)}
\end{array}
$$

The arrow $\rightarrow$ indicates the direction of the reactions as written. For the sake of clarity, the electrochemical reaction involves electrons and a chemical reaction does not.

The spontaneity of a reaction can be determined using the following criteria

$$\Delta G < 0 \quad \text{Spontaneous reaction at } T$$
$$\Delta G < 0 \quad \text{Non-spontaneous reaction at } T$$
$$\Delta G = 0 \quad \text{Equilibrium, no change with time at } T$$

As a matter of fact, all three reactions above are spontaneous as written if $\Delta G < 0$. Thermodynamically, the above reactions are intentionally included for the sake of clarity. For example, "oxidation" refers to loss of electrons and the neutral metal, such as copper Cu, becomes a cation (Cu^{2+}), "reduction" implies gaining electrons or changing chemical compound, and "phase change" signifies transformation of an atomic arrangement.

4.4 Phase Separation

The Gibbs energy diagram (G-X diagram) is a graphical representation of the $G = f(X_B)$ function at T and it has particular characteristics related to the concavity of the graph of $G_l, G_s = f(X_B)$ at T. For instance, the concave down shape of $G = f(X_B)$ function varies with temperature T as indicated in Fig. 4.7. The extreme conditions can be defined as $G_s(X_B) > G_l(X_B)$ at $T > T_f$ (Fig. 4.7a) and $G_s(X_B) < G_l(X_B)$ at $T < T_f$ (Fig. 4.7b).

Phase Separation For a solidifying hypothetical binary A-B alloy with nominal mole fraction X_o (Fig. 4.7c) and G_s, G_l have concave down shapes. The marked points in Fig. 4.7c indicate that

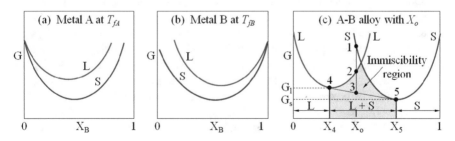

Fig. 4.7 Shape of $G = f(X_B)$ for two phases. (**a**) Stable liquid (L), (**b**) stable solid (S), and (**c**) alloy with nominal mole fraction X_o, solubility limits at X_4, X_5 and immiscibility region at $X_4 \leq X_B \leq X_5$

- **Point 1:** $G_l(X_o) = G_1(X_o)$ and only liquid phase exits at $T_l > T_f$
- **Point 2:** $G_s(X_o) = G_2(X_o)$ and only solid phase exits at $T_s < T_l$
- **Point 3:** $G_3(X_o) = G_4(X_B^s) + G_5(X_B^l)$ phase separation at $T_s < T > T_l$
- **Point 4 and 5:** These are connected by drawing a tangent line that acts as a tie-line (Lever rule) and $G_3(X_o)$ is the total Gibbs energy of mixing when phase separation occurs at $T_s < T > T_l$. Thus,

$$G_3(X_o) = G_4\left(X_B^s\right) + G_5\left(X_B^l\right) \quad \text{at } T_s < T > T_l \tag{4.9a}$$

$$G_3(X_o) = \left(\frac{X_B^l - X_o}{X_B^l - X_B^s}\right) G_s + \left(\frac{X_o - X_B^s}{X_B^l - X_B^s}\right) G_l \tag{4.9b}$$

$$G_3(X_o) = f_s G_s + f_l G_l \tag{4.9c}$$

$$f_s = \frac{X_B^l - X_o}{X_B^l - X_B^s} \quad \& \quad f_l = \frac{X_o - X_B^s}{X_B^l - X_B^s} \tag{4.9d}$$

Note that the f_s, f_l denote the phase fractions at point 3 and represent the well-known Lever rule. In particular, the Lever rule is commonly used to extract phase information from standard (equilibrium) binary phase diagrams for an A-B alloy with nominal composition X_o in mole fraction or C_o in weight percentage ($wt\%$ or %) at temperature T.

There are different types of binary phase diagrams with specific phase fields representing corresponding phases, which in turn have diverse atomic arrangements. Thus, phase diagrams are viewed as phase maps constructed using thermodynamic data obtained at atmospheric pressure.

For the sake of clarity, below are some particular characteristics of Gibbs energy diagram (Fig. 4.7c) related to binary phase diagrams.

- **Miscibility gap:** A particular A-B solution has a miscibility gap when it spontaneously decomposes into a mixture of two solutions with different concentrations, one of which is predominantly A-rich with some B atoms in solution, and the other predominantly B-rich with some A in solution. This physical behavior occurs at point 3. Subsequently, the $G = f(X_B)$ diagram is divided into three regions, which correspond to the mole fraction range $0 < X_B \leq X_4$ for the liquid, $X_5 \leq X_B \leq 1$ for the stable solid, and $X_4 \leq X_B \leq X_5$ for mixture of L and S. Moreover, note that the tangent line intercepts the minimum Gibbs energy, $G_{min} = G_l(X_4)$, for the liquid (L) and $G_{min} = G_s(X_5)$ for the solid (S) phases. Hence, X_4, X_5 are the solubility limits of B atoms in the L and S phases composed of A, B atoms or components.
- **Immiscibility region:** If a single-phase solution decomposes into two different phases, then the A,B components become immiscible. Nonetheless, the immiscibility region is located at a mole fraction range given as $X_4 \leq X_B \leq X_5$ between $G_{min}(X_4)$ and $G_{min}(X_5)$. It represents the immiscibility region, where

there exists a mixture of L and S phases. Moreover, if $G_l(X_4)$, $G_s(X_5)$ reach minimum values, then $dG/X_B = 0$ at X_4, X_5. Also, if $d^2G/X_B^2 = 0$, then $X_A = X_B$ at an inflection point.

- **Physical behavior:** The relevant physical behavior of the A-B solution is thermodynamically described by the Gibbs energy function, $G = f(X_B)$, as hypothetically illustrated in Fig. 4.7 at temperature T. In fact, the function $G = f(X_B)$ plots different types of curves at $0 \leq X_B \leq 1$ and temperature T, and reveals the physical behavior of the A-B mix upon freezing or melting. Hence, the Gibbs energy diagram or simply the G-X_B diagram, where $X_A + X_B = 1$.

- **Effect of temperature:** When the temperature $T = T_{fA}$ (Fig. 4.7a), Eq. (4.5) $\Delta G = \Delta H - T_{fA}\Delta S = \Delta G_A = \Delta H_f$ and $\Delta S = 0$ since $S_A = S_B$. Similarly, when $T = T_{fB}$ (Fig. 4.7b), $\Delta G = \Delta H - T_{fA}\Delta S = \Delta G_B = \Delta H_f$ and $\Delta S = 0$ since $S_A = S_B$. When $T_{fA} < T < T_{fB}$ (Fig. 4.7c), however, G_l and G_s curves intercept at some X_B and consequently, the minimum value of the $G_{\min,s} = f_s(X_B)$ curve is lower than that $G_{\min,l} = f_s(X_B)$ curve. In this case, both L and S phases are stable in the immiscibility gap at $X_4 \leq X_B \leq X_5$. The opposite physical behavior may occur when $G_{\min,l} < G_{\min,s}$.

- **Effect of composition:** When the mole fraction B is within the range $X_4 \leq X_B \leq X_5$, the phase separation indicates that the liquid composition is given by $X_4 = X_{l,B}$ and the solid becomes B-richer due to the fact that $X_5 = X_{s,B} < X_{l,B}$, and their fractions are given by the Lever rule, Eq. (4.9d). Therefore, $G_{\min,l} < G_{\min,s}$ as before since $\Delta S_{\min,l} > \Delta S_{\min,s}$. In other words, if ΔS increases, $\Delta G = \Delta H - T\Delta S$ decreases and vice versa.

- **Solubility Limit:** It is the maximum amount of a component that can be dissolved in a phase at a temperature T. For instance, carbon (C) has a limited solubility in iron (Fe) ferrite phase.

4.5 Binary Solutions

Consider the internal energy of atom pairs as the bonding energy, which takes into account the interatomic distances in ideal solutions with the enthalpy change $\Delta H = 0$ and regular (non-ideal) solutions with $\Delta H \neq 0$. Additional theoretical and analytical details on the subject matter can be found elsewhere ([2, p. 43], [3, p. 196] and [4, p. 81]).

In general, when solution containing N_A and N_B atoms in the gaseous state condenses into liquid or solid, one refers to this atomic process as a phase transformation. In this section, only metallic liquid and solid solutions are considered henceforth since the latter is the most useful phase in engineering applications.

According to the nature of the chemical bonding between atom pairs in a crystalline structure, a crystal is classified as ionic, covalent, and metallic. The latter is most relevant in this section for characterizing solidification of binary A-B alloys, such as Cu-Ni, Cu-Zn, Pb-Sn, and so forth.

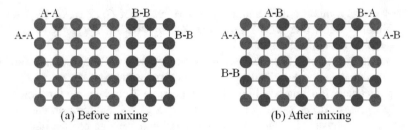

Fig. 4.8 Atomic arrangement (**a**) before mixing and (**b**) after mixing A and B atoms during phase transformation from the liquid phase at temperature T_f and pressure P

The general two-dimensional (2D) view of the concept of A-A, B-B, and A-B atom pairs (ij) in the solid-state phase is illustrated in Fig. 4.8. These arrangements are initially in the unmixed state (Fig. 4.8a) and only A-A and B-B atom pairs are shown.

After mixing (Fig. 4.8b) at a temperature T and pressure P, the A-B atom pair appears as part of the configurational arrangement of atoms. Normally, the pressure is kept constant during phase transformation, which is usually carried out directionally or unidirectionally under strict process control. Note that Fig. 4.8b shows a disordered planar atomic arrangement between A and B atoms. The assemblage of atom planes forms a 3-D space lattice known as a crystal, which in turn represents a crystalline solid structure.

This model serves as a guide to determine the coordination number (Z) of a particular atom in bond and related binding energy (or cohesive energy) U_{AA}, U_{BB}, or U_{AB}. In reality, these binding energies are also related to the mole fraction of each type of atoms and as a whole, the goal in here is to define a more common and practical thermal energy known as the enthalpy of mixing (ΔH_m). In practice, ΔH_m is used for characterizing the physical behavior of any system undergoing phase change at constant pressure (isobaric process). As a matter of fact, ΔH_m is also known as molar excess enthalpy and heat of mixing. These phrases are used interchangeably in the literature. Nonetheless, ΔH_m provides the meaningful concept of interatomic interaction within the mixture composed of elements capable of forming a solution, and it can be derived from calorimetric measurements at a series of temperatures.

4.5.1 Internal Energy

With regard to the concept of atom pairs (ij), i = A,B and j = A,B are denoted as A-A, B-B, and A-B and the internal energies (U_{ij}) for these type of atom pairs are denoted as U_{AA}, U_{BB}, and U_{AB}. Moreover, the relevant bond fractions (f_{ij}) are f_{AA}, f_{BB}, and f_{AB}. This leads to the total internal energy (U_m) of mixing defined by

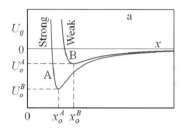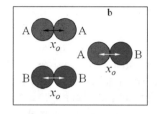

Fig. 4.9 (a) Internal energy U_{ij} as a function of interatomic distance x and (b) atom pairs for two hypothetical A and B atoms

$$U_m = \sum f_{ij}U_{ij} = f_{AA}U_{AA} + f_{BB}U_{BB} + f_{AB}U_{AB} \tag{4.10}$$

and the internal energy (U_{ij}) of the solid phase due to the atom pair model is the sum of the attractive and repulsive atomic energies

$$U_{ij} = -\left(\frac{a_1}{x^m}\right)_{\text{attraction}} + \left(\frac{a_2}{x^n}\right)_{\text{repulsive}} \tag{4.11}$$

where a_1, a_2 denote constants for atom attraction and repulsion, respectively, x denotes the interatomic distance, and m, n are material exponents. Figure 4.9a illustrates two plots for hypothetical materials A and B and Fig. 4.9b shows the equilibrium distance x_o for atom pairs (ij) in the solid state and the minimum internal energies $(U_o^A$ and $U_o^B)$ for equilibrium condition.

These plots suggest that

$$U_{ij}(x_o) = U_o^A < 0 \text{ at } x_o^a \quad \text{solid state}$$

$$U_{ij}(x_o) = U_o^B < 0 \text{ at } x_o^b \quad \text{solid state}$$

$$U_{ij}(x) \to 0 \text{ as } x \to \infty \quad \text{vapor state}$$

$$U_{ij}(x) \to \infty \text{ as } x \to 0 \quad \text{ionic state}$$

Additionally, atoms are at random in the liquid state and atoms are packed together in the solid state forming a three-dimensional space lattice, which in turn represents the atomic arrangement used for deducing the type solid geometry, such as face-centered cubic (FCC), body-centered cubic (BCC), hexagonal-closed packed (HCP), and so forth.

In materials science or physical metallurgy, the physical solid is a stable phase due to the chemical bonds between atoms being induced by an atomic force field. Hence, a crystal structure is a stable solid phase constituted by a packed atomic arrangement that has a potential energy, commonly known as the internal energy for atomic binding. In order for the A-B alloy to achieve equilibrium in the solid state, $U_{ij}(x_o) = U_o$ must be a minimum.

4.5.2 Entropy of Mixing

The goal in this section is to show how the Gibbs energy function $G = f(X_B)$ at temperatures T is used to obtain data for constructing a binary A-B phase diagram. This implies that a T-X_B diagram can be determined from G-X_B diagrams assuming an initially homogeneous A-B liquid solution.

First of all, the Gibbs energy G_1 for A and B components and the change of Gibbs energy ΔG due to mixing are

$$G_2 = G_1 + \Delta G = X_A G_A + X_B G_B + \Delta G \tag{4.12}$$

where ΔG is defined by Eq. (4.8b), $G_2 > G_1$ after mixing A and B atoms, and G_A, G_B are the standard or reference free energies. Nonetheless, both X_A, X_B and ΔG are defined below.

The entropy change of mixing (ΔS_m) for N_A and N_B number of atoms in the solidifying system can be defined using Boltzmann statistical definition of entropy. Thus, the physical behavior of the mixing can be determined as the Boltzmann entropy change

$$\Delta S_m = k_B \ln(\omega) = k_B \ln \frac{(N_A + N_B)!}{N_A! N_B!} \tag{4.13}$$

anywhere k_B denotes the Boltzmann's constant and ω denotes the probability parameter. Using the general Stirling approximation $\ln(N)! = N \ln(N) - N$ on Eq. (4.13) yields the configurational (no vibrational) entropy of mixing as

$$\Delta S_m = k_B [\ln(N_A + N_B)! - \ln(N_A)! - \ln(N_B)!] \tag{4.14a}$$

$$\Delta S_m = -k_B \left[N_A \ln\left(\frac{N_A}{N_A + N_B}\right) + N_B \ln\left(\frac{N_B}{N_A + N_B}\right) \right] \tag{4.14b}$$

For one mole of A-B mixture, $N_a = N_A + N_B$, $k_B = R/N_a$, $N_A = X_A N_a$ and $N_B = X_B N_o$, where N_o is the Avogadro's number. Thus, Eq. (4.14b) becomes

$$\Delta S_m = -R [X_A \ln(X_A) + X_B \ln(X_B)] \tag{4.15}$$

Combining Eqs. (4.14) and (4.15) yields G_2 for an ideal solution with $\Delta H_m = 0$ and ΔG_m for non-ideal (regular) solution with $\Delta H_m \neq 0$. Thus,

$$G_2 = X_A G_A + X_B G_B + RT [X_A \ln(X_A) + X_B \ln(X_B)] \tag{4.16a}$$

$$\Delta G_m = \Delta H_m + X_A G_A + X_B G_B$$
$$+ RT [X_A \ln(X_A) + X_B \ln(X_B)] \tag{4.16b}$$

where $\Delta H_m > 0$ for an endothermic mixing due to absorption of heat for melting and $\Delta H_m < 0$ for an exothermic mixing due to release of heat for solidification. Actually, enthalpy represents a thermodynamic quantity and measures the total heat content of a system.

Notice that $\Delta G_m = \Delta H_m - T\Delta S_m$ is the fundamental mathematical model being used to define Eq. (4.16b). Thus, thermodynamics of solidification treats the melt as an ideal or real solution and describes the Gibbs energy change of mixture ΔG_m as a function dependent on the characteristics of the mixture. In particular, Eq. (4.16b) describes the fundamental Gibbs equation relating equilibrium states due to the thermodynamics of reacting mixtures.

4.5.3 Enthalpy of Mixing

It is convenient now to classify alloy solid solutions as either ideal solid solutions ($\Delta H_m = 0$) or regular (non-ideal) solid solutions ($\Delta H_m \neq 0$), which is treated as an excess thermal energy (xs).

For one mole of disorder solution containing A and B atoms, the total mole fraction of the constituents (components) is $X_A + X_B = 1$ In this case, the bonding-energy factors (bond fractions) f_{ij} per type of atomic bonding are defined by multiplying the number of bonds $(Z/2)\,X_i$ and the number of atoms $N_o X_i$ in solution. Thus,

$$f_{AA} = \frac{1}{2}ZX_A\,(N_o X_A) = \frac{1}{2}ZN_o X_A^2 = \frac{1}{2}ZN_o\,(1 - X_B)^2 \tag{4.17a}$$

$$f_{BB} = \frac{1}{2}ZX_B\,(N_o X_B) = \frac{1}{2}ZN_o X_B^2 \tag{4.17b}$$

$$f_{AB} = \frac{1}{2}[ZX_B\,(N_o X_A) + (ZX_A)\,(N_o X_B)] = \frac{1}{2}ZN_o\,(2X_A X_B) \tag{4.17c}$$

$$= \frac{1}{2}ZN_o\,[2X_B\,(1 - X_B)]$$

where f_{AB} represents both A-B an B-A bonds, N_o denotes Avogadro's number, Z denotes the coordination number (CN) of an atom, and $1/2$ denotes the shared quantity in the A-B alloy. The bonding energy of mixing is the sum of the product $f_{ij}U_{ij}$ along with the mole fraction condition $X_A + X_B = 1$, where $i, j = A, B$

$$U_m = f_{AA}U_{AA} + f_{AB}U_{AB} + f_{BB}U_{BB} \tag{4.18a}$$

$$U_m = \frac{1}{2}ZN_o\left[(1 - X_B)^2\,U_{AA} + 2\Omega_m X_B\,(1 - X_B) + X_B^2 U_{BB}\right] \tag{4.18b}$$

and the interaction mixing parameter Ω_m, defined as an exchange binding energy term, takes the mathematical form

$$\Omega_m = \frac{1}{2} Z N_o \left[2 U_{AB} - (U_{AA} + U_{BB}) \right] \qquad (4.19)$$

This parameter is also known as the regular solution parameter and it can have three energy characters like

- $\Omega_m < 0$ for atomic attraction in regular solutions
- $\Omega_m > 0$ for atomic repulsion in regular solutions
- $\Omega_m = 0$ for ideal solutions independent of atomic arrangements

Further, the enthalpy of mixing ΔH_m and corresponding components are mathematically defined as

$$\Delta H_m = \Omega_m X_A X_B; \quad H_l = \Omega_l X_A X_B; \quad H_s = \Omega_s X_A X_B \qquad (4.20)$$

where the subscripts "l" and "s" stand for liquid and solid phases, respectively. Notice that regular solution parameter Ω_i is also a multiplier, as indicated by Eq. (4.20), that strongly depends on temperature and composition. Recall that $\Delta H_m > 0$, $\Delta H_m < 0$ or $\Delta H_m = 0$ (isothermal).

4.5.4   Gibbs Energy of Mixing

It is clear now that ΔG_m depends on the exchange binding energy term Ω_m and the disorder character of a regular solution through the entropy of mixing ΔS_m at constant pressure.

Inserting Eq. (4.20) into (4.14b) yields ΔG_m for a non-ideal solution

$$\Delta G_m = X_A G_A + X_B G_B + RT \left[X_A \ln(X_A) + X_B \ln(X_B) \right] + \Omega_m X_A X_B \qquad (4.21)$$

Now, Eq. (4.8b) along with (4.20) can be written as

$$\Delta G_m = \Delta H_m - T \Delta S_m \qquad (4.22a)$$

$$\Delta G_m = \Omega_m X_A X_B - T \Delta S_m \qquad (4.22b)$$

$$\Delta H_m = \Delta U_m + P \Delta V_m \qquad (4.22c)$$

According to the first law of thermodynamics for an isobaric process in a closed system, the enthalpy of mixing ΔH_m is defined by Eq. (4.22c) as an expression representing the heat of mixing at constant pressure. Specifically, the thermodynamic energy quantities ΔG_m, ΔH_m, and ΔS_m arise upon mixing at least two components at pressure P and temperature T. Thus, the thermodynamics of mixing indicates that there must be a change in internal energy ΔU_m and a change in volume ΔV_m of the mixture.

Furthermore, the Gibbs energy of the liquid and solid phases with reference state $G_A = G_B = 0$ and $X_A = 1 - X_B$ can be defined as $G_l, G_s = f(X_B, T)$ for plotting energy diagrams. This can be accomplished by letting $\Omega_m = 0$ in the liquid phase and $G_s = \Delta G_v + G_l$, where ΔG_v is defined by Eq. (4.8a).

For the sake of convenience, the Gibbs energy for the liquid and solid phases is written as

$$G_l = (1 - X_B) G_A + X_B G_B + \Omega_l X_B (1 - X_B) \tag{4.23a}$$
$$+ RT [X_B \ln X_B + (1 - X_B) \ln (1 - X_B)]$$

$$G_s = (1 - X_B) \Delta S_{fA} (T - T_{fA}) + X_B \Delta S_{fB} (T - T_{fB}) \tag{4.23b}$$
$$+ RT [X_B \ln X_B + (1 - X_B) \ln (1 - X_B)] + \Omega_s X_B (1 - X_B)$$

In particular, Ω_l, Ω_s denote the exchange binding energy terms for the regular solution and are essential for designing and optimizing individual thermodynamic systems using computational thermodynamics software packages, such as Thermo-Calc with reliable database. Nonetheless, if $\Omega_l, \Omega_s > 0$, then a repulsive interaction between atoms of components A and B occurs and if $\Omega_l, \Omega_s < 0$, an attractive interaction prevails in the mixture.

Accordingly, the Gibbs energy, as defined by Eq. (4.8b) or (4.22a), is directly related to the bonding energy and this may be the reason why it is commonly treated as the driving force for the solidification process of stable crystal structures at constant pressure.

For ideal solutions, $\Omega_m = 0$ and consequently, $\Delta H_m = 0$ (no evolution of heat). Conversely, regular solutions depend on the arrangement of atoms so that $\Delta H_m \neq 0$ and eventually, ΔG_m reaches a minimum. Moreover, Eq. (4.22c) gives $\Delta H_m \simeq \Delta U_m$ for condensed liquid or solid phases since $\Delta V_m \simeq 0$. Moreover, $\Delta G_m > 0$, $\Delta G_m < 0$ or $\Delta G_m = 0$ (equilibrium).

Example 4.2 Consider a binary A-B alloy in the solution state and assume that the enthalpy of mixing can be defined in terms of partial enthalpies (shown below) for components A and B, provided that $X_A + X_B = 1$ and

$$\Delta H_m = X_A \Delta H_A + X_B \Delta H_B \tag{E1}$$

with the linear functions $\Delta H_i = f(X_j)$ and $\Delta H_j = f(X_i)$

$$\Delta H_A = \Delta H_m + \frac{d(\Delta H_A)}{dX_A} X_B = \Delta H_m + \frac{d(\Delta H_A)}{dX_A}(1 - X_A) \tag{E2a}$$

$$\Delta H_B = \Delta H_m + \frac{d(\Delta H_A)}{dX_B} X_A = \Delta H_m + \frac{d(\Delta H_A)}{dX_B}(1 - X_B) \tag{E2b}$$

Based on this information, show that the partial enthalpy of mixing can be defined as $\Delta H_i = \Omega_m X_j^2$ and $\Delta H_j = \Omega_m X_i^2$, where $i = A$ and $j = B$.

Solution Start with Eq. (4.20) and take the derivatives with respect to X_A and X_B

$$\Delta H_m = \Omega_m X_A X_B = \Omega_m X_A (1 - X_A) = \Omega_m \left(X_A - X_A^2 \right) \tag{E3a}$$

$$\frac{d (\Delta H_m)}{d X_A} = \Omega_m (1 - 2X_A) \tag{E3b}$$

and

$$\Delta H_m = \Omega_m X_A X_B = \Omega_m (1 - X_B) X_B = \Omega_m \left(X_B - X_B^2 \right) \tag{E4a}$$

$$\frac{d (\Delta H_m)}{d X_B} = \Omega_m (1 - 2X_B) \tag{E4b}$$

Substituting Eqs. (E3b) and (E4b) into (E2a,b), respectively, yields the partial enthalpy of mixing for component A

$$\Delta H_A = \Delta H_m + X_B \frac{d (\Delta H_A)}{X_A} = \Omega_m X_A X_B + \Omega_m X_B (1 - 2X_A) \tag{E5a}$$

$$\Delta H_A = \Omega_m X_A (1 - X_A) + \Omega_m (1 - X_A) (1 - 2X_A) \tag{E5b}$$

$$= \Omega_m \left[X_A - X_A^2 + 1 - 2X_A - X_A + 2X_A^2 \right]$$

$$= \Omega_m (1 - X_A)^2 = \Omega_m X_B^2$$

Similarly,

$$\Delta H_B = \Delta H_m + X_A \frac{d (\Delta H_A)}{X_A} = \Omega_m X_A X_B + \Omega_m X_A (1 - 2X_B) \tag{E6a}$$

$$\Delta H_B = \Omega_m X_A (1 - X_A) + \Omega_m X_A [1 - 2 (1 - X_A)] \tag{E6b}$$

$$\Delta H_B = \Omega_m X_A^2 = \Omega_m (1 - X_B)^2$$

Notice that this thermodynamics of mixing example has only two components, which have the partial enthalpy of mixing connect through the real solution parameter Ω_m. Combining Eqs. (E6a) and (E6b) yields

$$\Omega_m = \frac{\Delta H_A}{X_B^2} = \frac{\Delta H_B}{X_A^2}$$

$$\Delta H_A = \Delta H_B \left(\frac{X_B}{X_A} \right)^2 = \Delta H_B \left(\frac{X_B}{1 - X_B} \right)^2$$

which are convenient and interchangeable equations.

4.6 Chemical Potential

In general, the chemical potential (μ) of a component is the rate of change of the Gibbs energy change with respect to the number of particles in solution ($\mu = \partial \Delta G / \partial N$). It measures the increment of the Gibbs energy (G) at a temperature T. Thus, the chemical potential derives from the changes in Gibbs energy and it is an energy that can be either absorbed or released during phase transformation.

Consider a binary A-B system undergoing solidification. Then, μ and G become directly related to the interactions of nearest-neighbors like A-A, B-B and unlike A-B atom pairs at relatively high temperatures. Hence, the state function of interest for $i = A, B$ components is the chemical potential of mixing

$$\mu_i = \mu_i^o + RT \ln(\gamma_i) \tag{4.24}$$

where μ_i^o denotes the chemical potential for a pure component i and γ_i denotes the activity coefficient related to the interatomic action between A and B atoms forming bonds. If $\gamma_i = 1$, the solution is ideal and if $\gamma_i > 1$, the solution is non-ideal, indicating a deviation from the ideal state.

Furthermore, Eq. (4.24) represents the partial Gibbs energy of mixing with $\mu_i = G_m$, which is related to the component chemical potentials (μ_A, μ_B). Thus,

$$G_m = X_A \mu_A + X_B \mu_B = (1 - X_B)\mu_A + X_B \mu_B \tag{4.25a}$$

$$\frac{\partial G_m}{\partial X_B} = \mu_B - \mu_A \tag{4.25b}$$

Combining Eqs. (4.25a,b) yields the chemical potentials representing the partial molar Gibbs energies

$$\mu_A = G_A = G_m - X_B \left(\frac{\partial G_m}{\partial X_B} \right) \tag{4.26a}$$

$$\mu_B = G_B = G_m + (1 - X_B) \left(\frac{\partial G_m}{\partial X_B} \right) \tag{4.26b}$$

and in general, the chemical potential function $\mu_i = f(a_i)$, where a_i denotes the chemical activity of a component $i = A, B$, can be defined as

$$\mu_i = G_i + RT \ln(a_i) \tag{4.27}$$

In consequence, the classical thermodynamical activity (a_i) equation describing any deviation from the ideal state is the Henry's law written as

$$a_i = \gamma_i X_i \tag{4.28}$$

Note that the variables in Eq. (4.28) are dimensionless and that $0 < X_i < 1$ in regular solutions. These are treated in a later section using the Gibbs–Duhem equation. Nonetheless, if $\gamma_i = 1$, then $a_i = X_i$ for a metallic ideal liquid solution. In general, the activity coefficient γ_i in Eq. (4.28) is the factor that determines the degree of deviation from the thermodynamically ideal behavior in a mixture of chemical substances, such as alloying elements. Moreover, Eq. (4.28) directly implies that the activity a_i is a measure of the effective concentration (X_i) of a species under non-ideal conditions. See Appendix 4A for activity calculations.

4.7 The Interaction Parameter

It is now convenient to define all the mixing functions in Eq. (4.22a) for a binary regular solution as

$$\Delta G_m = RT \, [X_A \ln (a_A) + X_B \ln (a_B)] \tag{4.29a}$$

$$\Delta H_m = RT \, [X_A \ln (\gamma_A) + X_B \ln (\gamma_B)] \tag{4.29b}$$

$$\Delta S_m = -RT \, [X_A \ln (X_A) + X_B \ln (X_B)] \tag{4.29c}$$

Combining Eqs. (4.20) and (4.29b) yields

$$\ln (\gamma_A) = \frac{\Omega_m}{RT} X_B^2 = \frac{\Omega_m}{RT} (1 - X_A)^2 = \omega (T) (1 - X_A)^2 \tag{4.30a}$$

$$\ln (\gamma_B) = \frac{\Omega_m}{RT} X_A^2 = \frac{\Omega_m}{RT} (1 - X_B)^2 = \omega (T) (1 - X_B)^2 \tag{4.30b}$$

$$\Omega_m = \frac{RT \ln (\gamma_B)}{X_A^2} = \frac{RT \ln (\gamma_B)}{(1 - X_B)^2} = \text{Constant at } T \tag{4.30c}$$

which is compared to Krupkowski's formalism [5] with $\omega (T) = \Omega_m / RT$ and

$$\ln (\gamma_B) = \omega (T) \left[\frac{1}{m-1} - \left(\frac{m}{m-1} \right) (1 - X_A)^{m-1} + (1 - X_A)^m \right] \tag{4.31}$$

If $m = 2$, then Eq. (4.31) reduces to (4.30b).

Example 4.3 Consider the activity coefficient data from electrochemical measurements at 527 °C for Zn-Cd alloys, where A = Cd and B = Zn. The goal is to verify that Eq. (4.31) gives a constant value for a regular solution and analyze ΔG_m, ΔH_m, and ΔS_m. Given data [6, p. 150]:

X_{Cd}	0.2	0.3	0.4	0.5
γ_{Cd}	2.153	1.817	1.544	1.352

Solution From Eq. (4.30a),

$$RT \ln (\gamma_{Cd}) = \Omega_m X_{Zn}^2 \tag{E1a}$$

$$\frac{RT \ln (\gamma_{Cd})}{X_{Zn}^2} = \frac{RT \ln (\gamma_{Cd})}{(1 - X_{Cd})^2} = \Omega_m \tag{E1b}$$

Calculations to the nearest integer values for Ω_m in J/mol units along with $R = 8.314$ J/mol K and temperature $T = 800$ K yield

X_{Zn}	X_{Cd}	$(1 - X_{Cd})^2$	γ_{Cd}	Ω_m
0.8	0.2	0.64	2.153	8000
0.7	0.3	0.49	1.817	8106
0.6	0.4	0.36	1.544	8025
0.5	0.5	0.25	1.352	8024

Note that Ω_m is virtually a constant at the given mole fractions of X_{Cd}. The average value is $\Omega_m = 8039$ J/mol. Thus, Eqs. (4.20) and (4.29) yield

$$\Delta H_m = 8039 X_A X_B \tag{E2a}$$

$$\Delta G_m = 8039 X_A X_B - T \Delta S_m \tag{E2b}$$

$$T \Delta S_m = 8039 X_A X_B - \Delta G_m \tag{E2c}$$

$$= 8039 X_A X_B - RT [X_A \ln (a_A) + X_B \ln (a_B)] \tag{E2d}$$

Therefore, these functions of mixing depend on $\Omega_m = 8039$ J/mol and the solution is regular. Accordingly, Eq. (E1b) can be used to determine γ_{Zn} and γ_{Cd}. Thus,

$$\gamma_{Zn} = \exp \left[\frac{\Omega_m (1 - X_{Zn})^2}{RT} \right] = \exp \left[1.2087 (X_{Cd})^2 \right] \tag{E3a}$$

$$\gamma_{Cd} = \exp \left[\frac{\Omega_m (1 - X_{Cd})^2}{RT} \right] = \exp \left[1.2087 (1 - X_{Cd})^2 \right] \tag{E3b}$$

Polynomial fit on the given data yields

$$\gamma_{Cd} = 2.942 - 4.155 X_{Cd} + 0.45 X_{Cd}^2 + 3 X_{Cd}^3 \quad \text{with} \quad R^2 = 0.992 \tag{E4}$$

Plotting Eqs. (E3a,b) along with the given data points yields
From Eqs. (4.28) and (E3), the activity expressions for Zn and Cd are

$$a_{Zn} = X_{Zn} \exp \left[1.2087 (X_{Cd})^2 \right] \tag{E5a}$$

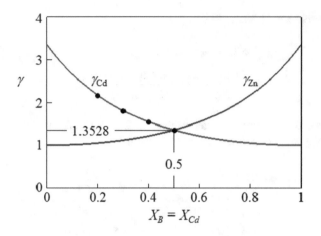

$$a_{Cd} = X_{Cd} \exp\left[1.2087\,(1 - X_{Cd})^2\right] \tag{E5b}$$

Example 4.4 Consider mixing a significant amount of Cu and Ni atoms to produce Cu-Ni alloys with $0 < X_B = X_{Ni} < 1$. Here, $A = Cu$ and $B = Ni$ so that A-B $= Cu$-Ni. (a) Derive Eqs. (4.17a,b,c) using the factor f_{ij} for A-A, B-B, and A-B type of bonds and (b) analyze Eqs. (4.18b) and (4.19) for individual pair of atoms before mixing. Calculate (c) Ω_{AA}, Ω_{BB}, and Ω_{AB} using the available latent heat of fusion for copper $\Delta H_{f,Cu} = 1628\,\text{J/cm}^3$ and for nickel $\Delta H_{f,Cu} = 2756\,\text{J/cm}^3$, and (d) Ω_m and U_{AB} after mixing Cu and Ni atoms. The activity coefficient equations (curve fitting expressions) at $870\,\text{K} \le T \le 1280\,\text{K}$ interval, as reported by Tanegashima et al. [7] using electrochemical and comparative studies, are

$$\ln(\gamma_{Cu}) = 0.345 + \left(1.33 \times 10^3\right)(1/T) \tag{E1a}$$

$$\ln(\gamma_{Ni}) = 0.691 + (604)(1/T) \tag{E1b}$$

(c) Find $\Omega_m = f(X_B)$ and $U_{AB} = g(X_B)$ functions for $0 \le X_B = X_{Ni} \le 0.5$ at $T = 870\,\text{K}$ and $1280\,\text{K}$ temperatures.

Solution

(a) Before mixing: For a pure metal, Eqs. (4.18b) and (4.19) with $i = A, B$ can be used to deduce the following:

$$U_{ii} \rightarrow \Delta H_{fi} \quad \& \quad \Omega_{ii} \rightarrow U_{ii} < 0 \quad \text{for } X_i = 1 \tag{E2a,b}$$

$$\Delta H_{fi} \rightarrow \Omega_{ii} < 0 \tag{E2c}$$

Thus,

$$\Delta H_{f,A} = \frac{1}{2} Z N_o U_{AA} \quad \text{and} \quad \Omega_{AA} = -\frac{1}{2} Z N_o U_{AA} \qquad \text{(E3a,b)}$$

$$\Delta H_{f,A} = -\Omega_{AA} \quad \& \quad U_{AA} = \frac{2\Delta H_{f,A}}{Z N_o} \qquad \text{(E3b,c)}$$

and

$$\Delta H_{f,B} = \frac{1}{2} Z N_o U_{BB} \quad \text{and} \quad \Omega_{BB} = -\frac{1}{2} Z N_o U_{BB} \qquad \text{(E4a,b)}$$

$$\Delta H_{f,B} = -\Omega_{BB} \quad \& \quad U_{BB} = -\frac{2\Delta H_{f,B}}{Z N_o} \qquad \text{(E4c)}$$

In the solid state, both Cu and Ni atoms form an FCC crystal structure with a coordination number of $Z = 12$ and different atomic radii; that is, $R_{Cu} = 0.128$ nm and $R_{Ni} = 0.125$ nm. Recall that the Avogadro's number (physical property) is $N_o = 6.022 \times 10^{23}$ atoms. Thus, Eqs. (E3c) and (E4c) yield the sought internal energies

$$U_{AA} = U_{Cu} = -\frac{2\left(1628 \text{ J/cm}^3\right)}{(12)\left(6.022 \times 10^{23}\right)} = -4.51 \times 10^{-22} \text{ J/cm}^3$$

$$U_{BB} = U_{Ni} = -\frac{2\left(2756 \text{ J/cm}^3\right)}{(12)\left(6.022 \times 10^{23}\right)} = -7.63 \times 10^{-22} \text{ J/cm}^3$$

and

$$\Omega_{AA} = \Omega_{Cu} = -\Delta H_{f,A} = \Delta H_{f,Cu} = -1628 \text{ J/cm}^3$$

$$\Omega_{BB} = \Omega_{Ni} = -\Delta H_{f,B} = \Delta H_{f,Ni} = -2756 \text{ J/cm}^3$$

Unit conversion: Multiply Ω_{Cu}, Ω_{Ni} by

$$(A_w/\rho)_{Cu} = (63.55 \text{ g/mol}) / \left(8.93 \text{ g/cm}^3\right) = 7.1165 \text{ cm}^3/\text{mol}$$

$$(A_w/\rho)_{Ni} = (58.69 \text{ g/mol}) / \left(8.80 \text{ g/cm}^3\right) = 6.6693 \text{ cm}^3/\text{mol}$$

so that

$$\Omega_{AA} = -11,586 \text{ J/mol}$$

$$\Omega_{BB} = -18,381 \text{ J/mol}$$

$$\frac{\Omega_{AA} + \Omega_{ABB}}{2} = -14,984 \text{ J/mol}$$

In fact, J/mol are most common units.

(b)   After mixing Cu and Ni: From Eq. (4.31b),

$$\Omega_m = \frac{RT \ln{(\gamma_B)}}{(1 - X_B)^2} = \frac{RT \ln{(\gamma_{Ni})}}{(1 - X_B)^2} \tag{E5a}$$

$$\Omega_m = \frac{RT}{(1 - X_B)^2} [0.691 + (604)(1/T)] \tag{E5b}$$

Equating Eqs. (E5a) and (4.19), and solving for U_{AB} yields

$$U_{AB} = \frac{1}{2}(U_{AA} + U_{BB}) + \frac{RT \ln{(\gamma_B)}}{ZN_o(1 - X_B)^2} \tag{E6}$$

By inspection, Eq. (E6) can be approximated by letting $N_o(1 - X_B)^2 >> 0$ and $RT \ln{(\gamma_B)}/[ZN_o(1 - X_B)^2] \to 0$.
     Substituting Eq. (E1b) into (E6) gives

$$U_{AB} = U_{Cu\text{-}Ni} = \frac{U_{AA} + U_{BB}}{2} + \frac{RT}{ZN_o(1 - X_B)^2}[0.691 + (604)(1/T)] \tag{E7}$$

Substitute the values of $R = 8.314 \text{ J/mol K}$, $Z = 12$ for Cu and Ni FCC structures, $N_o = 6.022 \times 10^{23}$ atoms/mol and $(\Omega_{AA} + \Omega_{BB})/2$ into Eqs. (E5b) and (E7) to get Ω_m and U_{AB} in J/mol units. Thus,

$$\Omega_m = \frac{8.314T}{(1 - X_B)^2}[0.691 + (604)(1/T)] \tag{E8a}$$

$$U_{AB} = -14,984 \text{ J/mol} + \frac{1.1509 \times 10^{-24}}{(1 - X_B)^2}[0.691 + (604)(1/T)] \tag{E8b}$$

Substituting the temperatures 870 and 1280 K along with $X_B = X_{Ni}$ yields two expressions for the interaction parameter

$$\Omega_m = \frac{10,020 \text{ J/mol}}{(1 - X_{Ni})^2} \quad \text{at } T = 870 \text{ K} \tag{E9a}$$

$$\Omega_m = \frac{12,375 \text{ J/mol}}{(1 - X_{Ni})^2} \quad \text{at } T = 1280 \text{ K} \tag{E9b}$$

and the sought functions for $U_{AB} = U_{Cu\text{-}Ni}$ are

$$U_{Cu-Ni} = (-14,984 \text{ J/mol}) + \frac{1.5943 \times 10^{-24} \text{ J/mol}}{(1 - X_{Ni})^2} \quad \text{at } T = 870 \text{ K} \tag{E10a}$$

$$U_{Cu-Ni} = (-14{,}984 \text{ J/mol}) + \frac{1.3384 \times 10^{-24} \text{ J/mol}}{(1 - X_{Ni})^2} \text{ at } T = 1280 \text{ K}$$

$$(\text{E10b})$$

In summary, Eq. (E9) clearly shows that $\Omega_m = f(X_{Ni})$ and Eq. (E10) yields $U_{Cu-Ni} \simeq -14{,}984$ J/mol for $0 < X_{Ni} < 1$. The reader is encouraged to plot Eq. (E9) for a graphical analysis of the function $\Omega_m = f(X_{Ni})$ at the given temperatures.

4.8   The Gibbs–Duhem Equation

In general, the thermodynamics of solutions can be described by the Gibbs–Duhem equation related to the energy of mixing Z_m, such as the Gibbs energy (G_m), enthalpy (H_m), entropy (S_m), and potential energy (μ_m). For an A-B solution, the Gibbs energy of mixing $(Z_m = G_m)$ is related to partial molar energies (G_A, G_B) and mole fractions (X_A, X_B)

$$G_m = X_A G_A + X_B G_B \tag{4.32a}$$

$$dG_m = X_A dG_A + G_A dX_A + X_B dG_B + G_B dX_B \tag{4.32b}$$

If $G_m = \sum X_i G_i$ for a binary A-B solution with $i = A, B$ components, then

$$dG_m = G_A dX_A + G_B dX_B \tag{4.33}$$

Combining Eqs. (4.32b) and (4.33) yields the Gibbs–Duhem equation

$$X_A dG_A + X_B dG_B = 0 \tag{4.34a}$$

$$dG_A = -\frac{X_B}{X_A} dG_B \tag{4.34b}$$

Fundamentally, the chemical potential μ_i is a measure of the driving force for a substance to undergo physical or chemical change within a thermodynamic system. Mathematically, the chemical potential is defined as the slope of the partial molar Gibbs state function $G = f(X_i)$ at constant temperature T and pressure P with $i = 1, 2, 3, \ldots$ components. Thus,

$$\mu_i = \left(\frac{\partial G_i}{\partial X_i} \right)_{T,P} \tag{4.35}$$

For a binary A-B system along with $X_1 = X_A$, $X_2 = X_B$, $\mu_1 = \mu_A$, $\mu_2 = \mu_B$, the total Gibbs energy and its infinitesimally change are, respectively

$$G = X_1\mu_1 + X_2\mu_2 \tag{4.36a}$$

$$dG = X_1 d\mu_1 + \mu_1 dX_1 + X_2 d\mu_2 + \mu_2 dX_2 \tag{4.36b}$$

$$dG = X_1 d\mu_1 + X_2 d\mu_2 \quad \text{(at constant } X_1, X_2\text{)} \tag{4.36c}$$

Combining Eqs. (4.36b) and (4.36c) at constant T and P, the Gibbs–Duhem equation can be written as

$$X_A d\mu_A + X_B d\mu_B = 0 \tag{4.37a}$$

$$d\mu_A = -\frac{X_B}{X_A} d\mu_B \tag{4.37b}$$

$$\mu_A = -\int \frac{X_B}{X_A} d\mu_B \tag{4.37c}$$

For a multicomponent system,

$$\sum_{i=1}^{N} X_i d\mu_i = 0 \tag{4.38a}$$

$$X_1 d\mu_1 + X_2 d\mu_2 + \ldots.. = 0 \tag{4.38b}$$

which is applied to binary A-B solutions throughout this section.

4.8.1   The Activity and Activity Coefficient Concepts

Defining the chemical potential in terms of the activity of components (a_i) provides (see Appendix 4A)

$$\mu_i = RT \ln (a_i) \tag{4.39a}$$

$$d\mu_i = RT d \ln (a_i) \tag{4.39b}$$

For a binary A-B solution with $i = A, B$ as the metal components, combine Eqs. (4.37b) and (4.39b) to get the Gibbs–Duhem equation in terms of activities (a_i)

$$d \ln (a_A) = -\frac{X_B}{X_A} d \ln (a_B) \tag{4.40a}$$

$$\int_0^{0<X_A<1} \ln (a_A) = -\int_0^{0<X_B<1} \frac{X_B}{X_A} d \ln (a_B) \tag{4.40b}$$

$$\ln(a_A) - \ln(1) = -\int_0^{0<X_B<1} \frac{X_B}{X_A} d\ln(a_B) \tag{4.40c}$$

$$\ln(a_A) = -\int_0^{0<X_B<1} \frac{X_B}{X_A} d\ln(a_B) \tag{4.40d}$$

This integral can be solved graphically or analytically once an equation for $\ln(a_B) = f(X_A)$ is known. Substituting the activity $a_i = \gamma_i X_i$, Eq. (4.28), into (4.40) yields the Gibbs–Duhem equation in another mathematical form

$$d\ln(\gamma_A X_A) = -\frac{X_B}{X_A} d\ln(\gamma_B X_B) \tag{4.41a}$$

$$d\ln(\gamma_A) + d\ln(X_A) = -\frac{X_B}{X_A}[d\ln(\gamma_B) + d\ln(X_B)] \tag{4.41b}$$

$$d\ln(\gamma_A) + \frac{dX_A}{X_A} = -\frac{X_B}{X_A}\left[d\ln(\gamma_B) + \frac{dX_B}{X_B}\right] \tag{4.41c}$$

$$d\ln(\gamma_A) + \frac{dX_A}{X_A} = -\frac{X_B}{X_A}d\ln(\gamma_B) - \frac{X_B}{X_A}\frac{dX_B}{X_B} \tag{4.41d}$$

$$d\ln(\gamma_A) + \frac{dX_A}{X_A} = -\frac{X_B}{X_A}d\ln(\gamma_B) - \frac{dX_B}{X_A} \tag{4.41e}$$

This integral defines the area under the curve described by the function $X_B/X_A = f(a_B)$. Subsequently, measuring this area solves the integral graphically. However, curve fitting the experimental data and obtaining a curve fitting equation for $X_B/X_A = f(a_B)$ also solves the integral with proper integral limits.

Substituting the activity $a_i = \gamma_i X_i$, Eq. (4.28), into (4.40d) yields the Gibbs–Duhem equation in another mathematical form

$$d\ln(\gamma_A) = -\frac{X_B}{X_A}d\ln(\gamma_B) \tag{4.42a}$$

$$\ln(\gamma_A) = -\int_0^{0<X_B<1} \frac{X_B}{X_A} d\ln(\gamma_B) \tag{4.42b}$$

This integral can be solved graphically or analytically once an equation for $\ln(\gamma_B) = f(X_A)$ is known.

4.8.2 The Alpha Function

In addition, determining the activity or activity coefficient of a component A knowing the value of the component B in a binary A-B solution at relatively high temperature T and pressure P is a common practice considered by researchers using the Gibbs–Duhem equation, such as Eq. (4.40d) or (4.42b). It is recognized by the scientific and academic communities that these equations introduce uncertainty in calculating a_A or γ_A from experimentally determined a_B or γ_B due to the asymptotic behavior of these variables. In order to improve the accuracy of a_A or γ_A, an α-function proposed by Darken-Gurry [8, p. 264] is used to modify the conventional Gibbs–Duhem equation.

Mathematically, manipulating Eq. (4.30) yields

$$\ln(\gamma_A) = \frac{\Omega_m X_B^2}{RT} = \frac{\Omega_m X_B (1 - X_A)}{RT} \tag{4.43a}$$

$$\ln(\gamma_A) = -\frac{X_A X_B \Omega_m}{RT} + \frac{X_B \Omega_m}{RT} \tag{4.43b}$$

$$\ln(\gamma_A) = -\frac{X_A X_B \Omega_m}{RT} + \frac{\Omega_m}{RT}(1 - X_A) \tag{4.43c}$$

$$\ln(\gamma_A) = -\frac{X_A X_B \Omega_m}{RT} - \frac{\Omega_m}{RT}(X_A - 1) \tag{4.43d}$$

$$\ln(\gamma_A) = -\frac{X_A X_B \Omega_m}{RT} - \int_1^{X_A < 1} \frac{\Omega_m}{RT} dX_A \tag{4.43e}$$

Again, from Eq. (4.30),

$$\ln(\gamma_B) = \frac{\Omega_m X_A^2}{RT} \tag{4.44a}$$

$$\Omega_m = \frac{RT \ln(\gamma_B)}{X_A^2} \tag{4.44b}$$

Insert Eq. (4.44b) into (4.43e) to get the Gibbs–Duhem equation in the form

$$\ln(\gamma_A) = -\frac{X_A X_B \Omega_m}{RT} - \int_1^{X_A < 1} \frac{\Omega_m}{RT} dX_A \tag{4.45a}$$

$$\ln(\gamma_A) = -\frac{X_A X_B}{RT} \frac{RT \ln(\gamma_B)}{X_A^2} - \int_1^{X_A < 1} \frac{1}{RT} \frac{RT \ln(\gamma_B)}{X_A^2} dX_A \tag{4.45b}$$

$$\ln(\gamma_A) = -\frac{X_A X_B \ln(\gamma_B)}{X_A^2} - \int_1^{0 < X_A < 1} \frac{\ln(\gamma_B)}{X_A^2} dX_A \tag{4.45c}$$

This integral can be solved graphically or analytically once an equation for $\ln(\gamma_B) = f(X_A)$ is known. Moreover, Eq. (4.45c) can be modified by introducing the alpha function (α-function) defined by Darken-Gurry as [8, Problem 10-8, p. 512]

$$\alpha_A = \frac{\ln(\gamma_A)}{X_B^2} \tag{4.46a}$$

$$\alpha_B = \frac{\ln(\gamma_B)}{X_A^2} \tag{4.46b}$$

which are alternative terms for determining accurate activity coefficients.

Combining Eqs. (4.45c) and (4.46b) gives another Gibbs–Duhem equation in terms of the alpha function (α-function)

$$\ln(\gamma_A) = -\alpha_B X_A X_B - \int_1^{0 < X_A < 1} \alpha_B \, dX_A \tag{4.47}$$

Furthermore, combining Eqs. (4.30) and (4.46) yields an expression relating the α-function for components A and B, and the interaction parameter Ω_m at a temperature T

$$\alpha_A = \alpha_B = \frac{\Omega_m}{RT} \tag{4.48}$$

An example can reveal some characteristics of the above Gibbs–Duhem equations and their particular accuracy for determining the activity (a_A) of the solvent A knowing the activity (a_B) of the solute B in a binary A-B solution at a relatively high temperature T and fixed pressure P.

Example 4.5 Consider the Kubaschewski and Alcock [9] mole fraction and pressure data set for Cu-Zn alloy solutions at $T = 1060\,°C$. Based on this information, (a) analytically calculate the activity of copper ($a_A = a_{Cu}$) in an alloy containing $X_{Cu} = 0.60$ and $X_{Zn} = 0.40$. Use the available Gibbs–Duhem equations for the four cases shown in the figures below. (b) Find the equation for the area under the curve in Figs. b,c to determine a_{Cu}. In this case, the approach to an analytical solution of the Gibbs–Duhem equation is via a curve fitting expression relating the activity a_{Cu} to mole fraction (composition) X_{Cu} or X_{Zn}. The data set is

X_{Zn}	1	0.45	0.30	0.20	0.15	0.10	0.05
P_{Zn} (atm)	4	1.28	0.60	0.24	0.12	0.0592	0.0289

where $P_{o,Zn} = 4\,atm$ (reference pressure) and $a_{Zn} = P_{Zn}/P_{o,Zn}$. The given plots are conveniently constructed beforehand since they are the basis for calculating the activity of copper at $X_{Zn} = 0.40$.

Solution The table below contains calculated data for plotting relevant relationships associated with the analysis of the Gibbs–Duhem equations cited previously.

(a) **Method A: The interaction parameter (Fig. a):** First of all, the pressure P_{Zn} and activity a_{Zn} trends for zinc are described by the curve fitting equations

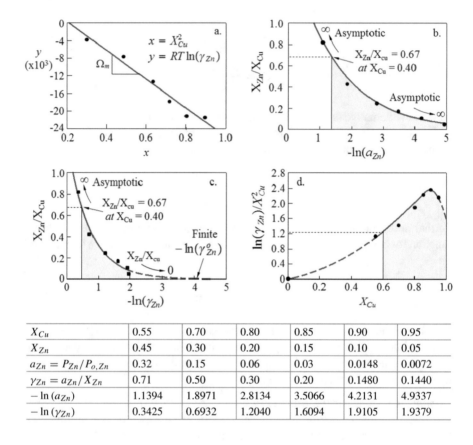

X_{Cu}	0.55	0.70	0.80	0.85	0.90	0.95
X_{Zn}	0.45	0.30	0.20	0.15	0.10	0.05
$a_{Zn} = P_{Zn}/P_{o,Zn}$	0.32	0.15	0.06	0.03	0.0148	0.0072
$\gamma_{Zn} = a_{Zn}/X_{Zn}$	0.71	0.50	0.30	0.20	0.1480	0.1440
$-\ln(a_{Zn})$	1.1394	1.8971	2.8134	3.5066	4.2131	4.9337
$-\ln(\gamma_{Zn})$	0.3425	0.6932	1.2040	1.6094	1.9105	1.9379

(along with the correlation coefficients R^2) shown below. From the activity equation, one obtains $\ln(a_{Zn}) = f(X_{Zn})$ and $d\ln(a_{Zn}) = g(X_{Zn})$ which are subsequently used to calculate the activity of copper, a_{Cu}, in the Cu-Zn alloy solution containing $X_{Zn} = 0.40$. Thus,

$$P_{Zn} = 4.2601\,(X_{Zn})^{1.7526} \text{ with } R^2 = 0.9862 \qquad (E1a)$$

$$a_{Zn} = 1.0659\,(X_{Zn})^{1.7534} \text{ with } R^2 = 0.9863 \qquad (E1b)$$

$$a_{Zn} = 1.0659\,(0.40)^{1.7534} = 0.21378 \qquad (E1c)$$

$$\ln(a_{Zn}) = \ln\left[1.0659\,(X_{Zn})^{1.7534}\right] \qquad (E1d)$$

$$d\ln(a_{Zn}) = \frac{1.7534}{X_{Zn}}\,dX_{Zn} \qquad (E1e)$$

From Eq.(4.30),

$$RT\ln(\gamma_{Cu}) = \Omega_m X_{Zn}^2 \qquad (E2a)$$

$$RT \ln (\gamma_{Zn}) = \Omega_m X_{Cu}^2 = \Omega_m (1 - X_{Zn})^2 \tag{E2b}$$

Plotting $RT \ln (\gamma_{Zn}) = f\left(X_{Cu}^2\right)$ or $y = f(x)$ yields Fig. a along with data points and the interaction parameter Ω_m as the slope of the straight line

$$y = -32{,}729x + 6878.7 \ \textit{with } R^2 = 0.9691 \tag{E3a}$$

$$\Omega_m = -32{,}729 \ \text{J/mol} \ \ \textit{(slope)} \tag{E3b}$$

From Eq. (E2a) at $X_{Cu} = 0.60$, $X_{Zn} = 0.40$, and $T = 1333$ K, the activity coefficient γ_{Cu} and the activity a_{Cu} of copper follow, respectively

$$RT \ln (\gamma_{Cu}) = \Omega_m X_{Zn}^2 \tag{E4a}$$

$$\ln (\gamma_{Cu}) = \frac{\Omega_m X_{Zn}^2}{RT} = \frac{(-32{,}729)\,(0.40)^2}{(8.314)\,(1333)} = -0.47251 \tag{E4b}$$

$$\gamma_{Cu} = \exp (-0.47251) = 0.62344 \tag{E4c}$$

$$a_{Cu} = \gamma_{Cu} X_{Cu} = (0.62344)\,(0.60) = 0.37406 \tag{E4d}$$

which can be taken as a reference activity of copper in the alloy containing $X_{Zn} = 0.40$.

Method B: Interaction Parameter and Activity Combining Eqs. (E2a,b) along with $a_{Cu} = \gamma_{Cu} X_{Cu}$ gives the activity for copper at composition $X_{Cu} = 0.60$ and $X_{Zn} = 0.40$

$$\ln (a_{Cu}) = \ln (X_{Cu}) + \frac{X_{Zn}^2}{X_{Cu}^2} \ln (a_{Zn}/X_{Zn}) \tag{E5a}$$

$$\ln (a_{Cu}) = \ln (0.60) + \left(\frac{0.40}{0.60}\right)^2 \ln (0.21378/0.40) \tag{E5b}$$

$$\ln (a_{Cu}) = -0.78928 \tag{E5c}$$

$$a_{Cu} = \exp (-0.78928) = 0.45417 \tag{E5d}$$

Notice that $a_{Cu} = 0.45417$ (E5d) is slightly higher than $a_{Cu} = 0.37406$ (E4d). At this moment, there is no experimental data for copper to compare to calculated a_{Cu} values.

Method C: Gibbs–Duhem Equation and Activity (Fig. b) Combining Eqs. (4.40d) and (E1e) with $a_A = a_{Cu}$ and $a_B = a_{Zn}$, the activity of copper is defined by

$$\ln (a_{Cu}) = - \int_0^{\ln(a_{Zn}) \text{ at } X_{Zn}=0.40} \frac{X_{Zn}}{X_{Cu}} d \ln (a_{Zn}) \tag{E6}$$

Notice that the $X_{Zn}/X_{Cu} = f[\ln(a_{Zn})]$ function shows an asymptotic behavior to both x and y axes. This implies that

$$a_{Zn} = \gamma_{Zn}X_{Zn} = 0, \ \ln(a_{Zn}) \to \infty \ at \ X_{Zn} = 0 \quad (x\text{-}axis) \tag{E7a}$$

$$\frac{X_{Zn}}{X_{Cu}} \to \infty \ as \ X_{Cu} \to 0 \ and \ X_{Zn} \to 1 \qquad (y\text{-}axis) \tag{E7b}$$

This is an approximation method since $\ln(a_{Zn})$ and X_{Zn}/X_{Cu} do not reach finite values, which are essential for evaluating the shaded area as the graphical solution of the integral. Consequently, this method is not recommended.

Instead, an analytical solution is considered using the corresponding curve fitting equation.

$$\ln(a_{Cu}) = -\int_0^{X_{Zn}=0.40} \frac{X_{Zn}}{X_{Cu}} d\ln(a_{Zn}) \tag{E8a}$$

$$d\ln(a_{Zn}) = \frac{1.7534}{X_{Zn}} dX_{Zn} \tag{E8b}$$

$$\ln(a_{Cu}) = -1.7534\int_0^{X_{Zn}=0.40} \frac{dX_{Zn}}{X_{Cu}} = -1.7534\int_0^{X_{Zn}=0.40} \frac{dX_{Zn}}{1-X_{Zn}} \tag{E8c}$$

$$\ln(a_{Cu}) = -1.7534\int_0^{X_{Zn}=0.40} \frac{dX_{Zn}}{1-X_{Zn}} = -0.89568 \ (area) \tag{E8d}$$

$$a_{Cu} = \exp(-0.89568) = 0.40833 \tag{E8e}$$

Note that

$$a_{Cu} = 0.37406[(E4d)] < a_{Cu} = 0.40833[(E8e)] < a_{Cu} = 0.45417[(E5d)] \tag{E9}$$

Method D: Gibbs–Duhem Equation and Activity Coefficient (Fig. c) The $X_{Zn}/X_{Cu} = f[-\ln(\gamma_{Zn})]$ function exhibits a finite value to the x-axis, but it shows an asymptotic behavior to the y-axis. Consequently, this method is recommended despite the fact that

$$\frac{X_{Zn}}{X_{Cu}} \to 0 \ and \ \ln(\gamma_{Zn}) \to \ln\left(\gamma_{Zn}^o\right) \ as \ X_{Cu} \to 1, \ X_{Zn} \to 0 \tag{E10a}$$

$$\frac{X_{Zn}}{X_{Cu}} \to \infty \ as \ X_{Cu} \to 0 \ and \ X_{Zn} \to 1 \tag{E10b}$$

Therefore, this is still an approximation method related to a higher degree of accuracy since $X_{Zn}/X_{Cu} \rightarrow 0$ and $-\ln(\gamma_{Zn}) \rightarrow -\ln(\gamma_{Zn}^o)$ reaches finite values. Nonetheless, a_{Cu} is calculated as shown below.

From Eq. (4.42b) with $\gamma_A = \gamma_{Cu}$ and $\gamma_B = \gamma_{Zn}$, the activity coefficient and the corresponding curve fitting equation are

$$RT \ln(\gamma_{Cu}) = \Omega_m X_{Zn}^2 \tag{E11a}$$

$$RT \ln(\gamma_{Zn}) = \Omega_m X_{Cu}^2 = \Omega_m (1 - X_{Zn})^2 \tag{E11b}$$

$$\ln(\gamma_{Zn}) = \frac{\Omega_m (1 - X_{Zn})^2}{RT} \tag{E11c}$$

$$d\ln(\gamma_{Zn}) = \frac{2\Omega_m (X_{Zn} - 1)}{RT} dX_{Zn} \tag{E11d}$$

Then,

$$\ln(\gamma_{Cu}) = -\int_0^{X_{Zn}=0.40} \frac{X_{Zn}}{X_{Cu}} d\ln(\gamma_{Zn}) = -\int_0^{X_{Zn}=0.40} \frac{X_{Zn}}{(1 - X_{Zn})} d\ln(\gamma_{Zn}) \tag{E12a}$$

$$\ln(\gamma_{Cu}) = -\int_0^{X_{Zn}=0.40} \frac{X_{Zn}}{(1 - X_{Zn})} \frac{2\Omega_m (X_{Zn} - 1)}{RT} dX_{Zn} \tag{E12b}$$

$$\ln(\gamma_{Cu}) = \int_0^{X_{Zn}=0.40} \frac{2\Omega_m X_{Zn}}{RT} dX_{Zn} = \int_0^{0.40} \frac{2(-32{,}729) X_{Zn}}{(8.314)(1333)} dX_{Zn} \tag{E12c}$$

$$\ln(\gamma_{Cu}) = -0.47251 \tag{E12d}$$

$$\gamma_{Cu} = \exp(-0.47251) = 0.62344 \tag{E12e}$$

$$a_{Cu} = \gamma_{Cu} X_{Cu} = (0.62344)(0.60) = 0.37406 \tag{E12f}$$

Method E: Gibbs–Duhem Equation and α-Function (Fig. d)   Figure d shows finite values at $X_{Cu} = 0$ and $X_{Cu} = 1$, leading to a more accurate result for a_{Cu}. According to the alloy composition in this example, the integration limits are indicated in Fig. d. From Eq. (4.47) with $\gamma_A = \gamma_{Cu}$ and $\gamma_B = \gamma_{Zn}$, the activity coefficient and related limits of integration are written as

$$\ln(\gamma_{Cu}) = -\alpha_{Zn} X_{Cu} X_{Zn} - \int_1^{X_{Cu}=0.60} \alpha_{Zn} dX_{Cu} \tag{E13}$$

Once the value of the alpha-function for zinc is determined, the solution of the integral follows.

From Eqs. (E2b) and (4.48),

$$\ln(\gamma_{Zn}) = \frac{\Omega_m (1 - X_{Zn})^2}{RT} \tag{E14a}$$

$$\alpha_{Zn} = \frac{\ln(\gamma_{Zn})}{X_{Cu}^2} = \frac{1}{(1 - X_{Zn})^2} \frac{\Omega_m (1 - X_{Zn})^2}{RT} \tag{E14b}$$

$$\alpha_{Zn} = \frac{\Omega_m}{RT} = -\frac{32{,}729}{(8.314)(1333)} = -2.9532 \tag{E14c}$$

Then,

$$\int_1^{X_{Cu}=0.60} \alpha_{Zn} dX_{Cu} = \int_1^{X_{Cu}=0.60} (-2.9532) dX_{Cu} = 1.1813 \tag{E15}$$

Inserting Eq. (E15) into (E13) yields

$$\ln(\gamma_{Cu}) = -\alpha_{Zn} X_{Cu} X_{Zn} - 1.1813 \tag{E16a}$$

$$\ln(\gamma_{Cu}) = -(-2.9532)(0.60)(0.40) - 1.1813 = -0.47253 \tag{E16b}$$

$$\gamma_{Cu} = \exp(-0.47253) = 0.62342 \tag{E16c}$$

$$a_{Cu} = \gamma_{Cu} X_{Cu} = (0.62342)(0.60) = 0.37405 \tag{E16d}$$

(b) **Method F: Area under the curve (Fig. b):** Finding a curve fitting function $y = f(x)$, where y maps to X_{Zn}/X_{Cu} and x maps to $-ln(a_{Zn})$, and setting the limits of integration yield

$$y = 1.7236 \exp(0.6797x) \tag{E16a}$$

$$X_{Zn}/X_{Cu} = 1.7236 \exp[0.6797 \ln(a_{Zn})] \tag{E16b}$$

$$X_{Zn}/X_{Cu} = 0.05761 \quad at \quad \ln(a_{Zn})_1 = -5 \tag{E16c}$$

$$X_{Zn}/X_{Cu} = 0.67 \quad at \quad \ln(a_{Zn})_2 = -1.3902 \tag{E16d}$$

The integral area and activity at $X_{Zn} = 0.40$ or $X_{Zn}/X_{Cu} = 0.67$ are

$$\ln(a_{Cu}) = -\int_{-5}^{-1.3902} 1.7236 \exp[0.6797 \ln(a_{Zn})] \, d \ln(a_{Zn}) = -0.90095 \tag{E17a}$$

$$a_{Cu} = \exp(-0.90095) = 0.40618 \tag{E17b}$$

If $\ln(a_{Zn})_1 = -\infty$, then $a_{Cu} = \exp(-0.9857) = 0.37318$

Method G: Area Under the Curve (Fig. c) Similarly, the curve fitting equation and limits of integration are

$$X_{Zn}/X_{Cu} = 1.3083 \exp[-1.4191 \ln(\gamma_{Zn})] \tag{E18a}$$

$$X_{Zn}/X_{Cu} \simeq 0 \text{ at } -\ln(\gamma_{Zn})_1 = 4 \tag{E18b}$$

$$X_{Zn}/X_{Cu} = 0.67 \text{ at } -\ln(\gamma_{Zn})_2 = 0.47158 \tag{E18c}$$

Then, the integral area and activity at $X_{Zn} = 0.40$ or $X_{Zn}/X_{Cu} = 0.67$ are

$$\ln(\gamma_{Cu}) = \int_4^{0.47158} 1.3083 \exp[-1.419 \ln(\gamma_{Zn})] \, d\ln(\gamma_{Zn}) \tag{E19a}$$

$$\ln(\gamma_{Cu}) = -0.46902 \tag{E19b}$$

$$\gamma_{Cu} = \exp(-0.46902) = 0.62562 \tag{E19c}$$

$$a_{Cu} = \gamma_{Cu} X_{Cu} = (0.62562)(0.60) = 0.37537 \tag{E19d}$$

Example 4.6 Consider the thermodynamic data set for a regular $Cr\text{-}Ti$ solutions at $T = 1250\,°C$ (1523 K) taken Upadhyaya–Dube example [6, p. 162].

X_{Cr}	0.09	0.190	0.270	0.370	0.470	0.670	0.780	0.890
γ_{Cr}	3.356	2.800	2.444	2.103	1.745	1.248	1.106	1.018
a_{Cr}	0.302	0.532	0.660	0.778	0.820	0.836	0.863	0.906
X_{Ti}	0.91	0.81	0.73	0.63	0.53	0.33	0.22	0.11

Data plotting and curve fitting have been done in several forms for convenience. The main task is to analytically determine the activity of titanium (a_{Ti}) in liquid $Cr\text{-}Ti$ alloys at $T = 1250\,°C$ (1523 K).

Solution

(a) **Non-linear curve fitting: Fig. a:** Below is the resultant curve fitting equation along with the correlation coefficient

$$\gamma_{Cr} = 3.7217 \exp(-1.539 X_{Cr}) \quad \text{with } R^2 = 0.9557 \tag{E1}$$

From Eq. (4.42b), the Gibbs–Duhem equation is

$$\ln(\gamma_1) = -\int_1^{X_{Ti}} \frac{X_2}{X_1} d\ln(\gamma_2) \tag{E2a}$$

$$\ln(\gamma_{Ti}) = -\int_1^{X_{Ti}} \frac{X_{Cr}}{X_{Ti}} d\ln(\gamma_{Cr}) \tag{E2b}$$

The solution of the integral in Eq. (E2b) is the area bounded by the curve described by the dimensionless function $X_{Cr}/X_{Ti} = f[-\ln(\gamma_{Cr})]$, where y maps to X_{Cr}/X_{Ti} and x to $-ln(\gamma_{Cr})$. Hence, $y = f(X)$. Instead of using a

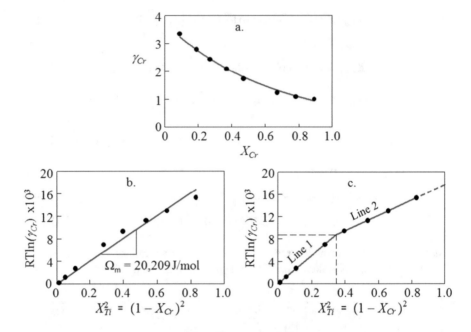

graphical solution, an analytical approach is chosen hereafter in order to show straightforward steps to comply with the main requirement in this part of the example.

Firstly, convert the curve fitting equation $\gamma_{Cr} = f(X_{Cr})$ into an ordinary differential $d \ln(\gamma_{Cr}) = df(X_{Cr})$ equation. Thus,

$$\gamma_{Cr} = 3.7217 \exp(-1.539X_{Cr}) \tag{E3a}$$

$$\ln(\gamma_{Cr}) = \ln\left[3.7217 \exp(-1.539X_{Cr})\right] \tag{E3b}$$

$$\frac{d \ln(\gamma_{Cr})}{dX_{Cr}} = -1.539 \tag{E3c}$$

$$d \ln(\gamma_{Cr}) = -1.539 dX_{Cr} \tag{E3d}$$

Inserting Eq. (E3d) into (E2b) yields the activity coefficient γ_{Ti} for titanium (Ti)

$$\ln(\gamma_{Ti}) = -\int \frac{X_{Cr}}{X_{Ti}} d \ln(\gamma_{Cr}) = 1.5390 \int \frac{X_{Cr}}{1 - X_{Cr}} dX_{Cr} \tag{E4a}$$

$$\ln(\gamma_{Ti}) = -1.5390 \left(X_{Cr} + \ln(X_{Cr} - 1)\right) \tag{E4b}$$

Secondly, using $X_{Cr} = 0.40$ and $X_{Ti} = 1 - X_{Cr} = 0.60$, Eq. (E4b) yields the real and imaginary parts of a complex number

$$\ln(\gamma_{Ti}) = -1.5390\,(0.40 + \ln(0.40 - 1)) \tag{E5a}$$

$$\ln(\gamma_{Ti}) = 0.17056 - 4.8349i \tag{E5b}$$

Equating the real parts of Eq. (E5b) gives

$$\ln(\gamma_{Ti}) = 0.17056$$

$$\gamma_{Ti} = \exp(0.17056) = 1.1860$$

$$a_{Ti} = \gamma_{Ti} X_{Ti} = (1.1860)(0.60) = 0.7116$$

This result $a_{Ti} = 0.7116$ compares with Upadhyaya–Dube [6, p. 166] reported value $a_{Ti} = 0.7344$ at $X_{Ti} = 0.60$ and $T = 1250\,°C$ (1523 K). The average percentage error is calculated as

$$\text{error} = \frac{0.7344 - 0.7116}{(0.7344 + 0.7116)/2} \times 100 = 3.15\%$$

(b) **Method A: Linear curve fitting (Fig. b):** This part is based on a classical approach using Eq. (4.30) as the relevant analytical model

$$RT \ln(\gamma_{Cr}) = \Omega_m X_{Ti}^2 \tag{E6}$$

Next, use the given data to plot $RT \ln(\gamma_{Cr}) = f\left(X_{Ti}^2\right)$ at $T = 1523\,K$ so that $y = RT \ln(\gamma_{Cr})$ and $x = X_{Ti}^2$. This means that the variable y maps to $RT \ln(\gamma_{Cr})$ and x to X_{Ti}^2 to fit the straight line equation with the slope (gradient) being the interaction parameter Ω_m. Thus, Eq. (E6) becomes

$$y = 20{,}209x \quad \text{with } R^2 = 0.9706 \tag{E7}$$

from which $\Omega_m = 20{,}209$ J/mol. Then,

$$RT \ln(\gamma_{Cr}) = 20{,}209 X_{Ti}^2 \tag{E8a}$$

$$\ln(\gamma_{Cr}) = \frac{20{,}209 X_{Ti}^2}{RT} \tag{E8b}$$

$$\ln(\gamma_{Ti}) = \frac{20{,}209 X_{Cr}^2}{RT} \tag{E8c}$$

At $X_{Cr} = 0.40$ and $X_{Ti} = 0.60$,

$$\ln(\gamma_{Cr}) = \frac{20{,}209 X_{Ti}^2}{RT} = \frac{(20{,}209)(0.60)^2}{(8.314)(1523)} = 0.57456$$

$$\gamma_{Cr} = \exp(0.57456) = 1.7763$$

$$a_{Cr} = \gamma_{Cr} X_{Cr} = (1.7763)\,(0.40) = 0.71052$$

and

$$\ln\left(\gamma_{Ti}\right) = \frac{(20{,}209)\,(0.40)^2}{(8.314)\,(1523)} = 0.25536$$

$$\gamma_{Ti} = \exp\,(0.25536) = 1.2909$$

$$a_{Ti} = \gamma_{Ti} X_{Ti} = (1.2909)\,(0.60) = 0.7745$$

This result $a_{Ti} = 0.7745$ also compares with Upadhyaya–Dube [6, p. 166] reported value $a_{Ti} = 0.7344$ at $X_{Ti} = 0.60$. The average percentage error is calculated as

$$\text{error} = \frac{0.7745 - 0.7344}{(0.7745 + 0.7344)\,/2} \times 100 = 5.32\%$$

Therefore, both linear and non-linear methods provide reasonable results.

(c) **Method B: Linear curve fitting (Fig. c):** Examining the trend of the plotted data points in Fig. c suggests that it can be divided into two data subsets, namely, Line 1 at $0.35 \le x = X_{Ti}^2 \le 1$ or $0.59 \le X_{Ti} \le 1$ and Line 2 at $0 \le X_{Ti}^2 \le 0.35$ or $0 \le X_{Ti} \le 0.59$. Below are the data subsets along with the curve fitting equations and their corresponding correlation coefficients. Thus,

$$RT \ln\left(\gamma_{Cr}\right) = f\left(X_{Ti}^2\right) \tag{E9a}$$

$$y = f\,(x) \tag{E9b}$$

and

Line 1		Line 2	
$0 \le x = X_{Ti} \le 0.59$		$0.59 \le x = X_{Ti} \le 1$	
x	y	x	y
0.2809	7049.70	0.8281	15,331.00
0.1089	2805.20	0.6561	13,037.00
0.0484	1275.70	0.5329	11,315.00
0.0121	225.89	0.3969	9412.70
$y = 25{,}203x$		$y = 13{,}735x + 3984.80$	
with $R^2 = 0.9995$		with $R^2 = 0.9998$	

The corresponding curve fitting equations are

$$y = 25{,}203x \text{ for } x = X_{Ti} \le 0.59 \ \text{ (Line 1)} \tag{E10a}$$

$$y = 13{,}735x + 3984.8 \text{ for } x = X_{Ti} \ge 0.59 \ \text{ (Line 2)} \tag{E10b}$$

From Eqs. (E10a) and (E10b), $(x, y) = (0.35, 8757.30)$ is the intercepting point for Line 1 and Line 2.

Line 1   For $X_{Cr} = 0.80$ and $X_{Ti} = 0.20$ at $T = 1523$ K,

$$RT \ln (\gamma_{Cr}) = 25{,}203 X_{Ti}^2 \tag{E11}$$

$$\ln (\gamma_{Cr}) = \frac{25{,}203 X_{Ti}^2}{RT} = \frac{(25{,}203)\,(0.20)^2}{(8.314)\,(1523)} = 0.07962$$

$$\gamma_{Cr} = \exp (0.07962) = 1.08290$$

$$a_{Cr} = \gamma_{Cr} X_{Cr} = (1.08290)\,(0.80) = 0.86632$$

and

$$\ln (\gamma_{Ti}) = \frac{(25{,}203)\,(0.80)^2}{(8.314)\,(1523)} = 1.2739$$

$$\gamma_{Ti} = \exp (1.2739) = 3.5748$$

$$a_{Ti} = \gamma_{Ti} X_{Ti} = (3.5748)\,(0.20) = 0.71496$$

Line 2   Extrapolating the linear equation $x = X_{Ti}^2 = 1$ yields

$$y = 13{,}735x + 3984.8 = 13{,}735\,(1) + 3984.8 = 17{,}720$$

$$y = RT \ln (\gamma_{Cr}) = 17{,}720 \text{ J/mol}$$

$$\ln (\gamma_{Cr}) = \frac{17{,}720}{RT} = \frac{17{,}720 \text{ J/mol}}{(8.314 \text{ J/mol K})\,(1523 \text{ K})} = 1.3994$$

Then, the Gibbs–Duhem integral at $X_{Cr} = 0.40$ and $X_{Ti} = 0.60$ yields

$$\ln (\gamma_{Ti}) = - \int_{\ln(\gamma_{Cr})_1}^{1} \frac{X_{Cr}}{X_{Ti}} d \ln (\gamma_{Cr}) = - \int_{1.3994}^{1} \left(\frac{0.40}{0.60} \right) d \ln (\gamma_{Cr}) \tag{E12}$$

$$\ln (\gamma_{Ti}) = 0.26627$$

$$\gamma_{Ti} = \exp (0.26627) = 1.3051$$

$$a_{Ti} = \gamma_{Ti} X_{Ti} = (1.3051)\,(0.60) = 0.78306$$

Thus, the average percentage error between the calculated $a_{Ti} = 0.78306$ and the experimental $a_{Ti} = 0.7344$ [6, p. 162] values is

$$\text{error} = \frac{0.78306 - 0.7344}{(0.78306 + 0.7344)\,/2} \times 100 = 6.41\%$$

Therefore, $a_{Ti} = 0.78306$ is considered to be within a reasonable marginal error since the experimental data indicates that the liquid Cr-Ti alloys at $T = 1523$ K show two linear behaviors.

(d) **Constant interaction parameter:** Assume that the interaction parameter Ω_m remains constant at $1523\,\text{K} \leq T < 1650\,\text{K}$. In such a case, one can predict $\gamma_{Ti,2}$ at T_2 knowing $\gamma_{Ti,1}$ at T_1. Thus,

$$RT_2 \ln\left(\gamma_{Ti,2}\right) = RT_1 \ln\left(\gamma_{Ti,1}\right) = \Omega_m X_{Ti}^2 \tag{E13a}$$

$$\ln\left(\gamma_{Ti,2}\right) = \frac{T_1}{T_2} \ln\left(\gamma_{Ti,1}\right) = \ln\left(\gamma_{Ti,1}\right)^{T_1/T_2} \tag{E13b}$$

$$\gamma_{Ti,2} = \left(\gamma_{Ti,1}\right)^{T_1/T_2} \tag{E13c}$$

If $\gamma_{Ti,1} = 1.3051$ at $T_1 = 1523$ K, then $\gamma_{Ti,2}$ at $T_2 = 1580$ K is

$$\gamma_{Ti,2} = \left(\gamma_{Ti,1}\right)^{T_1/T_2} = (1.3051)^{1523/1580} = 1.2926$$

4.9 Degree of Undercooling

It is appropriate now to derive the degree of undercooling ΔT for a pure metal (component) introduced in Eq. (4.7c). Consider melting pure solid A at T_f so that the reaction of pure metal A is simply $A_s(solid) = A_l(liquid)$ and the reaction constant K_A is $K_A = a_{A_l}/a_{A_s}$, where the activities are $a_{A_l} < 1$ and $a_{A_s} = 1$. Thus, the partial Gibbs energy for a diluted A-B alloy system is

$$\Delta G_{AB} = RT \ln\left(K_A\right) = RT \ln\left(a_{A_l}\right) = RT \ln\left(X_A\right) \tag{4.49a}$$

$$\Delta G_{AB} = RT \ln\left(1 - X_B\right) \tag{4.49b}$$

since $a_{A_l} = X_A$ as per Eq. (4.28) with $\gamma_A = 1$ (pure metal A). Adding Eqs. (4.7a) and (4.49b) approximates the total Gibbs energy for very diluted A-B alloys at a small undercooling $\Delta T = T_f - T$

$$\Delta G = \Delta G_A + \Delta G_{AB} \tag{4.50a}$$

$$\Delta G = \Delta H_{f,A}\left(\frac{\Delta T}{T_f}\right) + RT \ln\left(1 - X_B\right) \tag{4.50b}$$

However, if the A-B liquid and solid mix is in equilibrium at $T = T_f$, then $\Delta G = 0$ and the theoretical degree of undercooling ΔT for $X_A > 0$ and $X_B < X_A$ is written in a general form as

$$\Delta T = -\frac{RT_f^2 \ln (X_A)}{\Delta H_{f,A}} = -\frac{RT_f^2 \ln (1 - X_B)}{\Delta H_{f,A}} \tag{4.51}$$

If $\ln (1 - X_B) = -X_B - X_B^2/2 - X_B^3/3 \ldots \simeq -X_B$, then Eq. (4.51) for a diluted A-B alloy becomes

$$\Delta T = \frac{RT_f^2 X_B}{\Delta H_{f,A}} \tag{4.52}$$

which gives an approximated ΔT value since $\Delta H_{f,A} \simeq \Delta H_{f,AB}$.

Example 4.7 Calculate ΔT for an Ag-Pb alloy with a nominal composition $X_{Pb} = 0.01$, where $X_A = X_{Ag}$ and $X_B = X_{Pb}$. Given experimental data: $T_f = 962\,°C = 1235\,K$, $\rho_{Ag} = 10.49\,g/cm^3$ and $\Delta H_f = 965\,J/cm^3$. Compare ΔT with $\Delta T_{exp} = 10\,K$ [10, p. 211].

Solution For convenience, use one mole fraction $X_{Pb} = 0.01$ and convert the latent heat of fusion $\Delta H_f = 965\,J/cm^3$ into J/mol units. This is accomplished as follows:

$$\Delta H_f = \frac{\Delta H_f A_w}{\rho} = \left(\frac{965\,J/cm^3}{10.49\,g/cm^3}\right)\left(107.87\,\frac{g}{mol}\right) = 9923.2\,J/mol$$

Then, Eq. (4.52) gives

$$\Delta T = \frac{RT_f^2 X_B}{\Delta H_{f,A}} = \frac{(8.314\,J/mol\,K)\,(1235\,K)^2\,(0.01)}{9923.2\,J/mol} = 12.78\,K$$

4.10 Phase Stability Criteria

Following Gaskell's idea [11, p. 341], the criteria for phase stability is based on the derivatives of the Gibbs energy function $G = f(X_B)$ of mixing that provide fundamental information on the critical values of Ω_{lc}, Ω_{sc} and concavity of the graph of G_l, $G_s = f(X_B)$ at T.

From Eq. (4.23), the pertinent derivatives of $G = f(X_B)$ for the liquid and solid phases are

$$\frac{dG_l}{dX_B} = -G_A + G_B + \Omega_l(1 - 2X_B) - \frac{1}{X_B} \tag{4.53a}$$

$$\frac{dG_s}{dX_B} = -\Delta S_{fA}\left(T - T_{fA}\right) + \Delta S_{fB}\left(T - T_{fB}\right) \tag{4.53b}$$

$$+ RT\left[\ln X_B - \ln(1 - X_B)\right] + \Omega_s(1 - 2X_B)$$

$$\frac{d^2G_l}{dX_B^2} = -2\Omega_l - \frac{1}{X_B^2} \tag{4.53c}$$

$$\frac{d^2G_s}{dX_B^2} = RT\left[\frac{1}{X_B} + \frac{1}{1-X_B}\right] - 2\Omega_s \tag{4.53d}$$

$$\frac{d^3G_l}{dX_B^3} = -\frac{2}{X_B^3} \tag{4.53e}$$

$$\frac{d^3G_s}{dX_B^3} = RT\left[\frac{1}{(1-X_B)^2} - \frac{1}{X_B^2}\right] \tag{4.53f}$$

Equating these derivatives, Eqs. (4.53a–f), to zero yields

$$\frac{d^3G_s}{dX_B^3} = 0 \quad \rightarrow X_B = \frac{1}{2} \tag{4.54a}$$

$$\frac{d^2G_l}{dX_B^2} = 0 \quad \rightarrow \Omega_l = -\frac{1}{8} \text{ J/mol} \tag{4.54b}$$

$$\frac{d^2G_s}{dX_B^2} = 0 \quad \rightarrow \Omega_s = \frac{5}{4}RT \text{ in J/mol} \tag{4.54c}$$

$$\frac{dG_l}{dX_B} = 0 \quad \rightarrow G_B - G_A = \frac{15}{8} \text{ J/mol} \tag{4.54d}$$

$$\frac{dG_s}{dX_B} = 0 \quad \rightarrow \Delta S_{fA}\left(T - T_{fA}\right) = \Delta S_{fB}\left(T - T_{fB}\right) \text{ in J/mol} \tag{4.54e}$$

In addition, the second derivatives, d^2G_l/dX_B^2 and d^2G_s/dX_B^2, can be used to determine

- if a function $G = f(X_B)$ has a local minimum if $d^2G_l/dX_B^2 > 0$ or local maximum if d^2G_s/dX_B^2,), and
- if the function $G = f(X_B)$ has a concave up (convex) or concave down behavior.

In the final analysis, the second derivative of a function $G = f(X_B)$ measures the concavity of the graph for $G_l, G_s = f(X_B)$ at T and Ω_l, Ω_s. Nonetheless, d^2G_l/dX_B^2 and d^2G_s/dX_B^2 measure the rate of change of G_l and G_s at known T and Ω_l, Ω_s.

4.11 Construction of Binary Phase Diagrams

This section includes general features of equilibrium phase diagrams, also known as standard phase diagrams that correlate with the Gibbs energy curves described by the general function $G = f(X_B)$, where $X_B = 1 - X_A$. Moreover, a standard phase diagram is constructed at atmospheric pressure $P = 1$ atm and low freezing rates (dT/dt) under controlled experimental conditions. With regard to the skewness of the $G = f(X_B)$ curves, most are either uniform or altered U-shaped, as depicted in the plots given below, for individual liquid and solid phases. However, a mixture of phases requires a construction of a tangent line connecting the phase lines for defining the composition or mole fraction of the alloying element B with element A.

These standard phase diagrams are two-component maps that show the phase fields separated by phase-field lines and the general function that plots these phase-field lines is also of the form $T = T(X_B)$. As a result, one can interpret these maps in order to predict the temperature or temperature range for stable phases. Only a few $G = f(X_B)$ curves at T are shown below one isomorphous and one eutectic standard phase diagrams and subsequently, construct the standard phase diagrams.

Furthermore, simulations follow numerical methods for carrying out the construction of standard phase diagrams. This subject matter is out of the scope of this chapter.

4.11.1   Isomorphous Phase Diagrams

Consider mixing components A and B to develop Gibbs energy diagrams using $G = f(X_B)$ and subsequently, construct a standard isomorphous phase diagram, which is the most simple binary A-B phase diagram. Using the following data set found in Gaskell–Laughlin book [11, p. 356], the Gibbs energy change $\Delta G = \Delta H - T\Delta S$, as defined by Eq. (4.8b) for components A and B at constant pressure, are

$$\Delta G_A = (8000 - 10T)\left(10^{-3}\right) \quad \text{in kJ/mol} \tag{4.55a}$$

$$\Delta G_B = (12{,}000 - 10T)\left(10^{-3}\right) \quad \text{in kJ/mol} \tag{4.55b}$$

$$T_{fA} = 1200\,\text{K} \quad \& \quad T_{fB} = 1200\,\text{K} \tag{4.55c}$$

from which

$$\Delta H_A = 8000\,\text{J/mol and } \Delta H_B = 12{,}000\,\text{J/mol} \tag{4.56a}$$

$$\Delta S_{fA} = \Delta S_{fB} = 10\,\text{J/mol} \tag{4.56b}$$

and T_{fA}, T_{fB} denote the melting or freezing temperatures of components A and B, respectively. Apparently, this data set is for a hypothetical binary alloy. The idea now is to use this data set to plot $G = f(X_B)$ diagrams.

Based on the above information, Eq. (4.23) with $\Omega_l = \Omega_s = 0$ can be used to plot $G_l, G_s = f(X_B)$ at selected temperature T as illustrated in Fig. 4.10. Note that the plot in Fig. 4.10a is at $T > T_{fA}$, Fig. 4.10b is at $T_{fB} < T < T_{fA}$, and

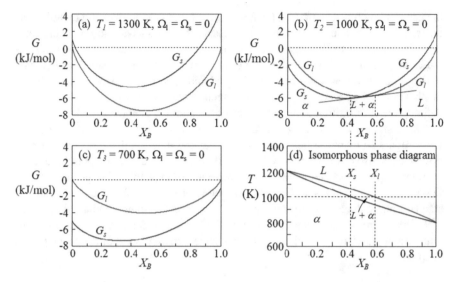

Fig. 4.10 Molar Gibbs energy diagrams at (**a**) 1300 K, (**b**) 1000 K, (**c**) 700 K, and (**d**) isomorphous phase diagram

Fig. 4.10c is at $T < T_{fB}$. These graphical representations of the Gibbs energies are included below side by side along with the hypothetical binary A-B alloy for proper interpretation of the theoretical approach being undertaken.

Notice that each of the Gibbs energy plots (Fig. 4.10) contains the assigned solution parameters, $\Omega_l = \Omega_s = 0$, for the liquid and solid phases. This is purposely carried out accordingly in order to simplify the graphs for an appropriate interpretation of the Gibbs energy as an important state function, which serves as a criterion for phase changes since it is not considered as a property of matter. It also serves as a convenient mathematical expression, $G = H - TS$ for assessing thermodynamic systems and making calculations easier to handle.

The interpretation of the above Gibbs energy curves is given below.

- Temperature T_1 (Fig. 4.9a): The entire $G = f(X_B)$ curve for the liquid, $G_l(X_B)$, lies below that for the solid, $G_s(X_B)$, at any composition X_B. This implies that this phase system is stable in the liquid (L) state or that the single homogeneous liquid is the stable phase at T_1 since $G_s > G_l$ for all X_B. Note that $G_s(X_B)$ and $G_l(X_B)$ are similar for very diluted A-B solution.
- Temperature T_2 (Fig. 4.9b): As the temperature decreases, the $G_l(X_B)$ and $G_s(X_B)$ curves intercept at $X_B = 0.5$, but the $G_l = f(X_B)$ and $G_s = g(X_B)$ have minimum values as depicted in Fig. 4.10a by the dashed vertical lines. Note that α is stable at $X_B \leq X_s$, α is stable in the liquid phase at $X_s \leq X_B \leq X_l$, and the liquid L is stable at $X_B \geq X_l$. Moreover, the minimum $G(X_B)$ values correspond to the point where the liquidus and solidus lines intercept at $T_2 = 1000$ K.

- Temperature T_3 (Fig. 4.9c): The $G_s = f(X_B)$ curve for the solid α is totally below the $G_l = f(X_B)$ curve for the liquid. Therefore, the single homogeneous solid α phase is entirely stable since $G_s < G_l$ for all X_B values. Eventually, the stable α phase cools off to room temperature for subsequent use in a relevant solid shape. The hypothetical A-B phase diagram in Fig. 4.10d shows the homogeneous α-phase region for the entire mole fraction range $0 \leq X_B \leq 1$.

4.11.2 Eutectic Phase Diagrams

This section is a continuation of the preceding thermodynamic treatment of phase changes during solidification, again, of a binary A-B eutectic system. As a result, the Gibbs energy graphical interpretation is slightly more complicated than before.

Consider the function $G = f(X_B)$, as defined by Eq. (4.23), and assume that the components A and B form a binary eutectic phase diagram. Available thermodynamic data:

$\Delta H_A = 8000$ J/mol	$\Delta S_{fA} = 10$ J/mol	$T_{fA} = 1200$ K
$\Delta H_B = 12{,}000$ J/mol	$\Delta S_{fB} = 10$ J/mol	$T_{fB} = 1200$ K
$\Omega_l = 0$	$\Omega_s = 1.5 \times 10^4$ J/mol	

Use this data set to plot $G_l = f(X_B)$ and $G_s = g(X_B)$ at temperatures T and constant pressure $P = 1$ atm $= 101.33$ kPa to construct a binary eutectic phase diagram. Figure 4.11 shows the resultant diagrams.

So far only two simple equilibrium phase diagrams are constructed using data provided by the function $G = f(X_B)$ for liquid and solid phases. More complicated equilibrium phase diagrams require a substantial amount of data from the Gibbs energy approach.

The reader should be aware of sophisticated numerical methods for mapping temperature versus composition $[T = f(C)]$, known as equilibrium phase diagram. For instance, the Ab initio method is a quantum mechanics and statistical analysis, and it may employ the Monte Carlo simulation technique in order to determine the Gibbs–Duhem energy data and related parameters at finite temperatures T. Subsequently, an optimization method follows for plotting the data as an equilibrium phase diagram.

Nevertheless, the Monte Carlo simulation is a computerized mathematical technique and the Gibbs–Duhem technique is an integration approach for determining partial properties.

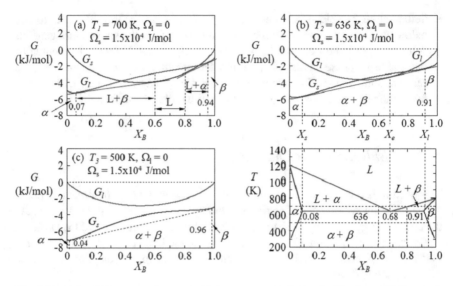

Fig. 4.11 Molar Gibbs energy diagrams. Gibbs energy at (**a**) 700 K, (**b**) 636 K, (**c**) 500 K, and (**d**) eutectic phase diagram

4.12 Nucleation Theory

Fundamentally, solidification is driven by a change in Gibbs energy per unit volume ($\Delta G_v = G_L - G_S$) between the liquid and the solid due to a change in temperature called *undercooling* defined as $\Delta T = T - T_f$, where T_f is the equilibrium freezing or melting temperature of the liquid metal. Subsequently, nano-sized particles (crystals) form randomly and coarsen as solidification proceeds at a *rate of change of the solid temperature* (dT/dt) in the order of $dT/dt = 10^{-3}\,°C/s$ for quasi-static and $dT/dt = 10^{6}\,°C/s$ for rapid phase transformation.

Furthermore, Fig. 4.12 schematically depicts the solidification models for a liquid metal, where Fig. 4.12a represents the onset of homogeneous nucleation in the form of a cluster of atoms treated as a suspended sphere (α) within the liquid metal at just below the equilibrium temperature T_f. The smallest particle (embryo) is also assumed to be the unit cell of the solidifying liquid.

Similarly, Fig. 4.12b illustrates the heterogeneous nucleation when a particle is in contact with a substrate β, such as the metallic mold wall or an impurity particle, known as grain refiner.

The notation related to the theoretical model shown in Fig. 4.12b is based on preexisting surfaces and the radius (r) of curvature of the droplet on a substrate β surface. Thus,

$$\gamma_{\alpha L} = \text{Solid–liquid interface surface energy}$$

$$\gamma_{\beta L} = \text{Agent–liquid interface surface energy}$$

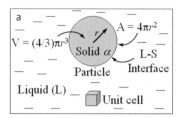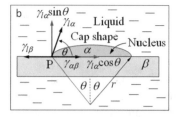

Fig. 4.12 Solidification models for (**a**) homogeneous nucleation of a sphere and (**b**) heterogeneous nucleation of a spherical cap

$$\gamma_{\alpha\beta} = \text{Solid–solid interface surface energy}$$

$$\theta = \text{Wet or contact angle}$$

In general, phase transformation is a liquid-to-solid (L-S) process related to energy release, and the kinetics of nucleation and growth of clusters of atoms, which are ideally modeled as perfect and hard spheres (Fig. 4.12a). As a result, a cluster contains an N number of unit cells and each unit cell contains a certain number of atoms, as determined in crystallography.

Conversely, Fig. 4.12b shows the cap shape model used to determine the interfacial surface energy at the interface between two phases. For the sake of clarity, the phrases "interfacial tension," "interfacial energy," and "surface energy" are interchangeably used throughout the vast literature as an interface property between two phases. Exclusively, liquid–solid and solid–solid interfaces (boundaries) during phase transformations prevail, in this book, as the main sources of surface energy (γ_s), which in turn is considered as an excess energy entity. Moreover, the initial stage of an interface is related to nucleation.

4.12.1 Homogeneous Nucleation of Metals

Consider a cluster of atoms susceptible to form a spherical particle as shown in Fig. 4.12a and assume that the particle radius is $r < r_c$, where r_c is a particle critical radius. In this case, the particle can be treated as an unstable spherical embryo and from Eq. (4.7b) along with a small undercooling ΔT, the corresponding Gibbs energy density (ΔG_v) for a growing unstable embryo can be approximated as

$$\Delta G_v = -\frac{\Delta H_f \Delta T}{T_f} = -\Delta S_f \Delta T \tag{4.57}$$

For homogeneous nucleation of a spherical solid phase, the total Gibbs energy change $\Delta G < 0$ is defined by

$$\Delta G = \Delta G_v + \Delta G_s = V \Delta G_v + A \gamma_{ls} \tag{4.58a}$$

$$\Delta G = -\frac{4}{3}\pi r^3 \Delta G_v + 4\pi r^2 \gamma_{ls} \tag{4.58b}$$

where ΔG is the Gibbs energy change per unit volume and that is released by solidification, and $\Delta G_s, \gamma_{ls}$ are the Gibbs energy and interfacial energy (surface energy) per unit area for the creation of a surface. Moreover, γ_{ls} is a temperature-independent and an isotropic energy term.

Substituting Eq. (4.57) into (4.58b) yields the homogeneous $\Delta G = \Delta G_{\text{hom}}$ and pertinent derivatives as

$$\Delta G = -\left(\frac{4}{3}\pi r^3\right)\frac{\Delta H_f \Delta T}{T_f} + \left(4\pi r^2\right)\gamma_{ls} \tag{4.59a}$$

$$\frac{d(\Delta G)}{dr} = -\left(4\pi r^2\right)\frac{\Delta H_f \Delta T}{T_f} + (8\pi r)\gamma_{ls} \tag{4.59b}$$

$$\frac{d^2(\Delta G)}{dr^2} = -(8\pi r)\frac{\Delta H_f \Delta T}{T_f} + (8\pi)\gamma_{ls} \tag{4.59c}$$

Letting $d(\Delta G)/dr = 0$ at $r = r_c$ gives the critical radius of the nuclei for solidification to proceed

$$r_c = \frac{2\gamma_{ls} T_f}{\Delta H_f \Delta T} = \frac{2\gamma_{ls}}{\Delta S_f \Delta T} = \frac{2\Gamma_{ls}}{\Delta T} \tag{4.60}$$

where Γ_{ls} denotes the Gibbs–Thomson effect. Substituting Eq. (4.60) into (4.59a) yields the critical Gibbs energy change for nuclei with $r = r_c$

$$\Delta G_c = \frac{16\pi \gamma_{ls}^3}{3\left(\Delta H_f\right)^2}\left(\frac{T_f}{\Delta T}\right)^2 \tag{4.61}$$

Note that Eqs. (4.59a,c) are $d^2(\Delta G)/dr^2 < 0$ and $\Delta G_c > 0$ at $r = r_c$. This means that $\Delta G_c(r_c)$ is a maximum and particle growth should occur if $r \geq r_c$. In addition, a particle with $r < r_c$ is referred to as an embryo because it is very small and consequently, it may dissolve into its original matrix. This means that the fundamental requirements for particle growth are governed by the Gibbs energy change $\Delta G_c(r_c) > 0$ and particle size $r \geq r_c$ at a certain undercooling ΔT, dependent on the pure metal being solidified.

For comparison purposes, Table 4.1 illustrates selected data taken from Askeland and Fulay [12, p. 262] for four elements. See Appendix 5A for changing engineering to scientific units.

Moreover, Eq. (4.59a) is plotted in Fig. 4.13 for copper (Cu). From Table 4.1 and Eq. (4.61), $\Delta G_c = 7.2442\ eV$ is the maximum energy at $r_c = 1.25\ nm$. Also shown

Table 4.1 Experimental data set [12, p. 262]

Elements	Crystal	T_f (°C)	ΔT (°C)	ρ (g/cm³)	ΔH_f (J/cm³)	γ_{ls} (J/cm²)
Cu	FCC	1085	236	8.96	1628	177×10^{-7}
Fe	FCC	1538	420	7.87	1737	204×10^{-7}
Ag	FCC	962	450	10.49	965	126×10^{-7}
Ni	FCC	1453	480	8.81	2756	255×10^{-7}

Fig. 4.13 Gibbs energy change profile using $\Delta G = f(r)$ for theoretical solidification of pure copper. Conversion: 1 $J = 6.242 * 10^{18}$ eV

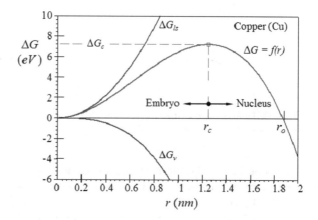

in Fig. 4.13 are the individual energy changes due to effects of particle surface area and volume; $G_{ls} = (4\pi r^2) \gamma_{ls}$ and $G_v = (4\pi r^2) H_s \Delta T / T_m$, respectively.

It is clear that $\Delta G = f(r)$ increases due to the randomly formation of clusters of atoms in the liquid. Eventually, these clusters reach critical and stable sizes as solidification proceeds.

Unstable Stage This occurs when $0 < \Delta G < \Delta G_{max}$ at $0 < r < r_c$ and therefore, nano-sized particles (known as embryos, clusters, and crystallites) dissolve; $\Delta G \to 0$ and $r \to 0$. For continuous nucleation, $\Delta G_c = \Delta G_{max}$ and the particles are treated as a nuclei having a critical radius $r = r_c$. Thus, ΔG_{max} is basically the maximum Gibbs energy change required for the formation of stable spherical nuclei and it is the energy barrier to the nucleation process or the thermodynamic driving force.

Stable Stage This is a stable particle growth process that is initiated at the critical state and proceeds when $\Delta G_c < \Delta G \le -\infty$ at $r_c \le r \le \infty$. Mathematically, $\Delta G \to 0$ as $r \to r_o$ and $\Delta G \to -\infty$ at $r > r_o$. The final solid is a crystalline structure having a particular grain morphology which is directly responsible for the properties of the final product.

Furthermore, homogeneous nucleation is characterized by a large undercooling ΔT to cause the formation of stable nuclei during directional or unidirectional solidification. Undoubtedly, the formation of nuclei requires conduction of heat

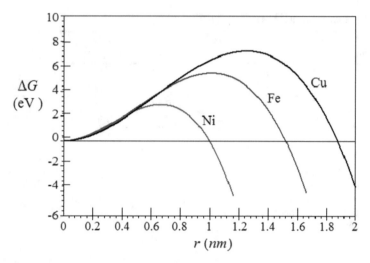

Fig. 4.14 Gibbs energy change profile using $\Delta G = f(r)$ for theoretical solidification of three pure elements as per data listed in Table 4.1

transfer at the liquid–solid interface, which represents an evolving boundary being dependent on time and the type of thermal resistance within the solidification domain Γ. As a result, microstructural morphology, in general, controls the mechanical and physical properties of the material. However, solidified materials with a mixture of different microstructural morphologies have different responses to mechanical deformation, electrochemical behavior in aggressive media, and the like.

In addition, Fig. 4.14 exhibits the total Gibbs energy change for the three pure elements listed in Table 4.1. Notice that $\Delta G_{c,Cu} > \Delta G_{c,Fe} > \Delta G_{c,Ni}$ and $r_{c,Cu} > r_{c,Fe} > r_{c,Ni}$. This is attributed to the magnitude of undercooling $\Delta T_{Cu} < \Delta T_{Fe} < \Delta T_{Ni}$. From Fig. 4.14 and Table 4.1, the lower ΔT, the higher ΔG_c and the larger r_c.

The effect of nucleation rate on the kinetics of phase transformation $L \rightarrow S$ is usually described by a time-temperature-transformation (TTT) diagram in the field of Materials Science or Physical Metallurgy.

In the field of heat transfer, however, phase transformation is characterized by a thermal energy balance associated with the type of thermal resistance between a mold and liquid metal or between the solidified melt and the remaining liquid metal. In this chapter, solidification heat transfer process is of interest with respect to thermal resistance cases related mainly to Fourier's law of thermal conduction and Newton's law of cooling.

It suffices to say that all heat transfer modes are present during solidification. However, conduction heat transfer appears to be the most relevant mechanism for freezing the melt since it derives from the evolved latent heat of fusion at the liquid–solid (L-S) solidification front (interface). This is the dominant source of fundamental importance for extracting heat from the melt. The principal approach for evaluating the solidification is related to the thermodynamics of nuclei formation in the melt using the Gibbs energy change since heat transfer theory does not provide any analytical means for such a purpose.

Example 4.8 Consider the homogeneous nucleation of pure iron (Fe) with a typical undercooling, $\Delta T = 420$ K, latent heat of fusion $\Delta H_f = 1737$ J/cm^3, surface energy $\gamma_{ls} = 204 \times 10^{-7}$ J/cm^2, and melting or freezing temperature $T_f = 1811$ K. Use this information to calculate (a) the critical radius (r_c), (b) the critical Gibbs energy change (ΔG_c) at $r = r_c$, (c) the number of unit cells (N_{cell}) and (d) the number of atoms (N_{atom}) in a nucleus. Moreover, iron at T_f has an FCC crystal structure with 4 *atoms* in the unit cell and lattice parameter $a = 0.287$ nm, which is normally determined using the X-ray technique.

Solution

(a) From Eq. (4.60), the critical radius is

$$r_c = \frac{2\gamma_{ls} T_f}{\Delta H_f \Delta T} = \frac{2\left(204 \times 10^{-7} \text{ J/cm}^2\right)(1811 \text{ K})}{\left(1737 \text{ J/cm}^3\right)(420 \text{ K})} = 1.01 \times 10^{-7} \text{ cm}$$

$$r_c = 1.10 \text{ nm}$$

(b) From Eq. (4.61), the critical Gibbs energy change at $r = r_c$ is

$$\Delta G_c = \frac{16\pi \gamma_{ls}^3}{3\left(\Delta H_f\right)^2}\left(\frac{T_f}{\Delta T}\right)^2 = \frac{16\pi \left(204 \times 10^{-7} \text{ J/cm}^2\right)^3}{3\left(1737 \text{ J/cm}^3\right)^2}\left(\frac{1811 \text{ K}}{420 \text{ K}}\right)^2$$

$$\Delta G_c = 8.77 \times 10^{-19} \text{ J}$$

(c) The volume of the unit cell (V_{cell}) and the volume of a sphere (V_s) are

$$V_{cell} = a^3 = (0.287 \text{ nm})^3 = 2.36 \times 10^{-2} \text{ nm}^3$$

$$V_s = \frac{4}{3}\pi r_c^3 = \frac{4}{3}\pi (1.01 \text{ nm})^3 = 4.32 \text{ nm}^3$$

Thus,

$$N_{cell} = \frac{V_s}{V_{cell}} = \frac{4.32 \text{ nm}^3}{2.36 \times 10^{-2} \text{ nm}^3/\text{cell}} = 183 \text{ cells}$$

(d) The total number of atoms (N_{atom}) in a nucleus is

$$N_{atom} = 4N_{cell} = 4(183) = 732 \text{ atoms}$$

Therefore, these results indicate that a spherical copper cluster with a volume $V_s = 4.32\,\text{nm}^3$ contains 183 unit cells, which in turn contain 732 atoms. Interestingly, assume a square nucleus or particle with an edge length defined as $d = 2r_c = 2.20\,\text{nm}$. Then, the particle volume becomes

$$V_s = d^3 = (2.20\,\text{nm})^3 = 10.648\,\text{nm}^3$$

Thus,

$$N_{cell} = \frac{V_s}{V_{cell}} = \frac{10.648\,\text{nm}^3}{2.36 \times 10^{-2}\,\text{nm}^3/\text{cell}} = 451\,\text{cells}$$

$$N_{atom} = 4N_{cell} = (4\,\text{atom/cell})\,(451\,\text{cells}) = 1804\,\text{atoms}$$

Example 4.9 Calculate (a) the number of unit cells, (b) the number of atoms in the critical nucleus (particle), (c) the entropy of fusion and (d) the Gibbs energy for solidification of copper under homogeneous nucleation. Copper has an FCC crystal structure with 4 atoms per unit cell, a critical particle radius of 1.2512 nm, a density of $8.96\,\text{g/cm}^3$, and a lattice parameter of $a = 0.362\,\text{nm}$.

Solution

(a) The FCC unit cell volume is

$$V_{cell} = a^3 = (0.362\,\text{nm})^3 = 4.74 \times 10^{-2}\,\text{nm}^3/\,\text{unit cell}$$

and the volume of a perfect spherical particle becomes

$$V_s = \frac{4}{3}\pi r_c^3 = \frac{4}{3}\pi\,(1.2512\,\text{nm})^3 = 8.20\,\text{nm}^3$$

Thus, the number of unit cells N_{cell} is

$$N_{cell} = \frac{V_s}{V_{cell}} = \frac{8.20\,\text{nm}^3}{4.7438 \times 10^{-2}\,\text{nm}^3/\text{unit cell}} = 173\,\text{cells}$$

(b) If there are four atoms in the FCC unit cell, then the number of atoms in the critical nucleus is

$$N_{atom} = 4N_{cell} = (4)\,(173) = 692\,\text{atoms}$$

Therefore, a copper cluster with $V_s = 8.20\,\text{nm}^3$ contains 173 unit cells and 692 atoms.

(c) From Eq. (4.6) and Table 4.1, the entropy of fusion is calculated with unusual units

$$\Delta S_f = \frac{\Delta H_f}{T_f} = \frac{1628\,\text{J/cm}^3}{(1085+273)\,\text{K}} = 1.20\,\text{J/(cm}^3\,\text{K)}$$

Divide this result by the given density value to get

$$\Delta S_f = \left[1.20\,\text{J/(cm}^3\,\text{K)}\right]/(8.96\,\text{g/cm}^3) \simeq 0.13\,\text{J/g K}$$

Using conventional units for the entropy of fusion yields

$$V_s = \frac{4}{3}\pi r_c^3 = \frac{4}{3}\pi \left(1.2512 \times 10^{-7}\,\text{cm}\right)^3 = 8.20 \times 10^{-21}\,\text{cm}^3$$

$$\Delta S_f = \left(1.20\,\text{J/cm}^3\,\text{K}\right)\left(8.20 \times 10^{-21}\,\text{cm}^3\right) = 9.84 \times 10^{-21}\,\text{J/K}$$

(d) From Eq. (4.5), the critical Gibbs energy change is

$$\Delta G_c = \Delta H_f - T_f \Delta S_f = (1628)\left(8.20 \times 10^{-21}\right)$$

$$- (1085+273)\left(9.84 \times 10^{-21}\right)$$

$$\Delta G_c = -1.312 \times 10^{-20}\,\text{J}$$

Note that a conversion of units is necessary in order to calculate the entropy of fusion and the Gibbs energy in proper units.

4.12.2 Heterogeneous Nucleation of Metals

Consider the classical heterogeneous nucleation model shown in Fig. 4.12b for solidification with the condition that α and β crystal structures are similar. In this case, the kinetics of heterogeneous nucleation can be characterized by the energy balance

$$\Delta G_{het} = \left(-\frac{\pi r^3}{3}\Delta G_v + 4\pi r^2 \gamma_{l\alpha}\right) f(\theta) \tag{4.62a}$$

$$\Delta G_{het} = \Delta G_{hom} f(\theta) \tag{4.62b}$$

where $f(\theta)$ is the shape factor (known as the geometric factor) takes the limit $0 \le f(\theta) \le 1$ and it is derived in Appendix 4B as [13, p. 288].

$$f(\theta) = \frac{1}{4}\left(2 - 3\cos\theta + \cos^3\theta\right) = \frac{1}{4}(2 + \cos\theta)(1 - \cos\theta)^2 \tag{4.63}$$

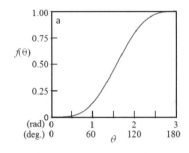
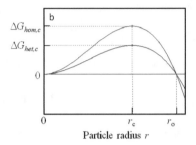

Fig. 4.15 (a) Profile of the shape factor $f(\theta)$ and (b) Gibbs energy change profiles for homogeneous and heterogeneous nucleation processes

The graphical representation of Eqs. (4.49) and (4.31a) is shown Fig. 4.15. For instance, Fig. 4.15a exhibits a sigmoidal behavior and Fig. 4.15b compares the Gibbs energy ΔG for homogeneous and heterogeneous cases.

Notice that $f(\theta)$ gives a sigmoidal curve (profile) evaluated at $0 \leq \theta \leq 180°$. The effect of $f(\theta)$ on ΔG_{het} is deduced as follows:

- If $f(\theta) = 1$, then $\Delta G_{het} = \Delta G_{hom}$ and homogeneous nucleation takes place. Therefore, the solidification process is based on an ideal nucleation mechanism. This implies that most nuclei formation occurs randomly within the liquid phase.
- For $f(\theta) = 0.5$, obviously, $\Delta G_{het} = 0.5\Delta G_{hom}$. This particular case indicates that nuclei formation occurs within the liquid, but the mold walls are the most susceptible sites for the least nuclei since they act as insoluble nucleation sites.
- If $f(\theta) = 0$, then Eq. (4.63) yields $\Delta G_{het} = 0$ and the model breaks down since the insoluble wetting agent (substrate β) has no effect on heterogeneous nucleation. One can conclude that the heterogeneous nucleation is a more realistic mechanism since the mold internal surfaces act as insoluble agents. Hence, $0 < f(\theta) < 1$ and $0 < \theta < \pi$ for heterogeneous nucleation. This is a key factor for adding nucleating agents to the melt.

4.12.3 Catalyst-Induced Solidification Process

Fundamentally, Fig. 4.15b exhibits $\Delta G = f(r)$ for homogeneous and heterogeneous nucleation mechanisms. As a matter of fact, $\Delta G_{het,c} < \Delta G_{hom,c}$ implies that the nucleation agent β (inoculant) acts as a weak barrier for heterogeneous nucleation to take place. Therefore, the nature of the nucleating agent induces the extent of undercooling prior to nucleation and controls the size of stable nuclei. Note that the stable critical radius r_c is theoretically the same for homogeneous and heterogeneous nucleation process. However, the amount of energy for nucleation is less for the latter process since β provides the nucleating sites and subsequently, enhances the solidification for producing small grain size microstructures.

In principle, the interface surface energy (γ_{ls}) between two phases (liquid and solid) is the basic energy entity for the nucleation of embryos, which eventually grow to form stable nuclei. Methods for measuring interface energies can be found elsewhere ([14, p. 107] and [15, p. 111]).

4.12.4 Nucleation Sites

Furthermore, heterogeneous nucleation represents a more realistic solidification process. The mold inner walls and/or suspended impurities act as nucleating agents (catalyst) in the liquid since they represent preexisting surfaces as preferred nucleation sites. This implies that heterogeneous undercooling (ΔT_{het}) is smaller than the homogeneous undercooling (ΔT_{hom}); $\Delta T_{het} < \Delta T_{hom}$.

Hence, heterogeneous nucleation is a catalyst-induced process that depends on the nucleus contact angle θ. The catalyst is known as the substrate or the grain refiner. For instance, adding grain refiners (such as TiB particles) to Al and Al-alloys is a common practice in the automotive and aerospace industries for refining the microstructure and improving properties. This can be achieved by controlling the mechanical agitation of the melt and the cooling rate.

The solidification response to adding grain refiners to the melt is to reduce the degree of undercooling ΔT of the liquidus temperature during solidification and minimizes the critical Gibbs energy ΔG_v.

4.12.5 Contact Area

In addition, the contact area between the solidifying melt and the impurities is the source for solidification in the form of stable nuclei, which are assumed to generate a uniform thickening solid with a L-S interface (boundary).

In order for a pure metal solidification to proceed, release of thermal energy called latent heat of fusion ΔH_f must occur in a continuous and uniform manner at $T \leq T_f$. For comparison, the latent heat of fusion ΔH_f^* for an alloy solidification is released at a temperature range $T_s < T_f^* < T_l$, where T_s, T_l denote the solidus and liquidus temperatures, respectively, at the alloy composition C_o.

The temperature difference between the liquid metal and the impurity surface areas causes a liquid undercooling ΔT needed for phase transformation $L \to S$ during solidification, making ΔT a driving force.

For most phase transformation cases, evaluating the derivative of Eq. (4.62a), $|d\Delta G_{het}/dr|_{r=r_c} = 0$, yields r_c as per Eq. (4.60) for a stable α particle, but $\Delta G_{het,c} < \Delta G_{hom,c}$ due to the effect of the shape factor; $0 < f(\theta) < 1$.

4.13   Steady-State Nucleation Rate

Consider the homogeneous nucleation process so that $\Delta G = f(r)$. In this case, the number of stable nuclei n_s (small solid particles having a crystalline structure in the solidifying liquid or melt) becomes a temperature-dependent function $n_s = f(T)$ with $r \geq r_c$ and the total enthalpy takes the form $n_s \Delta H$. The nuclei formation is a physical event described thermodynamically by the Gibbs energy change ΔG. From Eq. (4.8b),

$$\Delta G = n_s \Delta H - T \Delta S \tag{4.64}$$

The entropy change defined by Eq. (4.13) is now used to derive $n_s = f(T)$

$$\Delta S = k_B \ln(\omega) = k_B \ln \frac{(n_o + n_s)!}{n_o! n_s!} \tag{4.65a}$$

$$\Delta S = -k_B \left[n_s \ln \left(\frac{n_s}{n_o + n_s} \right) + n_o \ln \left(\frac{n_o}{n_o + n_s} \right) \right] \tag{4.65b}$$

where n_o denotes the number of atoms per volume in the liquid phase. Substituting Eq. (4.65b) into (4.64) yields

$$\Delta G = n_s \Delta H + k_B T \left[n_s \ln \left(\frac{n_s}{n_o + n_s} \right) + n_o \ln \left(\frac{n_o}{n_o + n_s} \right) \right] \tag{4.66}$$

If $d(\Delta G)/dn_s = 0$, then

$$\ln \left(\frac{n_s}{n_o + n_s} \right) = -\frac{\Delta G}{k_B T} \tag{4.67}$$

and if $n_o \gg n_s$, Eq. (4.67) along with $T < T_f$ that promotes a small undercooling $\Delta T > 0$ becomes

$$n_s = n_o \exp \left(-\frac{\Delta G}{k_B T} \right) \tag{4.68}$$

Here, the exponential function $\exp(-\Delta G/k_B T)$ can be treated as a phase transformation factor or as a thermodynamic energy factor needed to solidify a group of atoms n_o in the liquid into a solid particle (cluster) n_s with a critical spherical radius r_c. This implies that $\Delta G = \Delta G_c$ acts as the effective activation energy that should remain constant during the course of the liquid-to-solid phase transformation under the assumption of steady-state nucleation.

Linearizing Eq. (4.69) yields

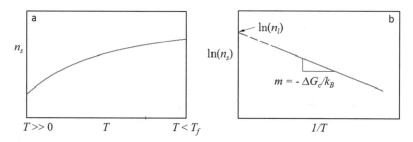

Fig. 4.16 Arrhenius plot for nucleation rate. (**a**) Non-linear behavior and (**b**) linearized Arrhenius equation

$$\ln(n_s) = \ln(n_o) - \left(\frac{\Delta G}{k_B}\right)\frac{1}{T} \tag{4.69}$$

Figure 4.16a shows the typical trend of an exponential function as defined by Eq. (4.68) and Fig. 4.16b plots Eq. (4.69) to find n_o and $\Delta G = \Delta G_c$ at a reciprocal temperature range $\Delta(1/T)$.

For $n_s = f(T) > 0$, groups of atoms in the liquid phase arrangement themselves randomly as solidification progresses under steady-state conditions and eventually, an atom attachment process transpire to what is known as nuclei of critical radius r_c predicted by Eq. (4.60) when $\Delta G = \Delta G_c$.

In addition to the preceding analysis, the number of atoms (n_o) per molar volume that can be attached to the nuclei is written as

$$n_o = \frac{N_o}{V_m} \tag{4.70a}$$

$$V_m = N_o V_{cell}/N_{atom} \simeq A_w/\rho \tag{4.70b}$$

where V_m denotes the molar volume, A_w denotes the atomic weight, and ρ denotes the mass density of the pure metal.

For nucleation of small clusters of atoms under a thermally activated process, the nucleation rate (I_s) model for a number of atoms n_s in a cluster of size $r \geq r_c$, incorporated henceforth, follows the Arrhenius-type equation as the approximated analytical approach, which is an energy-induced method. Originally, the Svante Arrhenius equation goes back to 1800s as a temperature-dependent expression for determining chemical reaction rates and subsequently, it has been adopted in many science fields as an important equation for finding the best fit to experimental data.

Thus, the kinetics of nucleation can be described by the nucleation rate (I_s) which strongly depends on the critical Gibbs energy ΔG_c for nuclei formation. This implies that the evolution of the number of nuclei (I_s) per unit volume with time at constant temperature can be modeled analytically using an Arrhenius-type equation of the form

$$I_s = \frac{dn_s}{dt} = I_o \exp\left(-\frac{\Delta G_c + Q_d}{k_B T}\right) \simeq I_o \exp\left(-\frac{\Delta G_c + Q_d}{k_B T}\right) \qquad (4.71\text{a})$$

$$\ln(I_s) = \ln(I_o) - \left(\frac{\Delta G_c}{k_B}\right)\frac{1}{T} \qquad (4.71\text{b})$$

where I_o is a constant related to vibration frequency of atoms ($1/s$) in the clusters per unit volume per unit time [$atom/(m^3\,s)$] as latent heat of fusion evolves during solidification, ΔG_c denotes the thermodynamic energy barrier for a cluster formation with a size r_c so that $\Delta G = \Delta G_c$ when $r = r_c$ at $\Delta T = (T_f - T) > 0$, Q_d is the activation energy for diffusion across the phase boundary.

As explained above, plotting Eq. (4.71b) provides useful insights on the amount of energy, namely ΔG_c, needed to solidify stable clusters and the values of I_o. Nevertheless, it can be concluded that the steeper the slope $(\Delta G_c/k_B)$, the higher the activation energy (ΔG_c) for cluster formation.

Accordingly, the proportionality constant I_o in Eq. (4.71a) can be defined in terms of atoms per unit volume per time $(atoms/m^3.s)$ under the assumption that the mold walls and nucleating agents are the potential sites for heterogeneous nuclei formation. Thus,

$$I_o = \frac{n_o k_B T}{h} \simeq v_o p_c n_o \qquad (4.72\text{a})$$

$$V_m = N_o V_{cell}/N_{atom} \qquad (4.72\text{b})$$

where $v_o = D_l/\delta^2$ denotes the atomic vibration frequency ($\simeq 10^{13}\ 1/s$) at T_f, D_l denotes the diffusion coefficient in the melt (m^2/s), δ denotes the atomic jump distance (m), p_c denotes the probability factor for attracting atoms to the nucleation sites (assumed $p_c \simeq 1$; otherwise $p_c < 1$), n_o denotes the particle density ($atoms/m^3$ or $particles/m^3$) in the melt or the density of nucleation sites at the melt–substrate interface ($sites/cm^2$ or $sites/m^2$), V_m denotes the molar volume of the liquid (cm^3/mol or m^3/mol), and V_{cell} volume of the unit cell (g/mol or kg/mol).

Substitute Eqs. (4.61) and (4.72a) into (4.71a) to get the nucleation rate $I_s = f(\Delta T)$ in more realistic form

$$I_s = \frac{n_o k_B T}{h} \exp\left[-\frac{16\pi \gamma_{ls}^3}{3k_B (\Delta H_f)^2}\frac{1}{T}\left(\frac{T_f}{\Delta T}\right)^2 - \frac{Q_d}{k_B T}\right] \qquad (4.73)$$

In fact, Eq. (4.73) clearly indicates that $I_s = f(T)$, but the driving force for nucleation is now the degree of undercooling $\Delta T = (T_f - T) > 0$. This means that phase transformation by the nucleation mechanism occurs at $T = T_f - \Delta T$.

Now, combining Eqs. (4.61) and (4.68) when $\Delta G = \Delta G_c$ yields a polynomial for nucleation at a temperature $T < T_f$

$$T^3 - 2T_f T^2 + T_f^2 T + \lambda = 0 \tag{4.74}$$

with

$$\lambda = \frac{16\pi \gamma_{ls}^3 T_f^2}{3\left(\Delta H_f\right)^2 k_B \ln\left(n_s/n_o\right)} \tag{4.75}$$

Thus, one of three roots of Eq. (4.74) is the theoretical temperature of the under-cooled liquid.

Example 4.10 Consider the nucleation of pure copper (Cu) at a slow cooling rate dT/dt under control conditions. Determine (a) the number of stable particles n_s per volume when the nuclei radius is 0.80 nm at the freezing temperature T_f, (b) the critical radius r_c and n_s at $\Delta T = 236$ K, and (c) r_c and n_s at $\Delta T = 300$ K. Will homogeneous nucleation take place at the freezing temperature T_f and at undercooling ΔT? Use the given and the relevant data from Table 4.1 to carry out all calculations. Recall that the stable nuclei n_s in the Cu-melt represent small solid particles having a crystalline structure in the solidifying liquid or melt and form randomly.

$$T_f = 1085\,°C = 1358\,K, \quad \Delta T = 236\,°C = 236\,K$$

$$\Delta S_m = 10\,J/mol\,K, \quad \Delta H_f = 1628\,J/cm^3$$

$$\gamma_{ls} = 177 \times 10^{-7}\,J/cm^2 = 0.177\,J/m^2$$

$$V_m = 7 \times 10^{-6}\,m^3/mol$$

Solution

(a) This part of the example problem assumes that there is no need to undercool the Cu-melt; $\Delta T = 0$. This means that solidification is assumed to start at the freezing temperature $T = T_f = 1358$ K. In such a case, the contribution of the Gibbs energy per volume is $G_v = 0$ and subsequently, ΔG depends on the surface energy. From Eq. (4.58b).

$$\Delta G = -\frac{4}{3}\pi r^3 \Delta G_v + 4\pi r^2 \gamma_{ls} = 4\pi r^2 \gamma_{ls}$$

$$\Delta G = 4\pi \left(0.80 \times 10^{-9}\,m\right)^2 \left(0.177\,J/m^2\right) = 1.42 \times 10^{-18}\,J$$

The constant n_o in Eq. (4.71) is

$$n_o = \frac{N_o}{V_m} = \frac{6.022 \times 10^{23}\,clusters/mol}{7 \times 10^{-6}\,cm^3/mol} = 8.60 \times 10^{28}\,clusters/m^3$$

From Eq. (4.68), the number of clusters having a radius of 0.8 nm is

$$n_s = n_o \exp\left(-\frac{\Delta G}{k_B T_f}\right)$$

$$n_s = \left(8.60 \times 10^{28} \text{ clusters/m}^3\right) \exp\left[-\frac{1.42 \times 10^{-18} \text{ J}}{\left(1.38 \times 10^{-23} \text{ J/K}\right)(1358 \text{ K})}\right]$$

$$n_s = 1.06 \times 10^{-4} \text{ clusters/m}^3$$

This result means that Cu is basically in liquid state since $n_s \simeq 0$. Therefore, no homogeneous nucleation takes place.

(b) For $\Delta T = 236$ K and $T < T_f$, Eq. (4.29) yields

$$r_c = \frac{2\gamma_{ls} T_f}{\Delta H_f \Delta T} = \frac{2\left(177 \times 10^{-7} \text{ J/cm}^2\right)(1358 \text{ K})}{\left(1628 \text{ J/cm}^3\right)(236 \text{ K})} = 1.25 \text{ nm}$$

From Eq. (4.61), the critical Gibbs energy change at $r = r_c$ is

$$\Delta G_c = \frac{16\pi \gamma_{ls}^3}{3\left(\Delta H_f\right)^2}\left(\frac{T_f}{\Delta T}\right)^2 = \frac{16\pi \left(177 \times 10^{-7} \text{ J/cm}^2\right)^3}{3\left(1628 \text{ J/cm}^3\right)^2}\left(\frac{1358 \text{ K}}{236 \text{ K}}\right)^2$$

$$\Delta G_c = 1.16 \times 10^{-18} \text{ J}$$

and at $T = T_f - \Delta T = 1358 - 236 = 1122$ K

$$n_s = n_o \exp\left(-\frac{\Delta G_c}{k_B T}\right)$$

$$n_s = \left(8.60 \times 10^{22} \text{ clusters/m}^3\right) \exp\left[-\frac{1.16 \times 10^{-18} \text{ J}}{\left(1.38 \times 10^{-23} \text{ J/K}\right)(1122 \text{ K})}\right]$$

$$n_s = 2.50 \times 10^{-10} \text{ clusters/m}^3$$

Therefore, homogeneous nucleation does not take place at $\Delta T = 236$ K because n_s is practically zero.

(c) From Eq. (4.29) with $\Delta T = 300$ K, the critical radius is

$$r_c = \frac{2\gamma_{ls} T_f}{\Delta H_f \Delta T} = \frac{2\left(177 \times 10^{-7} \text{ J/cm}^2\right)(1358 \text{ K})}{\left(1628 \text{ J/cm}^3\right)(300 \text{ K})} = 0.984 \text{ nm}$$

Again, from Eq. (4.30), the critical Gibbs energy change at $\Delta T = 300$ K is

$$\Delta G_c = \frac{16\pi \gamma_{ls}^3}{3\left(\Delta H_f\right)^2}\left(\frac{T_f}{\Delta T}\right)^2 = \frac{16\pi \left(177 \times 10^{-7}\,\text{J/cm}^2\right)^3}{3\left(1628\,\text{J/cm}^3\right)^2}\left(\frac{1358\,\text{K}}{300\,\text{K}}\right)^2$$

$$\Delta G_c = 7.18 \times 10^{-19}\,\text{J}$$

and at $T = T_f - \Delta T = 1358 - 300 = 1058$ K,

$$n_s = \left(8.60 \times 10^{22}\,\text{clusters/m}^3\right)\exp\left[-\frac{7.18 \times 10^{-19}\,\text{J}}{\left(1.38 \times 10^{-23}\,\text{J/K}\right)\left(1058\,\text{K}\right)}\right]$$

$$n_s = 37.79\,\text{clusters/m}^3 \simeq 38\,\text{clusters/m}^3$$

Consequently, nucleation takes place at $\Delta T = 300$ K since 38 clusters/cm^3 of Cu form, each having a radius of 0.984 nm, at $T = 1058$ K.

Example 4.11 This example calls for the minimum degree of undercooling ΔT and the minimum undercooled liquid temperature $T < T_f$ for the onset of nucleation. For simplicity, assume a homogeneous nucleation process in order to determine ΔT and T. Hereinafter, use both ΔT and T for heterogeneous nucleation and compare results.

For the solidification of pure aluminum (Al) with some properties taken from Dantzig-Rappaz [2, p. 275],

T_f	$= 933$ K	ρ	$= 2700\,\text{kg/m}^3$
γ_{ls}^*	$= 0.093\,\text{J/m}^2$	A_w	$= 26.98 \times 10^{-3}\,\text{kg/mol}$
$\rho \Delta S_f^*$	$= 1.02 \times 10^6\,\text{J/(m}^3\,\text{K)}$	N_{atoms}	$= 4\,\text{atoms}$
R	$= 1.43 * 10^{-10}\,\text{m}$	a	$= 2\sqrt{2}R = 4.04 \times 10^{-10}\,\text{m}$

Calculate (a) the proportionality constants n_o and I_o and the thermophysical properties ΔH_f, ΔS_f, and Γ_{ls}, (b) ΔT and T for nucleating 1 cluster/m^3 homogeneously and (c) r_c, N_c, ΔG_c, n_s, and I_s for homogeneous and heterogeneous nucleation using ΔT and T from part (b). Use the shape factor $\theta = \pi/2$ for heterogeneous nucleation and the probability factor $p_c = 1$.

Solution

(a) From Eq. (4.72b), the molar volume for Al-FCC structure along with the unit cell volume $V_{cell} = a^3$ is

$$V_m = \frac{N_o V_{cell}}{N_{atom}} = \frac{\left(6.022 \times 10^{23}\,\text{atoms/mol}\right)\left(4.04 * 10^{-10}\,\text{m}\right)^3}{4\,\text{atoms}}$$

$$V_m = 9.93 \times 10^{-6}\,\text{m}^3/\text{mol}$$

and from Eqs. (4.70a) and (4.72a) with $T = T_f = 933$ K, the proportionality constants are

$$n_o = \frac{N_o}{V_m} = \frac{6.022 \times 10^{23} \text{ atoms/mol}}{9.93 \times 10^{-6} \text{ m}^3/\text{mol}} = 6.06 \times 10^{28} \text{ atoms/m}^3$$

$$I_o = \frac{n_o k_B T_f}{h} = \frac{\left(6.06 \times 10^{28} \text{ atoms/m}^3\right)\left(1.38 \times 10^{-23} \text{ J/K}\right)(933 \text{ K})}{6.63 \times 10^{-34} \text{ J s}}$$

$$I_o = 1.18 \times 10^{42} \text{ atoms/(m}^3 \text{ s)} \quad \text{(typical value)}$$

Letting $v_o = 10^{13}$ 1/s and $p_c = 1$ Eq. (4.72a) yields

$$I_o \simeq v_o p_c n_o = \left(10^{13} \text{ 1/s}\right)(1)\left(5.80 \times 10^{28} \text{ atoms/m}^3\right)$$

$$I_o \simeq 6.06 \times 10^{41} \text{ atoms/(m}^3 \text{ s)} \quad \text{(approximate value)}$$

Both calculated I_o values are within the typical range $10^{41} \leq I_o \leq 10^{43}$ cited in the literature. Therefore, use the average $\overline{I}_o = 8.93 \times 10^{41}$ atoms/(m^3 s) for calculating I_s.

Thermophysical Properties The entropy of fusion and latent heat of fusion at $T_f = 933$ K are

$$\Delta S_f = \frac{1.02 \times 10^6 \text{ J/(m}^3 \text{ K)}}{2700 \text{ kg/m}^3} = 377.78 \text{ J/(kg K)}$$

$$\Delta H_f = T_f \Delta S_f = (933 \text{ K})\left(377.78 \frac{\text{J}}{\text{kg K}}\right) = 3.52 \times 10^5 \text{ J/kg}$$

$$\Delta H_f = \left(2700 \text{ kg/m}^3\right)\left(3.52 \times 10^5 \text{ J/kg}\right) = 9.50 \times 10^8 \text{ J/m}^3$$

and the Gibbs–Thomson coefficient is

$$\Gamma_{ls} = \frac{\gamma_{ls}}{\rho \Delta S_f} = \frac{0.093 \text{ J/m}^2}{1.02 \times 10^6 \text{ K m}} = 9.12 \times 10^{-8} \text{ K m}$$

The shape factor for heterogeneous nucleation with $\theta = \pi/2$ is

$$f\left(\theta = \pi/2\right) = \frac{1}{4}\left(2 - 3\cos\theta + \cos^3\theta\right) = 0.5$$

(b) *Finding the magnitudes of* λ, ΔT *and* T: From Eq. (4.75) with $n_s = $ 1 cluster/m^3 (minimum),

$$\lambda = \frac{16\pi \gamma_{ls}^3 T_f^2}{3\left(\Delta H_f\right)^2 k_B \ln\left(n_s/n_o\right)}$$ (E1)

$$\lambda = \frac{16\pi \left(0.093 \text{ J/m}^2\right)^3 \left(933 \text{ K}\right)^2}{3\left(9.50 \times 10^8 \text{ J/m}^3\right)^2 \left(1.38 \times 10^{-23} \text{ J/K}\right) \ln\left(1/\left(6.06 \times 10^{28}\right)\right)}$$

$$\lambda = -1.4213 \times 10^7 \text{ K}^3$$

Substitute λ and T_f into Eq. (4.74) to get

$$T^3 - 2T_f T^2 + T_f^2 T + \lambda = 0$$ (E2a)

$$T^3 - 1866T^2 + \left(8.7049 \times 10^5\right) T - 1.4213 \times 10^7 = 0$$ (E2b)

Solving this polynomial yields three roots

$$\{[T_1 = 1049.40 \text{ K}], [T_2 = 16.937 \text{ K}], [T_3 = 799.69 \text{ K}]\}$$

Root $T_1 \simeq 1049 \, K = 776\,°C$ is too high and therefore, it is not feasible for nucleation because $T_1 > T_f = 933$ K.

Root $T_2 \simeq 17 \, K = -256\,°C$ is much too low and therefore, it is not feasible for nucleation because $T_1 << T_f = 933$ K.

Root $T_3 \simeq 800 \, K = 527\,°C$ is a logical result and it is feasible for nucleation because $T_3 < T_f$ and ΔT is relatively small.

Nonetheless, using this root yields a reasonable and a small degree of undercooling

$$\Delta T = T_f - T_3 = 933 \text{ K} - 800 \text{ K} = 133 \text{ K}$$ (E3)

All relevant calculations are carried out below using $T = T_3 = 800$ K and $\Delta T = 133$ K. The results will yield interesting insights about the nucleation process and related theoretical background. Moreover, this theoretical approach suggests that $\Delta T = 133$ K is the minimum degree of undercooling needed for the onset of homogeneous nucleation of 1 cluster/volume containing N_c number of atoms, which in turn under slow and steady phase transformation acquire a preferred crystal structure at a random position within the liquid phase.

(c) Using Eq. (4.60) for both homogeneous and heterogeneous solidification at $\Delta T = 133$ K yields the theoretical radius of a cluster

$$r_c = \frac{2\Gamma_{ls}}{\Delta T} = \frac{2\left(9.12 \times 10^{-8} \text{ K m}\right)}{133 \text{ K}} = 1.37 \times 10^{-9} \text{ m}$$

$$r_c = 1.37 \text{ nm}$$

The number of atoms contained in a critical nucleus along with the constant $n_o = 6.06 \times 10^{28}$ atoms/m^3 is

$$N_c = \frac{N_o V_{sphere}}{V_m} = V_{sphere} n_o = \frac{4\pi}{3} r_c^3 n_o \tag{E4}$$

$$N_c = \frac{4\pi}{3} \left(\frac{1.8240 \times 10^{-7} \, \text{K m}}{\Delta T} \right)^3 \left(6.06 \times 10^{28} \, \text{atoms/m}^3 \right)$$

$$N_c = \frac{1.5404 \times 10^9 \, \text{atoms K}^3}{\Delta T^3} = \frac{1.54 \times 10^9 \, \text{atoms K}^3}{(133 \, \text{K})^3}$$

$$N_c = 655 \, \text{atoms}$$

The homogeneous driving force is

$$\Delta G_{c,\text{hom}} = \frac{16\pi \gamma_{ls}^3}{3 (\Delta H_f)^2} \left(\frac{T_f}{\Delta T} \right)^2 = \frac{16\pi \left(0.093 \, \text{J/m}^2 \right)^3}{3 \left(9.50 \times 10^8 \, \text{J/m}^3 \right)^2} \left(\frac{933 \, \text{K}}{\Delta T} \right)^2 \tag{E5}$$

$$\Delta G_{c,\text{hom}} = \frac{1.2999 \times 10^{-14} \, \text{J K}^2}{\Delta T^2} = \frac{1.2999 \times 10^{-14} \, \text{J K}^2}{(133 \, \text{K})^2}$$

$$\Delta G_{c,\text{hom}} = 7.35 \times 10^{-19} \, \text{J}$$

and for heterogeneous nucleation,

$$\Delta G_{c,het} = \Delta G_{c,\text{hom}} f(\theta) = \left(\frac{1.2999 \times 10^{-14} \, \text{J K}^2}{\Delta T^2} \right) (0.5) \tag{E6a}$$

$$\Delta G_{c,het} = \frac{6.4995 \times 10^{-15} \, \text{J K}^2}{\Delta T^2} = \frac{6.4995 \times 10^{-15}}{(133)^2} = 3.67 \times 10^{-19} \, \text{J} \tag{E6b}$$

Subsequently, verify that $n_{s,\text{hom}} = 1$ clusters/m^3 at $T = 800 \, \text{K}$

$$n_{s,\text{hom}} = n_o \exp \left(-\frac{\Delta G_{c,\text{hom}}}{k_B T} \right) = n_o \exp \left(-\frac{7.35 \times 10^{-19} \, \text{J}}{k_B T} \right) \tag{E7a}$$

$$n_{s,\text{hom}} = \left(6.06 \times 10^{28} \right) \exp \left[-\frac{7.35 \times 10^{-19} \, \text{J}}{\left(1.38 \times 10^{-23} \, \text{J/K} \right) (800 \, \text{K})} \right] \tag{E7b}$$

$$n_{s,\text{hom}} = n_{s,\text{hom}} = 1 \, \text{clusters/m}^3$$

and

$$n_{s,het} = n_o \exp\left(-\frac{\Delta G_{c,het}}{k_B T}\right) = n_o \exp\left(-\frac{3.67 \times 10^{-19}\,\text{J}}{k_B T}\right) \quad \text{(E8a)}$$

$$n_{s,het} = \left(6.06 \times 10^{28}\right) \exp\left(-\frac{3.67 \times 10^{-19}\,\text{J}}{\left(1.38 \times 10^{-23}\,\text{J/K}\right)(8\,\text{K})}\right) \quad \text{(E8b)}$$

$$n_{s,het} = 2 \times 10^{14}\ \text{clusters/m}^3$$

Similarly, using $\overline{I_o} = 8.93 \times 10^{41}$ atoms/(m^3 s) yields

$$I_{s,hom} = \overline{I_o} \exp\left(-\frac{\Delta G_{c,hom}}{k_B T}\right) = \overline{I_o} \exp\left(-\frac{7.35 \times 10^{-19}\,\text{J}}{k_B T}\right) \quad \text{(E9a)}$$

$$I_{s,hom} = \left(8.93 \times 10^{41}\right) \exp\left[-\frac{7.35 \times 10^{-19}\,\text{J}}{\left(1.38 \times 10^{-23}\,\text{J/K}\right)(1106\,\text{K})}\right] \quad \text{(E9b)}$$

$$I_{s,hom} = 1.09 \times 10^{21}\ \text{atoms/(m}^3\ \text{s)}$$

and

$$I_{s,het} = \overline{I_o} \exp\left(-\frac{\Delta G_{c,het}}{k_B T}\right) = \overline{I_o} \exp\left(-\frac{3.67 \times 10^{-19}\,\text{J}}{k_B T}\right) \quad \text{(E10a)}$$

$$I_{s,het} = \left(8.93 \times 10^{41}\right) \exp\left[-\frac{3.67 \times 10^{-19}\,\text{J}}{\left(1.38 \times 10^{-23}\,\text{J/K}\right)(1106\,\text{K})}\right] \quad \text{(E10b)}$$

$$I_{s,het} = 3.22 \times 10^{31}\ \text{atoms/(m}^3\ \text{s)}$$

Therefore, $\Delta G_{c,het} < \Delta G_{c,hom}$, and as expected, the following inequalities are true: $n_{s,het} \gg n_{s,hom}$ and $I_{s,het} \gg I_{s,hom}$ Plotting $r_c = f(\Delta T)$, $N_c = f(\Delta T)$, $\Delta G_c = f(\Delta T)$, $n_s = f(T)$ and so on is left out due to lack of space. Nonetheless, the plotting equations are

$$r_c = \left(1.824 \times 10^{-7}\right)/\Delta T$$
$$N_c = \left(1.5404 \times 10^{9}\right)/\Delta T^3$$
$$\Delta G_{c,hom} = \left(1.2999 \times 10^{-14}\right)/\Delta T^2$$
$$\Delta G_{c,het} = \left(6.4995 \times 10^{-15}\right)/\Delta T^2$$
$$n_{s,hom} = \left(6.06 \times 10^{28}\right) \exp\left(-53,261/T\right)$$
$$n_{s,het} = \left(6.06 \times 10^{28}\right) \exp\left(-26,594/T\right)$$
$$I_{s,hom} = \left(8.93 \times 10^{41}\right) \exp\left(-53,261/T\right)$$
$$I_{s,het} = \left(8.93 \times 10^{41}\right) \exp\left(-26,594/T\right)$$

4.14 Critical Temperature for Nuclei Formation

It has been established that solidification starts by formation of nuclei with volume V and area A at random positions in the melt and subsequently, crystal growth occurs from these nuclei, leading to the solidification reaction $L \rightarrow S$. First of all, a particle (crystal) with a radius $r < r_c$ is defined as an embryo and if it grows by additions of adjacent atoms from the melt, then it becomes a nucleus with $r = r_c$ with ΔG_c at a critical temperature T_c.

Now, assume a homogeneous nucleation process where the undercooling $\Delta T = T_f - T_c$ becomes the driving force for nuclei formation at T_c. Thus, the Gibbs excess energy $(-\Delta G_i)$ and the Gibbs molar energy $(-\Delta G_m)$ changes are mathematically approximated by

$$\Delta G_i = \frac{V}{V_m}(-\Delta G_m) + A\gamma_{ls} \tag{4.76a}$$

$$\Delta G_m = \frac{T_f - T_c}{T_c}(\Delta H_m) \tag{4.76b}$$

where $V_m = M/\rho$ denotes the molar volume (m³/kmol), M denotes the molar weight (kg/kmol), ρ denotes the particle density (kg/m³), and ΔH_m denotes the molar latent heat of fusion (kJ/kmol).

If ΔG_c is defined as the activation energy for formation of a nucleus with critical size r_c, then [16, p. 141]

$$\Delta G_c = \frac{16\pi}{3}\frac{V_m^2\gamma_{ls}^3}{(\Delta G_m)^2} \tag{4.77}$$

The average activation energy for forming a nucleus at T_c is written as

$$\Delta G_c = 60 k_B T_c \quad \text{(Energy barrier)} \tag{4.78}$$

Combining Eqs. (4.76b), (4.77), and (4.78) yields the critical temperature T_c equation

$$60 k_B T_c = \frac{16\pi}{3}\left(\frac{T_f}{T_f - T_c}\right)^2\frac{V_m^2\gamma_{ls}^3}{(\Delta H_m)^2} \tag{4.79a}$$

$$T_f^2 T_c - 2T_f T_c^2 + T_c^3 - \frac{4\pi V_m^2\gamma_{ls}^3 T_f^2}{45\Delta H_m^2} = 0 \tag{4.79b}$$

which is a convenient expression for predicting T_c for homogeneous nucleation of a pure metal.

Furthermore, for heterogeneous nucleation, as in inoculation of carbon steels with $FeTi$ particles, the Gibbs molar energy ΔG_m depends on the mole fraction of the added elements; X_{Ti} and X_C at the melt temperature T and at the critical temperature T_c. Additional details on this particular case can be found elsewhere [16, pp. 142–143].

Inoculation in casting metals and alloys is a practical method for controlling the solidification process and related microstructural features. Thus, the resultant microstructure controls mechanical behavior and to an extent, chemical or electro-chemical corrosion.

4.15 Solidification Methods

The solidification methods shown in Fig. 4.17 are common industrial practices, where the choice of sand or metallic mold material is an option for the producer. However, using metal molds implies that the liquid metal initially solidifies fast as a thin layer (metal skin) forming a gap at the interface due to mold expansion during absorption of heat and contraction (shrinkage) of the solidified layer during the release of heat (latent heat of fusion). As a result, an interface resistance develops and no physical bonding takes place at the metal–mold interface; otherwise, welding or soldering would occur.

4.15.1 Conventional Castings

The conventional casting technique is a time-consuming procedure, but it requires cost-effective sand molds and it is a common industrial practice for producing large amounts of alloys.

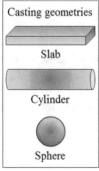
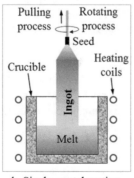
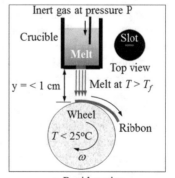

a. Conventional castings   b. Single crystal casting   c. Rapid casting

Fig. 4.17 Types of solidification techniques. (**a**) conventional quasi-static solidification of a slab, (**b**) quasi-static solidification of a single-crystal using the Czochralski process and (**c**) rapid solidification for producing ribbons. Here, ω denotes the rotating speed

This is a technique used for producing polycrystalline parts known as conventional castings, which may have simple (Fig. 4.17a) or intricate geometries.

In principle, the initial thermal contact at the start of solidification is influenced by the mold surface texture, gas interactions, casting geometric contours, and the like. Consequently, nucleation and growth stages in conjunction with the microstructural features, such as morphology and grain size, are strongly affected by the initial thermal metal-mold contact at the freezing temperature T_f and mainly by subsequent heat transfer conduction.

In general, solidification heat transfer is an energy transport phenomenon induced by the hot flowing molten metal into cold mold cavities. Then, the molten metal assumes the geometry within the mold and it begins to cool so that solidification initiates when it reaches the freezing temperature at $T = T_f$ or small undercooling ΔT. The latter is more likely to occur.

Accordingly, a thermal contact between the molten metal and the mold inner surfaces causes heat transfer due to a temperature difference ΔT. On the one hand, sand molds, however, must have some thermal conductivity coupled with porosity throughout the outer surfaces for gases and air to escape. On the other hand, ideal metallic molds do not have porosity, but a high thermal conductivity capacity for enhancing the heat removal during solidification.

4.15.2   Single Crystals

The single-crystal method (Fig. 4.17b) is based on a directional solidification process (DSP) for producing continuous monocrystalline solids known as single crystals. Figure 4.17b schematically shows the Czochralski process (Cz), discovered in 1916, that is nowadays used for manufacturing, in principle, impurity-free single crystals from a single nucleus at relatively slow upward speed (pulling rate), while the rod-seed and the crucible rotate in the opposite direction to minimize convection in the melt. Moreover, single crystals are monocrystalline solids with continuous lattice structures that lack grain boundaries and typically, they are produced as cylinders having variable dimensions.

In the Czochralski process, a seed rod is simultaneously pulled upwards and rotated in an inert environment in order to produce single crystals. This is an important directional solidification process for producing silicon single crystals used by the electronic industry. The pertinent analytical procedure related to directional solidification is included in a later chapter.

The production of single crystals is not to be confused with single-crystal X-ray diffraction. The latter is most commonly used for determining the type of unit cell and related dimensions.

4.15.3 Rapid Solidification

Rapid solidification is achievable at a larger liquid metal undercooling ΔT compared to conventional solidification. This extended undercooling causes the melt to solidify as an amorphous, a semi-amorphous or as a fine-grained microstructure. As a result, the metal casting is produced with enhanced properties.

Often amorphous alloys are produced in the form of thin ribbons, flakes, or wires by melt spinning, melt extraction, and the like. If a metallic powder is produced, then it is consolidated by hot extrusion, hot isostatic pressing (HIP) at relatively high temperatures or explosive consolidation. In any case, amorphous ribbons are subsequently chopped and consolidated. Figure 4.17c shows a schematic setup for such a purpose.

Furthermore, rapid solidification (RS) is a casting process carried out at high solidification or quenching rates. For comparison, the range of solidification rate for rapid and convention (slow) solidification techniques can be defined as indicated below

$$10^3 \,°C/s < dT/dt < 10^{12} \,°C/s \quad \text{(rapid)} \tag{4.80a}$$

$$10^{-6} \,°C/s < dT/dt > 10^3 \,°C/s \quad \text{(slow)} \tag{4.80b}$$

In fact, one must be concerned with the effect of solidification rate on microstructure evolution and mechanical properties. Thus, rapidly solidified alloys (RSA) can be classified as unique metallic or ceramic materials with superior mechanical and magnetic properties, and with corrosion-resistant capabilities.

In general, phase transformation at slow, intermediate, and rapid solidification velocity is an indirect manipulation process of atoms by decreasing the liquid temperature accordingly. Undoubtedly, more theoretical and experimental research on phase transformation phenomenon will lead to a deeper understanding of the liquid state of matter and subsequently, relevant analytical models will be developed on this subject matter. Specifically, rapid solidification can be achieved by a large of deep undercooling ΔT and a rapid extraction of heat.

4.16 Cooling Curves and Phase Diagrams

For solidifying a pure metal "A," the cooling curve shown in Fig. 4.18a is used to determine the temperature distribution related to the freezing process. The pouring and freezing temperatures denoted by T_p and T_f, respectively, are coupled with the phase diagram depicted in Fig. 4.18b.

A pure metal solidifies at a constant temperature T_f and subsequently, the solid phase cools to room temperature T_o having a possible equiaxed-grain microstructure. Nonetheless, Fig. 4.18a elucidates a two-phase transformation (liquid and

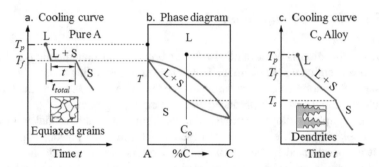

Fig. 4.18 Schematic cooling curves and related equilibrium phase diagram for a hypothetical A-C binary alloy. (**a**) Cooling curve for pure metal A showing the local solidification time t and total solidification time t_{total}, (**b**) phase diagram, and (**c**) cooling curve for an alloy with nominal composition C_o

solid), where t and t_{total} denote the local and the total solidification times,, respectively.

For a pure metal A and an A-C alloy with composition C_o, the pouring temperature is $T_p > T_f$ as indicated in the A-C equilibrium phase diagram (Fig. 4.18b). The sequence of solidification events can be summarized as follows:

- For freezing the melt, $Q_p = mc_{p,l}\Delta T_s$ where $\Delta T_s = T_p - T_f$
- For local solidification, $Q_f = m\Delta H_f$ at $T = T_f$ and $t > 0$
- For total solidification, $Q_s = mc_{p,l}\Delta T_s + m\Delta H_f$ when $T_p \to T_f$ at $t > 0$
- For cooling the solid, $Q_c = mc_{p,s}\Delta T_c$ when $T_f \to T_o = 25\,°C$ at $t \gg 0$

However, Fig. 4.16b,c are included in this section for comparison purposes since solidification of a C_o alloy is a two-phase transformation, and as expected, the analysis of transient solidification is rather complicated because the latent heat of fusion ΔH_f^* is released at a temperature range $T_s < T_f < T_l$.

In addition, the sketch shown in Fig. 4.18c is for a binary alloy that solidifies over a range of temperatures $\Delta T = T_f - T_s$. Notice the dendrites at the bottom of Fig. 4.18c. Dendrites are tree-like branching structures that grow along certain crystallographic directions as molten metal freezes.

Fundamentally, a two-phase mushy dendritic zone is formed during the solidification process. Actually, the nucleation and growth of crystals in deeply undercooled melts induce rapid formation of dendrites. This suggests that solidification proceeds by the movement of a dendritic interface within the mushy zone. Hence, dendritic solidification.

4.17 Specific Heat

The general definition of the specific heat c_p (J/kg K) or heat capacity (J/K) at constant pressure (P) is the amount of heat energy required to raise the temperature of unit mass $(m = 1\,\text{kg})$ of a material by one degree $(\Delta T = 1\,\text{K})$. Fundamentally, c_p is the slope or the first derivative of the enthalpy function $H = H(T)$, which is related to the heat transfer induced by the evolved latent heat of fusion (ΔH_f) at T_f. Thus, c_p, H, and G are

$$c_p = \left(\frac{dH}{dT}\right)_p = \left(\frac{\partial Q}{\partial T}\right)_p \qquad (4.81a)$$

$$dS = \frac{\partial Q}{T} = \frac{c_p dT}{T} \qquad (4.81b)$$

$$G = G(T) = H - TS \qquad (4.81c)$$

$$S = -(dG/dT)_p \qquad (4.81d)$$

The specific heat is temperature-dependent thermophysical property which can be described by a general function $c_p = f(T)$, but it is customary to let it be $c_p = c_p(T)$ since $f(T) = c_T$. Hence, an option on nomenclature. Commonly, the specific heat is commonly defined as a general polynomial

$$c_p = a_1 + a_2 T + a_3 T^2 + \ldots\ldots \qquad (4.82)$$

In addition, the physical behavior of the solidification process is characterized by $H = H(T)$ and $G = G(T)$ functions for the liquid and solid phases. This is schematically shown in Fig. 4.19 for a pure metal.

Fig. 4.19 Temperature-dependent energy. (**a**) Enthalpy $H = f(T)$ and (**b**) Gibbs energy $G = f(T)$

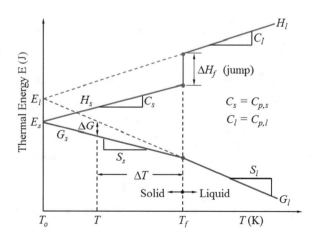

In principle, Eqs. (4.81c,d) are linear expressions applied to individual phases and as a result, on the one hand the $H = H(T)$ plot reveals an enthalpy jump defining ΔH_f and on the other hand, $G = G(T)$ shows a change in slope, which defines the entropy (Fig. 4.19), Eq. (4.81d), at the freezing temperature T_f. The extended liquidus lines for H and G intercept at a temperature T_o, where the thermal energy $E_l = H_l = G_l$. Similarly, $E_s = H_s = G_s$ at T_o. Moreover, $E_l > E_s$ since the liquid temperature is higher than that of the solid phase; otherwise, solidification would not occur.

4.18 Thermal Energy Diagrams for Metals

The fundamental relation between heat transfer and temperature for phase changes can be represented as a temperature-energy $(T\text{-}Q)$ diagram (Fig. 4.20) for easy interpretation of the amount of energy added to or extracted from a system. Notice the given equations for the thermal energy Q per region.

For the sake of clarity, assume that an amount of material is completely melted and chemically conditioned at a pouring temperature T_p. This typical case is indicated as point "P" in Figs. 4.19 and 4.20, from which the temperature range $\Delta T_s = T_p - T_f$ is called superheat. In practice, this is a common metallurgical condition prior to solidification.

Among the fundamental data required for a full analysis of the solidification process, the enthalpy profile as a function of temperature, $H = H(T)$, is of fundamental importance. For instance, Fig. 4.21 shows two schematic $H - T$ diagrams. Figure 4.21a clearly shows an enthalpy jump, defining the latent heat of fusion at the freezing temperature T_f, whereas Fig. 4.21b exhibits a similar behavior at a temperature range $T_s < T_f < T_l$.

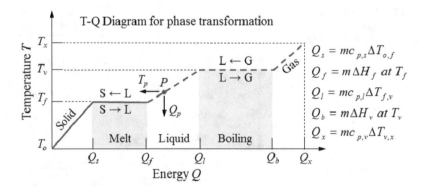

Fig. 4.20 Schematic temperature-energy (T-Q) diagram for phase transformation of a hypothetical metal or substance

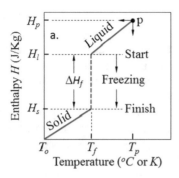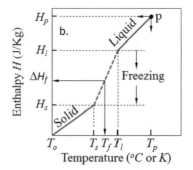

Fig. 4.21 Enthalpy diagram for freezing from $T_p \rightarrow T_f$. (**a**) Isothermal for pure metals and (**b**) non-isothermal for alloys and glassy substrates

Again, Fig. 4.21a schematically shows an isothermal freezing process at T_f for a pure metal, while Fig. 4.21b exhibits a non-isothermal process for an alloy that freezes at a temperature range $T_s \leq T_f \leq T_l$. Therefore, the latter is more complicated to analyze due to the formation of a planar or contoured L-S interface at a temperature $T_s < T_i < T_l$.

4.19 Summary

Nowadays, casting technology has evolved significantly for producing pure metals and their alloys using reliable sand or metal molds containing simple or intricate casting geometries. Undoubtedly, the mold thermophysical properties play a significant role in heat extraction during solidification.

The definition of the Gibbs energy change, $\Delta G = \Delta H - T \Delta S$, is treated as the driving force for solidification since $\Delta G < 0$ at $T < T_f$ is a general condition that describes the physical behavior of the solidifying melt. Thus, Gibbs energy function $G = f(X_B)$ for a binary A-B system when plotted at T provides significant information used to construct an equilibrium phase diagram (T-C diagram). Therefore, both G-X_B and T-C diagrams are essential for characterizing the solidification of binary alloys.

Furthermore, the classic theories of homogeneous and heterogeneous nucleation provide simple models for understanding solidification. Again, the Gibbs energy gives the basic analytical approach for determining the critical (maximum) energy change (ΔG_c) for obtaining the stable critical particle size (r_c) subjected to grow as solidification proceeds. From a macroscale point of view, the cooling curves for pure metals or alloys define the onset and end of solidification, and provide relevant data for constructing binary phase diagrams. Thus, $G = f(X_B)$ and $T = f(t)$ (cooling curve) are known methods for developing T-C equilibrium diagrams.

The concept of freezing the melt below its melting point is described by the degree of undercooling ΔT. Thus, a stable solidification process is conveniently characterized by a minimum value of the Gibbs energy change ΔG and a small value of ΔT. Nonetheless, a brief description of the solidification methods is conveniently introduced in this chapter since ΔT is an important variable that affects the final microstructural features of a casting geometry.

Furthermore, the theoretical background on the activity of binary A-B components in the liquid-state has been introduced as per Raoult's law, Henry's law, and Gibbs–Duhem equation. Specifically, an ideal solution contains an activity defined by the Raoult's law as $a_i = X_i$ at relatively high temperatures. Any deviation from this law is mathematically written as $a_i = \gamma_i X_i$, where $\gamma_i \neq 1$, and the melt is treated as a regular or non-ideal solution, which in turn is normally characterized by the Gibbs–Duhem equation.

4.20   Appendix 4A Calculation of Activities

4.20.1   Theoretical Basis of Raoult's and Henry's Laws

This appendix focuses on the activity of binary A-B components in the liquid-state as per Raoult's law, Henry's law, and Gibbs–Duhem equation. For binary A-B solutions at relatively high temperatures, Figs. 4A.1a,b schematically shows the behavior of the activity of components A and B being defined as $a_i = \gamma_i X_i$ (ideal dashed straight lines) and $a_i = f(X_i)$ (non-ideal or regular solid curves), where $i = A, B$.

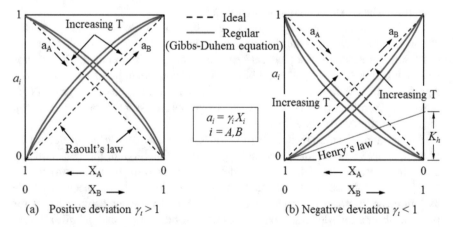

Fig. 4A.1 Schematic activity behavior and deviation from Raoult's law. (**a**) Positive deviations and (**b**) negative deviations

The **Raoult's law** dictates that $a_i = X_i = P_i/P_i^o$ with $\gamma_i = 1$ (dashed straight lines) for an ideal solution behavior. Here, P_i is the solvent pressure in the solution and P_i^o is the vapor pressure of pure component i. For regular or non-ideal solutions, a positive ($\gamma_i > 1$) or negative ($\gamma_i < 1$) deviation from the Raoult's law can be described by the Gibbs–Duhem equation at a temperature T. Notice that $a_i \rightarrow 1$ as $X_i \rightarrow 1$. Hence, A-A, B-B, and A-B atomic interactions are indirectly considered in regular solutions.

The **Henry's law** is applicable to infinite dilute solutions, where $X_i \rightarrow 0$. In this particular case, the Henry's law constant K_h can be determined graphically by plotting $P_B = f(X_B)$ in a similar manner as shown in Fig. 4A.1b. Moreover, $P_B = K_h X_B$ and $a_B = K_h X_B$ are straight lines intercepting the vertical axis at $X_B = 1$. In other words, the Henry's law is evaluated at $X_B \rightarrow 0$ and K_h is determined by extrapolating the Henry's law as illustrated in Fig. 4A.1b.

The **Gibbs–Duhem equation** is commonly applicable to regular solutions at the entire composition range so that the calculation of the solvent activity a_A knowing the solute activity a_B is possible. Nonetheless, the positive and negative deviations from the Raoult's law are represented by the solid curves in Fig. 4A.1a,b.

Example 4A.1 Assume a hypothetical pressure relationship such as $P_i = f(X_i)$ and combine Eqs. (4.27) and (4.37a) to show that $P_B = K_h X_B$, where K_h is the proportionality constant that resembles the vapor pressure by extrapolating Henry's law. What does K_h mean in Fig. 4A.1b?

Solution For a binary A-B solution at temperature T, Eq. (4.27) along with $a_A = P_A/P_A^o$ and $a_B = P_B/P_B^o$, where P_A^o, P_B^o are constant, at temperature T yields

$$\mu_A = \mu_A^o + RT \ln(a_A) = \mu_A^o + RT \ln\left(P_A/P_A^o\right) \tag{E1a}$$

$$\mu_B = \mu_B^o + RT \ln(a_B) = \mu_B^o + RT \ln\left(P_B/P_B^o\right) \tag{E1b}$$

from which

$$d\mu_A = RT \left[\frac{\partial \ln(P_A)}{\partial X_A}\right] dX_A \tag{E2a}$$

$$d\mu_B = RT \left[\frac{\partial \ln(P_B)}{\partial X_B}\right] dX_B \tag{E2b}$$

For $X_A + X_B = 1$ and $dX_A = -dX_B$, the Gibbs–Duhem equation, Eq. (4.37a), is written as

$$X_A d\mu_A + X_B d\mu_B = 0 \tag{E3}$$

Substituting Eqs. (E2a,b) into (E3) gives

$$X_A \left[\frac{\partial \ln(P_A)}{\partial X_A}\right] = X_B \left[\frac{\partial \ln(P_B)}{\partial X_B}\right] \tag{E4}$$

Letting the solvent and solute pressures be $P_A = P_A^o X_A$ and P_B, respectively, Eq. (E4) yields

$$X_A \left[\frac{\partial \ln (P_A)}{\partial X_A} \right] = X_A \left[\frac{\partial \ln (P_A^o X_A)}{\partial X_A} \right] = X_A \left[0 + \frac{1}{X_A} \right] = 1 \qquad \text{(E5a)}$$

$$X_B \left[\frac{\partial \ln (P_B)}{\partial X_B} \right] = 1 \qquad \text{(E5b)}$$

$$\partial \ln (P_B) = \frac{1}{X_B} \partial X_B \qquad \text{(E5c)}$$

Rearranging and integration Eq. (E5b) produces an integration constant C

$$\ln (P_B) = \int \frac{1}{X_B} \partial X_B = \ln (X_B) + C \qquad \text{(E6a)}$$

$$C = \ln [P_B / X_B] = \ln (K_h) \qquad \text{(E6b)}$$

from which the solute pressure (P_B) equation for infinitely dilute solutions with $X_B \to 0$ is written as

$$\ln (P_B) = \ln (X_B) + \ln (K_h) = \ln (K_h X_B) \qquad \text{(E7a)}$$

$$P_B = K_h X_B \quad \text{for } X_B \to 0 \qquad \text{(E7b)}$$

Here, on the one hand, K_h is regarded as the vapor pressure obtained by extrapolating the Henry's law in a similar manner as indicated in Fig. 4A.1b for the activity a_B. On the other hand, K_h in Fig. 4A.1b is now treated as a hypothetical activity coefficient of component B; $K_h = \gamma_h$.

4.20.2 Analytical Procedure

The analytical procedure outlined in this appendix provides the framework for calculating the chemical activity (a_A) of component A in a binary A-B alloy solution containing a nominal composition X_o at a temperature T. Thus, the goal is to show the usefulness of binary phase diagrams, such as those schematically shown in Figs. 4A.2a,b, and associated analytical thermodynamics for deriving a_A under the liquid-state condition at $T > T_c$, where T_c denotes the critical or liquidus temperature and T_f denotes the freezing or melting temperature of the A-B alloy.

Assume that the activity coefficient and the activity for a component A are $\gamma_{A,T} = \gamma_{A,T_c}$ and $a_A = \gamma_A X_A$, respectively. Letting $(X_A)_T = (X_A)_{T_c}$ and manipulating $\gamma_A = a_A / X_A$ yields

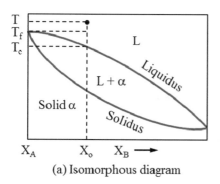

(a) Isomorphous diagram   (b) Eutectic diagram

Fig. 4A.2 Schematic A-B phase diagrams

$$\ln{(\gamma_A)}_T = \ln{(\gamma_A)}_{T_c} \tag{4A.1a}$$

$$RT\ln{(\gamma_A)}_T = RT_c\ln{(\gamma_A)}_{T_c} \tag{4A.1b}$$

$$RT\ln{(a_A)}_T - RT\ln{(X_A)}_T = RT_c\ln{(a_A)}_{T_c} - RT\ln{(X_A)}_{T_c} \tag{4A.1c}$$

$$\ln{(a_A)}_T = \frac{T_c}{T}\ln{(a_A)}_{T_c} + \frac{T-T_c}{T}\ln{(X_A)} \tag{4A.1d}$$

Fundamentally, the Gibbs energy change of mixing is

$$\Delta G = \Delta H - T\Delta S = \left(\Delta H_T - \Delta H_{T_c}\right) - \left(T_c\Delta S_{T_c} + T_f\Delta S_{T_f}\right) \tag{4A.2a}$$

$$\Delta G_m = \Delta H_m - \left(T_c - T_f\right)\Delta S_m \tag{4A.2b}$$

since the entropy change is assumed to be $\Delta S_{T_c} = \Delta S_{T_f}$, while the enthalpy change of mixing is denoted as ΔH_m for the moment.

Now, define ΔG_m in terms of specific heat capacities c_{p,T_f}, c_{p,T_c} at temperatures T_f, T_c and constant pressure P

$$\Delta G_m = \int_{T_c}^{T_f} \left(c_{p,T_f} - c_{p,T_c}\right)dT + T_o\Delta S_{m,T_c} \tag{4A.3}$$

$$- T_f\left[\Delta S_{m,T_f} + \int_{T_c}^{T_f} \frac{\left(c_{p,T_f} - c_{p,T_c}\right)}{T}dT\right]$$

These integrals are solved using a special condition.
   If $c_{p,T_f} = c_{p,T_c}$, then Eq. (4A.3) becomes

$$\Delta G_m = \left(T_o - T_f\right)\Delta S_m \tag{4A.4}$$

At equilibrium, $\Delta G_m = 0$ and $\Delta H_m = \Delta H_f$ so that

$$\Delta S_m = \frac{\Delta H_f}{T_f} \qquad (4A.5)$$

From Eqs. (4A.4) and (4A.5),

$$\Delta G_m = \frac{(T_o - T_f)\,\Delta H_f}{T_f} \qquad (4A.6a)$$

$$RT_c \ln (a_A)_{T_c} = \frac{(T_o - T_f)\,\Delta H_f}{T_f} \qquad (4A.6b)$$

$$\frac{T_c}{T} \ln (a_A) = -\frac{(T_f - T_o)\,\Delta H_f}{RTT_f} \qquad (4A.6c)$$

Combining Eqs. (4A.1d) and (4A.6c) gives the sought equation for the activity of component A in the binary A-B metallic solution at T [8, p. 51]

$$\ln (a_A)_T = -\frac{(T_f - T_o)\,\Delta H_f}{RTT_f} + \frac{T - T_c}{T} \ln (X_A) \qquad (4A.7)$$

For convenience, assume a binary A-B liquid solubility only at T and use $R = 1.987 \text{ cal/mol K}$ and $\ln(a_A)_T = 2.3026 \log(a_A)_T$ to modify Eqs. (4A.1d) and (4A.7). The resultant expressions are written as ([8, p. 51] and [17])

$$\log (a_A)_T = \frac{T_c}{T} \log (a_A)_{T_c} + \frac{T - T_c}{T} \log (X_A) \qquad (4A.8a)$$

$$\log (a_A)_T = -\frac{(T_f - T_c)\,\Delta H_f}{4.5753\,(TT_f)} + \frac{T - T_c}{T} \log (X_A) \qquad (4A.8b)$$

Furthermore, the corresponding solute vapor pressure (P_A) equation at T is written as a polynomial [9, pp. 358–377]

$$\log (P_A) = c_1 + c_2 T + c_3 \log (T) + c_4/T \qquad (4A.9)$$

For zinc (P_{Zn} in mm-Hg units) in Cu-Zn alloy solutions between the melting and boiling temperature range $419.5\,^\circ\text{C} \le T \le 907\,^\circ\text{C}$,

$$\log (P_{Zn}) = 12.34 - 1.255 \log (T) - 6620/T \quad \text{at } T_m \le T \le T_b \qquad (4A.10)$$

If $T = 1060\,^\circ\text{C} = 1333\,\text{K}$, then

$$\log (P_{Zn}) = 12.34 - (1.255) \log (1333) - 6620/1333$$

$$\log (P_{Zn}) = 3.4521$$

$$P_{Zn} = 10^{3.4521} = 2832 \, \text{mm-Hg}$$

$$P_{Zn} = (2832.0) \, (0.00131579) = 3.73 \, \text{atm}$$

which becomes $P_{Zn} \simeq 4 \, \text{atm}$ as shown in Example 4.5.

Example 4A.2 Consider a representative binary Cu-Zn alloy solution containing $X_A = X_{Cu} = 60\%$ at $= 0.60$. Use Eqs. (4A.7) and (4A.8b) to calculate a_{Cu} at (a) $T = 1060\,^{\circ}\text{C}$ (1333 K) and (b) $T = 1100\,^{\circ}\text{C}$ (1373 K). The pertinent phase diagram is shown below for two temperature cases.

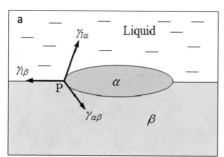
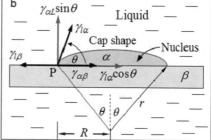

Solution

(a) From the phase diagram at $X_{Cu} = 60\%$ at $= 0.60$, the relevant temperatures are $T_f = 1083\,^{\circ}\text{C} = 1356 \, \text{K}$, $T_c = 900\,^{\circ}\text{C} = 1173 \, \text{K}$, and $T_o = 298 \, \text{K}$. And from Table 4.1, the latent heat of fusion is $\Delta H_f = 1628 \, \text{J/cm}^3$ which is converted to

$$\Delta H_f = \left(1628 \, \text{J/cm}^3\right) \left(1/8.96 \, \text{g/cm}^3\right) (63.546 \, \text{g/mol})$$

$$\Delta H_f = 11{,}546 \, \text{J/mol} = 2760 \, \text{cal/mol}$$

Using Eq. (4A.7) at $T_f > T = 1060\,^{\circ}\text{C} = 1333 \, \text{K}$ yields

$$\ln (a_{Cu})_T = -\frac{(T_f - T_o) \, \Delta H_f}{RTT_f} + \frac{T - T_c}{T} \ln (X_{Cu})$$

$$\ln (a_{Cu})_T = -\frac{(1356 - 298) \, (11{,}546)}{(8.314) \, (1333) \, (1356)} + \frac{(1333 - 1173)}{1333} \ln (0.60)$$

$$\ln (a_{Cu})_T = -0.87418$$

$$a_{Cu} = \exp (-0.87418) = 0.4172$$

From Eq. (4A.8b) along with $\Delta H_f = 2760 \, \text{cal/mol}$,

$$\log (a_{Cu})_T = -\frac{(T_f - T_c) \Delta H_f}{4.5753 T T_f} + \frac{T - T_c}{T} \log (X_{Cu})$$

$$\log (a_{Cu})_T = -\frac{(1356 - 1173)(2760)}{(4.5753)(1333)(1356)} + \frac{(1333 - 1173)}{(1333)} \log (0.60)$$

$$\log (a_{Cu})_T = -0.23063$$

$$a_{Cu} = 10^{-0.23063} = 0.58799$$

(b) Using Eq. (4A.7) at $T_f < T = 1100\,^\circ C = 1373\,K$ yields

$$\ln (a_{Cu})_T = -\frac{(T_f - T_o) \Delta H_f}{R T T_f} + \frac{T - T_c}{T} \ln (X_{Cu})$$

$$\ln (a_{Cu})_T = -\frac{(1356 - 298)(11,546)}{(8.314)(1373)(1356)} + \frac{(1373 - 1173)}{1373} \ln (0.60)$$

$$\ln (a_{Cu})_T = -0.86359$$

$$a_{Cu} = \exp (-0.86359) = 0.42165$$

Therefore, Eq. (4A.7) yields $(a_{Cu})_{1373\,K} > (a_{Cu})_{1333\,K}$ and it is consistent with the information given in Fig. 4A.1b.

4.21   Appendix 4B The Shape Factor

Figure 4B.1 shows the three-phase model for heterogeneous nucleation. Figure 4B.1a depicts a general elliptical nucleus α in contact with the liquid and substrate β and the directions of the surface energies at a triple junction "P". Figure 4B.1b, however, illustrates a simplified model along with trigonometric information. Nonetheless, the triple junction "P" imposes surface forces (surface tensions) that must be defined as per Fig. 4B.1b.

Fig. 4B.1 Three-phase model for heterogeneous nucleation. (**a**) General model for one nucleus and (**b**) modified model showing the directions of the surface tensions

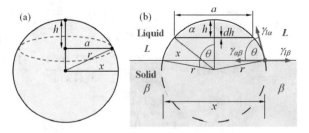

According to Newton's second law, the sum of the forces is $\sum F_i = 0$ at mechanical equilibrium. For heterogeneous nucleation, the mathematical definitions of individual forces based on trigonometry, as found in Kurz–Fisher book ([13] Appendix A3), are written as

$$\gamma_{\alpha\beta} = \gamma_{l\alpha} \cos (\theta) + \gamma_{l\beta} \cos (\theta) \tag{4B.1a}$$

$$\gamma_{l\alpha} \sin (\theta) = \gamma_{l\beta} \sin (\theta) \tag{4B.1b}$$

$$\gamma_{l\beta} = \gamma_{\alpha\beta} + \gamma_{l\alpha} \cos (\theta) \tag{4B.1c}$$

The energy balance at the triple junction suggests that the Gibbs energy change is due to the creation of the interface energy for nucleation and the energy gained by the substrate β. Mathematically,

$$\Delta G_\gamma = \left(A_{l\alpha}\gamma_{l\alpha} + A_{\alpha\beta}\gamma_{\alpha\beta}\right) - A_{\alpha\beta}\gamma_{l\beta} = A_{l\alpha}\gamma_{l\alpha} + A_{\alpha\beta}\left(\gamma_{\alpha\beta} - \gamma_{l\beta}\right) \tag{4B.2}$$

where $A_{l\alpha}, A_{\alpha\beta}, A_{\alpha\beta}$ denote surface areas and $\gamma_{l\alpha}, \gamma_{\alpha\beta}, \gamma_{l\beta}$ denote the surface energies. Combining Eqs. (4A.1c) and (4A.7) with $A_{\alpha\beta} = \pi r^2$ yields

$$\Delta G_\gamma = A_{l\alpha}\gamma_{l\alpha} - A_{\alpha\beta}\gamma_{l\alpha} \cos (\theta) = A_{l\alpha}\gamma_{l\alpha} - \pi r^2 \gamma_{l\alpha} \cos (\theta) \tag{4B.3a}$$

$$\Delta G_\gamma = \left[A_{l\alpha} - \pi r^2 \cos (\theta)\right] \gamma_{l\alpha} \tag{4B.3b}$$

The total Gibbs energy change for heterogeneous nucleation is the sum of Gibbs energy change (ΔG_f) per nucleus volume at the freezing temperature T_f and the Gibbs energy change (ΔG_γ) due to the formation of nucleus surface area

$$\Delta G_{het} = \Delta G_f + \Delta G_\gamma = V\Delta G_v + A\gamma_{l\alpha} \tag{4B.4a}$$

$$\Delta G_{het} = V_\alpha \Delta G_v + \left[A_{l\alpha} - \pi r^2 \cos (\theta)\right] \gamma_{l\alpha} \tag{4B.4b}$$

where $V = V_\alpha$ denotes the volume, $A = A_\alpha$ denotes the surface area of the nucleus α, and ΔG_f is written as

$$\Delta G_f = \Delta T \Delta S_f \tag{4B.5}$$

The remaining variables, V_α and $A_{l\alpha}$, are derived in the next section.

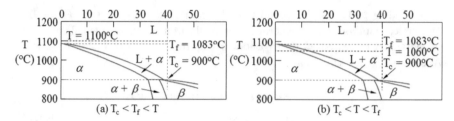

Fig. 4B.2 (a) Sphere and (b) spherical cap

4.21.1 Spherical Cap

Assume that the volume of a spherical cap (Spherical dome) in the region of a sphere is laying above a given plane. Actually, a spherical cap segment is a cutoff sphere along a plane.

Now, consider a sphere containing a thin disk (Fig. 4B.2a) and a spherical cap (Fig. 4B.2b) for deriving the shape factor $f(\theta)$ as a multiplier required to define the Gibbs energy ΔG_{het} for heterogeneous nucleation during solidification.

Alternatively, Fig. 4B.2b shows a general overview of the sphere on a substrate β surface and a thin α disk with radius a. This figure also illustrates the proper nomenclature for deriving the surface energy $\gamma_{l\alpha}$ and the volume V_a of the spherical cap. as per Eq. (4B.4b).

Using the spherical cap of an embryo is simply an approximation methodology that seems to capture the attention of the scientific and academic communities. In light of this, the surface areas of interest along with the thin disk radius $a = r\sin\phi$ are defined by

$$A_{\alpha\beta} = \pi x^2 = \pi r^2 \left[1 - \cos^2(\theta)\right] \tag{4B.6a}$$

$$A_{l\alpha} = \int_0^\theta 2\pi a r d\phi^2 \left[1 - \cos(\theta)\right] = 2\pi r^2 \left[1 - \cos(\theta)\right] \tag{4B.6b}$$

The area A_a, the differential thickness dh, and the differential volume dV_α of the α disk are

$$A_a = \pi r^2 \sin^2\theta \tag{4B.7a}$$

$$dh = (rd\theta)\sin(\theta) \tag{4B.7b}$$

$$dV_\alpha = A_a dh \tag{4B.7c}$$

Combining Eqs. (4B.5) and integrating the resultant expression yields the volume of the disk

$$V_\alpha = \int_0^\theta \left[\pi r^2 \sin^2(\phi)\right] [r\,d\phi \sin(\phi)] = \pi r^3 \int_0^\theta \sin^3(\phi)\,d\phi \qquad (4B.8a)$$

$$V_\alpha = \pi r^3 \left[\frac{2}{3} - \frac{3}{4}\cos\theta + \frac{1}{12}\cos 3\theta\right] \qquad (4B.8b)$$

where $V_\alpha = V_{het}$.

Use the trigonometric identity $\cos(3\theta) = 4\cos^3(\theta) - 3\cos(\theta)$ and the volume V_{hom} to get

$$V_{het} = \pi r^3 \left[\frac{2}{3} - \cos(\theta) + \frac{1}{3}\cos^3(\theta)\right] \qquad (4B.9a)$$

$$V_{het} = \frac{\pi r^3}{3}\left[2 - 3\cos(\theta) + \cos^3(\theta)\right] \qquad (4B.9b)$$

$$V_{hom} = \frac{4\pi r^3}{3} \qquad (4B.9c)$$

$$f(\theta) = \frac{V_{het}}{V_{hom}} = f_v = \text{Volume fraction} \qquad (4B.9d)$$

$$f(\theta) = \frac{2 - 3\cos(\theta) + \cos^3(\theta)}{4} = \frac{[2 + \cos(\theta)][1 - \cos(\theta)]^2}{4} \qquad (4B.9e)$$

which is the same expression defined by Eq. (4.63). Thus,

$$\Delta G_{het} = \left(-\frac{4\pi r^3}{3}\Delta G_v + 4\pi r^2 \gamma_{l\alpha}\right)\left[\frac{2 - 3\cos(\theta) + \cos^3(\theta)}{4}\right] \qquad (4B.10a)$$

$$\Delta G_{het} = \left(-\frac{\pi r^3}{3}\Delta G_v + 4\pi r^2 \gamma_{l\alpha}\right) f(\theta) \qquad (4B.10b)$$

$$\Delta G_{het} = \Delta G_{hom} f(\theta) \qquad (4B.10c)$$

$$f(\theta) = \frac{\Delta G_{het}}{\Delta G_{hom}} = \frac{V_{het}}{V_{hom}} \qquad (4B.10d)$$

Notice that Eqs. (4B.10b,c) are equivalent to (4.62a,b).

4.22 Problems

4.1 For the solidification of pure nickel (Ni) under homogeneous nucleation mechanism, calculate (a) the critical radius r_c and the critical Gibbs energy change ΔG_c (activation energy) for a stable nucleus having a perfect spherical shape. Use the data listed in Table 4.1. (b) If the atomic radius of nickel is $R = 0.125$ nm, then compute the number of unit cells N_{cell} and the number of atoms N_{atoms} in the critical nucleus. Assume that the nucleus is stable and spherical. [Solution: (a) $r_c = 0.67$ nm and (b) $N = 112$ atoms].

4.2 For the solidification of pure iron (Fe), calculate (a) the critical radius r_c and the critical Gibbs energy change ΔG_c for a stable nucleus having a perfect spherical shape under homogeneous nucleation mechanism. Use the data listed in Table 4.1. (b) If the atomic radius of nickel is $R = 0.124$ nm, then compute the number of unit cells N_{cell} and the number of atoms N_{atoms} in the critical nucleus. Assume that the nucleus is stable and spherical. [Solution: (a) $r_c = 1.01$ nm and (b) $N = 404$ atoms].

4.3 Show that $\Delta G = G_l - G_s = \Delta S_f \Delta T < 0$ for melting, provided that $\Delta H(T) = \Delta H(T_f)$ and $\Delta S(T) = \Delta S(T_f)$.

4.4 Show that $\Delta G = G_l - G_s = \Delta S_f \Delta T < 0$ for solidification, provided that $\Delta H(T) = \Delta H(T_f)$ and $\Delta S(T) = \Delta S(T_f)$.

4.5 Let $\Delta H_f = 8000$ J/mol and $\Delta S_f = 10$ J/mol K. Determine (a) if $L \rightarrow S$ occurs spontaneously at 1100 K and (b) the work done for such a phase transformation. [Solution: (b) $W = 3$ kJ/mol].

4.6 Let the fundamental Gibbs energy change equation for mixing A and B elements be defined by

$$\Delta G = RT [X_A \ln X_A + X_B \ln X_B + \Omega X_A X_B]$$

where Ω is a constant. Here, $X_A = 1 - X_B$ and consequently, there is symmetry. Based on this information, (a) find the critical values of $X_B = X_c$ and $\Omega = \Omega_c$ that lead to the symmetry inherent in this equation, (b) the critical ΔG_c and (c) plot $\Delta G / RT = f(X_B)$ to reveal that $\Delta G = \Delta G_c$ is a minimum at X_c.

4.7 If pure nickel (Ni) is solidified at a slow cooling rate dT/dt, determine (a) the number of stable particles n_s per volume when the nuclei radius is 0.80 nm at the freezing temperature T_f, (b) the critical radius r_c and n_s at $\Delta T = 300$ K, (c) r_c and n_s at $\Delta T = 480$ K, and (d) the number of unit cells and atoms for $\Delta T = 480$ K. Will homogeneous nucleation take place at the freezing temperature T_f and undercooling ΔT? Data:

$$T_f = 1726 \text{ K}, \quad \Delta S_m = 10 \text{ J/mol K}, \quad \Delta H_f = 2756 \text{ J/cm}^3$$

$$\gamma_{ls} = 0.255 \text{ J/m}^2, \quad V_o = A_w/\rho \simeq 7 \text{ cm}^3/\text{mol} = 7 \times 10^{-6} \text{ m}^3/\text{mol}$$

4.8 Consider the solidification of pure iron (Fe) at a slow cooling rate dT/dt. Determine (a) the number of stable particles n_s per volume when the nuclei radius is 0.80 nm at the freezing temperature T_f, (b) the critical radius r_c and n_s at $\Delta T = 400$ K and (c) r_c and n_s at $\Delta T = 420$ K, and (d) the number of unit cells and atoms for $\Delta T = 420$ K. Will homogeneous nucleation take place at the freezing temperature T_f and undercooling ΔT? Use the molar volume $V_o =$

7×10^{-6} m³/mol, the entropy of fusion $\Delta S_m = 10$ J/mol K and the relevant data from Table 4.1.

4.9 Assume that pure silver (Ag) solidifies at a slow cooling rate dT/dt. Determine (a) the number of stable particles n_s per volume when the nuclei radius is 0.80 nm at the freezing temperature T_f, (b) the critical radius r_c and n_s at $\Delta T = 200$ K and (c) r_c and n_s at $\Delta T = 300$ K, and (d) the number of unit cells and atoms for $\Delta T = 300$ K. Will homogeneous nucleation take place at the freezing temperature T_f and undercooling ΔT? Use the entropy of fusion $\Delta S_m = 10$ J/mol K and the relevant data from Table 4.1.

4.10 Consider the homogeneous nucleation of pure nickel (Ni) with a typical undercooling, $\Delta T = 480$ K, latent heat of fusion $\Delta H_f = 2756$ J/cm³, surface energy $\gamma_{ls} = 255 \times 10^{-7}$ J/cm², and melting or freezing temperature $T_f = 1726$ K. Use this information to calculate (a) the critical radius (r_c), (b) the critical Gibbs energy change (ΔG_c) at $r = r_c$, (c) the number of unit cells (N_{cell}), and (d) the number of atoms (N_{atom}) in a nucleus. Moreover, nickel at T_f has an FCC crystal structure with 4 atoms in the unit cell and an atomic radius $r = 0.125$ nm, which is normally determined using the X-ray technique.

4.11 This problem calls for the minimum degree of undercooling ΔT and the minimum undercooled liquid temperature $T < T_f$ for the onset of nucleation. For simplicity, assume a homogeneous nucleation process in order to determine ΔT and T. Hereinafter, use both ΔT and T for heterogeneous nucleation and compare results. Consider the solidification of pure copper $(Cu\text{-}FCC)$ with some properties taken from Askeland–Fulay [12, p. 262].

T_f	$= 1085\,°C$	ρ	$= 8940$ kg/m³
γ_{ls}^*	$= 0.177$ J/m²	A_w	$= 63.55 \times 10^{-3}$ kg/mol
$\Delta H_f^* =$	$= 1628$ J/cm³	N_{atom}	$= 4$ atoms per unit cell
R	$= 1.28 \times 10^{-10}$ m	a	$= 2\sqrt{2}R = 3.62 \times 10^{-10}$ m

Calculate (a) the proportionality constants n_o and I_o, the thermophysical properties ΔH_f, ΔS_f and Γ_{ls}, (b) ΔT and T for nucleating 1 cluster/m³ homogeneously and (c) r_c, N_c, ΔG_c, n_s and I_s for homogeneous and heterogeneous nucleation using ΔT and T from part (b). Use the shape factor $\theta = \pi/2$ for heterogeneous nucleation and the probability factor $p_c = 1$.

4.12 Assume that melting pure lead (Pb) under controlled conditions and consider the processes schematically shown in the T-S diagram given below. If the latent heat of melting (ΔH_f) of lead (Pb) is 4810 J/mol at 600 K melting temperature and 1-atm pressure, then calculate the entropy, enthalpy, and Gibbs energy changes when 1 mole of liquid spontaneously freezes at 590 K. This means that the amount of undercooling is a small absolute temperature change set to $\Delta T = 10$ K.

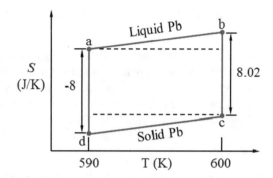

Here, the constant-pressure molar heat capacity of the liquid and the solid Pb phases and the entropy change cycle for solidifying Pb are adopted from Gaskell–Laughlin book given as [4, p. 87]

$$C_l = 32.40 - 3.10 \times 10^{-3} T \quad \text{(in J/K)}$$

$$C_s = 23.56 + 9.75 \times 10^{-3} T \quad \text{(in J/K)}$$

Calculate the relevant thermal energies, such as ΔS, ΔH, and ΔG for each process; ab, bc, cd, da. Then, compute the entropy generation ΔS_{gen} as the degree of irreversibility during solidification of lead under controlled conditions.

4.13 This problem calls for the minimum degree of undercooling ΔT and the minimum undercooled liquid temperature $T < T_f$ for the onset of nucleation. For simplicity, assume a homogeneous nucleation process in order to determine ΔT and T. Hereinafter, use both ΔT and T for heterogeneous nucleation and compare results. Consider the solidification of pure copper ($Fe\text{-}FCC$) with some properties taken from Askeland–Fulay [12, p. 262].

T_f	$= 1538\,°C$	ρ	$= 7870$ kg/m^3
γ_{ls}^*	$= 0.204$ J/m^2	A_w	$= 55.85 \times 10^{-3}$ kg/mol
ΔH_f^*	$= 1737$ J/cm^3	N_{atom}	$= 4$ atoms per unit cell
R	$= 1.24 \times 10^{-10}$ m	a	$= 2\sqrt{2}R = 3.51 \times 10^{-10}$ m

Calculate (a) the proportionality constants n_o and I_o, the thermophysical properties ΔH_f, ΔS_f, and Γ_{ls}, (b) ΔT and T for nucleating 1 cluster/m^3 homogeneously and (c) r_c, N_c, ΔG_c, n_s, and I_s for homogeneous and heterogeneous nucleation using ΔT and T from part (b). Use the shape factor $\theta = \pi/2$ for heterogeneous nucleation and the probability factor $p_c = 1$.

4.14 Consider the solidification process for a hypothetical material and assume the pressure-temperature relation as defined by the Clapeyron equation given below along with the molar volume V_m.

$$\frac{dP}{dT} = \frac{\Delta H_f}{\Delta V_m T_f} \quad \& \quad m = \rho V_m$$

Calculate the melting temperature T_f at $P = 150\,\text{kPa}$ using the following data

$\Delta H_f = 20\,\text{kJ/mol}$	$P_1 = 100\,\text{kPa}$	$T_1 = 1273\,\text{K}$
$\rho_s = 8000\,\text{kg/m}^3$	$\rho_l = 900\,\text{kg/m}^3$	
$V_s = 0.020\,\text{L/mol}$	$V_s = 0.018\,\text{L/mol}$	

4.15 Consider a hypothetical pure metal undergoing solidification by releasing $\Delta H_f = 8200\,\text{J/mol}$ at $1100\,\text{K}$. Determine (a) if $L \rightarrow S$ occurs spontaneously when $\Delta S_f = 11\,\text{J/mol K}$ and (b) the work done for such a phase transformation.

4.16 Consider a $Zn\text{-}Cd$ electrochemical solution at a temperature of $800\,\text{K}$ with mole fractions $X_{Zn} = 0.8$ and $X_{Cd} = 0.2$, and activity coefficient

$$\gamma_{Cd} = 2.942 - 4.155 X_{Cd} + 0.45 X_{Cd}^2 + 3 X_{Cd}^3$$

Calculate the (a) the interaction mixing parameter Ω_m, (b) the activity coefficient γ_{Zn}, (c) the activities a_{Zn} and a_{Cd}, (d) the thermodynamic energies of mixing ΔH_m, ΔG_m and ΔS_m.

4.17 Use the Gibbs–Duhem equation in the form of Eq. (4.42b) and assume that the following equation is valid above the melting temperature $T > T_f^*$

$$RT \ln (\gamma_2) = \Omega_m X_1^2$$

(a) For a binary alloy, show that

$$RT \ln (\gamma_1) = \Omega_m X_2^2$$

(b) For a suitable amount of liquid brass containing $X_1 = X_{Cu} = 0.60$, and $X_2 = X_{Zn} = 0.40$ mole fractions, calculate the activity coefficients ($\gamma_1 = \gamma_{Cu}$ and $\gamma_3 = \gamma_{Zn}$) and the corresponding activities (a_{Cu} and a_{Zn}) at $T = 1200\,\text{K}$. Assume that $\Omega_m = -38,300\,\text{J/mol}$ at $1000\,\text{K} \leq T \leq 1500\,\text{K}$.

4.18 Consider the given data set for $Al\text{-}Cu$ alloys at $1100\,^\circ\text{C}$ ($1373\,\text{K}$) taken from Upadhyaya–Dube example [6, p. 166]
(a) Perform curve fitting using the activity coefficient function $\gamma_{Cu} = f(X_{Cu})$, plot the given data and the resultant equation and (b) determine the γ_{Al} and a_{Al} values in an $Al\text{-}Cu$ alloy at $X_{Al} = 0.60$ and $X_{Cu} = 0.40$ mole fractions.

X_{Al}	0.10	0.20	0.30	0.40	0.50	0.60	0.70	0.80
X_{Cu}	0.90	0.80	0.70	0.60	0.50	0.40	0.30	0.20
a_{Cu}	0.86	0.61	0.34	0.18	0.08	0.05	0.02	0.01
γ_{Cu}	0.96	0.76	0.49	0.30	0.16	0.13	0.07	0.05

4.19 Assume that pure nickel Ni with a typical undercooling, $\Delta T = 450\,K$ undergoes homogeneous nucleation. For latent heat of fusion $\Delta H_f = 2756\,J/cm^3$, surface energy $\gamma_{ls} = 255 \times 10^{-7}\,J/cm^2$, and melting or freezing temperature $T_f = 1726\,K$, calculate (a) the critical radius (r_c), (b) the critical Gibbs energy change (ΔG_c) at $r = r_c$, (c) the number of unit cells (N_{cell}), and (d) the number of atoms (N_{atom}) in a nucleus. Moreover, nickel at T_f has an FCC crystal structure with 4 atoms in the unit cell and an atomic radius $r = 0.125\,nm$, which is normally determined using the X-ray technique.

4.20 Show that the Gibbs–Duhem equation can be written as

$$X_1 \frac{d \ln (\gamma_1)}{d X_1} = X_2 \frac{d \ln (\gamma_2)}{d X_2}$$

4.21 Use the given data set (taken from the thermodynamics world) (a) to determine the $\gamma_{Al} = f\,(X_{Al})$ and $a_{Al} = f\,(X_{Al})$ functions using least squares regression. The data is valid for liquid Al-Si alloys at $1150\,K \leq T \leq 1470\,K$. (b) Calculate Ω_m and a_{Al} at $X_{Si} = 0.18$ and $T = 1300\,K$.

X_{Si}	0	0.05	0.10	0.15	0.20	0.25	0.30	0.35	0.40
a_{Al}	1	0.92	0.83	0.77	0.69	0.62	0.55	0.49	0.43

4.22 The given schematic figures help visualize the aspects of thermodynamics for finding relationships between the activities a_i and the activity coefficients γ_i in binary A-B solutions at high temperatures. Manipulate Eqs. (4.21), (4.27) and (4.28) along with $X_A X_B = X_A^2 X_B + X_A X_B^2$ [17, p. 20] to derive Eq. (4.30) for the activity coefficients of components A and B.

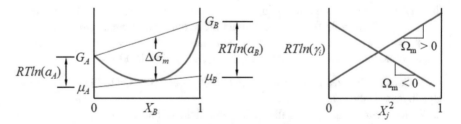

4.23 For a binary A-B solution, show that $\partial G_m / \partial X_B$ and $\partial G_m / \partial X_A$ are the linear slopes of the tangent lines (tangent rule) in Fig. 4.6 or 4.22, where ΔG_m denotes the Gibbs energy change of mixing and G_A, G_B denote the partial Gibbs energies.

4.24 For a binary Cu-Zn alloy solution containing $X_{Zn} = 0.40$ at $T = 1060\,°C$, analytically calculate the activity of copper, a_{Cu}, using the data given in Example 4.5 and a curve fitting equation related to the curve described by the relation $X_{Zn}/X_{Cu} = f[-ln(\gamma_{Zn})]$.

4.25 Consider the thermodynamic data set for regular $Cr\text{-}Ti$ alloy solutions at $T = 1250\,°C$ (1523 K) given in Example 4.6 to calculate (a) the shaded area under the curve described by the power function $X_{Cr}/X_{Ti} = b\,[\ln(\gamma_{Cr})]^c$, where b, c are constants and (b) the activity a_{Ti} using the Gibbs–Duhem equation at $X_{Cr} = 0.53$ and $X_{Ti} = 0.47$.

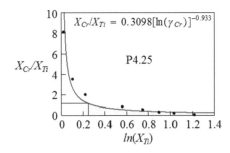
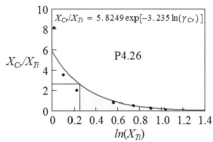

4.26 Consider the thermodynamic data set for regular $Cr\text{-}Ti$ alloy solutions at $T = 1250\,°C$ (1523 K) given in Example 4.6 to calculate (a) the shaded area under the curve described by the function $X_{Cr}/X_{Ti} = b\exp[c\ln(\gamma_{Cr})]$, where b, c are constants and (b) the activity a_{Ti} using the Gibbs–Duhem equation at $X_{Cr} = 0.37$ and $X_{Ti} = 0.63$.

References

1. N. Perez, Scientific method in analysis of local Montana semisynthetic sands and their suitability to ferrous foundry usage, M.S. Thesis, Montana Tech., Butte, Montana (1983)
2. J.A. Dantzig, M. Rappaz, *Solidification*, 2nd edn. (EPFL Press, Lausanne, 2016). Revised & Expanded
3. R.T. DeHoff, *Thermodynamics in Materials Science* (McGraw-Hill, New York, 1993)
4. N.A. Gokcen, *Statistical Thermodynamics of Alloys* (Plenum Press, New York, 1986)
5. A. Krupkowski, Activity coefficients in binary and many-component solutions. Bull. Int. Acad. Pol. Sci. Let. Vol. Ser. A **1**(1), 15–45 (1950)
6. G.S. Upadhyaya, R.K. Dube, *Problems in Metallurgical Thermodynamics and Kinetics* (Pergamon Press, New York, 1977)
7. T. Oishi, S. Tagawa, S. Tanegashima, Activity measurements of copper in solid copper-nickel alloys using copper-beta-alumina. Mater. Trans. **44**(6), 1120–1123 (2003)
8. L.S. Darken, R.W. Gurry, *Physical Chemistry of Metals* (McGraw Hill Book Company, New York, 1953)
9. O. Kubaschewski, C.B. Alcock, *Metallurgical Thermochemistry*, 5th edn. (Pergamon Press, Oxford, 1979)

10. D.V. Ragone, *Thermodynamics of Materials*, vol. I (Wiley, New York, 1995)
11. D.R. Gaskell, D.E. Laughlin, *Introduction to the Thermodynamics of Materials*, 6th edn. (Taylor & Francis Group, LLC/CRC Press, New York, 2018)
12. D.R. Askeland, P.P. Fulay, D.K. Bhattacharya, *Essentials of Materials Science and Engineering*, 2nd edn. (Cengage Learning, Stanford, 2010). ISBN-13: 978-0-495-43850-2
13. W. Kurz, D.J. Fisher, *Fundamentals of Solidification*, 3rd edn. (Trans Tech Publications, Zürich, 1992). ISBN 0-87849-522-3
14. H. Fredriksson, U. Akerlind, *Solidification and Crystallization Processing in Metals and Alloys* (Wiley, Chichester, 2012)
15. D.V. Ragone, *Thermodynamics of Materials*, vol. II (Wiley, New York, 1995)
16. H. Fredriksson, U. Akerlind, *Materials Processing during Casting* (Wiley, Chichester, 2006)
17. D.A. Porter, K. Easterling, M.Y. Sherif, *Phase Transformations in Metals and Alloys*, 3rd edn. (CRC Press/Taylor & Francis Group, LLC, Boca Raton, 2009)

Chapter 5
Planar Metal Solidification

5.1 Introduction

The theoretical background on phase-change heat transfer is directly related to melting and solidification processes used in casting practices, food and pharmaceutical processing, latent heat energy storage, ice making, welding, and so forth. For a pure material, the melting or solidification process occurs at a single temperature T_f, while a multi-phase material undergoes either process at a temperature range T_s (solid) $< T_f < T_l$ (liquid). Despite that melting and solidification problems are three-dimensional processes and are nowadays characterized using numerical methods for intricate geometries, the one-dimensional approach provides exact solutions and prevails as an elegant approach for teaching matter phase-change problems in a classroom.

Assume that the melting or solidification process occurs within a physical domain Γ containing the phase-change metal. Subsequently, the domain is divided into two semi-infinite regions by a sharp liquid–solid (L-S) interface with $\delta = 0$ thickness. However, if $\delta > 0$, then L-S interface is a mushy zone. The former case prevails in this chapter.

The goal in this chapter is to include a compilation of analytical solutions to heat equations for one-dimensional melting and solidification problems in rectangular coordinates with moving liquid–solid (L-S) interfaces. For rectangular coordinates, the concept of half-space region $x > 0$ is defined within a semi-infinite region $0 < x < \infty$ with a L-S interface located at $x = s(t)$.

5.2 Melting Problem in a Half-Space Region

The plots in Fig. 5.1 represent idealized profiles of the physical temperature fields $T(x, t)$ during plain melting and solidification. For instance, Fig. 5.1a illustrates

© Springer Nature Switzerland AG 2020
N. Perez, *Phase Transformation in Metals*,
https://doi.org/10.1007/978-3-030-49168-0_5

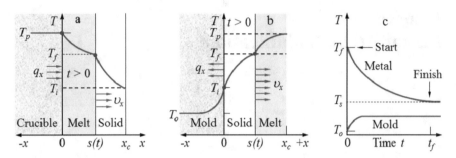

Fig. 5.1 Temperature profiles in a half-space $x > 0$ during (**a**) melting, (**b**) solidification, and (**c**) for freezing the melt and heating the mold

part of the crucible at constant temperature T_p, while heat flux q_x is supplied to the system to start the melting process at $x = 0$ and to finish it at location x_c in the positive x-direction. Figure 5.1b shows the idealized temperature profiles related to solidification of pure materials being superheated at T_p and Fig. 5.1c exhibits the temperature profile $T = f(t)$ in the mold and solid metal for the entire solidification processes.

Consider a solid-phase metal initially at $T_i < T_f$ in a confined semi-infinite planar region $0 < x < \infty$, where a continuous supply of heat flux q_x suddenly raises the solid temperature to $T_p > T_f$ (Fig. 5.1a). As a result, the onset of melting starts at the crucible–liquid interface $x = 0$. At times $t > 0$, the liquid–solid (L-S) interface with velocity v_x is located at $x = s(t)$ and the completion of the melting process is achieved when $s(t) = x_c$ at $t = t_f$. Moreover, $s(t)$ is the location of the L-S interface within the half-space region $0 < s(t) \leq x_c$ and it is called the characteristic length of the moving L-S interface in the positive x-direction. Additional references on melting can be found elsewhere [1–4].

Two-Phase Problem For one-dimensional analysis in the x-direction, the governing *heat equations* for the two-phase problem, continuity of temperature at the L-S interface and the Fourier heat flux balance ($q_l - q_s = q_f$) for melting a slab in a half-space $x > 0$ in a semi-infinite region $0 < x < \infty$ are, respectively

$$\frac{\partial T_l}{\partial t} = \alpha_l \frac{\partial^2 T_l}{\partial x^2} \quad \text{for } 0 < x < s(t), t > 0 \quad \text{(melt)} \tag{5.1a}$$

$$\frac{\partial T_s}{\partial t} = \alpha_s \frac{\partial^2 T_s}{\partial x^2} \quad \text{for } x < s(t), t > 0 \quad \text{(solid)} \tag{5.1b}$$

$$T_l = T_s = T_f \quad \text{at } x = s(t) \quad \text{(L-S interface)} \tag{5.1c}$$

$$k_s \frac{\partial T_s}{\partial x} - k_l \frac{\partial T_l}{\partial x} = \rho_l \Delta H_f \frac{ds(t)}{dt} \quad \text{at } x = s(t) \text{ (energy balance)} \tag{5.1d}$$

where q_f denotes the heat flux (W/m^2) due to release of latent heat of fusion (specific enthalpy at T_f), ρ_l denotes the density (kg/m^3), k_l, k_s denotes the thermal conductivity (W/m K), c_p denotes the specific heat capacity at constant pressure (J/kg K), ΔH_f denotes the latent heat of fusion ΔH_f (J/kg), $ds\,(t)\,/dt$ denotes the rate of melting, and T denotes the absolute temperature. Actually, ΔH_f represents the amount of thermal energy that must be supplied to overcome the binding energy in the solid phase during melting or extracted from the liquid phase to allow atoms to form a solid phase during solidification. Nonetheless, Eq. (5.1d) is known as the Stefan condition.

In general, the thermal diffusivity α (m^2/s), the thermal inertia (thermal effusivity) γ [J/ (m^2 K$\sqrt{s}$)], the heat flux q, the rate of removed energy (dQ/dt) and the amount of removed energy (Q) applicable to liquid and solid phases, and mold region are, respectively, given by

$$\alpha = \frac{k}{\rho c_p} \quad \& \quad \gamma = k/\sqrt{\alpha} = \sqrt{k\rho c_p} \tag{5.2a}$$

$$q = k\frac{\partial T}{\partial x} \tag{5.2b}$$

$$\frac{dQ}{dt} = -Ak\frac{\partial T}{\partial x} \tag{5.2c}$$

$$Q = -Ak \int \left(\frac{\partial T}{\partial x}\right) dt \tag{5.2d}$$

where Eqs. (5.2) contain the thermophysical properties (Appendix 5A). Furthermore, finding the analytical solutions of the heat partial differential equations (PDEs), Eq. (5.1), is possible if the PDEs admit similarity solutions by reducing them to ordinary differential equations (ODE's).

Using the separation of variables method found in Appendix 5B yields the general temperature equations for the liquid and solid phase

$$T_l = C_1 + C_2 \operatorname{erf}\left(\frac{x}{2\sqrt{\alpha_l t}}\right) \quad \& \quad T_s = C_3 + C_4 \operatorname{erfc}\left(\frac{x}{2\sqrt{\alpha_s t}}\right) \tag{5.3}$$

which satisfy the heat equations. Before finding the constants C_1, C_2, C_3, C_4, it is necessary to introduce some general mathematical definitions when $x = s(t)$ so that

$$n = \frac{x}{2\sqrt{\alpha t}} \quad \text{(similarity variable)} \tag{5.4a}$$

$$\lambda = \frac{s\,(t)}{2\sqrt{\alpha t}} \quad \text{for } \alpha = \alpha_l \text{ (liquid); } \alpha = \alpha_s \text{ (solid)} \tag{5.4b}$$

$$t = \frac{s\,(t)^2}{4\lambda^2 \alpha} \quad \text{for } \alpha = \alpha_l \text{ (liquid); } \alpha = \alpha_s \text{ (solid)} \tag{5.4c}$$

$$s(t) = 2\lambda\sqrt{\alpha t} \quad \text{for } \alpha = \alpha_l \text{ (liquid)}; \alpha = \alpha_s \text{ (solid)} \tag{5.4d}$$

$$\frac{ds(t)}{dt} = \lambda\sqrt{\frac{\alpha}{t}} \quad \text{for } \alpha = \alpha_l \text{ (liquid)}; \alpha = \alpha_s \text{ (solid)} \tag{5.4e}$$

where n_l, n_s (similarity variables) and λ (solution characteristic) denote dimensionless parameters, t denotes the melting time, $s(t)$ denotes the characteristic length and the location of the L-S interface, and $ds(t)/dt$ denotes the rate of the L-S interface motion (interface velocity) or the rate of melting in the positive x-direction. Among all the variables defined above, the most important ones are the n and $s(t)$. The former is part of the solution of the heat equation and the latter describes the position of the moving interface between the liquid and solid phases during melting or solidification.

Finding the constants C_1, C_2:

$$T_l = T_l(x,t) = C_1 + C_2 \operatorname{erf}\left(\frac{x}{2\sqrt{\alpha_l t}}\right) \quad \text{(liquid)} \tag{5.5a}$$

BC1 $T_l = T_l(0,t) = T_p$ \hfill (5.5b)

$$T_p = C_1 + C_2 \operatorname{erf}(0) \quad \rightarrow \quad C_1 = T_p \tag{5.5c}$$

BC2 $T_l = T_l[s(t),t] = T_f$ \hfill (5.5d)

$$T_f = T_p + C_2 \operatorname{erf}\left(\frac{s(t)}{2\sqrt{\alpha_l t}}\right) = T_p + C_2 \operatorname{erf}(\lambda) \tag{5.5e}$$

$$C_2 = -\frac{(T_p - T_f)}{\operatorname{erf}(\lambda)} \tag{5.5f}$$

$$T_l = T_p - \frac{(T_p - T_f)}{\operatorname{erf}(\lambda)} \operatorname{erf}\left(\frac{x}{2\sqrt{\alpha_l t}}\right) \tag{5.5g}$$

Finding the constants C_3, C_4:

$$T_l = T_l(x,t) = C_1 + C_2 \operatorname{erf}\left(\frac{x}{2\sqrt{\alpha_l t}}\right) \quad \text{(liquid)} \tag{5.6a}$$

BC1 $T_l = T_l(0,t) = T_p$ \hfill (5.6b)

$$T_p = C_1 + C_2 \operatorname{erf}(0) \quad \rightarrow \quad C_1 = T_p \tag{5.6c}$$

BC2 $T_l = T_l[s(t),t] = T_f$ \hfill (5.6d)

$$T_f = T_p + C_2 \operatorname{erf}\left(\frac{s(t)}{2\sqrt{\alpha_l t}}\right) = T_p + C_2 \operatorname{erf}(\lambda) \tag{5.6e}$$

$$C_2 = -\frac{(T_p - T_f)}{\operatorname{erf}(\lambda)} \tag{5.6f}$$

$$T_l = T_p - \frac{(T_p - T_f)}{\operatorname{erf}(\lambda)} \operatorname{erf}\left(\frac{x}{2\sqrt{\alpha_l t}}\right) \tag{5.6g}$$

From Eqs. (5.5g) and (5.6g),

$$\left(\frac{\partial T_l}{\partial x}\right)_{s(t)} = -\frac{(T_p - T_f)}{\text{erf}(\lambda)\sqrt{\pi\alpha_l t}}\exp\left(-\lambda^2\right) \tag{5.7a}$$

$$\left(\frac{\partial T_s}{\partial x}\right)_{s(t)} = -\frac{(T_f - T_i)}{\text{erfc}\left(\lambda\sqrt{\alpha_l/\alpha_s}\right)\sqrt{\pi\alpha_s t}}\exp\left(-\lambda^2\alpha_l/\alpha_s\right) \tag{5.7b}$$

Liquid phase at $T_f \leq T_l \leq T_p$ *and solid phase at* $T_i \leq T_s \leq T_f$: Substituting Eqs. (5.7a) and (5.7b) along with (5.4e) into (5.1d) gives the transcendental equation for the melting process [2, p. 48]

$$\frac{\lambda\,\text{erf}(\lambda)}{\exp\left(-\lambda^2\right)} + \frac{\gamma_s}{\gamma_l}\frac{c_l\left(T_f - T_i\right)}{\sqrt{\pi}\Delta H_f}\frac{\exp\left[\lambda^2\left(1 - \alpha_l/\alpha_s\right)\right]\text{erf}(\lambda)}{\text{erfc}\left(\lambda\sqrt{\alpha_l/\alpha_s}\right)} = \frac{c_l\left(T_p - T_f\right)}{\sqrt{\pi}\Delta H_f} \tag{5.8}$$

Liquid phase at $T_f \leq T_l \leq T_p$ *and solid phase at* $T_i = 0$: In this case, Eq. (5.8) for the melting process becomes [1, p. 289]

$$\frac{\lambda\,\text{erf}(\lambda)}{\exp\left(-\lambda^2\right)} + \frac{\gamma_s}{\gamma_l}\frac{c_l T_f}{\sqrt{\pi}\Delta H_f}\frac{\exp\left[\lambda^2\left(1 - \alpha_l/\alpha_s\right)\right]\text{erf}(\lambda)}{\text{erfc}\left(\lambda\sqrt{\alpha_l/\alpha_s}\right)} = \frac{c_l\left(T_p - T_f\right)}{\sqrt{\pi}\Delta H_f} \tag{5.9}$$

One-Phase Melting Problem of a Slab In this circumstance, Eq. (5.8) with $T_i = T_f$ becomes [3, p. 405]

$$\lambda\exp\left(\lambda^2\right)\text{erf}(\lambda) = \frac{c_l\left(T_p - T_f\right)}{\sqrt{\pi}\Delta H_f} = \frac{S_{t,l}}{\sqrt{\pi}} \tag{5.10}$$

where $S_{t,l}$ denotes the Stefan number for the liquid phase. The preceding solutions are useful in melting problems since $S_{t,l}$ is the principal parameters in Eq. (5.10) that depends on material properties (Appendix 5A) and temperature differences.

Example 5.1 Consider melting copper in a crucible using one-phase melting problem. If the slab thickness is 0.08 m, then calculate (a) the melting time at half and full space, and (b) the melting temperature for a pouring temperature of $T_p = 1183\,°C$. Data: $\Delta H_f = 205,000\,J/kg$, $k_l = 162\,W/m\,K$, $\rho_l = 8020\,kg/m^3$, $c_l = 570\,J/kg\,K$, $\alpha_l = 3.54 \times 10^{-5}\,m^2/s$ and $T_f = 1083\,°C$.

Solution

(a) Calculate the following constant when the liquid copper is at its pouring temperature $T_p = 1183\,°C$:

$$\frac{c_l\left(T_p - T_f\right)}{\sqrt{\pi}\Delta H_f} = \frac{(570\,J/kg\,K)\,(1183 - 1083)\,K}{\sqrt{\pi}\,(205,000\,J/kg)} = 0.15687$$

Using Eq. (5.10),

$$\lambda \exp\left(\lambda^2\right) \operatorname{erf}(\lambda) = \frac{c_l\left(T_p - T_f\right)}{\sqrt{\pi}\,\Delta H_f} = 0.15657$$

Solving this transcendental equation for the melting process yields λ as

$$0 = \lambda \exp\left(\lambda^2\right) \operatorname{erf}(\lambda) - 0.15657$$

$$\lambda = 0.35689$$

From Eq. (5.4c) with half-space region $s\,(t) = 0.08/2 = 0.04\,\text{m}$ and full space $s\,(t) = x = 0.08\,\text{m}$, the melting times are, respectively

$$t_{1/2} = \frac{s\,(t)^2}{4\lambda^2 \alpha_l} = \frac{0.04^2}{4\,(0.35689)^2\,(3.54 \times 10^{-5})} = 88.71\,\text{s} = 1.48\,\text{min}$$

$$t = \frac{s\,(t)^2}{4\lambda^2 \alpha_l} = \frac{0.08^2}{4\,(0.35689)^2\,(3.54 \times 10^{-5})} = 354.85\,\text{s} = 5.91\,\text{min}$$

(b) From Eq. (5.5g) with $s\,(t) = 0.08/2 = 0.04\,\text{m}$ (half-space region), the melting temperature is

$$T_{l_{1/2}} = T_p - \frac{\left(T_p - T_f\right)}{\operatorname{erf}(\lambda)} \operatorname{erf}\left(\frac{x}{2\sqrt{\alpha_l t_{1/2}}}\right)$$

$$= 1183\,°\text{C} - \frac{(1183 - 1083)\,°\text{C}}{\operatorname{erf}(0.35689)} \operatorname{erf}\left[\frac{4 \times 10^{-2}\,\text{m}}{2\sqrt{(3.54 \times 10^{-5}\,\text{m}^2/\text{s})\,(88.71\,\text{s})}}\right]$$

$$= 1083\,°\text{C}$$

Therefore, the melting time is not linear and as expected, the slab liquid temperature is exactly at $T_l = T_f = 1083\,°\text{C}$ and $s(t) = 4\,\text{cm}$. Similarly, $T_l = T_f = 1083\,°\text{C}$ at $s\,(t) = x = 8\,\text{cm}$.

5.3 Solidification in Half-Space Regions

5.3.1 Heat Transfer Through Planar Surfaces

In principle, heat transfer between plane (|) and contoured (convex $\smile$ and concave $\frown$) mold inner surfaces can be assumed similar [5–8]. Figure 5.2a illustrates a rectangular casting and Fig. 5.2b shows its cross-sectional plane during unidirec-

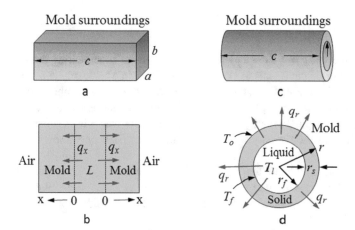

Fig. 5.2 Sketches of simple casting geometries and heat flux directions. (**a**) Rectangular. (**b**) Cross section. (**c**) Cylinder. (**d**) Annular ring

tional solidification. For one-dimensional solidification heat transfer, the heat flux q_x crosses the liquid–mold (*L-m*) interface at $x > 0$ and flows in the x-direction at the freezing temperature (T_f).

Similarly, Fig. 5.2c shows a cylinder-type casting and Fig. 5.2d exhibits its annular cross-sectional area, illustrating the heat flux q_r flow in the radial direction at T_f. For spherical solidification, q_r also flows in the radial direction (Fig. 5.2d). Here, T_m denotes the mold temperature.

For one-dimensional solidification of a slab, the analysis of steady heat flow across a uniform plane wall is simple because the flow area A does not change during the flow direction. Thus, the temperature $T(x, t)$ is a function of the Cartesian coordinate x and time t. In the case of cylindrical or spherical solidification, the surface area for heat flow continuously changes in the radial direction along with the temperature field $T(r, t)$ [2].

For a complete thermal analysis, the heat fluxes q_x, q_{ls} or the rate of thermal energy quantities $dQ_t/dt, dQ_{ls}/dt = -A(q_x, q_{ls})$ will suffice to give the amount of energy associated with a two-phase (liquid and solid) solidification problem, where A is the assumed smooth and flat *L-S* interface contact area. Accordingly, the notation *L-S* denotes $L \to S$, transformation from liquid to solid at T_f.

Furthermore, the thermal conductivity strongly depends on the heat interaction between two surfaces during solidification: mold–liquid metal and solid–liquid interface, where "solid" stands for solidified melt and "interaction" represents some degree of heat transfer resistance by the phase absorbing the thermal energy at the two-phase boundary, which evolves at T_f and $t > 0$.

The source of heat transfer is the evolution of latent heat of fusion ΔH_f at the *L-S* interface. At this moment, assume that solidification is fundamentally controlled by a planar interface $s(t)$ at T_f.

5.3.2 Solidification with Superheating

In principle, $\Delta T_s = (T_p - T_f) > 0$ is a measure of superheating the liquid metal. The entire mold cavity has to be completely filled in a relatively short time prior to the onset of solidification at $x > 0$, $t = 0$, and $T = T_f$; otherwise, premature solidification may take place in the gating or riser system and consequently, an incomplete casting geometry may be attained. Moreover, a high superheat saturates gases, induces the formation of metal oxides, and melt penetration into the mold surface.

Superheat is related to pouring rate (volumetric rate) and heat dissipation. Fast pouring rates induce melt turbulence and erosion within the mold cavity. Conversely, too low pouring rates cause premature solidification before filling the mold completely.

Once the melt completely fills the mold cavity in a very short time interval at T_p, the temperature of the mold inner walls instantaneously increases to T_p at $t = 0$ and subsequently, solidification starts at T_f and the heat flux q_x flows in the negative x-direction through the solid and mold walls.

5.3.3 Solidification with Supercooling

The magnitude of the supercooling is defined as ΔT_u and it is the driving force required for nucleation of a solid phase. In principle, the solidification heat transfer is treated as an idealized phase-change process, which starts at $x = 0$, the ideal initial mold–liquid smooth and defect-free interface position. Explicitly, the mold is considered as a chemically nonreactive material at high temperatures so that solidification is uniform and continuous along the evolved L-S interface at $x = s(t) > 0$.

Furthermore, two different supercooling conditions can be highlighted at a macroscale ([2, p. 85] and [9, p. 429], respectively)

$$T_f - \frac{\Delta H_f}{c_l} < T_i < T_f \tag{5.11a}$$

$$\Delta T = (T_f - T_i) < \frac{\Delta H_f}{c_l} \tag{5.11b}$$

where $T_i < T_f$ is the melt initial temperature prior to phase change and c_l is the specific heat of the liquid phase. Eventually, the L-S interface (solidification front) forms at a location $x = s(t)$ for planar solidification or $r = s(t)$ for radial solidification process.

In this case, thermal heat flows into the supercooled liquid while the solid thickens at the freezing temperature T_f in the positive x-direction within the limits of the half-space dimension; $s(t) \to x_c$ at $t > 0$.

Fundamentally, the microstructural features of the solidified melt are affected by the degree of supercooling, rate of cooling, magnetic field, and agitation of the melt. Therefore, the forthcoming sections on solidification highlight the conventional mathematical modeling of solidification based on the heat transfer theory at a macroscale [1–10]. The fundamental theory of heat transfer is based on the concept of thermal resistance of a phase due to a certain temperature gradient dT/dx. It suffices to say that one-dimensional heat transfer provides reliable analytical results for determining the temperature distribution equations in the mold, solid, and liquid phases.

5.4 Three-Phase Thermal Resistances

Consider a pure metal undergoing solidification in a half-space region $0 \leq x \leq x_c$ and assume that the temperature profiles are as schematically shown in Fig. 5.3.

The goal in this section is to derive a general transcendental equation that couples the mold, solid, and liquid as a solidification domain free of convective heat transfer effect on the solidification process and variable thermophysical properties (Appendix 5A).

5.4.1 One-Dimensional Heat Equations

For an ideal mold–solid contact surface area, the governing heat equations, continuity of temperature at the L-S interface, and the Fourier heat flux balance for solidification are, respectively, generalized and presented as a collection of dimensional group of equations. Thus, for one-dimensional analysis of the solidification process

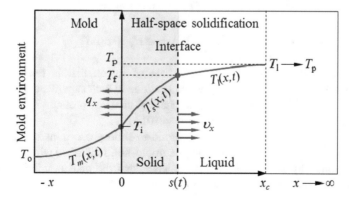

Fig. 5.3 Coupled mold–solid resistance problem with negligible liquid temperature gradient

$$\frac{\partial T_m}{\partial t} = \alpha_m \frac{\partial^2 T_m}{\partial x^2} \quad \text{for } -\infty < x \leq 0, t > 0 \tag{5.12a}$$

$$\frac{\partial T_s}{\partial t} = \alpha_s \frac{\partial^2 T_s}{\partial x^2} \quad \text{for } 0 < x \leq s(t), t > 0 \tag{5.12b}$$

$$\frac{\partial T_l}{\partial t} = \alpha_l \frac{\partial^2 T_l}{\partial x^2} \quad \text{for } s(t) \leq x < \infty, t > 0 \tag{5.12c}$$

$$T_s = T_l = T_f \quad \text{at } x = s(t) \tag{5.12d}$$

$$k_m \frac{\partial T_m}{\partial x} = k_s \frac{\partial T_s}{\partial x} = \rho_s \Delta H_f \frac{ds(t)}{dt} \quad \text{at } x = 0 \tag{5.12e}$$

$$k_s \frac{\partial T_s}{\partial x} - k_l \frac{\partial T_l}{\partial x} = \rho_s \Delta H_f \frac{ds(t)}{dt} \quad \text{at } x = s(t), T_p = T_f \tag{5.12f}$$

$$k_s \frac{\partial T_s}{\partial x} - k_l \frac{\partial T_l}{\partial x} = \rho_s \Delta H_s \frac{ds(t)}{dt} \quad \text{at } x = s(t), T_p > T_f \tag{5.12g}$$

Firstly, the heat transfer is based on the concept of thermal resistance of a phase due to a temperature gradient $\partial T_i/\partial x$, where $i = m, s, l$ and secondly, the Fourier heat flux balance for each phase is governed by the solidification velocity, $ds(t)/dt$, at the interface location $s(t)$ and time $t > 0$.

5.4.2   Latent Heat of Solidification

The latent heat of solidification (ΔH_s) is a special relationship between superheating and freezing states of the liquid phase. For a motionless superheated melt in a mold cavity, the effective latent heat of fusion is intentionally called the latent heat of solidification and it is mathematically defined as

$$\Delta H_s = \Delta H_f + \int_{T_f}^{T_p} c_l dT \quad \text{(with } c_{pl} = c_l\text{)} \tag{5.13a}$$

$$\Delta H_s = \Delta H_f + c_l (T_p - T_f) = \Delta H_f + c_l \Delta T_s \tag{5.13b}$$

where ΔH_f denotes, again, the latent heat of fusion that is released at T_f, $c_l (T_p - T_f)$ denotes the sensible specific heat due to melt superheating effect, and c_l denotes the specific heat of the liquid phase at constant pressure P. Commonly, experiments are carried out at $P = 1\, atm = 101$ kPa.

Firstly, as a first approximation, the liquid-to-solid phase transformation starts when ΔH_f is released on a smooth and flat mold–solid interface with the heat flux balance defined by Eq. (5.12f), which indicates that the melt is initially at T_f or $T_p \simeq T_f$ and releases ΔH_f at $x = s(t)$.

Secondly, metals are normally cast with some degree of superheat defined henceforth as $\Delta T_s = T_p - T_f$, where $T_p > T_f$, and consequently, the heat flux balance is now defined by Eq. (5.12g), which includes ΔH_s to compensate for the effect of superheating the melt before the actual phase transformation begins at a certain predictable rate of solid formation denoted by $ds(t)/dt$.

Superheating the melt is an inevitable physical event during conventional casting procedures for conditioning the melt chemistry in a suitable ladle before casting. Eventually, the mold initially absorbs this superheated energy when the melt freezes from T_p to T_f at $x = 0$ into the mold enclosure and the released thermal energy by the liquid during phase transformation $(L \rightarrow S)$ at T_f. The latter thermal energy is known in the literature as the latent heat of fusion (ΔH_f), which is the enthalpy change of formation for phase transformation. Hence, $\Delta H_s = \Delta H_{super} + \Delta H_f$ when the melt is stationary in the mold cavity at the pouring temperature $T_p > T_f$.

Most analytical solutions available in the literature consider the melt is initially stationary at T_f which implies that the melt pouring temperature T_p, in principle, is its own freezing temperature T_f.

Furthermore, during solidification, ΔH_f is released while the L-S interface evolves at $0 < s(t) < \infty$ in a time interval $0 < t \leq t_s$ for the local solidification time or $0 < t \leq t_f$ for the total solidification time due to the superheat effect. Hence, t_s is the real solidification time at $T = T_f$ and $t_f = t_{cool} + t_s$ is the apparent solidification time since it takes into account the effect of superheat since the melt has to freeze before it transforms to solid.

Moreover, the solidification coordinates adapted herein requires that the heat flux (q_x) must flow in the negative x-direction through the solid, crosses the mold–solid interface and continuous flowing through the mold wall until it reaches the mold surroundings at a uniform temperature T_o.

It suffices to say that the latent heat of fusion ΔH_f for a pure metal or ΔH_f^* for an alloy is a quantity of thermal energy associated with phase transformation. For instance, it is absorbed during melting and released during solidification at the location of the characteristic length $s(t) > 0$ at time $t > 0$.

5.4.3 Analytical Solutions of the Heat Equations

Consider the solidification process of pure metals in sand and metal molds, and assume that the heat equations admit similarity solutions. Mathematically, one-dimensional analysis is considered henceforth since it will suffice as a suitable analytical approach to solve solidification problems.

Solidification, in general, is a materials process associated with one specific moving boundary problem (MBP) within a two-phase semi-infinite region, where $0 < x < \infty$ and time $t > 0$. In this case, the moving boundary is called the solidification front or simply the L-S interface, which divides the solid and liquid phases. Such a problem is treated as a transient and non-linear solidification because the moving interface is strongly dependent on the rate of released latent heat of fusion. Transient solidification is known as a Stefan problem, which can analytically be solved by the classical Neumann similarity solution.

Mold Region The coordinate system shown in Fig. 5.3 puts the mold in the negative side of the x-direction. For a thick enough metal mold, modeled as a one-dimensional semi-infinite system $(-\infty < x \leq 0)$ with interface thermal resistance, the general temperature field equation with boundary conditions is

$$T_m = C_1 + C_2 \, \text{erfc}\left(\frac{-x}{2\sqrt{\alpha_m t}}\right) \quad \text{(mold)} \tag{5.14a}$$

$$\text{BC1 } T_m = T_m\,(-\infty, t) = T_o \tag{5.14b}$$

$$T_o = C_1 + C_2 \, \text{erfc}\,(\infty) = C_1 \tag{5.14c}$$

$$\text{BC2 } T_m = T_m\,(0, t) = T_i \tag{5.14d}$$

$$T_i = T_o + C_2 \, \text{erfc}\,(0) = T_o + C_2 \tag{5.14e}$$

$$C_2 = T_i - T_o \tag{5.14f}$$

where the error function yields $\text{erfc}\,(\infty) = 0$ and $\text{erfc}\,(0) = 1$. Substituting C_1, C_2 into Eq. (5.14a) yields

$$T_m = T_o + (T_i - T_o)\, \text{erfc}\left(\frac{-x}{2\sqrt{\alpha_m t}}\right) \quad \text{(mold)} \tag{5.15a}$$

$$\frac{\partial T_m}{\partial x} = \frac{T_i - T_o}{\sqrt{\pi \alpha_m t}} \exp\left(\frac{x^2}{4\alpha_m t}\right) \tag{5.15b}$$

$$\left(\frac{\partial T_m}{\partial x}\right)_{x=0} = \frac{T_i - T_o}{\sqrt{\pi \alpha_m t}} \tag{5.15c}$$

Solid Phase Similarly, the temperature equation and the boundary conditions for the solid phase are, respectively

$$T_s = T\,(x, t)_s = D_1 + D_2 \, \text{erf}\left(\frac{x}{2\sqrt{\alpha_s t}}\right) \quad \text{(solid)} \tag{5.16a}$$

$$\text{BC1 } T_s = T_s\,(0, t) = T_i \tag{5.16b}$$

$$T_i = D_1 + D_2 \, \text{erf}\,(0) \;\; \to \;\; D_1 = T_i \tag{5.16c}$$

$$\text{BC2 } T_s = T_s\,[s\,(t), t] = T_f \tag{5.16d}$$

$$T_f = T_i + D_2 \, \text{erf}\left(\frac{s\,(t)}{2\sqrt{\alpha_s t}}\right) = T_i + D_2 \, \text{erf}\,(\lambda) \tag{5.16e}$$

$$D_2 = \left(T_f - T_i\right)/\text{erf}\,(\lambda) \tag{5.16f}$$

Substituting D_1, D_2 into Eq. (5.16a) gives the temperature field equation and subsequently, the rate of freezing and the temperature gradient equations

$$T_s = T_i + \frac{(T_f - T_i)}{\mathrm{erf}(\lambda)} \, \mathrm{erf}\left(\frac{x}{2\sqrt{\alpha_s t}}\right) \quad \text{(solid)} \tag{5.17a}$$

$$\frac{\partial T_s}{\partial t} = -\frac{(T_f - T_i)\, x}{2t\sqrt{\pi \alpha_s t}\, \mathrm{erf}(\lambda)} \exp\left(-\frac{x^2}{4\alpha_s t}\right) \tag{5.17b}$$

$$\frac{\partial T_s}{\partial x} = \frac{(T_f - T_i)}{\sqrt{\pi \alpha_s t}\, \mathrm{erf}(\lambda)} \exp\left(-\frac{x^2}{4\alpha_s t}\right) \tag{5.17c}$$

In order to compute solid temperature gradient $\partial T_s / \partial x$ as per Eq. (5.17c), λ has to be evaluated as a priori.

It is now appropriate to set a new group of equations. That is, evaluating the solid temperature gradient $\partial T_s / \partial x$ at $x = 0$ and $x = s(t)$, and defining the dimensionless parameter λ and the rate of the evolved characteristic length, $ds(t)/dt$, respectively, gives

$$\left(\frac{\partial T_s}{\partial x}\right)_{x=0} = \frac{(T_f - T_i)}{\sqrt{\pi \alpha_s t}\, \mathrm{erf}(\lambda)} \tag{5.18a}$$

$$\left(\frac{\partial T_s}{\partial x}\right)_{s(t)} = \frac{(T_f - T_i)}{\sqrt{\pi \alpha_s t}\, \exp\left(\lambda^2\right) \mathrm{erf}(\lambda)} \quad \text{at } x = s(t) \tag{5.18b}$$

$$\lambda = \frac{s(t)}{2\sqrt{\alpha_s t}} \tag{5.18c}$$

$$s(t) = 2\lambda\sqrt{\alpha_s t} \tag{5.18d}$$

$$t_\lambda = t = \frac{s(t)^2}{4\lambda^2 \alpha_s} \tag{5.18e}$$

$$\frac{ds(t)}{dt} = \frac{\lambda\sqrt{\alpha_s}}{\sqrt{t}} \tag{5.18f}$$

Liquid Phase Similarly, the temperature equation and the boundary conditions for the liquid phase under superheating effects are, respectively

$$T_l = T(x, t)_l = E_1 + E_2\, \mathrm{erfc}\left(\frac{x}{2\sqrt{\alpha_l t}}\right) \quad \text{(liquid)} \tag{5.19a}$$

$$\text{BC1 } T_l = T_l(0, t) = T_p \tag{5.19b}$$

$$T_p = E_1 + E_2\, \mathrm{erfc}(\infty) \tag{5.19c}$$

$$E_1 = T_p \tag{5.19d}$$

$$\text{BC2 } T_l = T_l[s(t), t] = T_f \tag{5.19e}$$

$$T_f = T_p + E_2 \operatorname{erfc}\left(\frac{s\,(t)}{2\sqrt{\alpha_l t}}\frac{\sqrt{\alpha_s}}{\sqrt{\alpha_s}}\right) = T_p + E_2 \operatorname{erfc}\left(\lambda\sqrt{\frac{\alpha_s}{\alpha_l}}\right) \qquad (5.19f)$$

$$E_2 = \frac{\left(T_f - T_p\right)}{\operatorname{erfc}\left(\lambda\sqrt{\alpha_s/\alpha_l}\right)} \qquad (5.19g)$$

Substituting the constants E_1, E_2 into Eq. (5.19a) yields the fully defined liquid-phase temperature field equation along with the temperature gradient equation evaluated at the location $s\,(t)$ of the L-S interface.

Thus,

$$T_l = T_p - \frac{\left(T_p - T_f\right)}{\operatorname{erfc}\left(\lambda\sqrt{\alpha_s/\alpha_l}\right)} \operatorname{erfc}\left(\frac{x}{2\sqrt{\alpha_l t}}\right) \quad \text{(liquid)} \qquad (5.20a)$$

$$\frac{\partial T_l}{\partial t} = \frac{\left(T_p - T_f\right)x}{\sqrt{\pi \alpha_l t}\,\operatorname{erfc}\left(\lambda\sqrt{\alpha_s/\alpha_l}\right)} \exp\left(-\frac{x^2}{4\alpha_l t}\right) \qquad (5.20b)$$

$$\left(\frac{\partial T_l}{\partial x}\right)_{s(t)} = -\frac{\left(T_p - T_f\right)}{\sqrt{\pi \alpha_l t}\,\operatorname{erfc}\left(\lambda\sqrt{\alpha_s/\alpha_l}\right)} \exp\left(-\frac{x^2}{4\alpha_s t}\frac{\alpha_s}{\alpha_l}\right) \qquad (5.20c)$$

$$\left(\frac{\partial T_l}{\partial x}\right)_{s(t)} = -\frac{\left(T_p - T_f\right)\exp\left(-\lambda^2\alpha_s/\alpha_l\right)}{\sqrt{\pi \alpha_l t}\,\operatorname{erfc}\left(\lambda\sqrt{\alpha_s/\alpha_l}\right)} \qquad (5.20d)$$

Let the mold–solid interface temperature be T_i at $x = 0$ so that the heat flux balance gives T_i without any degree of superheating

$$q_m = q_s \qquad (5.21a)$$

$$k_m \left(\frac{\partial T_m}{\partial x}\right)_{x=0} = k_s \left(\frac{\partial T_s}{\partial x}\right)_{x=0} \qquad (5.21b)$$

$$\frac{k_m\,(T_i - T_o)}{\sqrt{\pi \alpha_m t}} = \frac{k_s\left(T_f - T_i\right)}{\sqrt{\pi \alpha_s t}\,\operatorname{erf}(\lambda)} \qquad (5.21c)$$

$$\gamma_m\,(T_i - T_o) = \frac{\gamma_s\left(T_f - T_i\right)}{\operatorname{erf}(\lambda)} \qquad (5.21d)$$

$$T_i = \frac{\gamma_s T_f + \gamma_m T_o\,\operatorname{erf}(\lambda)}{\gamma_s + \gamma_m\,\operatorname{erf}(\lambda)} \quad \text{(mold-solid)} \qquad (5.21e)$$

Inserting Eqs. (5.18b), (5.18c) and (5.20d) into (5.12f) and collecting dimensionless terms gives the general transcendental equation related to mold, solid, and liquid thermal resistances containing γ_m, γ_s, γ_l [10]. Thus,

$$\frac{S_t}{\sqrt{\pi}} = \left[\lambda\exp\left(\lambda^2\right) - \frac{\gamma_l}{\gamma_s}\frac{c_s\left(T_p - T_f\right)}{\sqrt{\pi}\Delta H_s}B\left(\lambda\right)\right]\left[\operatorname{erf}(\lambda) + \frac{\gamma_s}{\gamma_m}\right] \qquad (5.22a)$$

$$f(\lambda) = \left[\lambda \exp\left(\lambda^2\right) - \frac{\gamma_l}{\gamma_s}\frac{c_s\left(T_p - T_f\right)}{\sqrt{\pi}\,\Delta H_s}B(\lambda)\right]\left[\mathrm{erf}(\lambda) + \frac{\gamma_s}{\gamma_m}\right] \tag{5.22b}$$

$$f(\lambda) = \frac{S_t}{\sqrt{\pi}} \tag{5.22c}$$

with

$$B(\lambda) = \frac{\exp\left[(1 - \alpha_s/\alpha_l)\,\lambda^2\right]}{\mathrm{erfc}\left(\lambda\sqrt{\alpha_s/\alpha_l}\right)} \quad \& \quad S_t = \frac{c_s\left(T_f - T_o\right)}{\Delta H_f} \tag{5.23}$$

Appendix 5C illustrates the analytical procedure for deriving the transcendental expression Eq. (5.22a) as (5C.17a) or (5C.17b), which is confined to a two-region solidification domain as a Stefan condition for solidification with solid and liquid thermal resistances only.

Once λ is numerically evaluated from Eq. (5.22a), computations of desired variables become feasible. Nonetheless, Eq. (5.22) is a collection of dimensionless groups; γ_l/γ_s denotes the effusivity ratio, $c_s\left(T_p - T_f\right)/\Delta H_f$ denotes the dimensionless Stefan number due to superheating $\Delta T_s = T_p - T_f$, $B(\lambda)$ denotes the dimensionless ratio responsible for the non-linear behavior solidification from T_p to T_f, and $f(\lambda)$ denotes the monotonic dimensionless function that never decreases as λ increases. However, $f(\lambda)$ depends on the Stefan number S_t as indicated by Eq. (5.22c).

Inserting Eqs. (5.18b) and (5.20d) into the heat flux (q_x) balance (5.12g) and integrating the resultant expression yields the solidification time t_q for a solidified thickness $s(t)$ under the assumption that solidification is controlled by heat conduction

$$t_q = \frac{\pi}{4}\left[\frac{\rho_s\Delta H_s\,\mathrm{erf}(\lambda)\,\mathrm{erfc}\left(\lambda\sqrt{\alpha_s/\alpha_l}\right)}{B_1(\lambda) + B_2(\lambda)}\right]^2 s(t)^2 \tag{5.24a}$$

$$B_1(\lambda) = \gamma_s\left(T_f - T_i\right)\exp\left(-\lambda^2\right)\mathrm{erfc}\left(\lambda\sqrt{\alpha_s/\alpha_l}\right) \tag{5.24b}$$

$$B_2(\lambda) = \gamma_l\left(T_p - T_f\right)\exp\left(-\lambda^2\alpha_s/\alpha_l\right)\mathrm{erf}(\lambda) \tag{5.24c}$$

Continuing the analysis of solidification reveals the effect of superheating the liquid metal on the transcendental equation and other variables. For instance, the thermophysical data set for pure aluminum found in Dantzig–Rappaz book [10, p. 170] is intentionally adapted here for characterizing the effect of superheating on the solidification time as per Eqs. (5.24a) and (5.18d). In this particular case, let the characteristic length and the pouring temperature range be $s(t) = 0.04\,\mathrm{m}$ and $660\,^\circ\mathrm{C} \leq T_p \leq 850\,^\circ\mathrm{C}$, and use the selected data set listed in Table 5.1 to characterize the solidification process for pure aluminum at a macroscale.

Table 5.1 Thermophysical data for pure aluminum [10, p. 170]

Material	T (°C)	k (W/m K)	ρ (kg/m^3)	c_p (J/kg K)	ΔH_f (J/kg)
Graphite mold	25	100	2200	1700	
Solid Al		211	2555	1190	3.98×10^5
Liquid Al	700	91	2368	1090	

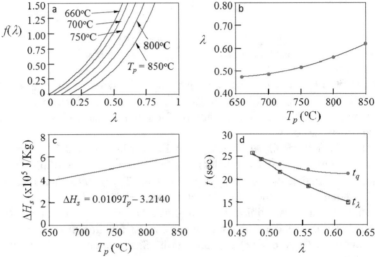

Fig. 5.4 Effect of melt superheating on solidification of pure aluminum. (**a**) The general transcendental equation, (**b**) the characteristic solution, (**c**) the latent heat of solidification, and (**d**) the solidification time. Pouring temperature range $660\,°C \leq T_p \leq 850\,°C$

Note that ΔH_f can be replaced by ΔH_s in Eq. (5.22a) to evaluate $f(\lambda)$ as a superheat-dependent dimensionless function as shown in Fig. 5.4a for $660\,°C \leq T_p \leq 850\,°C$. The $f(\lambda)$ curves are displaced downwards as T_p increases. Figure 5.4b shows an increasing dimensionless parameter as a function of pouring temperature, $\lambda = f(T_p)$.

Additionally, $\Delta H_s = f(T_p)$ shows a linear behavior (Fig. 5.4c) as per Eq. (5.13b) and the solidification times t_λ and t_q decrease with increasing λ (Fig. 5.4d) and indirectly with increasing T_p. However, high superheating temperatures may induce casting problems as mentioned above. Nonetheless, t_λ and t_q deviate from each other very significantly at $T_p > T_f$. The only common point for $t_\lambda = t_q$ is at $T_p = T_f$, meaning lack of superheating effects. It is assumed now that the solidification time t_q is the most appropriate for characterizing the freezing or solidification process since it is derived from the heat flux balance defined by Eq. (5.12g).

With regard to Fig. 5.4c, plotting $\Delta H_s = f(T_p)$ is a convenient graphical representation of the released latent heat of solidification. The resultant increasing linear trend is indicative of the influence of superheating on $\Delta H_s = f(T_p)$. This

Table 5.2 Results related to Fig. 5.4

T_p (°C)	ΔH_s ($\times 10^5$ J/kg)	λ	T_i (°C)	t_q (s)	t_λ (s)	t_q/t_λ
660	3.98	0.47370	485.30	25.69	25.69	1.00
700	4.42	0.48715	482.25	24.29	24.28	1.00
750	4.96	0.51591	476.00	21.66	23.07	0.94
800	5.51	0.55926	467.21	18.43	22.19	0.83
850	6.05	0.62041	456.04	14.98	21.12	0.71

physically means that the higher (T_p), the faster the release of $\Delta H_s = f(T_p)$ at the L-S interface (solidification front).

For convenience, the above graphical results are listed in Table 5.2 for pure aluminum solidification.

This data set is purposely included in this section as a reference for curve fitting. So the reader may want to have it available when needed.

For the sake of clarity, the resultant curve fitting expressions are

$$\Delta H_s \left(\times 10^5\right) = (0.0109 \text{ J/kg }°\text{C}) \, T_p - 3.2140 \text{ J/kg} \tag{a}$$

$$\Delta H_s = (1090 \text{ J/kg }°\text{C}) \, T_p - 3.214 \times 10^5 \text{ J/kg} \tag{b}$$

$$\frac{d\left(\Delta H_s\right)}{dT_p} = 1090 \text{ J/kg }°\text{C} \tag{c}$$

Notice that the slope, $d\left(\Delta H_s\right)/dT_p = 1090$ J/kg °C, represents the rate of release of latent heat of fusion.

5.5   Mold and Solid Thermal Resistances

Consider a coupled mold–solid interface at $x = 0$ with a new ideal planar (smooth and flat) interface at T_i. In fact, Eq. (5.21e) does not apply to this case. Inserting Eq. (5.17c) at $x = s(t)$ into (5.12f) along with (5.2a) and (5.18c), and range $T_i \leq T \leq T_f$ yields the transcendental equation related to the solid phase

$$\lambda \exp\left(\lambda^2\right) \text{erf}(\lambda) = \frac{c_s\left(T_f - T_i\right)}{\sqrt{\pi}\Delta H_f} \tag{5.25}$$

from which T_i takes a second definition

$$T_f - T_i = \sqrt{\pi}\lambda \exp\left(\lambda^2\right) \text{erf}(\lambda)\left(\frac{\Delta H_f}{c_s}\right) \tag{5.26a}$$

$$T_i = T_f - \sqrt{\pi}\lambda \exp\left(\lambda^2\right) \text{erf}(\lambda)\left(\frac{\Delta H_f}{c_s}\right) \tag{5.26b}$$

Now, combining Eqs. (5.24a) and (5.26a) T_i takes a third definition

$$T_i = T_o + \sqrt{\pi} \lambda \exp\left(\lambda^2\right) \frac{\Delta H_f}{c_s} \frac{\gamma_s}{\gamma_m} \tag{5.27a}$$

$$T_i = T_o + \sqrt{\pi} \lambda \exp\left(\lambda^2\right) \frac{\Delta H_f}{c_s} \sqrt{\frac{k_s \rho_s c_s}{k_m \rho_m c_m}} \tag{5.27b}$$

Substitute Eq. (5.27b) into Eq. (5.25) to get the transcendental equation in the following two forms [5, p. 341], [6, p. 413], [7, p. 413] and [10, p. 172]:

$$\lambda \exp\left(\lambda^2\right) \left[\mathrm{erf}\left(\lambda\right) + \sqrt{\frac{k_s \rho_s c_s}{k_m \rho_m c_m}} \right] = \frac{c_s \left(T_f - T_o\right)}{\sqrt{\pi} \Delta H_f} \tag{5.28a}$$

$$\lambda \exp\left(\lambda^2\right) \left[\mathrm{erf}\left(\lambda\right) + \frac{\gamma_s}{\gamma_m} \right] = \frac{S_t}{\sqrt{\pi}} \tag{5.28b}$$

Once λ is numerically evaluated, computations of variables become feasible.

Example 5A.2 For producing a copper casting in a steel mold having a 0.04-m thick cavity, calculate (a) the solidification time t and the solidification velocity dx_s/dt. Assume a superheat of 20 °C. Data for copper casting: $T_f = 1083\,°C$, $k_s = 400\,W/m\,K$, $\rho_s = 7525\,kg/m^3$, $c_s = 429\,J/kg\,K$, $\alpha_s = 1.2391 \times 10^{-4}\,m^2 s$ and $\Delta H_f = 205{,}000\,kJ/kg$. Data for metal mold: $T_o = 25\,°C$, $k_m = 7689\,W/m\,K$, $\rho_m = 35\,kg/m^3$, $c_m = 669\,J/kg\,K$, and $\alpha_m = 6.8041 \times 10^{-6}\,m^2 s$.

Solution

(a) Iteration solves Eq. (5.28a) for λ. Thus,

$$\lambda \exp\left(\lambda^2\right) \left[\mathrm{erf}\left(\lambda\right) + \sqrt{\frac{k_s \rho_s c_s}{k_m \rho_m c_m}} \right] = \frac{c_s \left(T_f - T_o\right)}{\sqrt{\pi} \Delta H_f}$$

$$\lambda \exp\left(\lambda^2\right) \left[\mathrm{erf}\left(\lambda\right) + 2.6781\right] = 1.2491$$

$$\lambda \exp\left(\lambda^2\right) \left(\mathrm{erf}\left(\lambda\right) + 2.6781\right) - 1.2491 = 0$$

$$\lambda = 0.35834$$

From Eq. (5.18d) with $s\,(t) = 0.04/2 = 0.02\,m$ and (5.18e), the solidification time t and the solidification velocity dx_s/dt are

$$t = \frac{s\,(t)^2}{4\lambda^2 \alpha_s} = \frac{(0.02)^2}{4\,(0.34774)^2 \left(1.2391 \times 10^{-4}\right)} = 6.29\,s$$

$$\frac{dx_s}{dt} = \lambda \sqrt{\frac{\alpha_s}{t}} = (0.35834) \sqrt{\frac{1.2391 \times 10^{-4}}{6.29}} = 1.59 \times 10^{-4} \text{ m/s}$$

Example 5A.3 (Metal Mold) Assume that a 0.04-m thick iron slab casting is to be produced in a carbon steel mold and that there are temperature gradients in the mold and solid only. Use the thermophysical properties given below to calculate (a) the solidification time t, and the solidification velocity dx_s/dt, (b) the heat flux at $x = 0$ and, (c) plot the temperature distribution at $-0.04 \leq x \leq 0$ m for the mold and $0 \leq x \leq 0.02$ m for the solid. Assume superheats 0, 20, and 100 °C. Data for iron casting: $T_f = 1535$ °C, $k_s = 73$ W/m K, $\rho_s = 7530$ kg/m^3, $c_s = 728$ J/kg K, and $\Delta H_f = 247.20$ kJ/kg. Data for metal mold: $T_o = 25$ °C, $k_m = 35$ W/m K, $\rho_m = 7689$ kg/m^3, and $c_m = 669$ J/kg K.

Solution

(a) Only one set of calculations are included below. From Eq. (5.2a), the thermal diffusivity terms are

$$\alpha_s = \frac{k_s}{\rho_s c_s} = 1.3317 \times 10^{-5} \text{ m}^2\text{s}$$

$$\alpha_m = \frac{k_m}{\rho_m c_m} = 6.8041 \times 10^{-6} \text{ m}^2\text{s}$$

The latent heat of solidification due to a superheat, defined as $\Delta T_s = T_p - T_f$, with $c_l \simeq c_s$ is

$$T_p = 1635 \,^\circ\text{C} = 1908 \text{ K}$$

$$\Delta T_s = T_p - T_f = 100 \,^\circ\text{C} = 100 \text{ K}$$

$$\Delta H_s = \Delta H_f + c_l (T_p - T_f) = 247{,}200 + 728 (1908 - 1808)$$

$$\Delta H_s = 320{,}000 \text{ J/kg}$$

Iteration solves Eq. (5.28a) for λ. Here, ΔH_f is replaced by ΔH_s to compensate for the superheat on solidification

$$\lambda \exp\left(\lambda^2\right) \left[\text{erf}(\lambda) + \sqrt{\frac{k_s \rho_s c_s}{k_m \rho_m c_m}} \right] = \frac{c_s (T_f - T_o)}{\sqrt{\pi} \Delta H_s}$$

$$\lambda \exp\left(\lambda^2\right) [\text{erf}(\lambda) + 1.4909] = 1.9381$$

$$\lambda \exp\left(\lambda^2\right) (\text{erf}(\lambda) + 1.4909) - 1.9381 = 0$$

$$\lambda = 0.62265$$

Then, the interface temperature, Eqs. (5.26b) or (5.27b), is

$$T_i = T_f - \sqrt{\pi}\lambda \exp\left(\lambda^2\right) \operatorname{erf}\left(\lambda\right) \left(\frac{\Delta H_s}{c_s}\right) = 1090.80\,°C$$

$$T_i = T_o + \sqrt{\pi}\lambda \exp\left(\lambda^2\right) \frac{\Delta H_s}{c_s} \sqrt{\frac{k_s \rho_s c_s}{k_m \rho_m c_m}} = 1090.80\,°C$$

From Eq. (5.18d) with $s(t) = 0.04/2 = 0.02$m, the solidification time t and the solidification velocity dx_s/dt are

$$t = \frac{s(t)^2}{4\lambda^2 \alpha_s} = \frac{(0.02)^2}{4\,(0.62265)^2\,(1.3317 \times 10^{-5})} = 19.37\,s$$

$$\frac{dx_s}{dt} = \lambda\sqrt{\frac{\alpha_s}{t}} = 5.16 \times 10^{-4} \text{m/s} = 1857.60\,\text{mm/h}$$

(b) From Eq. (5.21c) at $x = 0$ and $\Delta T_s = 100\,°C = 100\,K$, the heat flux flowing through the solid is

$$q_s = \frac{k_s\left(T_f - T_i\right)}{\operatorname{erf}\left(\lambda\right)\sqrt{\pi \alpha_s t}} = 1.83 \times 10^6 \text{ W/m}^2 = 1.83\,\text{MW/m}^2$$

and that crossing the mold–solid interface, into the mold and eventually dissipates in the mold surroundings is

$$q_m = \frac{k_m\left(T_i - T_o\right)}{\sqrt{\pi \alpha_m t}} = 1.83 \times 10^6 \text{ W/m}^2 = 1.83\,\text{MW/m}^2$$

(c) From Eqs. (5.15a) and (5.17a), the temperature equations are

$$T_m = T_o + (T_i - T_o)\operatorname{erfc}\left(\frac{-x}{2\sqrt{\alpha_m t}}\right) \quad \text{(mold)}$$

$$T_m = 25\,°C + (1065.80\,°C)\operatorname{erfc}\left(-43.553x\right)$$

$$T_s = T_i + \frac{(T_f - T_i)}{\operatorname{erf}\left(\lambda\right)}\operatorname{erf}\left(\frac{x}{2\sqrt{\alpha_s t}}\right) \quad \text{(solid)}$$

$$T_s = 1090.80\,°C + (714.79\,°C)\operatorname{erf}\left(31.132x\right)$$

Plotting these equations at $\Delta T_s = 0$, 20 and $100\,°C$ reveals the effect of superheating the melt at a particular holding time, which is not included in the current solidification analysis. Note the opposite behavior of the temperature profile in the mold and solid.

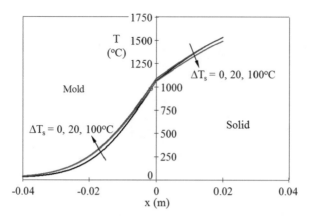

Therefore, the imposed superheating range $0 < \Delta T_s \leq 100\,°C$ on the temperature fields in the steel mold and the solid (solidified melt) regions can be considered insignificant since $\Delta H_s \simeq \Delta H_f$. This may be the reason most authors use ΔH_f instead of ΔH_s in evaluating the solidification process of pure metals under controlled conditions.

5.5.1   Sand Mold Thermal Resistance

A theoretical study of factors related to controlled rate of heat flow from the solidifying melt in green sand molds with and without chilling effects, and metallic molds lead to finding the analytical solution of the governing heat equation, Eq. (5.29a), for simple symmetrical slab-shaped, cylindrical, and spherical castings. This is the fundamental approach adopted in this book in order to convey to the reader facts associated with thermal conditions for an optimum solidification process. However, solidification problems due to thermal-contact resistance at the mold–metal interface and complicated casting geometries are amenable to analysis by a suitable iterative procedure (numerical method). Truthfully, casting in sand molds is a conventional foundry practice because it is cost effective and the molding sand can be recycled to an extent.

Consider casting a metal slab into a low-thermal conductivity sand mold and assume that mold resistance to heat transfer during solidification is controlled by the sand temperature gradient $\partial T_m/\partial x$. In this case, Fig. 5.5 schematically shows the corresponding temperature profiles for the mold, solid, and liquid phases. This implies that $(\partial T_m/\partial x) \gg (\partial T_s/\partial x) \simeq (\partial T_l/\partial x) \simeq 0$ and the convective heat transfer is negligible.

In this case, the governing heat equation, the continuity of temperature, and the heat flux balance are, respectively

Fig. 5.5 Mold resistance and negligible solid and liquid thermal resistances

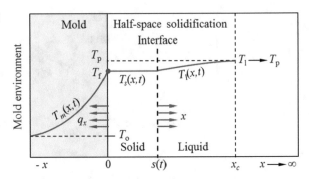

$$\frac{\partial T_m}{\partial t} = \alpha_m \frac{\partial^2 T_m}{\partial x^2} \quad \text{for } -\infty < x < 0, t > 0 \tag{5.29a}$$

$$T_s = T_l = T_f \quad \text{at } x = s\,(t) \tag{5.29b}$$

$$k_m \frac{\partial T_m}{\partial x} = \rho_s \Delta H_f \frac{ds\,(t)}{dt} \quad \text{at } x = s\,(t) \tag{5.29c}$$

The solution of Eq. (5.29a) is defined by (5.15a) above, but in this section T_i is replaced by T_f so that

$$T_m = T_o + \left(T_f - T_o\right) \operatorname{erfc}\left(\frac{-x}{2\sqrt{\alpha_m t}}\right) \quad \text{(mold)} \tag{5.30a}$$

$$\frac{\partial T_m}{\partial x} = \frac{T_f - T_o}{\sqrt{\pi \alpha_m t}} \exp\left(\frac{-x}{2\sqrt{\alpha_m t}}\right) \tag{5.30b}$$

$$\left(\frac{\partial T_m}{\partial x}\right)_{x=0} = \frac{T_i - T_o}{\sqrt{\pi \alpha_m t}} \tag{5.30c}$$

These equations are now suitable for redefining and simplifying the Stefan problem, mathematically defined by Eq. (5.29c).

Combining Eqs. (5.29c) and (5.30c) yields the solidification velocity, and integrating it gives the characteristic length $s\,(t)$ and the solidification time t at $x = s\,(t)$ given by

$$\frac{ds\,(t)}{dt} = \frac{1}{\sqrt{\pi}} \left(\frac{T_f - T_o}{\rho_s \Delta H_f}\right) \left(\frac{k_m}{\sqrt{\alpha_m t}}\right) \tag{5.31a}$$

$$s\,(t) = \frac{2}{\sqrt{\pi}} \left(\frac{T_f - T_o}{\rho_s \Delta H_f}\right) \left(\frac{k_m}{\sqrt{\alpha_m}}\right) \sqrt{t} \tag{5.31b}$$

$$s\,(t) = \frac{2}{\sqrt{\pi}} \frac{\gamma_m \left(T_f - T_o\right)}{\rho_s \Delta H_f} \sqrt{t} \tag{5.31c}$$

$$t = \frac{\pi \alpha_m}{4} \left[\frac{\rho_s \Delta H_f}{k_m \left(T_f - T_o\right)}\right]^2 s\,(t)^2 \tag{5.31d}$$

The Amount of Thermal Energy According to Fourier's law, the heat flux q_x crossing the mold–solid interface and the rate of thermal energy dQ/dt through the mold are, respectively

$$q_x = -k_m \left(\frac{\partial T}{\partial x} \right)_{x=0} \tag{5.32a}$$

$$\frac{dQ}{dt} = -Ak_m \left(\frac{\partial T}{\partial x} \right)_{x=0} = \frac{Ak_m \left(T_f - T_o \right)}{\sqrt{\pi \alpha_m t}} \tag{5.32b}$$

Actually, q_x denotes the amount of heat that crosses a unit area per unit time and A denotes the interface surface area. Integrating Eq. (5.32b) yields the amount of thermal energy through the mold in the negative x-direction

$$Q_t = \frac{Ak_m \left(T_f - T_o \right)}{\sqrt{\pi \alpha_m}} \int_0^{t_s} \frac{dt}{\sqrt{t}} \tag{5.33a}$$

$$Q_t = \frac{2Ak_m \left(T_f - T_o \right)}{\sqrt{\pi \alpha_m}} \sqrt{t} \tag{5.33b}$$

For solidifying a liquid metal of mass m, the amount of thermal energy by conduction (Q_f) due to the release of latent heat of fusion ΔH_f is written as

$$Q_f = m \Delta H_f = \rho_s V \Delta H_f \tag{5.34}$$

where V is the casting volume. Equating Eqs. (5.33b) and (5.34) yields exactly (5.31b) in terms of volume to area ratio

$$s(t) = \frac{V}{A} = \frac{2}{\sqrt{\pi}} \left(\frac{T_f - T_o}{\rho_s \Delta H_f} \right) \left(\frac{k_m}{\sqrt{\alpha_m}} \right) \sqrt{t} \tag{5.35}$$

At this time, the mathematical definition $s(t) = V/A$ in Eq. (5.35) can be referred to as the Chvorinov's thermal modulus used in the foundry industry based on metal/sand molding and casting. Particularly, this volume-to-area (V/A) casting ratio is the basis for developing the Chvorinov's method, which can be treated as a special foundry rule in relation to the solidification time. This is accomplished in the next section below in a general form.

5.5.2 Casting Modulus and Chvorinov's Rule

This section includes details on the well-known Chvorinov's rule related to the solidification time t and casting modulus $s(t) = M = V/A$. The main objective of the Chvorinov's rule is to predict and compare the solidification time of the casting parts. In principle, at least one riser is designed to feed liquid metal to the casting geometry and gating system. As a result, defect-free castings can be obtained under control casting variables and mold thermophysical properties.

For any casting geometry, the Chvorinov's casting modulus $s(t)$ is the thickness or radius of the solidified melt, which in turn, is the length of the L-S boundary (dual-phase front) that has advanced into the liquid metal phase at a velocity defined as $ds(t)/dt$.

Solving Eq. (5.35) for t yields the Chvorinov's rule for a semi-infinite casting $(0 < x < \infty)$ [5, p. 332], [6, p. 331–332] and [7, p. 404]

$$t = \frac{\pi \alpha_m}{4} \left[\frac{\rho_s \Delta H_f}{k_m (T_f - T_o)} \right]^2 \left(\frac{V}{A} \right)^2 \tag{5.36a}$$

$$t = B \left(\frac{V}{A} \right)^2 = Bs(t)^2 \tag{5.36b}$$

$$B = \frac{\pi \alpha_m}{4} \left[\frac{\rho_s \Delta H_f}{k_m (T_f - T_o)} \right]^2 \tag{5.36c}$$

where B is the Chvorinov's constant or simply the mold constant [8]. Again, $M = V/A$ is the *casting modulus* with V as the volume of the casting geometry and A as the contact surface area of the metal–mold interface. Additionally, B, as defined by Eq. (5.36c), is the simplest closed-form equation when compared with expressions found in [4, p. 99].

In addition, the casting modulus $M = s(t)$ for rectangular, cylindrical, and spherical geometries can also be defined by Wlodawer [11, pp. 5–6]

$$M = s(t) = \frac{V}{A} = \frac{A}{P} = \frac{\text{Cross-sectional area}}{\text{Perimeter of cross section}} \tag{5.37a}$$

Rectangular Casting with Thickness $x = a$

$$s(t) = \frac{V}{A} = \frac{abc}{2bc} = \frac{a}{2} \tag{5.37b}$$

Cylindrical Casting with Radius $r = a$

$$s(t) = \frac{V}{A} = \frac{\pi r^2 c}{2\pi rc} = \frac{r}{2} \tag{5.37c}$$

Spherical Casting with Radius $r = a$

$$s(t) = \frac{V}{A} = \frac{(4/3) \pi r^3}{4\pi r^2} = \frac{r}{3} \tag{5.37d}$$

According to the Chvorinov's rule, solidification is considered complete when the casting modulus reaches the position of the characteristic length, $s(t) = M$, which denotes half the thickness of a rectangular casting geometry or the radius of a cylinder and sphere [4, p. 99] and [12, p. 37].

Example 5A.4 (Sand Mold) Consider a 4-cm cubic sand–mold cavity being filled with a hypothetical alloy at T_p and assume that solidification is complete in 3 min. Calculate the solidification time as per Chvorinov's rule for casting a 2-cm radius and 6-cm long cylinder and for a sphere with a 2-cm radius.

Solution From Eq. (5.36b) with $a = 4$ cm,

$$t_{cube} = 3 \text{ min and } s(t)_{cube} = V/A = a/6 = 0.33 \text{ cm}$$

$$t_{cube} = B\left(\frac{V}{A}\right)^2 = B\left(\frac{a^3}{6a^2}\right)^2 = \frac{a^2 B}{36} = Bs(t)^2$$

$$B = \frac{36t_{cube}}{a^2} = \frac{36(3 \text{ min})}{(4 \text{ cm})^2} = 6.75 \text{ min}/\text{cm}^2$$

For a cylinder,

$$s(t)_{cylinder} = \frac{V}{A} = \frac{r}{2} = 1 \text{ cm}$$

$$t_{cylinder} = Bs(t)^2 = \left(6.75\frac{\text{min}}{\text{cm}^2}\right)(1 \text{ cm})^2 = 6.75 \text{ min}$$

For a sphere,

$$s(t)_{sphere} = \frac{V}{A} = \frac{r}{3} = \frac{2}{3} \simeq 0.67 \text{ cm}$$

$$t_{sphere} = Bs(t)^2 = \left(6.75\frac{\text{min}}{\text{cm}^2}\right)\left(\frac{2}{3}\text{ cm}\right)^2 = 3 \text{ min}$$

Therefore, the cylinder solidifies slower because $t_{cube} = t_{sphere} < t_{cylinder}$.

Example 5A.5 (Sand Mold) Consider casting a pure copper slab ($4 \times 6 \times 10$ cm^3) into a sand mold. Assume the effect of superheat at $\Delta T_s = 10\,°C$ and calculate (a) the total solidification time t, (b) the solidification velocity $ds(t)/dt$, (c) the heating rate $\partial T_m/\partial t$ and mold thermal gradient $(\partial T_m/\partial x)_{x=0}$, (d) the amount of thermal energy Q and heat flux q_x through the mold, and (e) the Chvorinov's constant B. Also assume that the mold external temperature remains at 25 °C. From Appendix 5A,

Copper (Cu)
$T_f = 1356\,\text{K}$
$\rho_s = 7525\,\text{kg/m}^3$
$c_s = 429\,\text{J/kg K}$
$k_s = 400\,\text{W/m K}$
$\Delta H_f = 205{,}000\,\text{J/kg}$

Sand mold
$T_o = 298\,\text{K}$
$\rho_m = 1550\,\text{kg/m}^3$
$c_m = 986\,\text{J/kg K}$
$k_m = 0.40\,\text{W/m K}$

Solution

(a) Pouring temperature: $T_p = T_f + \Delta T_s = 1083 + 10 = 1093\,^\circ\text{C} = 1366\,\text{K}$. The thermal diffusivity, Eq. (5.2a), with $W = $ J/s units is

$$\alpha_m = \frac{k_m}{\rho_m c_m} = \frac{0.40\,\text{J/s/m K}}{(1550\,\text{kg/m}^3)\,(986\,\text{J/kg K})} = 2.6173 \times 10^{-7}\,\text{m}^2/\text{s}$$

For $x = 0.04\,\text{m}$ and $s\,(t) = x/2 = 0.02\,\text{m}$, the latent heat of solidification that accounts for any superheating the liquid and the solidification time are, respectively

$$\Delta H_s = \Delta H_f + c_s \Delta T_s = 205{,}000 + (429)\,(10) = 209{,}290\,\text{J/kg}$$

$$t = \frac{\pi \alpha_m}{4} \left[\frac{\rho_s \Delta H_s}{k_m\,(T_f - T_o)} \right]^2 s\,(t)^2 = 1138.70\,\text{s}$$

$$t \simeq 19\,\text{min} = 0.32\,\text{h}$$

(b) From Eq. (5.31a), the solidification velocity being influenced by the mold thermophysical properties is

$$\frac{ds\,(t)}{dt} = \frac{1}{\sqrt{\pi}} \left(\frac{T_f - T_o}{\rho_s \Delta H_s} \right) \left(\frac{k_m}{\sqrt{\alpha_m t}} \right) = 8.78 \times 10^{-6}\,\text{m/s} = 31.61\,\text{mm/h}$$

(c) From Eqs. (5.30a) and (5.30c), the mold heating rate and the mold temperature gradient are, respectively

$$\frac{\partial T_m}{\partial t} = \frac{(T_f - T_o)\,x}{2t\,\sqrt{\pi \alpha_m t}} \exp\left(-\frac{x^2}{4\alpha_m t} \right) = 0.16\,\text{K/s}$$

$$\left(\frac{\partial T_m}{\partial x} \right)_{x=0} = -\frac{(T_f - T_o)}{\sqrt{\pi \alpha_m t}} = -34{,}576\,\text{K/m}$$

(d) For an area $A = 2bc = 2\,(6 \times 10) = 120\,\text{cm}^2 = 0.012\,\text{m}^2$, Eqs. (5.33b) and (5.32a) give

$$Q = \frac{2Ak_m\left(T_f - T_o\right)}{\sqrt{\pi\alpha_m}}\sqrt{t} = 3.78 \times 10^5\,\text{J} = 378\,\text{kJ}$$

$$q_m = -k_m\left(\frac{\partial T_m}{\partial x}\right) = -\,(0.40\,\text{W/m K})\,(-34{,}576\,\text{K/m})$$

$$q_m = 13{,}830\,\text{W/m}^2 \simeq 13.83\,\text{kW/m}^2$$

$$q_m = 13.83\,\text{kJ/m}^2\,\text{s}$$

Notice that the heat flux q_m is analogous to diffusion flux. Both fluxes have the same units.

(e) From Eq. (5.36c), the Chvorinov's constant needed to compute the solidification time is

$$B = \frac{\pi\alpha_m}{4}\left[\frac{\rho_s\Delta H_s}{k_m\left(T_f - T_o\right)}\right]^2 = 2.8468 \times 10^6\,\text{s/m}^2$$

$$B = 4.7447\,\text{min}\,/\text{cm}^2$$

Using Eq. (5.36b) with the planar interface located at $s\,(t) = 2\,\text{cm}$ gives the predicted solidification time of the slab

$$t_{\text{slab}} = Bs\,(t)^2 = \left(4.7447\,\text{min}\,/\text{cm}^2\right)(2\,\text{cm})^2$$

$$t_{\text{slab}} = 19\,\text{min}$$

Therefore, Eq. (5.36b) verifies that the solidification time is 19 min as per Chvorinov's rule. This simple example indicates that the Chvorinov's rule is applicable to simple casting geometries.

5.5.3   Thermal Resistance in Solids

For solidification of a pure metal in a **water-cooled (chill) metal mold**, a high thermal resistance develops in the solid (solidified melt) due to a temperature field profile $T_s\,(x, t)$ schematically shown in Fig. 5.3. In this particular case, the solid thermal resistance R_s predominates over other thermal resistances during solidification because $\partial T_m/\partial x \simeq \partial T_l/\partial x \to 0$ and $\partial T_s/\partial x \gg 0$.

For a half-space region $x > 0$, the set of equations related to this thermal resistance case are

$$\frac{\partial T_s}{\partial t} = \alpha_s\frac{\partial^2 T_s}{\partial x^2} \quad \text{for } -\infty < x < 0, t > 0 \tag{5.38a}$$

$$T_s = T_l = T_f \quad \text{at } x = s\,(t) \tag{5.38b}$$

$$-k_s \left(\frac{\partial T_s}{\partial x} \right)_{s(t)} = \rho_s \Delta H_f \frac{ds\,(t)}{dt} \quad \text{at } x = s\,(t) = 2\lambda \sqrt{\alpha_s t} \tag{5.38c}$$

and the solution of Eq. (5.38a) is defined by (5.17a), which is recast here with $T_i = T_o$

$$T_s = T_o + \frac{\left(T_f - T_o\right)}{\text{erf}\,(\lambda)} \, \text{erf}\left(\frac{x}{2\sqrt{\alpha_s t}} \right) \tag{5.39}$$

This expression provides the temperature distribution within the solid phase and it is amenable to be analyzed as $T_s \, f\,(X)$ at a fixed t.

5.5.3.1 Moving Boundary Theory

The general moving L-S boundary (MB) (known as free boundary, melting front, or freezing front) theory is now used for modeling the liquid-to-solid (L-S) phase transformation, referred to as phase-change in the literature.

Accordingly, this is a two-phase subsystem within the solidification domain Γ that embraces the mold material and the casting geometry. Moreover, the L-S boundary position $s\,(t) \rightarrow x_c$ is driven by a liquid temperature gradient (no supercooling below T_f).

For convenience, assume that solidification occurs by a moving L-S boundary located at $x = s(t) > 0,\, t > 0$ and the local solidification temperature is the freezing temperature T_f. This means that there is not an imposed undercooling (supercooling) on the liquid metal (melt) and solidification proceeds at T_f. Actually, this is a simplification of solidification since there must be, at least, a small undercooling $\Delta T = \left(T_f - T\right) > 0$ for heterogeneous nucleation of small crystals. The concept of undercooling is discussed and characterized in later chapters as the driving force for solidification.

From Eq. (5.39) at $x = s\,(t)$,

$$\left(\frac{\partial T_s}{\partial x} \right)_{s(t)} = -\frac{\left(T_f - T_o\right)}{\sqrt{\pi \alpha_s t} \exp\left(\lambda^2\right) \text{erf}\,(\lambda)} \tag{5.40}$$

Upon inserting Eq. (5.40) into (5.38c) and collecting terms yields the transcendental equation similar to (5.10)

$$\lambda \exp\left(\lambda^2\right) \text{erf}\,(\lambda) = \frac{c_s \left(T_f - T_o\right)}{\sqrt{\pi} \Delta H_f} = \frac{S_t}{\sqrt{\pi}} \tag{5.41a}$$

$$f\,(\lambda) = \lambda \exp\left(\lambda^2\right) \text{erf}\,(\lambda) \tag{5.41b}$$

Fig. 5.6 Dimensionless transcendental non-linear function $f(\lambda)$

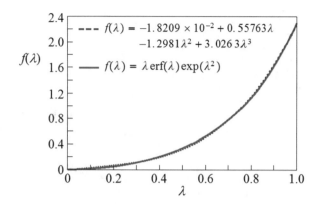

Plotting Eq. (5.41b) yields a non-linear trend similar to a fitted polynomial as shown in Fig. 5.6.

Once the right-hand side of Eq. (5.41a) is known, λ is evaluated numerically or graphically. Subsequently, additional calculations follow in order to fully character-ize the solidification process using conventional casting procedures.

5.5.3.2 Quasi-Static Approximation (QSA) Theory

For the classical Stefan problem, λ can be approximated using the exponential and error functions as power series

$$\exp\left(\lambda^2\right) = 1 + \frac{\lambda^2}{1!} + \frac{\lambda^4}{2!} + \frac{\lambda^6}{3!} + \ldots \simeq 1 \tag{5.42a}$$

$$\text{erf}\,(\lambda) = \frac{2}{\sqrt{\pi}}\left(\lambda - \frac{\lambda^3}{3 \times 1!} + \frac{\lambda^5}{5 \times 2!} \ldots\right) \simeq \frac{2\lambda}{\sqrt{\pi}} \tag{5.42b}$$

$$\lambda \exp\left(\lambda^2\right)\text{erf}\,(\lambda) \simeq \frac{2\lambda^2}{\sqrt{\pi}} \quad \text{for small } \lambda \tag{5.42c}$$

Combining Eqs. (5.41a) and (5.42c) yields the solution characteristic λ as

$$\lambda = \sqrt{\frac{c_s\left(T_f - T_o\right)}{2\Delta H_f}} \tag{5.43}$$

Thus, merging Eqs. (5.28a) and (5.43), and (5.18d) gives

$$S_t = 2\lambda^2 \tag{5.44a}$$

$$\lambda = \sqrt{S_t/2} \tag{5.44b}$$

Fig. 5.7 Comparison between the quasi-static approximation (QSA) and the moving boundary (MB) theories for the Stefan number S_t

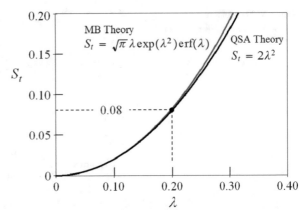

Plotting Eqs. (5.41b) and (5.44a) reveals a non-linear trend as shown in Fig. 5.7. According to the scale used in Fig. 5.7, the Stefan number S_t, as defined by the MB and QSA theories, compares at $\lambda \leq 0.20$ since $S_t < 0.1$. For $S_t > 0.1$, the QSA theory is considered inadequate to describe the heat transfer process during unidirectional solidification [9, pp. 91–92].

The objective now is to identify the significance of MB and QSA theories and rank them in order of importance for predicting the solidification time. It turns out that if $\lambda \leq 0.2$, these theories are in agreement, and the use of either theory is justified as long as $S_t \rightarrow 0$ and $\lambda \rightarrow 0$ and according to the scale in Fig. 5.7, the point $(S_t, \lambda) = (0.2, 0.08)$ serves as an upper limit for these theories to be in agreement.

Note that emphasis has been given to one-dimensional solidification heat transfer by conduction for single-phase materials. Heat transfer with or without convection and radiation effects is an option for the moment. Nonetheless, heat transfer by conduction prevails as the dominant mode for heat flow in conventional casting procedures.

5.5.3.3   Internal Resistance

In addition, an electric analogy with Ohm's law to heat flow by conduction used in solidification dictates that the temperature range ΔT is also a driving force for phase transformation ($L \rightarrow S$) being restricted by an internal resistance R_x written as [7, p. 243]

$$R_x = \frac{\Delta T}{q_{s(t)}} = \frac{T_f - T_o}{q_{s(t)}} \tag{5.45}$$

where the heat flux crossing the L-S interface located at $s(t)$ is

$$q_{s(t)} = -k_s (dT_s/dx)_{s(t)} \tag{5.46a}$$

$$q_{s(t)} = \frac{k_s (T_f - T_o)}{\sqrt{\pi \alpha_s t} \exp(\lambda^2) \operatorname{erf}(\lambda)} \tag{5.46b}$$

During solidification, a phase transformation ($L \rightarrow S$) imparts a two-phase boundary motion. For this reason, the evolved latent heat of solidification, ΔH_s, is justifiably a thermal energy in motion and the theory of heat transfer sets the mathematical background for determining the solidification parameters previously defined above and in preceding chapter.

With regard to Eq. (5.46b), the heat flux, $q_{s(t)}$, evaluated at the L-S solidification front, $s(t) > 0$, strongly depends on the value of the characteristic solution λ and the temperature difference $\Delta T = T_f - T_o$.

Example 5A.6 (Metal Mold) Use the moving boundary (MB) theory to determine (a) the Stefan number S_t, (b) solidification time t and the solidification velocity $ds(t)/dt$, (c) the temperature gradient $(\partial T/\partial x)_{x=0}$ and, (d) the heat flux $q_{s(t)}$ for casting a 0.04 m-thick copper slab into a carbon steel mold at a pouring temperature of 1093 °C. Compare results using the QSA theory. Use the data set for pure copper (Cu) from Appendix 5A.

Copper (Cu)	Carbon steel mold
$T_f = 1356$ K	$T_o = 298$ K
$\rho_s = 7525$ kg/m^3	$\rho_m = 7689$ kg/m^3
$c_s = 429$ J/kg K	$c_m = 669$ J/kg K
$k_s = 400$ W/m K	$k_m = 35$ W/m K
$\Delta H_f = 205{,}000$ J/kg	

Solution

Moving Boundary (MB) Theory
The pouring temperature:

$$T_p = T_f + \Delta T_s = 1083 + 10 = 1093\,°C = 1366\,K$$

Thermal diffusivity using Eq. (5.2a):

$$\alpha_m = \frac{k_m}{\rho_m c_m} = 2.6173 \times 10^{-7}\,m^2/s$$

$$\alpha_s = \frac{k_s}{\rho_s c_s} = 1.2391 \times 10^{-4}\,m^2/s$$

(a) For a half-space region, $x = 0.04\,\mathrm{m}$ and $s\,(t) = x/2 = 0.02\,\mathrm{m}$ and Eq. (5.13b) with compensated superheat effect gives

$$\Delta H_s = \Delta H_f + c_s \Delta T_s = 205{,}000 + (429)\,(10) = 209{,}290\,\mathrm{J/kg}$$

From Eq. (5.28),

$$S_t = \frac{c_s\,(T_f - T_o)}{\Delta H_s} = \sqrt{\pi}\lambda \exp\left(\lambda^2\right)\mathrm{erf}\,(\lambda) = 2.17$$

(b) From Eq. (5.41a),

$$\lambda \exp\left(\lambda^2\right)\mathrm{erf}\,(\lambda) = \frac{c_s\,(T_f - T_o)}{\sqrt{\pi}\,\Delta H_s} = 1.2235$$

$$\lambda \exp\left(\lambda^2\right)\mathrm{erf}\,(\lambda) - 1.2235 = 0$$

$$\lambda = 0.82287$$

From Eqs. (5.18d) and (5.18e), respectively,

$$t = \frac{s\,(t)^2}{4\lambda^2 \alpha_s} = \frac{(0.02)^2}{4\,(0.82287)^2\,(1.2391 \times 10^{-4})} = 1.19\,\mathrm{s}$$

$$\frac{ds\,(t)}{dt} = \lambda\sqrt{\frac{\alpha_s}{t}} = (0.82287)\sqrt{\frac{1.2391 \times 10^{-4}}{1.19}} = 8.40 \times 10^{-3}\,\mathrm{m/s}$$

Combining Eqs. (5.38c) and (5.40) yields

$$\frac{ds\,(t)}{dt} = \left[\frac{1}{\sqrt{\pi \alpha_s t}\,\exp\left(\lambda^2\right)\mathrm{erf}\,(\lambda)}\right]\left[\frac{k_s\,(T_f - T_o)}{\rho_s\,\Delta H_s}\right]$$

$$\frac{ds\,(t)}{dt} = 8.40 \times 10^{-3}\,\mathrm{m/s}$$

(c) From Eq. (5.40),

$$\left(\frac{\partial T_s}{\partial x}\right)_{s(t)} = -\frac{(T_f - T_o)}{\sqrt{\pi \alpha_s t}\,\mathrm{erf}\,(\lambda)}\exp\left(-\lambda^2\right) = -34{,}534\,\mathrm{K/m} = -33{,}060\,^\circ\mathrm{C/m}$$

(d) The heat flux density at the L-S interface, $s\,(t) = 0.02\,\mathrm{m}$, is calculated using Eq. (5.46b)

$$q_{s(t)} = \frac{k_s \left(T_f - T_o\right)}{\sqrt{\pi \alpha_s t} \exp\left(\lambda^2\right) \operatorname{erf}\left(\lambda\right)} = 1.322 \times 10^7 \, \text{W/m}^2 = 13.22 \, \text{MW/m}^2$$

$$q_f = \rho_s \Delta H_s \frac{ds\left(t\right)}{dt} = (7525)\,(209{,}290)\left(8.40 \times 10^{-3}\right) = 13.22 \, \text{MW/m}^2$$

Quasi-Static Approximation (QSA) Theory   Using Eq. (5.43) yields

$$\lambda = \sqrt{\frac{c_s \left(T_f - T_o\right)}{2\Delta H_s}} = 1.0413$$

and from (5.44a), $S_t = 2\lambda^2 = 2\,(1.0413)^2 = 2.17$. Then,

$$t = \frac{s\left(t\right)^2}{4\lambda^2 \alpha_s} = \frac{(0.02)^2}{4\,(1.5494)^2\left(1.1 \times 10^{-5}\right)} = 0.74 \, \text{s}$$

$$\frac{ds\left(t\right)}{dt} = \lambda \sqrt{\frac{\alpha_s}{t}} = (1.0413) \sqrt{\frac{1.2391 \times 10^{-4}}{0.74}} = 1.35 \times 10^{-2} \, \text{m/s}$$

Therefore, $t\,(\text{MB}) > t\,(\text{QSA})$ and $[ds\,(t)\,/dt]_{\text{MB}} < [ds\,(t)\,/dt]_{\text{QSA}}$ are different, but the conventional Stefan number remains the same, $S_t = 2.17$.

Example 5A.7 (Metal Mold) Consider pouring an amount of copper melt at $1183\,^\circ\text{C}$ into a cooled metal mold having a slab geometry 0.08-m thick. Model the solidification process as a stable and planar L-S boundary that moves into the melt very slowly in the half-space region $x > 0$ until it reaches a location $x = s\,(t) = 0.04\,\text{m}$. **Case 1:** Assumed that a thermal conduction resistance to heat transfer develops in the solidified melt and that the melt cools to a temperature $T_f = 1083\,^\circ\text{C}$ and solidifies at an insignificant temperature gradient (no supercooling $\Delta T_u = \left(T_f - T_u\right) \to 0$). Determine (a) the solidification time t and the rate of solidification $ds\,(t)\,/dt$, (b) the temperature gradients $(\partial T_s/\partial x)_{s(t)}$ and $(\partial T_l/\partial x)_{s(t)}$, (c) the heat fluxes, and (d) plot the temperature field equations $T_s\,(x,t)$ and $T_l\,(x,t)$ when $s\,(t) = 0.03\,\text{m}$. **Case 2:** Now, assume that the melt undergoes a significant supercooling $\Delta T_u = \left(T_f - T_u\right) \gg 0$ and as a result, solid and liquid temperature gradients develop during the solidification process. Analyze this case and determine the solidification time. Use the data for pure copper from Appendix 5A.

Solution From Appendix 5A,
From Eq. (5.2a), the thermal diffusivity terms are

$$\alpha_m = \frac{k_m}{\rho_m c_m} = 6.8041 \times 10^{-6} \, \text{m}^2\text{s}$$

$$\alpha_s = \frac{k_s}{\rho_s c_s} = 1.7590 \times 10^{-4} \, \text{m}^2\text{s}$$

Mold	Pure copper	
$T_o = 25\,°C$	$T_p = 1183\,°C$	$T_f = 1083\,°C$
$\rho_m = 7689\,kg/m^3$	$\rho_l = 8960\,kg/m^3$ at T_f	$\rho_s = 8033\,kg/m^3$
$c_m = 669\,J/kg\,K$	$c_l = 118\,J/kg\,K$	$c_s = 92\,J/kg\,K$
$k_m = 35\,W/m\,K$	$k_l = 163\,W/m\,K$	$k_s = 130\,W/m\,K$
	$\Delta H_f = 209,000\,Jkg$	

$$\alpha_l = \frac{k_l}{\rho_l c_l} = 1.5417 \times 10^{-4}\,m^2 s$$

Use Eq. (5.13b) to calculate the latent heat of solidification due to a superheat $\Delta T_s = T_p - T_f$ is

$$T_p = 1183\,°C = 1456\,K$$

$$\Delta T_s = T_p - T_f = 100\,°C = 100\,K$$

$$\Delta H_s = \Delta H_f + c_l\,(T_p - T_f) = 209,000 + 728\,(1183 - 1083)$$

$$\Delta H_s = 281,800\,J/kg$$

Subsequently, use of ΔH_s compensates for the superheating effect on solidification. Thus, evaluate λ using Eq. (5.28a)

$$\lambda \exp\left(\lambda^2\right)\left[\operatorname{erf}(\lambda) + \sqrt{\frac{k_s \rho_s c_s}{k_m \rho_m c_m}}\right] = \frac{c_s\,(T_f - T_o)}{\sqrt{\pi}\,\Delta H_s}$$

$$\lambda \exp\left(\lambda^2\right)[\operatorname{erf}(\lambda) + 0.7305] = 0.19488$$

$$\lambda \exp\left(\lambda^2\right)(\operatorname{erf}(\lambda) + 0.7305) - 0.19488 = 0$$

$$\lambda = 0.19726$$

Additionally, evaluate the interface temperature T_i at $x = 0$ between the mold and solid planar interface without convective heat transfer effects. Thus, Eq. (5.26b) gives

$$T_i = T_f - \sqrt{\pi}\,\lambda \exp\left(\lambda^2\right)\operatorname{erf}(\lambda)\left(\frac{\Delta H_s}{c_s}\right) = 838.34\,°C$$

Case 1: No Supercooling Effect of the Melt
(a) From Eq. (5.41a),

$$\lambda \exp\left(\lambda^2\right)\operatorname{erf}(\lambda) - \frac{c_s\,(T_f - T_o)}{\sqrt{\pi}\,\Delta H_s} = 0$$

$$\lambda \exp\left(\lambda^2\right) \mathrm{erf}\,(\lambda) - 0.19488 = 0$$

$$\lambda = 0.39438$$

From (5.18d) and (5.18e), respectively, with $s\,(t) = 0.03\,\mathrm{m}$

$$t = \frac{s\,(t)^2}{4\lambda^2\alpha_s} = \frac{(0.03)^2}{4\,(0.39438)^2\,(1.7590 \times 10^{-4})} = 8.22\,\mathrm{s}$$

$$\frac{ds\,(t)}{dt} = \lambda\sqrt{\frac{\alpha_s}{t}} = (0.39438)\sqrt{\frac{1.7590 \times 10^{-4}}{8.22}} = 1.82 \times 10^{-3}\,\mathrm{m/s}$$

Combining Eqs. (5.38c) and (5.40) yields an expression that verifies the $ds\,(t)/dt$ calculated above

$$\frac{ds\,(t)}{dt} = \left[\frac{1}{\sqrt{\pi\alpha_s t}\,\exp\left(\lambda^2\right)\mathrm{erf}\,(\lambda)}\right]\left[\frac{k_s\,(T_f - T_o)}{\rho_s\,\Delta H_s}\right]$$

$$\frac{ds\,(t)}{dt} = 1.82 \times 10^{-3}\,\mathrm{m/s}$$

(b) From Eq. (5.40), the temperature gradients in the solid phase at $s\,(t) = 0.03\,\mathrm{m}$ are

$$\left(\frac{\partial T_s}{\partial x}\right)_{s(t)} = -\frac{(T_f - T_o)}{\sqrt{\pi\alpha_s t}\,\exp\left(\lambda^2\right)\mathrm{erf}\,(\lambda)} = -31{,}767\,\mathrm{K/m} = -31{,}767\,^\circ\mathrm{C/m}$$

$$\left(\frac{\partial T_l}{\partial x}\right)_{s(t)} \rightarrow 0 \quad (\text{assumed})$$

(c) The heat flux balance ($q_s = q_f$) at the $L\text{-}S$ interface, $s\,(t) = 0.03\,\mathrm{m}$, is calculated using Eq. (5.46b)

$$q_s = -k_s\left(\frac{\partial T_s}{\partial x}\right)_{s(t)} = \frac{k_s\,(T_f - T_o)}{\sqrt{\pi\alpha_s t}\,\exp\left(\lambda^2\right)\mathrm{erf}\,(\lambda)} \simeq 4.12\,\mathrm{MW/m^2}$$

$$q_f = \rho_s\,\Delta H_s\frac{ds\,(t)}{dt} = (8033)\,(281{,}800)\left(1.82 \times 10^{-3}\right) = 4.12\,\mathrm{MW/m^2}$$

(d) The solid and liquid temperature field equations up to $s\,(t) = 0.03\,\mathrm{m}$ are

$$T_s = T_i + \frac{T_f - T_i}{\mathrm{erf}\,(\lambda)}\,\mathrm{erf}\left(\frac{x}{2\sqrt{\alpha_s t}}\right) = 838.34 + (578.42)\,\mathrm{erf}\,(13.149x)$$

$$T_l = T_f = 1083\,^\circ\mathrm{C} \quad (\text{at interface})$$

Then, T_s at $0 \leq x \leq 0.03$ m and T_l at $0.03 \leq x \leq 0.04$ m intervals are shown in the figure at the end of this example.

Case 2: Supercooling Effect on the Melt

(a) In this case, the order of calculations is altered for convenience. From Table 4.1, the undercooling temperature range is $\Delta T_u = 236\,°C$ and the supercooled liquid temperature is

$$\Delta T_u = T_f - T_u$$
$$T_u = T_f - \Delta T_u = 1083\,°C - 236\,°C = 847\,°C$$

which is approximately 22% below the equilibrium freezing temperature T_f. Nonetheless, this a significant undercooling value.

(b) For $s(t) = 0.03$ m, Eqs. (5.15a), (5.17a), and (5.20a) at $t = 8.22$ s yield the temperature field equations as

$$T_m = T_o + (T_i - T_o)\,\text{erfc}\left(\frac{-x}{2\sqrt{\alpha_m t}}\right) = 25 + (813.34)\,\text{erfc}\,(-66.857x)$$

$$T_s = T_i + \frac{(T_f - T_i)}{\text{erf}(\lambda)}\,\text{erf}\left(\frac{x}{2\sqrt{\alpha_s t}}\right) = 838.34 + (578.42)\,\text{erf}\,(13.149x)$$

$$T_l = T_u - \frac{(T_u - T_f)}{\text{erfc}\left(\lambda\sqrt{\alpha_s/\alpha_l}\right)}\,\text{erfc}\left(\frac{x}{2\sqrt{\alpha_l t}}\right) = 847 + (428.\,05)\,\text{erfc}\,(14.045x)$$

(c) From the above temperature equations, the heat flux terms are

$$q_s = k_s \left(\frac{\partial T_s}{\partial x}\right)_{s(t)} = (130)\,(7345.20) = 0.95\,\text{MW/m}^2$$

$$q_l = k_l \left(\frac{\partial T_s}{\partial x}\right)_{s(t)} = (163)\,(-5680.30) = -0.93\,\text{MW/m}^2$$

$$q_f = q_s - q_l = 0.95 + 0.93 = 1.88\,\text{MW/m}^2$$

(d) Using $q_f = \rho_s \Delta H_s\,[ds(t)/dt]$, the solidification time is

$$\frac{ds(t)}{dt} = \frac{q_f}{\rho_s \Delta H_s} = \frac{1.88 \times 10^6\,\text{W/m}^2}{(8033)\,(281{,}800)} = 8.31 \times 10^{-4}\,\text{m/s}$$

$$t = 0.03/\left(8.31 \times 10^{-4}\right) = 36.10\,\text{s}$$

The temperature distributions for T_m, T_s, and T_l are given below.

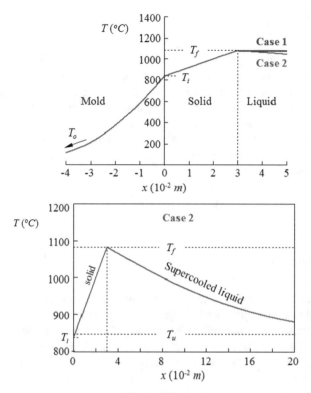

Therefore, expanding this region shows the non-linearity of the supercooling as indicated by the bottom figure.

5.6   Phase-Change Numerical Methods

Mathematical modeling simple and intricate solidification problems can be feasible through the use of software packages that handle complicated algebra systems for solving ODEs and PDEs, and developing an optimization procedure. Hence, computer algebra systems, such as MAPLE, MATLAB, and LINDO, are valuable tools for scientists and engineers. The literature is abundant in planar solidification numerical solutions of simple and intricate casting geometries.

In particular, the reader is encouraged to consult excellent review papers on cylindrical and/or spherical melting and solidification numerical methods, where cylindrical and spherical solidification processes include comparative analytical and numerical results [2, p. 210–211] and [13, 14].

5.7   Solidification Under Magnetic Fields

In general, applied magnetic fields on the solidification process may have positive or negative effects on the metal structure due to the presence of thermoelectric magnetic convection. Characterizing the electromagnetic solidification processes includes the effects of weak or strong magnetic fields, electromagnetic vibration and alternative electric current (AC), and so forth. Obviously, a type of magnetic field may distort the L-S interface surface area and influence the freezing temperature of pure metals and subsequently, affect to an extent the final microstructural features of metals and alloys.

The reader is also encouraged to search the literature for publications that characterize the effects of magnetic fields on solidification heat transfer [15–17].

5.8   Summary

Theoretically, the solidification heat transfer is a complicated process to model, but a quasi-static approximation (QSA) theory for the thermal field, known as a Stefan problem, is related to the moving L-S boundary (MB) theory. Thus, the mathematical solutions for predicting the solidification time t_f, solidification velocity $v = dx/dt$, temperature gradient dT/dx, and the amount of thermal energy Q removed from the casting through the mold wall were derived based on the heat transfer and evolved QSA and MB theories.

The simple analytical solutions and example problems on solidification of pure metals included in this chapter serve as a strong foundation for studying phase-change heat transfer. Different mathematical modeling on alloy solidification is also available and fortunately, the reader interested on this topic has a vast literature field to search for specific publications.

In conclusion, there are solutions to one-phase and two-phase heat transfer problems available in the literature. Among the cited references in this chapter, Rubinstein book [18] and Tarzia [19, Chapter 20] are excellent resources for additional theoretical background on a compilation of phase-change solidification problems, usually solved using the half-space concept in semi-infinite regions and the internal thermal resistance approach. Among important variables used to analyze the solidification process, the solidification time predicted by the Chvorinov's rule and by the heat transfer theory are analytically compared.

5.9   Appendix 5A Thermophysical Properties

This appendix contains a convenient unit conversion table and a short numerical database on thermophysical properties of some pure metals.

Conversion Table

Variable/Property	Eng. units	Scientific units
Mass m	$1\,lb_m$	$0.453592\,kg$
Thermal energy Q	1 Btu	1.05506 kJ
Heat of fusion ΔH_f	$1\,Btu/lb_m$	2.326 kJ/kg
Density ρ	$1\,lb_m/ft^3$	$16.0185\,kg/m^3$
Thermal conductivity k	1 Btu/ (hr ft °F)	1.7307 (J/m s K = W/mK)
Thermal conductivity k	1 cal/s cm °C	4.18 W/m K
Thermal diffusivity α	$1\,ft^2/h$	$2.5807 \times 10^{-5}\,m^2/s$
Heat capacity c_p	1 Btu/lbm °F	4184 J/kg K
Heat capacity c_p	1 cal/g °C	4188 J/kg K

Average thermophysical properties for some metals at T_f and foundry molds. The subscripts l, s, and m denote liquid, solid, and mold, respectively. For the mold materials in the table given below, $\rho_s = \rho_m$, $c_s = c_m$, and $k_s = k_m$.

Material	T_f^a	ΔH_f^a	ρ_s^b	ρ_l^b	c_s^b	c_l^b	k_s^b	k_l^b
Ag	961	104	10,312	9329	240	310	362	175
Al	660	397	2700	2378	216	259	211	91
Cu	1083	209	8960	8033	92	118	130	163
Fe	1535	247	7870	7035	108	187	34	33
Mg	650	349	1740	1589	110	316	145	79
Ni	1450	292	8890	7890	131	157	70	60
Si	1412	1877	2329	2469	210	231	25	56
Steel mold			7689		669		35	
$\overline{C}$ mold^c			1800		720		65	
Sand mold			1550		986		0.40	

Units of tabulated thermophysical data:
$^a T_f$ in °C and ΔH_f in $\times 10^3$ kJ/kg units
$^b \rho_s$, ρ_l in kg/m^3, c_s, c_s in J/kg K and k_s, k_l in W/m K
$^c \overline{C}$ stands for graphite

Example 5A.1 For silver (Ag), change $\Delta H_f = 965$ J/cm³ $\rightarrow$ ΔH_f in J/mol

$$\Delta H_f = \frac{\Delta H_f\,(\text{J/cm}^3)}{\rho\,(\text{g/cm}^3)} A_w\,(\text{g/mol}) \quad \text{units are parenthesis}$$

$$\Delta H_f = \left(\frac{965\,\text{J/cm}^3}{10.49\,\text{g/cm}^3} \right) \left(107.87\,\frac{\text{g}}{\text{mol}} \right) = 9923.2\,\text{J/mol}$$

5.10 Appendix 5B Separation of Variables

The solution to the one-dimensional heat equation is based on the transformation of a partial differential equation (PDE) to ordinary differential equations (ODE). This can be achieved if the temperature of the liquid phase is defined as $T(x, t) = f(x) g(t)$. Thus, the transformation of PDE to ODE is achieved by equating the function that depends on time t on one side and the one depending on x on the other side. Subsequently, the solution will contain some constants that must be evaluated according to the chosen IC and BC.

For one-dimensional analysis, the heat equation is

$$\frac{\partial T}{\partial t} = \alpha \frac{\partial^2 T}{\partial x^2} \qquad (5B.1)$$

The heat equation, Eq. (4B.1), can be approximated using unidimensional analysis and its solution can be derived by converting PDE to ODE by letting $T = T(x, t) = f(n)$ with n defined by

$$n = \frac{x}{2\sqrt{\alpha t}} \qquad (5B.2)$$

$$\frac{dn}{dx} = \frac{1}{2\sqrt{\alpha t}} \qquad (5B.3)$$

$$\frac{dn}{dt} = -\frac{x}{4\sqrt{\alpha t^3}} = -\frac{x}{4t\sqrt{\alpha t}} \qquad (5B.4)$$

Using the chain rule of differentiation yields

$$\frac{\partial T}{\partial t} = \frac{dT}{dn} \frac{\partial n}{\partial t} = -\frac{x}{4t\sqrt{\alpha t}} \frac{dT}{dn} \qquad (5B.5a)$$

$$\frac{\partial T}{\partial x} = \frac{dT}{dn} \frac{\partial n}{\partial x} = \frac{1}{2\sqrt{\alpha t}} \frac{dT}{dn} \qquad (5B.5b)$$

$$\frac{\partial^2 T}{\partial x^2} = \frac{d^2 T}{dn^2} \left(\frac{\partial n}{\partial x}\right)^2 + \frac{dT}{dn} \frac{\partial^2 n}{\partial x^2} = \left(\frac{1}{2\sqrt{\alpha t}}\right)^2 \frac{d^2 T}{dn^2} \qquad (5B.5c)$$

$$\frac{\partial^2 T}{\partial x^2} = \frac{1}{4Dt} \frac{d^2 T}{dn^2} \qquad (5B.5d)$$

where $\partial^2 n / \partial x^2 = 0$. Substitute Eqs. (5B.5a) and (5B.5d) into (5B.1) to get the sought ODE

$$-\frac{x}{4t\sqrt{\alpha t}} \frac{dT}{dn} = \frac{D}{4Dt} \frac{d^2 T}{dn^2} \qquad (5B.6a)$$

$$-2n\frac{dT}{dn} = \frac{d^2T}{dn^2} \qquad\qquad (5B.6b)$$

For convenience, let $f = dT/dn$ and $df = d^2T/dn^2$ so that Eq. (5B.6b) becomes

$$-2nf = df \qquad\qquad (5B.7a)$$

$$\frac{df}{f} = -2n \qquad\qquad (5B.7b)$$

$$\int \frac{df}{f} = -2 \int ndn \qquad\qquad (5B.7c)$$

$$\ln(f) = -n^2 + h \qquad\qquad (5B.7d)$$

where h is a constant of integration. Solving for f yields

$$f = \exp\left(-n^2 + h\right) = A\exp\left(-n^2\right) \qquad\qquad (5B.8a)$$

$$\frac{dT}{dn} = A\exp\left(-n^2\right) \qquad\qquad (5B.8b)$$

$$dT = A\exp\left(-n^2\right)dn \qquad\qquad (5B.8c)$$

In order to integrate Eq. (5B.8c) and solve for A one needs to describe the heat problem and set the initial condition (IC) and boundary conditions (BC) as the limits of integration.

For an isotropic material as a semi-infinite $(0 < x < \infty)$ medium,

$$\text{IC} \rightarrow T = T_o \text{ for } n = \infty \text{ at } x > 0, \ t = 0 \qquad\qquad (5B.9a)$$

$$\text{BC} \rightarrow T = T_f \text{ for } n = 0 \text{ at } x = 0, \ t > 0 \qquad\qquad (5B.9b)$$

$$\text{BC} \rightarrow T = T_o \text{ for } n = \infty \text{ at } x = \infty, t > 0 \qquad\qquad (5B.9c)$$

Integrating Eq. (5B.8c) using the BC given by Eq. (5B.9c) yields the constant A written as

$$\int_{T_o}^{T_f} dT = A \int_0^\infty \exp\left(-n^2\right)dn = A\frac{\sqrt{\pi}}{2} \qquad\qquad (5B.10a)$$

$$A = \frac{2\left(T_f - T_o\right)}{\sqrt{\pi}} \qquad\qquad (5B.10b)$$

Integrate Eq. (5B.8c) again with IC defined by (5B.9a) to get

$$\int_{T_o}^{T} dT = A \int_0^n \exp\left(-n^2\right) dn \tag{5B.11a}$$

$$T - T_o = \left(T_f - T_o\right) \frac{2}{\sqrt{\pi}} \int_n^{\infty} \exp\left(-n^2\right) dn \tag{5B.11b}$$

This integral does not have an exact solution, but it defines the Gauss complementary error function, erfc (n), of sigmoid shape.

Furthermore, the solution of Eq. (5B.11b) can be written in two mathematical forms

$$T - T_o = \left(T_f - T_o\right) \text{erfc}\,(n) \tag{5B.12a}$$

$$T - T_o = \left(T_f - T_o\right) [1 - \text{erf}\,(n)] \tag{5B.12b}$$

For convenience, Eq. (5B.12) is slightly modified in order to include the definition of the similarity variable n in the error function and the complementary error function. Thus,

$$T - T_o = \left(T_f - T_o\right) \text{erfc}\left(\frac{x}{2\sqrt{\alpha t}}\right) \tag{5B.13a}$$

$$T - T_o = \left(T_f - T_o\right) \left[1 - \text{erf}\left(\frac{x}{2\sqrt{\alpha t}}\right)\right] \tag{5B.13b}$$

$$T = T_o + \left(T_f - T_o\right) \text{erfc}\left(\frac{x}{2\sqrt{\alpha t}}\right) \tag{5B.13c}$$

where

$$\text{erf}\,(n) = \frac{2}{\sqrt{\pi}} \int_0^n \exp\left(-n^2\right) dn \tag{5B.14a}$$

$$= \frac{2}{\sqrt{\pi}} \int_o^n \sum_{k=0}^{\infty} \frac{(-1)^k n^{2k}}{k!} dn \tag{5B.14b}$$

$$\text{erf}\,(n) = \frac{2}{\sqrt{\pi}} \left(n - \frac{n^3}{3} + \frac{n^5}{10} - \frac{n^7}{42} + \ldots\right) \tag{5B.14c}$$

This concludes the analytical procedure.

5.10.1 Similarity Solution

Below is the procedure for deriving the general similarity solution for the heat equation. This is just a similarity transformation of a partial differential equation (PDE), and its boundary conditions and initial condition to an ordinary differential equation (ODE) by combining two independent variables (x and t) into a single independent variable n. In fact, x and t are known as independent invariants.

For one-dimensional heat equation, the Hellums–Churchill method is used here with general initial condition (IC) and boundary conditions (BC)

$$\frac{\partial T}{\partial t} = \alpha \frac{\partial^2 T}{\partial x^2} \quad \text{for } x > 0, t > 0 \tag{5B.15a}$$

$$T(x, t) = T(x, 0) = T_{IC} \quad \text{for } t = 0 \tag{5B.15b}$$

$$T(x, t) = T(0, t) = T_{BC} \quad \text{for } t = 0 \tag{5B.15c}$$

where T_{IC} denotes the initial condition temperature and T_{BC} denotes the boundary condition temperature. Now, making PDE, IC, and BC dimensionless one can easily find the solution to Eq. (5B.15a).

Thus,

$$T^* = \frac{T - T_{IC}}{T_d}; \quad x^* = \frac{x}{x_o}; \quad t^* = \frac{t}{t_o} \tag{5B.16a}$$

$$\frac{\partial T^*}{\partial t^*} = \left(\frac{\alpha t_o}{x_o^2}\right) \frac{\partial^2 T^*}{\partial x^{*2}} \tag{5B.16b}$$

$$T^*(x^*, 0) = 0 \tag{5B.16c}$$

$$T^*(0, t^*) = \frac{T_{BC} - T_{IC}}{T_d} = 1 \tag{5B.16d}$$

$$T_d = T_{BC} - T_{IC} \tag{5B.16e}$$

Let

$$\frac{\alpha t_o}{x_o^2} = \frac{1}{4} \tag{5B.17a}$$

$$n = \frac{x}{\sqrt{4\alpha t}} \tag{5B.17b}$$

$$T(x, t) = T(n) \tag{5B.17c}$$

where two independent variables, $x/\sqrt{t}$, are made into a single independent variable n. Thus,

$$\frac{\partial n}{\partial x} = \frac{1}{\sqrt{4\alpha t}} \quad \& \quad \frac{\partial n}{\partial t} = -\frac{n}{2t} \tag{5B.18}$$

and

$$\frac{\partial T}{\partial t} = \frac{dT}{dn}\frac{\partial n}{\partial t} = -\frac{n}{2t}\frac{dT}{dn} \quad \text{(Chain rule)} \tag{5B.19a}$$

$$\frac{\partial T}{\partial x} = \frac{1}{\sqrt{4\alpha t}}\frac{dT}{dn} \tag{5B.19b}$$

$$\frac{\partial^2 T}{\partial x^2} = \frac{1}{4\alpha t}\frac{d^2 T}{dn^2} \tag{5B.19c}$$

Substituting Eqs. (5B.19a), (5B.19c) into (5B.15a) and upon transformation of the PDE into an ODE yields

$$\frac{d^2 T}{dn^2} + 2n\frac{dT}{dn} = 0 \tag{5B.20a}$$

$$\frac{\partial^2 T^*}{\partial n^2} + 2n\frac{\partial T^*}{\partial x} = 0 \tag{5B.20b}$$

with boundary condition (BC) and initial condition (IC), respectively

$$T^*(n) = 1 \text{ for } n = 0 \text{ at } x = 0, \ t > 0 \tag{5B.21a}$$

$$T^*(n) = 0 \text{ for } n \to \infty \text{ at } x > 0, \ t = 0 \tag{5B.21b}$$

Let $v = dT^*/dn$ so that

$$0 = \frac{dv}{dn} + 2nv \tag{5B.22a}$$

$$\frac{dv}{v} = d(\ln v) = -2ndn \tag{5B.22b}$$

$$v = C_1 \exp\left(-n^2\right) \tag{5B.22c}$$

Then, the dimensionless temperature function along with BC and IC becomes

$$T^*(n) = C_1 \int_0^n \exp\left(-n^2\right) + C_2 \tag{5B.23a}$$

$$T^*(n) = T^*(0) = 1 \ \to C_2 = 1 \tag{5B.23b}$$

$$T^* (n) = T^* (\infty) = 0 \tag{5B.23c}$$

$$0 = C_1 \int_0^\infty \exp\left(-n^2\right) + 1 \tag{5B.23d}$$

$$C_1 = -\frac{1}{\int_0^n \exp\left(-n^2\right)} \tag{5B.23e}$$

Thus, Eq. (5B.23a) becomes the

$$T^* (n) = -\frac{\int_0^n \exp\left(-n^2\right)}{\int_0^\infty \exp\left(-n^2\right)} + 1 = 1 - \text{erf}(n) = \text{erfc}(n) \tag{5B.24a}$$

$$T^* (x, t) = \text{erfc}(n) = \text{erfc}\left(\frac{x}{\sqrt{4\alpha t}}\right) \tag{5B.24b}$$

Combining Eqs. (5B.16a), (5B.16c) and (5B.24b) yields the temperature similarity solution by transforming two independent variables, x and t, into one single similarity variable n.

$$\text{erfc}(n) = \frac{T - T_{IC}}{T_d} = \frac{T - T_{IC}}{T_{BC} - T_{IC}} \tag{5B.25a}$$

$$T - T_{IC} = (T_{BC} - T_{IC})\,\text{erfc}(n) \tag{5B.25b}$$

Notice that T_{BC} and T_{IC} are the temperatures due to the boundary condition (BC) and initial condition (IC), respectively.

Substitute Eq. (5B.17b) into (5B.25b) to get the dimensional temperature field equation as

$$T - T_{IC} = (T_{BC} - T_{IC})\,\text{erfc}\left(\frac{x}{2\sqrt{\alpha t}}\right) \tag{5B.26}$$

If $T_{IC} = T_o$ is the initial mold temperature and $T_{BC} = T_f$ is the melt freezing temperature, then Eq. (5B.26) takes the form

$$T - T_o = \left(T_f - T_o\right)\text{erfc}\left(\frac{x}{2\sqrt{\alpha t}}\right) \tag{5B.27}$$

Solving Eq. (5B.27) for T yields the field temperature equation written as

$$T = T(x, t) = T_o + \left(T_f - T_o\right)\text{erfc}\left(\frac{x}{2\sqrt{\alpha t}}\right) \tag{5B.28}$$

from which $\partial T/\partial x$ and $\partial T/\partial t$ are easily determined

$$\frac{\partial T}{\partial x} = -\frac{(T_f - T_o)}{\sqrt{\pi \alpha t}} \exp\left(-\frac{x^2}{4\alpha t}\right) \tag{5B.29a}$$

$$\frac{\partial T}{\partial t} = (T_f - T_o)\left(\frac{x}{2t\sqrt{\pi \alpha t}}\right)\exp\left(-\frac{x^2}{4\alpha t}\right) \tag{5B.29b}$$

$$\frac{\partial T}{\partial t} = \frac{(T_f - T_o)\,x}{2t\sqrt{\pi \alpha t}}\exp\left(-\frac{x^2}{4\alpha t}\right) \tag{5B.29c}$$

where $\partial T/\partial x$ denotes the one-dimensional temperature gradient in the x-direction and $\partial T/\partial t$ denotes the rate of freezing or cooling of the molten material.

5.11   Appendix 5C Transcendental Equation

The subsequent analytical procedure defines the transcendental equation related to solid and liquid thermal resistances as schematically shown by the temperature profiles in Fig. 5.1b. This case is considered as the Neumann's solution to a Stefan problem when the liquid metal is initially at $T > T_f$ in the region $x > 0$. Thus, $T = T_p$ in a practical casting work is defined as the pouring temperature.

Consider the solid temperature (T_s) equation as the general solution of the heat equation, the boundary and initial conditions written as

$$T_s = C_1 + C_2\,\mathrm{erf}\left(\frac{x}{2\sqrt{\alpha_s t}}\right) \quad \text{(solid)} \tag{5C.1a}$$

$$\frac{\partial T_s}{\partial t} = \alpha_s\frac{\partial^2 T_s}{\partial x^2} \quad \text{at } 0 < x < s\,(t)\,,\ t > 0 \tag{5C.1b}$$

$$T_s\,(x,t) = T_o \quad \text{at } x = 0,\ t > 0 \tag{5C.1c}$$

$$T_s\,(x,t) = C_1 + C_2\,\mathrm{erf}\left(\frac{x}{2\sqrt{\alpha_s t}}\right) = T_o + C_2\,\mathrm{erf}\left(\frac{x}{2\sqrt{\alpha_s t}}\right) \tag{5C.1d}$$

and

$$T_l = C_3 + C_4\,\mathrm{erf}\left(\frac{x}{2\sqrt{\alpha_s t}}\right) \quad \text{(liquid)} \tag{5C.2a}$$

$$\frac{\partial T_l}{\partial t} = \alpha_l\frac{\partial^2 T_l}{\partial x^2} \quad \text{at } s\,(t) < x < \infty,\ t > 0 \tag{5C.2b}$$

$$T_l\,(x,t) = T_p \quad \text{at } x > 0,\ t = 0 \tag{5C.2c}$$

$$T_l\,(x,t) = C_3 + C_4\,\mathrm{erfc}\left(\frac{x}{2\sqrt{\alpha_l t}}\right) = T_p + C_4\,\mathrm{erfc}\left(\frac{x}{2\sqrt{\alpha_l t}}\right) \tag{5C.2d}$$

Here, $T_l = T_p > T_f$, where T_p is the pouring temperature. The coupling conditions at $x = s(t)$ are

$$T_s(x, t) = T_l(x, t) = T_f \quad \text{at } x = s(t), \ t > 0 \tag{5C.3a}$$

$$k_s \frac{\partial T_s}{\partial x} - k_l \frac{\partial T_l}{\partial x} = \rho \Delta H_f \frac{ds(t)}{dt} \quad \text{at } x = s(t), \ t > 0 \tag{5C.3b}$$

Substituting Eqs. (5C.1d) and (5C.2d) at $x = s(t)$ into (5C.3a) yields

$$T_o + C_2 \operatorname{erf}\left(\frac{s(t)}{2\sqrt{\alpha_s t}}\right) = T_p + C_4 \operatorname{erfc}\left(\frac{s(t)}{2\sqrt{\alpha_l t}}\right) = T_f \tag{5C.4a}$$

$$T_o + C_2 \operatorname{erf}(\lambda) = T_p + C_4 \operatorname{erf}\left(\lambda\sqrt{\alpha_s/\alpha_l}\right) = T_f \tag{5C.4b}$$

where

$$\lambda = \frac{s(t)}{2\sqrt{\alpha_s t}} \quad \& \quad s(t) = 2\lambda\sqrt{\alpha_s t} \tag{5C.5a}$$

$$\frac{s(t)}{2\sqrt{\alpha_l t}} = \frac{s(t)}{2\sqrt{\alpha_l t}} \frac{\sqrt{\alpha_s}}{\sqrt{\alpha_s}} = \frac{s(t)}{2\sqrt{\alpha_s t}} \frac{\sqrt{\alpha_s}}{\sqrt{\alpha_l}} = \lambda\sqrt{\frac{\alpha_s}{\alpha_l}} \tag{5C.5b}$$

From Eq. (5C.4a),

$$C_2 = \frac{T_f - T_o}{\operatorname{erf}(\lambda)} \quad \& \quad C_4 = \frac{T_f - T_p}{\operatorname{erfc}(\lambda)} \tag{5C.6}$$

Then, Eqs. (5C.1d) and (5C.2d) become

$$T_s(x, t) = T_o + \frac{T_f - T_o}{\operatorname{erf}(\lambda)} \operatorname{erf}\left(\frac{x}{2\sqrt{\alpha_s t}}\right) \tag{5C.7a}$$

$$T_l(x, t) = T_p + \frac{T_f - T_p}{\operatorname{erfc}\left(\lambda\sqrt{\alpha_s/\alpha_l}\right)} \operatorname{erfc}\left(\frac{x}{2\sqrt{\alpha_l t}}\right) \tag{5C.7b}$$

(b) Take $\partial T_s(x, t)/\partial x$ and $\partial T_l(x, t)/\partial x$,

$$\frac{\partial T_s(x, t)}{\partial x} = \frac{(T_f - T_o)}{\operatorname{erf}(\lambda)}\left(\frac{1}{\sqrt{\pi \alpha_s t}}\right) \exp\left(-\frac{x^2}{4\alpha_s t}\right) \tag{5C.8a}$$

$$\frac{\partial T_l(x, t)}{\partial x} = \frac{T_f - T_p}{\operatorname{erfc}\left(\lambda\sqrt{\alpha_s/\alpha_l}\right)}\left(\frac{1}{\sqrt{\pi \alpha_l t}}\right) \exp\left(-\frac{x^2}{4\alpha_l t}\right) \tag{5C.8b}$$

Substituting Eqs. (5C.8a,b) into (5C.3b) yields

$$\rho \Delta H_f \frac{ds\,(t)}{dt} = k_s \frac{\partial T_s}{\partial x} - k_l \frac{\partial T_l}{\partial x} \quad \text{at } x = s\,(t)\,, \ t > 0 \tag{5C.9a}$$

$$\rho \Delta H_f \frac{ds\,(t)}{dt} = k_s \frac{(T_f - T_o)}{\text{erf}\,(\lambda)} \left(\frac{1}{\sqrt{\pi \alpha_s t}} \right) \exp\left(-\frac{x^2}{4\alpha_s t} \right) \tag{5C.9b}$$

$$- k_l \frac{T_f - T_p}{\text{erfc}\,\left(\lambda \sqrt{\alpha_s / \alpha_l}\right)} \left(\frac{1}{\sqrt{\pi \alpha_l t}} \right) \exp\left(-\frac{x^2}{4\alpha_l t} \right)$$

but, $x = s\,(t)$ so that

$$\rho \Delta H_f \frac{ds\,(t)}{dt} = \frac{k_s \left(T_f - T_o \right) \exp\left[-s\,(t)^2 / (4\alpha_s t) \right]}{\sqrt{\pi \alpha_s t}} \tag{5C.10}$$

$$- \frac{k_l \left(T_f - T_p \right) \exp\left[-s\,(t)^2 / (4\alpha_l t) \right]}{\sqrt{\pi \alpha_l t}} \ \frac{}{\text{erfc}\,\left(\lambda \sqrt{\alpha_s / \alpha_l}\right)}$$

Substituting Eq. (5C.5a) into (5C.10) yields

$$\rho \Delta H_f \frac{ds\,(t)}{dt} = \frac{k_s \left(T_f - T_o \right) \exp\left(-\lambda^2 \right)}{\sqrt{\pi \alpha_s t}} \ \frac{}{\text{erf}\,(\lambda)} \tag{5C.11}$$

$$- \frac{k_l \left(T_f - T_p \right) \exp\left(-\lambda^2 \alpha_s / \alpha_l \right)}{\sqrt{\pi \alpha_l t}} \ \frac{}{\text{erfc}\,\left(\lambda \sqrt{\alpha_s / \alpha_l}\right)}$$

Integrating yields

$$\int \frac{dt}{\sqrt{\pi \alpha_s t}} = \frac{2\sqrt{t}}{\sqrt{\pi \alpha_s}} \tag{5C.12a}$$

$$\sqrt{\pi} \rho \Delta H_f s\,(t) = \frac{2k_s \left(T_f - T_o \right) \sqrt{t}}{\sqrt{\alpha_s}} \ \frac{\exp\left(-\lambda^2 \right)}{\text{erf}\,(\lambda)} \tag{5C.12b}$$

$$- \frac{2k_l \left(T_f - T_p \right) \sqrt{t}}{\sqrt{\alpha_l}} \ \frac{\exp\left(-\lambda^2 \sqrt{\alpha_s / \alpha_l} \right)}{\text{erfc}\,\left(\lambda \sqrt{\alpha_s / \alpha_l}\right)}$$

and

$$\sqrt{\pi} \rho \Delta H_f = \frac{2\sqrt{t}}{s\,(t)\sqrt{\alpha_s}} \ \frac{k_s \left(T_f - T_o \right)}{\text{erf}\,(\lambda)} \exp\left(-\lambda^2 \right) \tag{5C.13a}$$

$$- \frac{2\sqrt{t}}{s\,(t)\sqrt{\alpha_l}} \ \frac{k_l \left(T_f - T_p \right)}{\text{erfc}\,\left(\lambda \sqrt{\alpha_s / \alpha_l}\right)} \exp\left(-\lambda^2 \alpha_s / \alpha_l \right)$$

$$\sqrt{\pi}\rho\Delta H_f = \frac{2\sqrt{\alpha_s t}}{s\left(t\right)\alpha_s}\frac{k_s\left(T_f - T_o\right)}{\mathrm{erf}\left(\lambda\right)}\exp\left(-\lambda^2\right) \tag{5C.13b}$$

$$- \frac{2\sqrt{\alpha_s t}}{s\left(t\right)\sqrt{\alpha_s}\sqrt{\alpha_l}}\frac{k_l\left(T_f - T_p\right)}{\mathrm{erfc}\left(\lambda\sqrt{\alpha_s/\alpha_l}\right)}\exp\left(-\lambda^2\alpha_s/\alpha_l\right)$$

Use the definition of thermal diffusivity for the solid and liquid phases written as $\alpha_s = k_s/\rho_s c_s$ and $\alpha_l = k_l/\rho_l c_l$ so that $c_s = k_s/\rho_s\alpha_s$ and let $\rho = \rho_s = \rho_l$ at the freezing temperature T_f. Manipulating Eq. (5C.13b) using these definitions yields a convenient expression written as

$$\sqrt{\pi}\Delta H_f = \frac{k_s\left(T_f - T_o\right)}{\lambda\rho_s\alpha_s\,\mathrm{erf}\left(\lambda\right)}\exp\left(-\lambda^2\right) \tag{5C.14}$$

$$- \frac{k_l\left(T_f - T_p\right)\exp\left(-\lambda^2\alpha_s/\alpha_l\right)}{\lambda\rho_s\sqrt{\alpha_s}\sqrt{\alpha_l}\,\mathrm{erfc}\left(\lambda\sqrt{\alpha_s/\alpha_l}\right)}$$

Again, manipulate Eq. (5C.14) to get

$$\frac{\sqrt{\pi}\lambda\Delta H_f}{\left(T_f - T_o\right)} = \frac{c_s}{\mathrm{erf}\left(\lambda\right)}\exp\left(-\lambda^2\right) \tag{5C.15a}$$

$$- \frac{\sqrt{\alpha_s}}{\sqrt{\alpha_s}}\frac{k_l}{\rho_s\sqrt{\alpha_s}\sqrt{\alpha_l}}\frac{\left(T_f - T_p\right)\exp\left(-\lambda^2\alpha_s/\alpha_l\right)}{\left(T_f - T_o\right)\mathrm{erfc}\left(\lambda\sqrt{\alpha_s/\alpha_l}\right)}$$

$$\frac{\sqrt{\pi}\lambda\Delta H_f}{\left(T_f - T_o\right)} = \frac{c_s\exp\left(-\lambda^2\right)}{\mathrm{erf}\left(\lambda\right)} - \frac{k_l\sqrt{\alpha_s}}{\rho_s\alpha_s\sqrt{\alpha_l}}\frac{\left(T_f - T_p\right)\exp\left(-\lambda^2\sqrt{\alpha_s/\alpha_l}\right)}{\left(T_f - T_o\right)\mathrm{erfc}\left(\lambda\sqrt{\alpha_s/\alpha_l}\right)} \tag{5C.15b}$$

Using $\rho_s\alpha_s = k_s/c_s$ in Eq. (5C.15b) yields

$$\frac{\sqrt{\pi}\lambda\Delta H_f}{\left(T_f - T_o\right)} = \frac{c_s\exp\left(-\lambda^2\right)}{\mathrm{erf}\left(\lambda\right)} - \frac{c_s k_l}{k_s}\sqrt{\frac{\alpha_s}{\alpha_l}}\frac{T_f - T_p}{T_f - T_o}\frac{\exp\left(-\lambda^2\alpha_s/\alpha_l\right)}{\mathrm{erfc}\left(\lambda\sqrt{\alpha_s/\alpha_l}\right)} \tag{5C.16}$$

At last, the transcendental equation is

$$\frac{\exp\left(-\lambda^2\right)}{\mathrm{erf}\left(\lambda\right)} + \frac{k_l}{k_s}\sqrt{\frac{\alpha_s}{\alpha_l}}\frac{T_f - T_p}{T_f - T_o}\frac{\exp\left(-\lambda^2\alpha_s/\alpha_l\right)}{\mathrm{erfc}\left(\lambda\sqrt{\alpha_s/\alpha_l}\right)} = \frac{\sqrt{\pi}\lambda\Delta H_f}{c_s\left(T_f - T_o\right)} \tag{5C.17a}$$

$$\frac{\exp\left(-\lambda^2\right)}{\mathrm{erf}\left(\lambda\right)} - \frac{\gamma_l}{\gamma_s}\frac{T_f - T_p}{T_f - T_o}\frac{\exp\left(-\lambda^2\alpha_s/\alpha_l\right)}{\mathrm{erfc}\left(\lambda\sqrt{\alpha_s/\alpha_l}\right)} = \frac{\sqrt{\pi}\lambda}{S_t} \tag{5C.17b}$$

Therefore, this transcendental equation for a two-region solidification domain is as complicated as the expression defined by Eq. (5.22a) for a three-region solidification domain in a half-space region $x > 0$.

5.12 Problems

5.1 Derive Eq. (5.1b) using the Fourier's law of heat conduction and the first law of thermodynamics.

5.2 Compare the solidification time using the Chvorinov's rule for casting a copper spheroid with $a = 2$ cm and $c = 4$ cm, and a copper sphere with $r = 2$ cm in a sand mold with $B = 1.951$ min/cm^2. Which casting geometry solidifies first? Why?

5.3 Consider a sand casting procedure for producing a flat $20 \times 10 \times 3$ cm^3 AISI 1030 carbon steel flat plate. Assume that the pouring and freezing temperatures of the steel are $1520\,^{\circ}\mathrm{C}$ and $1510\,^{\circ}\mathrm{C}$, respectively, and that the mold constant is 3 min/cm^2. Based on this information, determine the solidification time for (a) the plate and (b) a 4-cm diameter sphere.

5.4 Plot the given data on a log–log scale and determine n and B as per Chvorinov's rule. Is $n = 2 \pm 0.05$?

Geometry	$s(t) = M = V/A$ (cm)	t (min)
Cube	0.500	0.63
Sphere	0.67	1.11
Plate	1.03	2.68
Cylinder	1.33	4.44
Spheroid	1.51	5.67

5.5 Assume that the Czochralski method is used to produce single-crystal silicon ingots at a pulling velocity $v = 0.5$ mm/s. In this solidification process, Q_{S-L} denotes the amount of heat released by the solid to the fluid phase and Q_k denotes the amount of energy removed from the solid–liquid interface A_s by conduction through the ingot. Assume that $\rho_s = \rho_l = 2400$ kg/m^3, $T_f = 1687$ K, $\Delta H_f = 1{,}877{,}000$ J/kg, $k_s = 25$ W/m K, and $L = 20$ cm. Calculate the L-S interface temperature T_i and the amount of heat flux density q_k removed from it.

5.6 Consider casting a 0.04-m thick copper slab into a mold with high thermal conductivity and assume that the copper solid develops a high thermal resistance during solidification. The thermophysical properties are $\rho_s = 7525$ kg/m^3, $k_s = 400$ W/m K, $\alpha_s = 1.24 \times 10^{-4}$ m^2/s, and $\Delta H_f = 205$ kJ/kg at $T_f = 1083\,^{\circ}\mathrm{C}$. If initial mold temperature is $T_o = 25\,^{\circ}\mathrm{C}$, calculate (a) the solidification time t, (b) the solidification velocity $v = dx_s/dt$, and (c) the temperature gradient and the heat flux $(dT/dx)_{x_s}$ and the heat flux q_{x_s} at x_s.

5.7 Assume that a 0.04-m thick iron slab casting is to be produced in carbon steel mold and that there are temperature gradients in the mold and solid. Use the thermophysical properties given below to calculate (a) the solidification time t_s, and the solidification velocity dx_s/dt and (b) plot the temperature distribution for the mold and for the solid up to x_s. Assume a pouring temperature of 1555 °C so that the superheat is 20 °C. Data for iron casting: $T_f = 1535$ °C, $k_s = 73$ W/m K, $\rho_s = 7530$ kg/m^3, $c_s = 728$ J/kg K, and $\Delta H_f = 247.20$ kJ/kg. Data for the steel mold: $T_o = 25$ °C, $k_m = 35$ W/m K, $\rho_m = 7689$ kg/m^3, and $c_m = 669$ J/kg K.

5.8 Consider casting a 0.08-m thick iron slab in a 0.04-m thick copper mold walls. Using the given data compute (a) the solidification time t and the solidification velocity $ds(t)/dt$ if $(\partial T/\partial x)_{slab} > 0$ and $(\partial T/\partial x)_{mold} > 0$. Assume negligible solid–mold thermal resistance and superheat, and constant air environmental temperature at 25 °C. (b) Plot $T = f(x, t_s)$ for the iron slab up to $x \leq s(t)$. Explain. Data for iron casting: $T_f = 1535$ °C, $k_s = 73$ W/m K, $\rho_s = 7530$ kg/m^3, $c_s = 728$ J/kg K, and $\Delta H_f = 247.20$ kJ/kg. Data for the steel mold: $T_o = 25$ °C, $k_m = 35$ W/m K, $\rho_m = 7689$ kg/m^3, and $c_m = 669$ J/kg K.

5.9 Consider casting a constant metal volume $(0.02 \, \text{m}^3)$ in the shape of a cube with a length c, a cylinder with radius r and $D = H$ (equal diameter and height), and a sphere with radius r. Use the Chvorinov's rule to determine which casting shape solidifies faster. Let the Chvorinov's constant be $B = 20{,}000$ min /m^2.

5.10 (a) Consider a cylindrical casting with height H and diameter D. Based on this information minimize the casting modulus $M = V/A$ so that the solidification time as per Chvorinov's rule is also minimized. This implies that $V_{min} = f(r) < V_{actual} = f(r, H)$ so that $D = H$. (b) Let $r = 0.04$ m and $B = 20{,}000$ min/m^2, and calculate t_{actual} for $D = 1.5H$ and t_{min} for $D = H$.

5.11 If it takes 4 min to solidify a 5-cm cube in a sand mold, then calculate how long will it take to solidify a 2-cm radius and 6-cm long cylinder. Use the Chvorinov's rule.

5.12 If it takes 3.5 min to solidify a 2-cm radius and 6-cm long cylinder, then calculate how long will it take to solidify a 5-cm cube in a sand mold. Use the Chvorinov's rule.

5.13 Assume that a 0.04-m thick copper slab casting is to be produced in a steel mold and that there are temperature gradients in the mold and solid. Use the thermophysical properties given below to calculate (a) the solidification time t, and the solidification velocity dx_s/dt, (b) the heat flux q_s crossing the mold–solid interface at $x = 0$, and (c) plot the temperature distribution at $-0.04 \leq x \leq 0$ m for the mold and $0 \leq x \leq 0.02$ m for the solid. Assume superheats 0, 20 and 100 °C. Data for copper casting: $T_f = 1083$ °C, $k_s = 400$ W/m K, $\rho_s = 7525$ kg/m^3,

$c_s = 429$ J/kg K, and $\Delta H_f = 205{,}000$ kJ/kg. Data for metal mold: $T_o = 25\,^\circ$C, $k_m = 7689$ W/m K, $\rho_m = 35$ kg/m^3, and $c_m = 669$ J/kg K.

5.14 Assume that solidification occurs uniformly by a moving straight interface located at $x = s(t)$ and apply the boundary condition $T_s = T_s[s(t),t] = T_f$ to Eq. (5.17a) to show that

$$s(t) = 2\lambda\sqrt{\alpha_s t}$$

for a similarity solution of the solid heat equation.

5.15 Consider Neumann's solution for solidification of a pure metal using liquid and solid thermal resistances. Assume that the governing equations and boundary conditions are

$$\frac{\partial T_s}{\partial t} = \alpha_s \frac{\partial^2 T_s}{\partial x^2} \quad \text{for } 0 < x < s(t), t > 0$$

$$\frac{\partial T_l}{\partial t} = \alpha_l \frac{\partial^2 T_l}{\partial x^2} \quad \text{for } x > s(t), t > 0$$

$$T_s(x,t) = T_s(x,t) = T_f$$

$$k_s \frac{\partial T_s}{\partial x} - k_l \frac{\partial T_l}{\partial x} = \rho_s \Delta H_s \frac{ds(t)}{dt}$$

and

$$\text{BC } T_s = T_s(0,t) = T_i \quad \& \quad T_l = T_l[s(t),t] = T_f$$

$$T_s = T_s[s(t),t] = T_f \quad \& \quad T_l = T_l(\infty,t) = T_p$$

$$\text{IC } T_s = T_s(x,0) = T_i \quad \& \quad T_l = T_l(x,0) = T_p$$

Show that

$$T_s = T_i + \frac{T_f - T_i}{\text{erf}(\lambda)}\,\text{erf}\left(\frac{x}{2\sqrt{\alpha_s t}}\right)$$

$$T_l = T_p + \frac{T_f - T_p}{\text{erfc}\left(\lambda\sqrt{\alpha_s/\alpha_l}\right)}\,\text{erfc}\left(\frac{x}{2\sqrt{\alpha_l t}}\right)$$

and

$$\frac{\exp\left(-\lambda^2\right)}{\text{erf}(\lambda)} - \frac{k_l}{k_s}\sqrt{\frac{\alpha_s}{\alpha_l}}\frac{T_p - T_f}{T_f - T_i}\frac{\exp\left(-\lambda^2\alpha_s/\alpha_l\right)}{\text{erf}\left(\lambda\sqrt{\alpha_s/\alpha_l}\right)} = \frac{\sqrt{\pi}\lambda\Delta H_s}{c_s(T_f - T_i)}$$

5.16 Consider the solidification of a semi-infinite slab of aluminum (Al) into a graphite mold. Use the given data [10] below to carry out all required calculations.

Initially, the mold and liquid *Al* are at temperatures indicated in the table, and the *Al* freezing temperature is $T_f = 660\,°\mathrm{C}$.

Material	T (°C)	k (W/m K)	ρ (kg/m^3)	c_p (J/kg K)	ΔH_f (J/kg)
Mold	25	100	2200	1700	
Solid *Al*		211	2555	1190	3.98×10^5
Liquid *Al*	700	91	2368	1090	

(a) Assume that there exist temperature gradients in the mold, solid *Al* and liquid *Al* regions, and that the phase transformation process at hand is a two-phase solidification problem along with the mold thermal resistance. Set up the governing equations for each region, solve the corresponding heat equations (similarity solutions), and calculate the mold–solid interface temperature T_i, the solidification time t and the solidification velocity $ds\,(t)\,/dt$. (b) Let the mold thickness be $x_m = 8s\,(t)$ and the half-space be $s\,(t) = 0.04$ m. Plot the temperature profiles for $T_m(x, t)$, $T_s(x, t)$ and $T_l(x, t)$.

5.17 Assume that the Chvorinov's constant is 4.50 min /cm^2. How long will it take to solidify a (2 cm) $\times$ (20 cm) $\times$ (30 cm) flat steel plate in an enclosed sand mold?

References

1. H.S. Carslaw, J.C. Jaeger, *Conduction of Heat in Solids*, 2nd edn. (Oxford University Press, Clarendon, 1959)
2. V. Alexiades, A.D. Solomon, *Mathematical Modeling of Melting and Freezing Processes* (Hemisphere Publishing Corporation, Washington, 1993). ISBN 1-56032-125-3
3. M. Necati Ozisik, *Heat Conduction* (Wiley, New York, 1993)
4. B. Gebhart, *Heat Conduction and Mass Diffusion* (McGraw-Hill, New York, 1993)
5. G.H. Geiger, D.R. Poirier, *Transport Phenomena in Metallurgy* (Addison-Wesley, Reading, 1973)
6. D.R. Poirier, G.H. Geiger, *Transport Phenomena in Materials Processing* (The Minerals, Metals & Materials Society, A Publication of TMS, Springer, Cham, 2016)
7. D.R. Gaskell, *An Introduction to Transport Phenomena in Materials Engineering*, 2nd edn. (Momentum Press, LLC, New Jersey, 2013)
8. D.M. Stefanescu, *Science and Engineering of Casting Solidification* , 3rd edn. (Springer, Switzerland, 2015). ISBN 978-3-319-15692-7
9. M.E. Glicksman, *Principles of Solidification: An Introduction to Modern Casting and Crystal Growth Concepts* (Springer, New York, 2011)
10. J.A. Dantzig, M. Rappaz, *Solidification*, 2nd edn. (Revised & Expanded, EPFL Press, Lausanne, 2016)
11. R. Wlodawer, *Directional Solidification of Steel Castings* (Translated by L.D. Hewitt and English Translation Edited by R.V. Riley) (Pergamon Press, New York, 1966)
12. R.A. Flinn, *Fundamentals of Metal Casting* (Addison-Wesley, Reading, 1963)
13. H. Hu, S.A. Argyropoulos, *Mathematical Modelling of Solidification and Melting: A Review*. Modelling and Simulation in Materials Science and Engineering, vol. 4, no. 4 (IOP Publishing, Bristol, 1996), pp. 371–396

14. J. Caldwell, D.K.S. Ng, *Mathematical Modelling: Case Studies and Projects* (Kluwer Academic Publishers, New York, 2004)
15. G.M. Poole, Mathematical modeling of solidification phenomena in electromagnetically stirred melts. Dissertation, The University of Alabama, Tuscaloosa, 2014
16. W.J. Boettinger, F.S. Biancaniello, S.R. Coriell, Solutal convection induced macrosegregation and the dendrite to composite transition in off-eutectic alloys. Metallurgical Trans. A **12**(2), 321–327 (1981)
17. C.J. Li, H. Yang, Z.M. Ren, W.L. Ren, Y.Q. Wu, On nucleation temperature of pure aluminum in magnetic fields. Progr. Electromag. Res. Lett. **15**, 45–52 (2010)
18. L.I. Rubinstein, *The Stefan Problem* (Translations of the American Monographs), vol. 27 (American Mathematical Society, Providence, 1971)
19. D.A. Tarzia, Explicit and approximated solutions for heat and mass transfer problems with a moving interface, in *Advanced Topics in Mass* Transfer, edited by M. El-Amin, intechweb.org (InTech, Rijeka, 2011), pp. 339–484

Chapter 6
Contour Metal Solidification

6.1 Introduction

This chapter includes a group of analytical solutions to heat equations for one-dimensional melting and solidification problems in cylindrical coordinates with moving liquid–solid (L-S) interfaces. Recall that rectangular coordinates require the half-space region $x > 0$ to be defined in a semi-infinite region $0 < x < \infty$ with a L-S interface located at the characteristic length $x = s\,(t)$. Similarly, $r = s\,(t)$ denotes the location of the L-S interface in cylindrical coordinates. Accordingly, a single-phase (liquid or solid) or two-phase (liquid and solid) analysis is known as a Stefan problem or Neumann solution.

For melting and solidification in cylindrical coordinates, solutions to phase-change problems are based on temperature field equations in terms of the exponential integral function $E_i\,(-\mu)$, where $\mu = r/4\alpha t$ denotes the similarity dimensionless variable and the liquid-to solid (L-S) interface is located at a characteristic length $r = s\,(t)$ and $t > 0$. For comparison purposes, solutions to these phase-change problems in rectangular coordinates are introduced in Chap. 5 with the temperature field equations being based on the error function $\mathrm{erf}\,(n)$ for the liquid and the complementary error function $\mathrm{erfc}\,(n)$ for the solid phases, where $n = x/4\alpha t$ is the similarity dimensionless variable, α is the thermal diffusivity (m^2/s), and t is the solidification time. Again, the temperature function $T = f\,(n)$, such as $\mathrm{erf}\,(n)$ is the similarity solution function to a partial differential equation (PDE).

6.2 Interface Thermal Resistance

This is the case for solidification in **metal molds**, in which a high mold–solid interface thermal resistance develops due to an air gap (air layer) interface schemat-

© Springer Nature Switzerland AG 2020
N. Perez, *Phase Transformation in Metals*,
https://doi.org/10.1007/978-3-030-49168-0_6

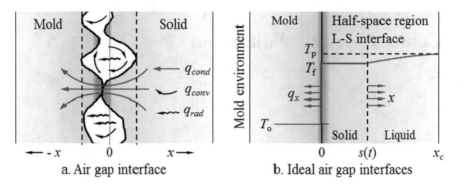

Fig. 6.1 Air gap at the mold–solid interface. (**a**) Actual and (**b**) ideal

ically shown in Fig. 6.1a as a rough interface ([1, p. 52] and [2, p. 674]). This is a representation of an actual interface with a small thickness where conductive, convective, and radioactive heat transfer modes are present during solidification. Notice the nomenclature used to represent the heat flux modes. Figure 6.1b, on the other hand, represents the ideal interface with smooth and flat surfaces.

Assume now that the mold–solid interface due to an air gap is the predominant thermal resistance during solidification. In fact, the air gap prevents fusion (welding or soldering) at the mold–solidified melt interface.

The air gap forms as a result of mold expansion due to the absorption of thermal energy (heat) and shrinkage of the solidified layer (solid skin) at an interface temperature T_i. Moreover, the air gap thickness is normally defined as the ratio of the thermal conductivity to the heat transfer coefficient; $\delta = k_s / h$, where h is the heat transfer coefficient.

The amount of thermal energies for solidifying a liquid metal of mass m by convection and conduction (see Eq. 5.34) due to the release of latent heat of fusion ΔH_f is written as

$$Q_t = Ah \, (T_i - T_o) \, t \qquad \text{(convection)} \tag{6.1a}$$

$$Q_f = m \Delta H_f = \rho_s V \Delta H_f \quad \text{(conduction)} \tag{6.1b}$$

Combining Eqs. (6.1a) and (6.1b) yields the characteristic length as per conductive heat transfer

$$s \, (t) = \frac{V}{A} = \frac{h \, (T_i - T_o) \, t}{\rho_s \Delta H_f} \tag{6.2}$$

where T_i is the interface temperature and $s(t) = V/A$ in Eq. (6.2) has the same physical meaning as (5.35). These expressions define the position of the solidification front at $t > 0$ due to conductive heat transfer and convective heat transfer modes, respectively. Yet heat transfer by conduction due to the evolution of latent heat of fusion ΔH_f at $s(t)$ appears to be the dominant heat extraction mechanism during solidification in sand and metallic molds.

The heat flux $q_{s(t)}$ due to conduction at $t > 0$ and $q_{x=0}$ due to convection are defined as per Newton's law of cooling and Fourier's law of heat conduction

$$q_{x=0} = h\,(T_i - T_o) \quad \text{(Newton's law)} \tag{6.3a}$$

$$q_{s(t)} = \frac{k_s\,(T_f - T_i)}{s\,(t)} \quad \text{(Fourier's law)} \tag{6.3b}$$

Solving for T_i from Eq. (6.3b) yields

$$T_i = T_f - \frac{s\,(t)\,q_{s(t)}}{k_s} \tag{6.4}$$

Now, substitute Eq. (6.4) into (6.3a) to get

$$q_{x=0} = \frac{k_s}{s\,(t)} \left[h\,(T_f - T_o) - \frac{h s\,(t)\,q_{s(t)}}{k_s} \right] \tag{6.5}$$

In addition, the heat flux due to evolution of latent heat of fusion ΔH_f at the equilibrium temperature T_f is defined by

$$q_{s(t)} = \rho_s \Delta H_f \frac{ds\,(t)}{dt} \tag{6.6}$$

Letting $q_{x=0} = q_{s(t)}$ in Eq. (6.5) due to a linear temperature profile, solving for $q_{s(t)}$, and combining the resultant expression with Eq. (6.6) yields the Stefan thermal energy balance based on the Newton's and Fourier's laws

$$\rho_s \Delta H_f \frac{ds\,(t)}{dt} = \frac{T_f - T_o}{1/h + s\,(t)/k_s} \tag{6.7}$$

If $1/h \gg s\,(t)/k_s$ and $s\,(t) \to 0$ (thin enough layer), then the initial heat transfer with convection dominates the solidification process. Eventually, $q_{s(t)}$ dissipates through the mold walls reaching the environmental surroundings.

If $1/h \ll s\,(t)/k_s$ and $s\,(t) > 0$ (thick enough layer), then $q_{s(t)}$ is due to conduction and the rate of solidification is controlled and limited by the internal resistance of the solidified melt.

If $1/h = s\,(t)/k_s$, it defines the heat flux as $q_{s(t)} = 0.5 k_s\,(T_f - T_o)/s\,(t)$ or $q_{s(t)} = 0.5 h\,(T_f - T_o)$ at the interface, where the thermal resistance may be defined as $R_{inter} = \Delta T/q_{x=0}$.

For one-dimensional quasi-static solidification, the solidification rate or velocity can be estimated by solving Eq. (6.7) for $ds\,(t)/dt$. Mathematically,

$$\frac{ds\,(t)}{dt} = \left(\frac{T_f - T_o}{\rho_s \Delta H_f}\right)\left(\frac{1}{1/h + s\,(t)\,/k_s}\right) \tag{6.8}$$

Integrating Eq. (6.8) yields

$$\int_0^{s(t)} (1/h + x/k)\,dx = \left(\frac{T_f - T_o}{\rho_s \Delta H_f}\right)\int_0^t dt \tag{6.9a}$$

$$\frac{s\,(t)}{h} + \frac{s\,(t)^2}{2k_s} = \left(\frac{T_f - T_o}{\rho_s \Delta H_f}\right) t \tag{6.9b}$$

Rearranging Eq. (6.9b) gives

$$s\,(t) = k_s \left[\sqrt{\frac{1}{h^2} + \frac{2}{k_s}\frac{(T_f - T_o)}{\rho_s \Delta H_f}t} - \frac{1}{h}\right] \tag{6.10a}$$

$$t = \left(\frac{\rho_s \Delta H_f}{T_f - T_o}\right)\left[\frac{s\,(t)}{h} + \frac{s\,(t)^2}{2k_s}\right] \tag{6.10b}$$

Significantly, Eq. (6.10) links conductive and convective heat transfer modes for defining the position of the interface $s(t)$ at a solidification time t and the solidification time t at a position $s(t)$ within the solidification domain Γ.

In principle, these thermal heat modes are connected in a dimensionless form called the Biot number B_i, defined in the next section along with the Fourier number F_o.

6.2.1   Stefan Dimensionless Thermal Energy Balance

Now, following the ideas of Glicksman [3, p. 72] by multiplying Eq. (6.8) by $c_s L^2/k_s$ and rearranging the resultant expression yields the dimensionless interfacial thermal energy balance as

$$\frac{df_s}{dF_o} = \frac{S_t}{B_i^{-1} + f_s} \tag{6.11}$$

where the B_i defines the ratio of conduction and convection thermal resistances, F_o denotes dimensionless parameter that arises due to heat conduction and L denotes the slab thickness. For convenience, below are given a group of dimensionless parameters and variables that are useful in one form or another during the analysis or solidification. Thus,

$$f_s = \frac{s(t)}{L} \qquad \text{(fraction of solid)} \qquad (6.12a)$$

$$F_o = \left(\frac{\alpha_s}{L^2}\right) t \qquad \text{(Fourier number)} \qquad (6.12b)$$

$$S_t = \frac{c_p(T_f - T_o)}{\Delta H_f} \qquad \text{(Stefan number)} \qquad (6.12c)$$

$$B_i = \frac{hL}{k_s} \qquad \text{(Biot number)} \qquad (6.12d)$$

$$\tau = S_t F_o \qquad \text{(dimensionless time)} \qquad (6.12e)$$

Assume that S_t remains constant in Eq. (6.12e) so that $d\tau = S_t dF_o$. Solve for S_t and substitute the resultant expression into Eq. (6.11) and integrate to get the solid fraction equation written as

$$f_s = \sqrt{\frac{1}{B_i^2} + 2\tau} - \frac{1}{B_i} \qquad (6.13)$$

This expression, Eq. (6.13), is plotted in Fig. 6.2 for some values of B_i. Notice that the function $f_s = g(\tau)$ shifts upwards and the slope $df_s/d\tau$ increases with increasing B_i. Therefore, the solidification rate, defined as $df_s/d\tau$, increases with increasing B_i. This can be attributed to a decrease in the thermal conductivity k_s term in the denominator of Eq. (6.12d), provided that the heat transfer coefficient h remains constant.

Now, merging Eq. (6.13) with $B_i \to \infty$ and (6.12e) and combining the resultant expression with (5.44b) yields

$$F_o = \frac{f_s^2}{2S_t} \quad \& \quad F_o = \left(\frac{f_s}{2\lambda}\right)^2 \qquad (6.14)$$

Fig. 6.2 Fraction of solid vs. dimensionless time with some Biot numbers

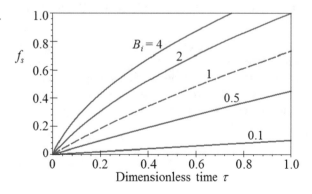

Conceptually, the Stefan, Biot, and Fourier numbers reduce the independent variables in solving heat transfer problems analytically and they represent the transient dimensionless parameters in heat transfer field.

Example 6.1 Assume that pure copper undergoes 50% solidification of the total liquid mass in a planar semi-infinite space of thickness x_c and total length $L = 0.04 \, \text{m}$. Let $S_t = 0.50$ (solid phase) and $\alpha_s = 3.4490 \times 10^{-4} \, \text{m}^2/\text{s}$. Calculate the Biot number using the moving boundary (MB) theory.

Solution For $f_s = 0.5$ and $S_t = 0.50$, Eq. (5.41a) yields

$$0 = \lambda \exp(\lambda) \operatorname{erf}(\lambda) - S_t/\sqrt{\pi} \quad \Rightarrow \quad \lambda = 0.41748$$

From Eqs. (5.4c) and (6.12b) with $s(t) = f_s L = 0.50 \times 0.04 = 0.02 \, \text{m}$,

$$t = \frac{s(t)^2}{4\lambda^2 \alpha_s} = \frac{(0.02 \, \text{m})^2}{4 \, (0.41748)^2 \, (3.4490 \times 10^{-4} \, \text{m}^2/\text{s})} = 1.66 \, \text{s}$$

$$F_o = \frac{\alpha_s t}{L^2} = \frac{(3.4490 \times 10^{-4} \, \text{m}^2/\text{s}) \, (1.66 \, \text{s})}{(0.04 \, \text{m})^2} = 0.36$$

Thus, Eq. (6.13) with $\tau = S_t F_o$, the Biot number yields

$$f_s = \sqrt{\frac{1}{B_i^2} + 2\tau} - \frac{1}{B_i} = \sqrt{\frac{1}{B_i^2} + 2\,(S_t F_o)} - \frac{1}{B_i} = \sqrt{\frac{1}{B_i^2} + 2\,(0.50)\,(0.36)} - \frac{1}{B_i}$$

$$0 = \sqrt{\frac{1}{B_i^2} + 0.36} - \frac{1}{B_i} - 0.50$$

$$B_i = 9.09$$

6.3 Heat Transfer on Contoured Surfaces

A contoured casting requires a two-dimensional analysis in cylindrical coordinates. According to Fig. 5.2d, the internal thermal resistance to radial heat transfer crosses the thickening annular ring during transient solidification. The general heat conduction equation, its analytical solution and the solidification velocity in the radial direction ($r = z$) are, respectively, [4, p. 405]

$$\frac{\partial T}{\partial t} = \alpha_m \left[\frac{\partial^2 T}{\partial r^2} + \frac{\beta}{r} \left(\frac{\partial T}{\partial r} \right)^2 \right] \tag{6.15a}$$

$$s(t) = \frac{V}{A} = \frac{(T_f - T_o)}{\rho_s \Delta H_f} \left[\left(\frac{2k_m}{\sqrt{\pi \alpha_m}} \right) t^{1/2} + \left(\frac{\beta k_m}{2r} \right) t \right] \tag{6.15b}$$

$$\frac{ds(t)}{dt} = \frac{(T_f - T_o)}{\rho_s \Delta H_f} \left(\frac{k_m}{\sqrt{\pi \alpha_m t}} + \frac{\beta k_m}{2r} \right) \tag{6.15c}$$

where $\beta = 1$ for a cylinder and $\beta = 2$ for a sphere in semi-infinite molds for which the radius is $r^2 = x^2 + y^2$.

The heat flux components and the heat flux balance $q_f = q_r$ are

$$q_f = -\rho_s \Delta H_f \frac{ds(t)}{dt} \tag{6.16a}$$

$$q_r = -k_m \frac{\partial T}{\partial r} \tag{6.16b}$$

$$-\rho_s \Delta H_f \frac{ds(t)}{dt} = -(T_f - T_o) \left(\frac{k_m}{\sqrt{\pi \alpha_m t}} + \frac{\beta k_m}{2r} \right) \tag{6.16c}$$

with

$$\frac{\partial T}{\partial r} = \frac{\rho_s \Delta H_f}{k_m} \frac{ds(t)}{dt} = (T_f - T_o) \left(\frac{1}{\sqrt{\pi \alpha_m t}} + \frac{\beta}{2r} \right) \tag{6.17}$$

Integrating Eq. (6.17) yields the near-field temperature equation

$$T = T_o + (T_f - T_o) \left[\frac{r}{\sqrt{\pi \alpha_m t}} + \frac{\beta}{2} \ln(r) \right] \tag{6.18}$$

With respect to an electric analogy with Ohm's law to heat flow by conduction, the internal thermal resistance is written as [4, p. 244]

$$R_r = \frac{\Delta T}{q_r} = \frac{T_f - T_o}{q_r} \tag{6.19}$$

In addition, one is tempted to simply state that $R_r \propto 1/q_r$ or $q_r \propto 1/R_r$ at a fixed ΔT. However, it is most appropriate to indicate that the latter proportionality suggests that thermal conduction is high at low thermal resistance and vice versa. Mathematically, $q_r \to 0$ as $R_r \to \infty$ or $q_r \to \infty$ as $R_r \to 0$.

Physically, $q_r > 0$ is attributed to the vibration of atoms that generates kinetics of energy in the solid phase. Thus, heat transfer can also be characterized as heat vibrational motion of atoms and eventually, heat (q_r) flows through the solid to the mold walls, through the mold material to the environment. In theory, heat transfer by conduction is governed by Fourier's law.

Example 6.2 (Sand Mold) Consider the solidification of copper into a slab ($4 \times 6 \times 10 \, \text{cm}^3$), cylindrical and spherical casting geometries. Include the effect of superheat

$\Delta T = 10\,°C = 283\,K$ on solidification. Let the slab thickness be $x = 4\,cm$ and, the radius of both cylinder and sphere be $r = 2\,cm$. Thus, the common dimensional characteristic lengths are $x_c = 4/2 = 2\,cm$ for the slab, $x_c = r/2$ for the cylinder with $c = 6\,cm$ long, and $x_c = r/3$ for the sphere. Based on this information, determine the final solidification time (t) and the solidification rate dx_s/dt for each casting geometry using (a) the heat transfer approach and (b) the Chvorinov's rule. Compare results and determine which casting geometry solidifies first. (c) Plot $T = f(x, t)$ for the slab casting geometry. Use the data for pure copper (Cu), plot the temperature field equation, Eqs. (5.20a), at $-0.08 \leq x \leq 0\,m$ and (5.29) at $0 \leq x \leq 0.02\,m$ for the slab, and explain the results.

Copper (Cu)
$T_f = 1356\,K = 1083\,°C$
$\rho_s = 7525\,kg/m^3$
$c_s = 429\,J/(kg\,K)$
$k_s = 400\,W/(m\,K)$
$\Delta H_f = 205{,}000\,J/kg$

Sand mold
$T_o = 298\,K = 25\,°C$
$\rho_m = 1550\,kg/m^3$
$c_m = 986\,J/(kg\,K)$
$k_m = 0.40\,W/(m\,K)$

Solution

(a) **Heat Transfer Approach**

Slab The pouring temperature $T_p = T_f + \Delta T = 1366\,K$ induces a significant superheating effect on the liquid phase. Thus,

$$\Delta H_s = \Delta H_f + c_s \Delta T_s = 205{,}000 + (429)(283) = 326{,}410\,J/kg$$

$$\alpha_s = \frac{k_s}{\rho_s c_s} = \frac{400\,W/m\,K}{(7525\,kg/m^3)(429\,J/kg\,K)} = 1.2391 \times 10^{-4}\,m^2/s$$

$$\alpha_m = \frac{k_m}{\rho_m c_m} = \frac{0.40\,W/m\,K}{(1550\,kg/m^3)(986\,J/kg\,K)} = 2.6173 \times 10^{-7}\,m^2/s$$

$$t = \frac{\pi \alpha_m}{4}\left[\frac{\rho_s \Delta H_s}{k_m (T_f - T_o)}\right]^2 \quad s(t)^2 = 2769.80\,s \simeq 46.16\,min = 0.77\,h$$

$$\frac{ds(t)}{dt} = \frac{1}{\sqrt{\pi t}}\left(\frac{T_f - T_o}{\rho_s \Delta H_s}\right)\left(\frac{k_m}{\sqrt{\alpha_m}}\right) = 3.61 \times 10^{-6}\,m/s = 13\,mm/h$$

Cylinder From Eq. (6.15b) with $\beta = 1$, $r = 0.02\,m$, and $s(t) = r/2 = 0.01\,m$ the solidification time is

$$0 = \frac{T_f - T_o}{\rho_s \Delta H_s}\left[\left(\frac{2k_m}{\sqrt{\pi \alpha_m}}\right)\sqrt{t} + \left(\frac{\beta k_m}{2r}\right)t\right] - s(t)$$

$$0 = \left(4.3074 \times 10^{-6}\right)t + \left(3.8002 \times 10^{-4}\right)\sqrt{t} - 0.01$$

$$t = 450.02\,s \simeq 7.50\,min$$

Using Eq. (6.15c) yields the solidification velocity

$$\frac{ds\,(t)}{dt} = \frac{T_f - T_o}{\rho_s \Delta H_s} \left(\frac{k_m}{\sqrt{\pi \alpha_m t}} + \frac{\beta k_m}{2r} \right) = 1.33 \times 10^{-5}\,\text{m/s} = 47.88\,\text{mm/h}$$

Sphere From Eq. (6.15b) with constant variables $\beta = 2$, $r = 0.02\,\text{m}$, and $s\,(t) = r/3 = 6.67 \times 10^{-3}\,\text{m}$,

$$0 = \frac{T_f - T_o}{\rho_s \Delta H_s} \left[\left(\frac{2k_m}{\sqrt{\pi \alpha_m}} \right) \sqrt{t} + \left(\frac{\beta k_m}{2r} \right) t \right] - s\,(t)$$

$$0 = \left(8.6148 \times 10^{-6} \right) t + \left(3.8002 \times 10^{-4} \right) \sqrt{t} - 0.00667$$

$$t = 180.92\,\text{s} = 3.02\,\text{min}$$

and from Eq. (6.15c),

$$\frac{ds\,(t)}{dt} = \frac{T_f - T_o}{\rho_s \Delta H_s} \left(\frac{k_m}{\sqrt{\pi \alpha_m t}} + \frac{\beta k_m}{2r} \right) = 2.2741 \times 10^{-5}\,\text{m/s} = 81.87\,\text{mm/h}$$

Therefore, the above results show that $t_{sphere} < t_{cylinder} < t_{bar}$ for a radius $r = 2\,\text{cm}$. Accordingly, the casting geometry has a significant influence on the solidification time.

(b) **The Chvorinov's rule**

Slab From Eq. (5.36c),

$$B = \frac{\pi \alpha_m}{4} \left[\frac{\rho_s \Delta H_s}{k_m \left(T_f - T_o \right)} \right]^2 = 6.9246 \times 10^6\,\text{s/m}^2 = 11.5410\,\text{min/cm}^2$$

Using characteristic length $s\,(t) = a/2 = 4/2 = 2\,\text{cm}$, and Eq. (5.36b) yields the solidification time

$$t = Bs\,(t)^2 = \left(11.5410\,\text{min/cm}^2 \right) (2\,\text{cm})^2 = 46.16\,\text{min}$$

$$t = 46.16\,\text{min} \quad \text{Chvorinov's rule}$$

$$t = 46.16\,\text{min} \quad \text{Heat transfer approach}$$

Cylinder For $s\,(t) = r/2 = 1\,\text{cm}$ and $b = 6\,\text{cm}$, Eq. (5.36b) gives the solidification times

$$t = Bs\,(t)^2 = \left(11.5410\,\text{min/cm}^2 \right) (1\,\text{cm})^2 = 11.54\,\text{min}$$

$$t = 11.54\,\text{min} \quad \text{Chvorinov's rule}$$

$$t = 7.50\,\text{min} \quad \text{Heat transfer approach}$$

Sphere   For $s(t) = r/3 = 2/3$ cm, Eq. (5.36b) yields

$$t = B(x_c)^2 = \left(11.5410\,\text{min/cm}^2\right)(2/3\,\text{cm})^2 = 5.13\,\text{min}$$

$$t = 5.13\,\text{min}\quad \text{Chvorinov's rule}$$

$$t = 3.02\,\text{min}\quad \text{Heat transfer approach}$$

Therefore, the effects of casting geometry on the solidification time are evident since $t_{sphere} < t_{cylinder} < t_{slab}$.

(c) **Plot:** From Eq. (5.27a),

$$\lambda \exp\left(\lambda^2\right)\left[\text{erf}(\lambda) + \sqrt{\frac{k_s \rho_s c_s}{k_m \rho_m c_m}}\right] = \frac{c_s\left(T_f - T_o\right)}{\sqrt{\pi}\,\Delta H_s}$$

$$\lambda \exp\left(\lambda^2\right)[\text{erf}(\lambda) + 45.96] - 0.78452 = 0$$

$$\lambda = 1.7058 \times 10^{-2}$$

From (5.26b),

$$T_i = T_o + \sqrt{\pi}\lambda \exp\left(\lambda^2\right)\frac{\Delta H_s}{c_s}\sqrt{\frac{k_s \rho_s c_s}{k_m \rho_m c_m}} = 1355.60\,\text{K} = 1082.60\,^\circ\text{C}$$

Eqs. (5.15a) and (5.17a) with $t = 2769.80$ s (slab) yield

$$T_m = T_o + (T_i - T_o)\,\text{erfc}\left(\frac{-x}{2\sqrt{\alpha_m t}}\right) = 25 + 1057.60\,\text{erfc}\,(-18.57x)$$

$$T_s = T_i + \frac{\left(T_f - T_i\right)}{\text{erf}(\lambda)}\,\text{erf}\left(\frac{x}{2\sqrt{\alpha_s t}}\right) = 1082.60 + 20.784\,\text{erf}\,(0.85348x)$$

Therefore, this temperature plot indicates that (1) the mold temperature at the extreme left-hand side, where $L = 2s(t) = 4(2\,\text{cm}) = 8\,\text{cm}$, is maintained at $25\,^\circ\text{C}$ during the solidification time of 19 min, (2) the non-linear temperature profile represents a strong resistance to heat transfer flow in the negative x-direction, and (3) the solid shows a small temperature gradient in the chosen scale for the half-space region. In addition, the temperature gradients are

$$\frac{\partial T_m}{\partial x} = 22,161\exp\left(-344.84x^2\right)$$

$$\frac{\partial T_s}{\partial x} = 20.016\exp\left(-0.728\,43x^2\right)$$

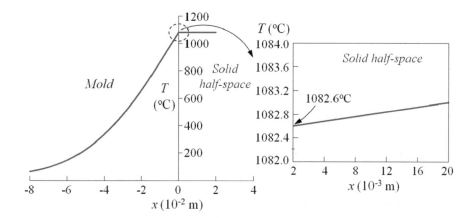

Further analysis of this example is summarized as follows:

- $\partial T_m/\partial x$ increases significantly as $x \rightarrow -0.08$ m due to the nature of the sand mold being a porous and heterogeneous material.
- $\partial T_s/\partial x$ decreases moderately as $x \rightarrow 0.02$ m, which is the centerline of the slab.
- Solidification is complete when $s(t) = x_c = 0.02$ m.

Lastly, the Chvorinov's rule and the heat transfer approach differ on calculated solidification times for the cylinder and sphere geometries.

6.3.1 Exponential Integral Function

This section deals with the generalized exponential integral $Ei(x)$, which arises in the physics and engineering fields. The solidification problem at hand is related to cylindrical and spherical casting geometries. The subsequent analysis is similar to the slab analysis presented in previous chapters.

Consider that the solidification problem, at an elapsed time $t > 0$, is described as a volume of material divided by a moving boundary defined as a L-S interface. This leads to a two-phase solidification domain composed of two regions with their own density and thermophysical properties [5].

Below are the theoretical solutions for cylindrical and spherical solidification modeling found in Paterson's paper [5], Stefanescu's book [6, pp. 295–296], and Hu and Argyropoulos' paper [7]. These references give some insights on the L-S boundary motion and the importance of the exponential integral in solving a transcendental equation for its root λ in the above casting geometries.

Cylindrical Geometry In the absence of a heat sink, the transcendental equation
is written as [6, p. 295]

$$\lambda^2 \exp\left(\lambda^2\right) \operatorname{Ei}\left(-\lambda^2\right) + \frac{c_l\left(T_f - T_i\right)}{\Delta H_f} = 0 \tag{6.20}$$

where the exponential integral function is of the form

$$\operatorname{Ei}(x) = -\int_{-x}^{\infty} \frac{\exp(-z)}{z} dz = \gamma_e + \ln(x) + \sum_{n=1}^{\infty} \frac{(-x)^n}{nn!} \tag{6.21}$$

Here, γ_e is the Euler–Mascheroni constant. More precisely, the position of the L-
S interface $r = r(t) = s(t)$ at $t > 0$ and the interface velocity $dr(t)/dt$ are
commonly written as

$$r(t) = 2\lambda\sqrt{\alpha_l t} \tag{6.22a}$$

$$\frac{dr(t)}{dt} = \lambda\sqrt{\frac{\alpha_l}{t}} \tag{6.22b}$$

Spherical Geometry The temperature field in the liquid phase equation for this
casting geometry is [6, pp. 295–296]

$$T_l(r, t) = T_o + \frac{2\lambda\left(T_p - T_f\right)X}{\exp\left(-\lambda^2\right) - \lambda\sqrt{\pi}\operatorname{erfc}(\lambda)} \tag{6.23a}$$

$$X = \left(\frac{\sqrt{\alpha_l t}}{r}\right)\exp\left(\frac{r^2}{4\alpha_s t}\right) - \left(\frac{\sqrt{\pi}}{2}\right)\operatorname{erfc}\left(\frac{r^2}{4\alpha_s t}\right) \tag{6.23b}$$

and the transcendental equation becomes

$$2\lambda^2 \exp\left(\lambda^2\right)\left[\exp\left(-\lambda^2\right) - \lambda\sqrt{\pi}\operatorname{erfc}(\lambda)\right] = \frac{S_{t,l}}{2} \tag{6.24}$$

where the Stefan number for the liquid phase is

$$S_{t,l} = \frac{c_l\left(T_p - T_f\right)}{\Delta H_f} \quad \text{for } 0 < S_t < 1 \tag{6.25}$$

Again, S_t is considered sufficiently small in the quasi-static solidification field,
the L-S moving boundary results from the phase transformation process due to
the evolution of latent heat of fusion ΔH_f. At this moment, a planar interface
(planar solidification front) prevails during quasi-static solidification of cylindrical,
spherical, and even rectangular geometries.

6.4 Single-Crystal Solidification

Only the Czochralski (Cz) process is considered in this section for developing the mathematical formulation related to the manufacturing of semiconductors using the pulling rate technique. The sketch shown in Fig. 4.17b represents the crystal-melt-crucible arrangement for producing cylindrical single crystals at a constant rate of solidification referred to as the vertical pulling rate $v_p = dz/dt$.

In reality, the Czochralski process is nowadays used to manufacture high quality cylindrical crystals, which are vertically pulled out of a crucible placed in an induction furnace. The resultant crystal is an intrinsic silicon ingot known as undoped silicon (without added impurities intentionally) or extrinsic silicon ingot referred to as doped silicon (with added impurities intentionally).

Furthermore, the solidifying melt takes on the crystal structure of the seed, and the rotating and pulling rates determine the diameter and length of a particular silicon ingot. A typical diameter range is in the order of 10–30 cm.

In one-dimensional heat transfer analysis, the coupling of conductive, convective, and radiative heat transfer modes in the crystal growth process is a boundary-value problem. The general governing heat flux balance, including the pulling rate term, initial condition (IC), and boundary conditions (BC's), is written as [8]

$$\frac{\partial T}{\partial t} = \left(\alpha_s \frac{\partial^2 T}{\partial z^2}\right)_{cond} - \left(\frac{2h}{r}(T - T_o)\right)_{conv} \tag{6.26a}$$

$$- \left(\frac{2}{r}q_r\right)_{rad} - \left(\rho_s c_s v_p \frac{\partial T}{\partial z}\right)_{pull}$$

$$\text{IC: } T = T(z, 0) = T_f \tag{6.26b}$$

$$\text{BC's: } T = T(0, t) = T_f \ \& \ T = T(z, t) = T_o \tag{6.26c}$$

$$\frac{dT}{dz} = \frac{\varepsilon\sigma\left(T^4 - T_o^4\right)}{k_s} - \frac{h(T - T_o)}{k_s} \tag{6.26d}$$

where ε denotes the emissivity, $\sigma = 5.67 \times 10^{-8}$ (W/m^2 K^4) denotes the Stefan–Boltzmann constant. This expression, Eq. (6.26), is subjected to approximations, but the main purpose henceforth is to analytically define v_p as simple as possible.

For steady-state heat transfer, the transition heat transfer term is $\partial T/\partial t = 0$ and subsequently, Eq. (6.26) reduces to an ordinary differential equation for a flat growth interface [8]

$$0 = \left(\alpha_s \frac{d^2 T}{dz^2}\right)_{cond} - \left[\frac{2h}{r}(T - T_o)\right]_{conv} \tag{6.27}$$

$$- \left(\frac{2}{r}q_r\right)_{rad} - \left(\rho_s c_s v_p \frac{dT}{dz}\right)_{pull}$$

The radial temperature gradient is defined as [9]

$$k_s \frac{dT}{dr} = -h\,(T - T_c) - \varepsilon_s \sigma \left(T^4 - T_o^4\right) \tag{6.28}$$

where T_c denotes the crucible wall temperature.

6.4.1   Conductive–Radioactive Heat Transfer

In the Czochralski process (Fig. 4.17b) the pulling rate ($\upsilon_p = dz/dt$) is directly related to the length of the crystal (with a desired radius) and speed of rotation (rpm) which are controlled during the growth process; otherwise, forced convection in the melt causes solidification problems.

The theoretical maximum pulling rate in the Czochralski process can also be predicted by using the heat flux balance

$$q_s = -k_s \left(\frac{dT}{dz}\right)_s \tag{6.29a}$$

$$q_l = -k_l \left(\frac{dT}{dz}\right)_l - \upsilon_p \rho_s \Delta H_f \tag{6.29b}$$

Equate ($q_s = q_l$) and rearrange Eqs. (6.29a) and (6.29b) to get the Stefan condition for the evolving or moving L-S interface

$$k_s \left(\frac{dT}{dz}\right)_s - k_l \left(\frac{dT}{dz}\right)_l = \upsilon_p \rho_s \Delta H_f \tag{6.30}$$

Manipulating Eq. (6.30) yields the pulling rate or solidification velocity as

$$\upsilon_p = -\frac{1}{\rho_s \Delta H_f} \left[k_s \left(\frac{dT}{dz}\right)_s - k_l \left(\frac{dT}{dz}\right)_l\right] \tag{6.31}$$

In addition, the Stefan–Boltzmann law describes the heat loss due to radiation as

$$dQ_{rad} = (2\pi r dz)\,\varepsilon \sigma T^4 \tag{6.32a}$$

$$\frac{dQ_{rad}}{dz} = (2\pi r)\,\varepsilon \sigma T^4 \tag{6.32b}$$

where dQ_{rad}/dz denotes the radiation energy gradient, $2\pi r dz$ denotes the radiating surface area of an increment length of the ingot, ε denotes the emissivity for radioactive heat transfer from the solid crystalline surface to the furnace surroundings.

From Fourier's law, the conductive heat energy and its gradient are

$$Q_{cond} = \left(\pi r^2\right) k_s \frac{dT}{dz} \tag{6.33a}$$

$$\frac{dQ_{cond}}{dz} = \left(\pi r^2\right) k_s \left[\frac{d^2 T}{dz^2} + \frac{dT}{dz}\frac{dk_s}{dz}\right] = \left(\pi r^2\right) k_s \frac{d^2 T}{dz^2} \tag{6.33b}$$

Here, dQ_{cond}/dz denotes the thermal energy gradient, πr^2 denotes the cross-sectional area of the ingot conducting the heat, k_s denotes the thermal conductivity of the solidified melt, and dT/dz denotes the temperature gradient.

Now, equating (6.32b) and (6.33b) yields

$$\frac{d^2 T}{dz^2} - \frac{2\varepsilon\sigma}{rk_s} T^4 = 0 \tag{6.34a}$$

$$\frac{d^2 T}{dz^2} - \frac{2\varepsilon\sigma}{rk_l T_f} T^5 = 0 \tag{6.34b}$$

since $k_s = k_l T_f / T$ at T_f. The solution of Eq. (6.34b) (Appendix 6A) is

$$\frac{dT}{dz} = -\frac{1}{2}\left(\frac{3rk_l T_f}{8\varepsilon\sigma}\right)^{1/4}\left[z + \left(\frac{3rk_l}{8\varepsilon\sigma T_f^3}\right)^{1/2}\right]^{-3/2} \tag{6.35a}$$

$$T = \left(\frac{3rk_l T_f}{8\varepsilon\sigma}\right)^{1/4}\left[z + \left(\frac{3rk_l}{8\varepsilon\sigma T_f^3}\right)^{1/2}\right]^{-1/2} \tag{6.35b}$$

$$\left(\frac{dT}{dz}\right)_{z=0} = \sqrt{\frac{2\varepsilon\sigma k_l T_f^5}{3r}} \tag{6.35c}$$

Substituting Eq. (6.35c) into (6.31) gives (Eq. (6A.4c), Appendix 6A)

$$v_p = \frac{1}{\rho_s \Delta H_f}\left(\frac{2\varepsilon\sigma k_l T_f^5}{3r}\right)^{1/2} \tag{6.36}$$

Figure 6.3 shows some $v_p = dz/dt$ experimental data obtained by hot-zone designs, coded as standard designs (STD1 and STD2) and as high-performance design (HPD1) [10]. Here, z denotes the crystal length.

Example 6.3 (Czochralski Process) For solidification of a silicon single-crystal ingot using the Czochralski process (Cz), calculate (a) the pulling rate v_p and the time t to produce one 2-m long ingot with $r = 100$ mm radius using the coupled conductive–reactive heat transfer. (b) This part calls for conductive heat transfer

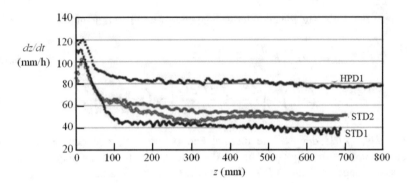

Fig. 6.3 Experimental pulling speed for silicon single crystals. Legend 1, 2, 3 stand for different design conditions [10]

in the radial direction. Thus, approximate the solidification velocity υ_r and the solidification time t in the radial direction by letting $\upsilon_p \rightarrow \upsilon_r$ and $dT/dz \rightarrow dT/dr$ in Eq. (6.31). Assume that the liquid silicon condition is at $T_l = 1850\,\text{K} = 1577\,°\text{C}$, $T_g = 700\,\text{K} = 427\,°\text{C}$ (inert gas). Given data:

$T_f = 1685\,\text{K}$	$T_o = 600\,\text{K}$	$T_l = 1850\,\text{K}$
$\rho_s = 2400\,\text{kg/m}^3$	$\rho_l = 2400\,\text{kg/m}^3$	$k_s = 25\,\text{W/m K}$
$c_s = 239\,\text{J/kg K}$	$c_l = 1000\,\text{J/kg K}$	$k_l = ?$
$\Delta H_f = 1{,}877{,}000\,\text{J/kg}$	$\sigma = 5.67 \times 10^{-8}\,\text{W/(m}^2\,\text{K}^4)$	
$h = 7\,\text{W/m}^2\,\text{K}$	$\varepsilon_s = 0.75$	$\varepsilon_l = 0.20$

Solution

(a) For $r = 0.1\,\text{m}$ and $z = L = 2\,\text{m}$ and

$$k_l = \frac{k_s T_f}{T_l} = \frac{(25)\,(1685)}{1850} = 22.77\,\text{W/m K}$$

Using Eq. (6.36) gives

$$\upsilon_p = \frac{1}{\rho_s \Delta H_f} \left(\frac{2\varepsilon_s \sigma k_l T_f^5}{3r} \right)^{1/2} = 6.57 \times 10^{-5}\,\text{m/s} = 236.52\,\text{mm/h}$$

$$t = \frac{L}{\upsilon_p} = \frac{2000\,\text{mm}}{236.52\,\text{mm/h}} = 8.46\,\text{h}$$

(b) Combining Eq. (6.31) with $\upsilon_p \rightarrow \upsilon_r$, $K_s(dT/dz)_s = k_s(dT/dr)$ and $(dT/dz)_1 = 0$, and (6.28) approximates the solidification velocity and the solidification time in the radial direction

$$v_r = -\frac{k_s}{\rho_s \Delta H_f}\frac{dT}{dr} = \frac{1}{\rho_s \Delta H_f}\left[h\left(T_f - T_g\right) + \varepsilon_s \sigma\left(T_f^4 - T_o^4\right)\right]$$

$$v_r = 7.64 \times 10^{-5}\,\text{m/s} = 275.04\,\text{mm/h}$$

$$t = \frac{r}{v_r} = \frac{100\,\text{mm}}{275.04\,\text{mm/h}} = 0.36\,\text{h} = 21.60\,\text{min}$$

Additionally, the temperature distribution within the cylinder can be approximated using Eq. (6.35b). Thus,

$$T = \left(\frac{3rk_l T_f}{8\varepsilon_s \sigma}\right)^{1/4}\left[z + \left(\frac{3rk_l}{8\varepsilon\sigma T_f^3}\right)^{1/2}\right]^{-1/2}$$

$$T = \left[\frac{3r\,(22.77\,\text{W/m K})\,(1685\,\text{K})}{8\,(0.75)\,\left(5.67 \times 10^{-8}\,\text{W/}\left(\text{m}^2\,\text{K}^4\right)\right)}\right]^{1/4}$$

$$\times\left[2\,m + \frac{3r\,(22.77\,\text{W/m K})}{8\,(0.75)\,\left(5.67 \times 10^{-8}\,\text{W/}\left(\text{m}^2\,\text{K}^4\right)\right)(1685\,\text{K})^3}\right]^{-1/2}$$

$$T = \frac{762.67 r^{1/4}}{\left(4.1971 \times 10^{-2}r + 2\right)^{1/2}}$$

which gives

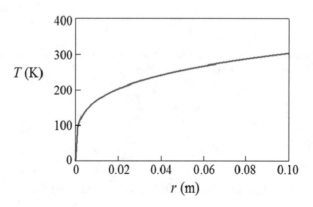

6.4.1.1 Near-Field Temperature Distribution

Following Poirier–Geiger analytical procedure [11, pp. 356–358], the conduction heat transfer in the hot crystal can be described by energy equation in cylindrical coordinates

$$\frac{\partial T}{\partial t} - \upsilon_p \frac{\partial T}{\partial z} = \alpha \left[\frac{\partial^2 T}{\partial z^2} + \frac{1}{r} \frac{\partial}{\partial r} \left(r \frac{\partial T}{\partial r} \right) \right] \tag{6.37}$$

Assuming a flat L-S interface at $z \rightarrow 0$ and a constant crystal radius, one-dimensional analysis under steady-state condition implies that $\partial T / \partial t = 0$ and $\partial T / \partial r = 0$. As a result, Eq. (6.37) is adopted to the solid (s) and liquid (l) phases along with the Stefan condition and boundary condition (BC) $T = T_f = T_l$ at $z = 0$

$$\alpha_s \frac{\partial^2 T_s}{\partial z^2} + \upsilon_p \frac{\partial T_s}{\partial z} = 0 \tag{6.38a}$$

$$\alpha_l \frac{\partial^2 T_l}{\partial z^2} + \upsilon_p \frac{\partial T_l}{\partial z} = 0 \tag{6.38b}$$

$$k_s \frac{dT_s}{dz} - k_l \frac{dT_l}{dz} = \upsilon_p \rho_s \Delta H_f \tag{6.38c}$$

where Eq. (6.30) has been renumbered as (6.38c) for convenience. The solutions of Eqs. (6.38a, 6.38b) for a near-field condition ($z \rightarrow 0$) are [11, p. 357]

$$T_s = T_f + \frac{\alpha_s}{\upsilon_p} \frac{dT_s}{dz} \left[1 - \exp\left(-\frac{z\upsilon_p}{\alpha_s} \right) \right] \tag{6.39a}$$

$$T_l = T_f + \frac{\alpha_l}{\upsilon_p} \frac{dT_l}{dz} \left[1 - \exp\left(-\frac{z\upsilon_p}{\alpha_l} \right) \right] \tag{6.39b}$$

Mathematically, the far-field condition yields $z \rightarrow \infty$, $\exp(-\infty) = 0$ and by deduction, $dT_l/dz \rightarrow 0$. Thus, Eqs. (6.39) becomes

$$T_s = T_f + \frac{\alpha_s}{\upsilon_p} \frac{dT_s}{dz} \tag{6.40a}$$

$$T_l = T_f + \frac{\alpha_l}{\upsilon_p} \frac{dT_l}{dz} = T_f \tag{6.40b}$$

Furthermore, using Eq. (6.31) or (6.38c) one can determine that

$$\upsilon_p = \frac{1}{\rho_s \Delta H_f} \left[k_s \left(\frac{dT}{dz} \right)_s - k_l \left(\frac{dT}{dz} \right)_l \right] > 0 \tag{6.41a}$$

$$k_s \frac{dT_s}{dz} > k_l \frac{dT_l}{dz} \tag{6.41b}$$

$$\frac{dT_l}{dz} < \frac{k_s}{k_l} \frac{dT_s}{dz} \tag{6.41c}$$

Using k_s, k_l values from Example 6.3 gives $v_p > 0$ if

$$dT_l/dz < (25/22.77)\, dT_s/dz = (1.0979)\, dT_s/dz \tag{6.42}$$

Moreover, this condition, Eq. (6.41b) or (6.41c), is very important for industrial production of silicon (Si) single crystals using the Cz process in the order of $v_p = 2 \times 10^{-5}$ m/s $= 72$ mm/h [11, p. 358].

6.5 Summary

Another aspect of solidification is the one-dimensional melting and solidification analyses in cylindrical coordinates. Hence, contoured solidification prevails within a cylindrical half-space in semi-infinite regions, leading to the well-known Stefan problems and Neumann solutions.

Of particular interest is the introduction to interface thermal resistance theory, from which the Stefan dimensionless thermal energy balance describes the obtainable fraction of a solid related to the Fourier number (F_o), Stefan number (S_t), and Biot number (B_i). Subsequently, the heat transfer on contoured surfaces follows a radial direction and it induces a contoured interface velocity along with a thermal resistance described by Ohm's law.

Cylindrical and spherical geometries comprise the contoured solidification theory, analytically described by an exponential integral function, which is more complicated to solve than the error function approach for planar solidification. The most common contoured solidification techniques are known as the Czochralski (Cz) and the Bridgman processes for producing cylindrical single crystals. Each technique has its own characteristics linked to the reliability of the solidifying system. Moreover, the theory of coupled conductive–radioactive heat transfer is introduced in order to link the Stefan condition to the Stefan–Boltzmann law for predicting the heat loss due to radiation.

6.6 Appendix 6A: Maximum Pulling Rate

The ODE to be solved, Eq. (6.34b), is

$$\frac{d^2T}{dz^2} = \frac{2\varepsilon\sigma}{rk_lT_f}T^5 \tag{6A.1}$$

Let $a = 2\varepsilon\sigma / \left(r k_l T_f \right)$ and $k_s T = k_f T_f$, where $k_f = k_l$, so that

$$\frac{d^2T}{dz^2} = aT^5 \tag{6A.2a}$$

$$\left(2\frac{dT}{dz} \right) \frac{d^2T}{dz^2} = \left(2\frac{dT}{dz} \right) aT^5 \tag{6A.2b}$$

$$\left(\frac{dT}{dz} \right)^2 = \frac{aT^6}{6} + C_1 \tag{6A.2c}$$

$$\frac{dT}{dz} = \sqrt{\frac{aT^6}{6} + C_1} \tag{6A.2d}$$

If $T = 1$ and $dT/dz = 1/\sqrt{3}$, then $C_1 = 0$ and the solution of Eq. (6A.2d) is

$$T = \left(\frac{3rk_lT_f}{8\varepsilon\sigma} \right)^{1/4} \left[z + \left(\frac{3rk_l}{8\varepsilon\sigma T_f^3} \right)^{1/2} \right]^{-1/2} \tag{6A.3a}$$

$$a_1 = \left(\frac{3rk_lT_f}{8\varepsilon\sigma} \right)^{1/4} \quad \& \quad a_2 = \left(\frac{3rk_l}{8\varepsilon\sigma T_f^3} \right)^{1/2} \tag{6A.3b}$$

$$T = a_1 (z + a_2)^{-1/2} \tag{6A.3c}$$

$$\frac{dT}{dz} = -\frac{1}{2}\frac{a_1}{(z+a_2)^{\frac{3}{2}}} = -\frac{1}{2}\left(\frac{3rk_lT_f}{8\varepsilon\sigma} \right)^{1/4} \left[z + \left(\frac{3rk_l}{8\varepsilon\sigma T_f^3} \right)^{1/2} \right]^{-3/2} \tag{6A.3d}$$

At $z = 0$,

$$\frac{dT}{dz} = -\frac{1}{2}\frac{a_1}{(z+a_2)^{\frac{3}{2}}} = -\frac{1}{2}\left(\frac{3rk_lT_f}{8\varepsilon\sigma} \right)^{1/4} \left[\left(\frac{3rk_l}{8\varepsilon\sigma T_f^3} \right)^{1/2} \right]^{-3/2} \tag{6A.4a}$$

$$\frac{dT}{dz} = -\frac{1}{2}\left(\frac{3rk_lT_f}{8\varepsilon\sigma} \right)^{1/4} \left(\frac{3rk_l}{8\varepsilon\sigma T_f^3} \right)^{-3/4} = -\left(\frac{2\varepsilon\sigma k_l T_f^5}{3r} \right)^{1/2} \tag{6A.4b}$$

$$v_p = -\frac{1}{\rho_s \Delta H_f}\frac{dT}{dx} = \frac{1}{\rho_s \Delta H_f}\left(\frac{2\varepsilon\sigma k_l T_f^5}{3r} \right)^{1/2} \tag{6A.4c}$$

which is the sought solution.

6.7 Problems

6.1 Calculate the solidification time for casting pure iron (Fe) into a graphite mold using one 0.04-m thick plate, a cylinder, and a sphere with 0.04-m diameter. Use data from Appendix 5A.

6.2 Consider casting a 0.04-m radius copper cylinder with negligible heat transfer interface at a steady-state solidification. The thermophysical properties are $\rho_s = 8033 \, \text{kg/m}^3$, $k_m = 0.89903 \, \text{W/m K}$, $\alpha_m = 4.9532 \times 10^{-7} \, \text{m}^2/\text{s}$, and $\Delta H_f = 205{,}400 \, \text{J/kg}$ at $T_f = 1083 \, ^\circ\text{C}$. If $T_o = 25 \, ^\circ\text{C}$, calculate the solidification time t, the solidification rate $ds\,(t)\,/dt$, the temperature gradient $\partial T/\partial r$, and the freezing rate $\partial T/\partial t$.

6.3 Assume that a hypothetical pure copper undergoes 50% solidification of the total liquid mass in a planar semi-infinite space of thickness $L = 0.06$ m. Let $B_i \to \infty$, $S_t = 0.46$ (solid phase) and thermal diffusivity of the liquid and solid phases at the L-S interface

$$\alpha_l = \frac{k_l}{\rho_l c_l} = 1.7193 \times 10^{-4}$$

$$\alpha_s = \frac{k_s}{\rho_s c_s} = 3.4490 \times 10^{-4} \, \text{m}^2/\text{s}$$

Determine the Fourier's number (F_o) as a measure of dimensionless solidification time and the dimensional solidification time (t) using (a) the quasi-static approximation (QSA) and (b) the moving boundary (MB) or Neumann's theories. (c) Derive $S_t = 2\lambda^2$ and calculate its value. Explain.

6.4 Consider that pure copper (Cu) undergoes 60% solidification of the total liquid mass in a planar semi-infinite space of thickness $L = 0.04$ m. Let $S_t = 0.50$ (solid phase) and $\alpha_s = 3.4490 \times 10^{-4} \, \text{m}^2/\text{s}$. Calculate the Fourier number using the moving boundary (MB) theory.

6.5 Consider the solidification process of pure copper (Cu) in a planar semi-infinite space of thickness $L = 0.04$ m. Let $S_t = 0.50$ (solid phase) and $\alpha_s = 3.4490 \times 10^{-4} \, \text{m}^2/\text{s}$. Calculate the Fourier number F_o using the moving boundary (MB) theory for 40% solidification.

6.6 Consider the solidification of copper into a slab ($4 \times 6 \times 10 \, \text{cm}^3$), cylindrical ($r = 2$ cm, $c = 6$ cm long), and spherical ($r = 2$ cm) casting geometries using sand molds. Include the effect of superheat $\Delta T_s = 20 \, ^\circ\text{C} = 293$ K on solidification. Calculate the final solidification time (t) using (a) the heat transfer approach along with the solidification rate $ds(t)/dt$ for each casting geometry and (b) the Chvorinov's rule. Compare results and determine which casting geometry solidifies first. (c) Plot $T = f(x, t)$ for the slab casting geometry. Use the data for pure copper (Cu) given in Example 6.2, plot the temperature field equation, Eqs. (5.20a), at $-0.08 \leq x \leq 0$ m and (5.29) at $0 \leq x \leq 0.02$ m for the slab, and explain the results.

6.7 For solidification of a silicon single-crystal ingot using the Czochralski process (Cz), calculate (a) the pulling rate v_p and the time t to produce one 3-m long ingot with $r = 200$ mm radius using the coupled conductive–reactive heat transfer. (b) This part calls for conductive heat transfer in the radial direction. Thus, approximate the solidification velocity v_r and the solidification time t in the radial direction by letting $v_p \rightarrow v_r$ and $dT/dz \rightarrow dT/dr$ in Eq. (6.31). Assume that the liquid silicon and the inert gas are at $T_l = 1850$ K $= 1577\,°C$ and $T_g = 700$ K $= 427\,°C$, respectively. Use the data given in Example 6.3.

6.8 Suppose that the solidification of copper into a slab $(4 \times 6 \times 10\,\text{cm}^3)$, cylindrical $(r = 2\,\text{cm}, c = 6\,\text{cm long})$, and spherical $(r = 2\,\text{cm})$ casting geometries using sand molds is carried out including the effect of superheat at $\Delta T = 30\,°C = 303$ K. Calculate the final solidification time (t) and the solidification rate dx_s/dt for each casting geometry using (a) the heat transfer approach and (b) the Chvorinov's rule. Compare results and determine which casting geometry solidifies first. (c) Plot $T = f(x, t)$ for the slab casting geometry. Use the data for pure copper (Cu) given in Example 6.2, plot the temperature field equation, Eqs. (5.20a), at $-0.08 \le x \le 0$ m and (5.29) at $0 \le x \le 0.02$ m for the slab, and explain the results.

6.9 For solidification of a silicon single-crystal ingot using the Czochralski process (Cz), calculate (a) the pulling rate v_p and the time t to produce one 2-m long ingot with $r = 200$ mm radius using the coupled conductive–reactive heat transfer. (b) This part calls for conductive heat transfer in the radial direction. Thus, approximate the solidification velocity v_r and the solidification time t in the radial direction by letting $v_p \rightarrow v_r$ and $dT/dz \rightarrow dT/dr$ in Eq. (6.31). Assume that the liquid silicon and the inert gas are at $T_l = 1850$ K $= 1577\,°C$ and $T_g = 700$ K $= 427\,°C$, respectively. Use the data given in Example 6.3 to carry out the required calculations.

6.10 Consider the Czochralski process (Cz) for solidification of a silicon single-crystal ingot in an inert gas environment at $T_g = 700$ K $= 427\,°C$. Calculate (a) the pulling rate v_p and the time t to produce one 1.5-m long ingot with $r = 200$ mm radius using the coupled conductive–reactive heat transfer. (b) This part calls for conductive heat transfer in the radial direction. Thus, approximate the solidification velocity v_r and the solidification time t in the radial direction by letting $v_p \rightarrow v_r$ and $dT/dz \rightarrow dT/dr$ in Eq. (6.31). Assume that the liquid silicon and the inert gas are at $T_l = 1850$ K $= 1577\,°C$. Use the data given in Example 6.3.

6.11 Let a suitable amount of pure copper undergo 70% solidification in a planar semi-infinite space of thickness x_c and length $L = 0.04$ m. Assume that the Stefan number and the thermal diffusivity for the solid phase are $S_t = 0.50$ and $\alpha_s = 3.4490 \times 10^{-4}\,\text{m}^2/\text{s}$, respectively. Based on this information, calculate (a) the Fourier number (F_o) using the moving boundary (MB) theory included in Chap. 5, (b) the Biot number (B_i), (c) the heat transfer coefficient (h) for natural convection, and (d) the heat flux $q_{x=0}$ at the mold wall $(x = 0)$. Neglect any effect of superheating on solidification and assume that the mold is initially at $T_o = 25\,°C$.

6.12 Consider an amount of pure copper undergoing 80% solidification in a planar semi-infinite space of thickness x_c and length $L = 0.04$ m. Assume that the Stefan number and the thermal diffusivity for the solid phase are $S_t = 0.50$ and $\alpha_s = 3.4490 \times 10^{-4}$ m²/s, respectively. Based on this information, calculate (a) the Fourier number (F_o) using the moving boundary (MB) theory included in Chap. 5, (b) the Biot number (B_i), (c) the heat transfer coefficient (h) for natural convection, and (d) the heat flux $q_{x=0}$ at the mold wall ($x = 0$). Neglect any effect of superheating on solidification and assume that the mold is initially at $T_o = 25\,°C$.

6.13 Suppose that an amount of pure copper is undergoing 90% solidification in a planar semi-infinite space of thickness x_c and length $L = 0.04$ m. Assume that the Stefan number and the thermal diffusivity for the solid phase are $S_t = 0.50$ and $\alpha_s = 3.4490 \times 10^{-4}$ m²/s, respectively. Based on this information, calculate (a) the Fourier number (F_o) using the moving boundary (MB) theory included in Chap. 5, (b) the Biot number (B_i), (c) the heat transfer coefficient (h) for natural convection, and (d) the heat flux $q_{x=0}$ at the mold wall ($x = 0$). Neglect any effect of superheating on solidification and assume that the mold is initially at $T_o = 25\,°C$.

6.14 Presume that an amount of pure copper is undergoing 95% solidification in a planar semi-infinite space of thickness x_c and length $L = 0.04$ m. Assume that the Stefan number and the thermal diffusivity for the solid phase are $S_t = 0.50$ and $\alpha_s = 3.4490 \times 10^{-4}$ m²/s, respectively. Based on this information, calculate (a) the Fourier number (F_o) using the moving boundary (MB) theory included in Chap. 5, (b) the Biot number (B_i), (c) the heat transfer coefficient (h) for natural convection, and (d) the heat flux $q_{x=0}$ at the mold wall ($x = 0$). Neglect any effect of superheating on solidification and assume that the mold is initially at $T_o = 25\,°C$.

6.15 Assume that an amount of pure copper undergoes 100% solidification in a planar semi-infinite space of thickness x_c and length $L = 0.04$ m. Assume that the Stefan number and the thermal diffusivity for the solid phase are $S_t = 0.50$ and $\alpha_s = 3.4490 \times 10^{-4}$ m²/s, respectively. Based on this information, calculate (a) the Fourier number (F_o) using the moving boundary (MB) theory included in Chap. 5, (b) the Biot number (B_i), (c) the heat transfer coefficient (h) for natural convection, and (d) the heat flux $q_{x=0}$ at the mold wall ($x = 0$). Neglect any effect of superheating on solidification and assume that the mold is initially at $T_o = 25\,°C$.

6.16 Assume that an amount of pure copper ($T_f = 1083\,°C$) undergoes solidification in a planar semi-infinite space of thickness x_c and length $L = 0.04$ m, where the sand mold is initially at $T_o = 25\,°C$. Use the following data set:

$S_t = 0.50$	$F_o = 1.43$	$B_i = 4.65$
$\alpha_s = 3.449 \times 10^{-4}$ m²/s	$h = 15{,}620$ W/(m² K)	$k_s = 400$ W/m K
$\rho_s = 7525$ kg/m³	$c_s = 429$ J/kg K	$\Delta H_f = 205{,}000$ J/kg

Determine (a) the solid fraction f_s. Is solidification complete at f_s? (b) the characteristic length s (t), and the solidification time t and (c) the heat flux q_x at the mold–solid interface $x = 0$.

6.17 Consider a complete solidification of copper in a planar semi-infinite space of thickness x_c and length $L = 0.05$ m. Assume that the Stefan number and the thermal diffusivity for the solid phase are $S_t = 0.52$ and $\alpha_s = 3.45 \times 10^{-4}\,\mathrm{m^2/s}$, respectively. Based on this information, calculate (a) the Fourier number (F_o) using the moving boundary (MB) theory included in Chap. 5, (b) the Biot number (B_i), (c) the heat transfer coefficient (h) for natural convection, and (d) the heat flux $q_{x=0}$ at the mold wall $(x = 0)$. Neglect any effect of superheating on solidification and assume that the mold is initially at $T_o = 25\,^\circ\mathrm{C}$. Use any additional data from Example 6.2.

6.18 Consider that pure nickel (Ni) undergoes 60% solidification of the total liquid mass in a planar semi-infinite space of thickness $L = 0.06$ m. Let $S_t = 0.50$ (solid phase), $T_f = 1453\,^\circ\mathrm{C}$, and $\alpha_s = 6.12 \times 10^{-6}\,\mathrm{m^2/s}$. Calculate the Fourier number using the moving boundary (MB) theory.

6.19 Consider the solidification of a silicon single-crystal ingot using the Czochralski process (Cz), calculate (a) the pulling rate υ_p and the time t to produce one 2.80-m long ingot with $r = 200$ mm radius using the coupled conductive–reactive heat transfer. (b) This part calls for conductive heat transfer in the radial direction. Thus, approximate the solidification velocity υ_r and the solidification time t in the radial direction by letting $\upsilon_p \rightarrow \upsilon_r$ and $dT/dz \rightarrow dT/dr$ in Eq. (6.31). Assume that the liquid silicon and the inert gas are at $T_l = 1850\,\mathrm{K} = 1577\,^\circ\mathrm{C}$ and $T_g = 700\,\mathrm{K} = 427\,^\circ\mathrm{C}$, respectively. Use the data given in Example 6.3.

6.20 For one-dimensional Czochralski process (Cz process) under conductive heat transfer only along the vertical z-axis, calculate (a) the vertical pulling crystal velocity υ_p and the solidification time t for $\partial T_l/\partial z = 2\partial T_s/\partial z = 800\,\mathrm{K/m}$ for producing a 2-m long cylindrical single crystal with a radius $r = 0.1$ m. Use the heat balance expression described by Eq. (6.30). (b) If $\partial T_s/\partial z = 400\,\mathrm{K/m}$ and $\partial T_l/\partial z \simeq 0$, then calculate υ_p and t. Is conductive heat transfer appropriate to predict the solidification time for a Cz specimen with a moving L-S interface? Explain.

6.21 Use the data set given in Example 6.3 to calculate (a) the maximum limit of $\partial T_l/dz$ so that the pulling velocity is $\upsilon_p > 0$ during the steady-state production of a silicon single-crystal ingot with a radius r and a length $L \gg r$ under a constant solid temperature gradient $\partial T_s/\partial z = 500\,\mathrm{K/m}$, (b) υ_p when $\partial T_l/\partial z = 200\,\mathrm{K/m}$.

6.22 Calculate (a) the maximum limit of $\partial T_l/dz$ so that the pulling velocity is $\upsilon_p > 0$ during the steady-state production of a silicon single-crystal ingot with a radius r and a length $L \gg r$ under a constant solid temperature gradient $\partial T_s/\partial z = 1000\,\mathrm{K/m}$, (b) υ_p when $\partial T_l/\partial z = 800\,\mathrm{K/m}$. Use the data set given in Example 6.3.

References

1. B. Gebhart, *Heat Conduction and Mass Diffusion* (McGraw-Hill, New York, 1993)
2. H. Biloni, W.J. Boettinger, Solidification, Chapter 8, in *Physical Metallurgy*, vol. 1, 4th revised and enhanced edition, ed. by R.W. Cahn, P. Haasen (Elsevier Science BV, North-Holland, 1996)
3. M.E. Glicksman, *Principles of Solidification: An Introduction to Modern Casting and Crystal Growth Concepts* (Springer Science+Business Media, LLC, New York, 2011)
4. D.R. Gaskell, *An Introduction to Transport Phenomena in Materials Engineering*, 2nd edn. (Momentum Press, LLC, New Jersey, 2013)
5. S. Paterson, *Propagation of a Boundary of Fusion*. Glasgow Math. J. **1**(01), 42–47 (1952)
6. D.M. Stefanescu, *Science and Engineering of Casting Solidification*, 3rd edn. (Springer International Publishing, Cham, 2015). ISBN 978-3-319-15692-7
7. H. Hu, S.A. Argyropoulos, Mathematical modeling of solidification and melting: a review. Modeling Simul. Mater. Sci. Eng. **4**(4), 371–396 (1996)
8. S.N. Rea, P.S. Gleim, Large area Czochralski silicon. Texas Instrument Final Report No. 03-77-23, ERDA/JPL-954475-77/4, Texas Instruments, Inc., Dallas, TX (1977)
9. A. Benmeddour, S. Meziani, Numerical study of thermal stress during different stages of silicon Czochralski crystal growth. Rev. Energ. Renouv. Algeria **12**(4), 575–584 (2009)
10. C.-W. Lan, C.-K. Hsieh, W.-C. Hsu, Czochralski silicon crystal growth for photovoltaic applications, in *Crystal Growth of Si for Solar Cells: Advances in Materials Research*, vol. 14, ed. by K. Nakajima, N. Usami (Springer, Berlin, 2009), pp. 25-39
11. D.R. Poirier, G.H. Geiger, *Transport Phenomena in Materials Processing* (Springer, Cham, 2016). The Minerals, Metals & Materials Society, A Publication of TMS

Chapter 7
Alloy Solidification I

7.1 Introduction

This chapter is devoted to solidification of substitutional and interstitial binary alloys of solute composition C_o. Accordingly, the solidification heat transfer is analytically characterized as a thermal conduction or as a coupled heat and mass transport process at the solidified melt boundary $s(t)$. Hence, mathematical modeling of solidification henceforth is based on finding the analytical solutions to one-dimensional heat equations, which admit similarity solutions in a half-space region $x > 0$ within a semi-infinite domain $0 \leq x < \infty$.

Only rectangular casting geometries are considered for introducing the analytical procedure for solving solidification problems related to thermal resistances in the mold, solid, and the solidifying melt at a particular rate of interface advancement known as the solidification velocity, which is a measure of the rate of stable or unstable planar-front solidification.

Mathematically, the analytical solutions provide valuable insights into the solidification process associated with solute boundary layers that build up at the advancing liquid–solid interface, which is treated as a smooth and flat solidification front. The seeking similarity solutions depend on the assumptions considered in a particular solidification model. Typically, the temperature and solute concentration field equations are initially derived for determining the effect of temperature and related solute concentration gradients on the solidification process.

7.2 Binary Equilibrium Phase Diagrams

The analysis of the solidification process may require the inclusion of solute microsegregation directly related to a binary phase diagram using the Lever rule and the selected Gulliver–Scheil microsegregation models for determining the

© Springer Nature Switzerland AG 2020
N. Perez, *Phase Transformation in Metals*,
https://doi.org/10.1007/978-3-030-49168-0_7

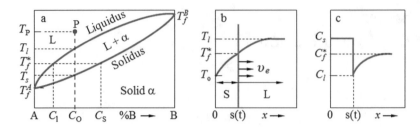

Fig. 7.1 Hypothetical equilibrium isomorphous phase diagrams. (**a**) *A-B* alloy, (**b**) temperature profile and (**c**) solute concentration profile showing a jump at the interface position $s(t)$

composition of solute in the solid and liquid phases at the solidifying freezing temperature T_f^*. Actually, melting and solidification of pure metals and their alloys are phase transformation (phase-change) processes. Once an amount of material is in the liquid state (known as the melt), a closed control of the chemical composition and removal of slag that floats on the melt is a practical procedure in the foundry industry.

Consider the schematic "isomorphous" system shown in Fig. 7.1a as the simplest equilibrium phase diagram for a hypothetical *A-B* binary alloy. This particular equilibrium phase diagram is a temperature-concentration (T-C diagram) graphical representation of the solubility of A and B in the liquid and solid states. In addition, the term "equilibrium" implies slow phase transformation, referred to as equilibrium solidification, and the equilibrium phase diagram is used to predict the type of solid one obtains after solidification.

Furthermore, assume that an amount of melt of a binary alloy with nominal composition C_o is initially at the pouring temperature $T_p > T_l$ (point "P" in Fig. 7.1a), chemically conditioned and slowly poured into a relatively cold mold cavity at $T_o \ll T_p$. Subsequently, (1) the melt freezes from T_p to T_l by liberating heat so that $L_p \to L + \alpha$ and $T_p \to T_l$ and (2) phase transformation (solidification) starts as an exothermic reaction, $L \to \alpha$, due to the evolved latent heat of fusion ΔH_f^* at the equilibrium temperature range $T_s < T_f^* < T_l$. Recall that T_f^A and T_f^B are the equilibrium freezing temperature for pure metals A and B, respectively.

In practice, the initially liquid *A-B* binary alloy with a uniform solute composition C_o is at the pouring temperature T_p so that the effect of superheat is imposed on solidification time and velocity. Moreover, Fig. 7.1b schematically shows the alloy freezing temperature $T_f^* = T_f^*(C_l)$, where C_l denotes the solute composition at the assumed moving stable planar L-S interface boundary located at $x = s(t)$ and $T_s < T_f^* < T_l$. Figure 7.1c exhibits the solute concentration profile with a solute concentration jump at $s(t)$.

The fundamental concept of stable planar L-S interface (planar front) is dealt with in a later section. For the moment, assume that there exists a stable interface in the absence of constitutional supercooling due to the compositional effect on the liquidus temperature.

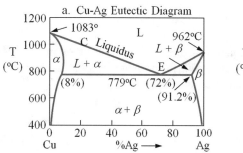
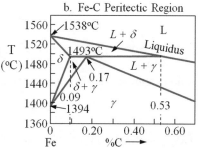

Fig. 7.2 (a) Copper-silver (*Cu-Ag*) eutectic system and (b) peritectic region of the iron-carbon (*Fe-C*) phase diagram. Nomenclature: % means weight percent

Furthermore, most commercial alloys have complicated phase diagrams. For the sake of simplicity, eutectic (Fig. 7.2a) and peritectic (Fig. 7.2b) regions of equilibrium phase diagrams are henceforth considered appropriate for providing additional details on binary alloy solidification. For instance, Fig. 7.2a shows the substitutional *Cu-Ag* eutectic system and Fig. 7.2b illustrates the peritectic region of the interstitial *Fe-C* phase diagram. The former indicates that *Cu* and *Ag* are completely soluble in the liquid state, but they are insoluble, to an extent, since they are immiscible (not mixable or incapable of mixing) in the solid state because they form different crystal structures such as α and β. Here, E is the eutectic point.

Note that the liquidus lines intercept at the eutectic point E, where L, α, and β coexist and the reversible ($\rightleftharpoons$) phase transformation occurs as per the invariant reaction

$$L \overset{Cu\text{-}57Ag}{\rightleftharpoons} \alpha + \beta \quad \text{at } T_E = 779\,^{\circ}\text{C} \qquad (7.1)$$

which occurs at the eutectic temperature (T_E) in a similar manner as that for pure *Cu* and iron *Fe*

$$L \overset{Cu}{\rightleftharpoons} \alpha \quad \text{at } T_f = 1083\,^{\circ}\text{C} \qquad (7.2a)$$

$$L \overset{Fe}{\rightleftharpoons} \delta \quad \text{at } T_f = 1538\,^{\circ}\text{C} \qquad (7.2b)$$

In general, the formation of *L-S* interface has previously been defined as a moving boundary problem during solidification of a pure metal to obtain a single solid phase at a single or fixed freezing temperature T_f, as indicated by Eq. (7.2). Similarly, eutectic solidification also occurs at a single eutectic temperature T_E and the solidified melt is a two-phase solid (α-β) as denoted in Eq. (7.1). Hence, a two-phase solidification.

On the other hand, Fig. 7.2b depicts the peritectic region where carbon is the interstitial (small) element in the iron crystalline structure and the peritectic invariant reaction for interstitial solid solution is

$$L + \delta \overset{Fe\text{-}0.17C}{\rightleftharpoons} \gamma \quad \text{at } T_{\text{per}} = 1493\,^{\circ}C \qquad (7.3)$$

Further details on phase transformation of binary alloys will be referred to as eutectic solidification, specifically under equilibrium conditions. Conversely, rapid solidification is a deviation from the equilibrium process.

7.3 Microsegregation Models

This section includes simple analytical procedures related to solid and liquid fractions using microsegregation models, such as (1) the linear Lever rule for determining the proportions of liquid and solid phases present in a binary alloy at $T_s < T_f^* < T_l$. This model considers complete mixing in both liquid and solid phases. (2) The non-linear Lever rule known as the Gulliver–Scheil model or Scheil equation, which considers complete mixing only in the liquid. However, back-diffusion models based on microsegregation with diffusion in the solid phase during solidification are available in the literature. A list of suitable models can be found elsewhere [1, p. 457].

In this section, only binary A-B phase diagrams are considered (Fig. 7.3), where the element "A" is the solvent and "B" is solute in any binary A-B alloy.

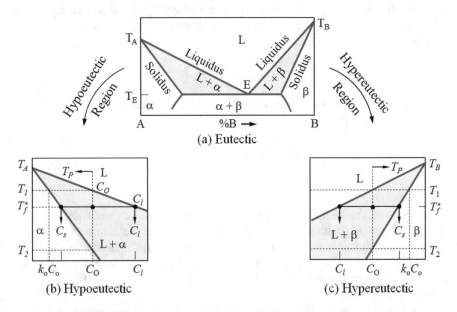

Fig. 7.3 (a) Hypothetical equilibrium eutectic phase diagram, (b) partial hypoeutectic region and (c) partial hypereutectic region

First of all, Fig. 7.3a is a schematic binary A-B phase diagram illustrating the *hypoeutectic* and *hypereutectic* regions used for defining the mushy zone as the $L + \alpha$ region for a *hypoeutectic alloy* and the $L + \beta$ region for a *hypereutectic alloy*. Here, α and β are the respective solid phases in the phase diagram. For instance, Fig. 7.3b depicts the hypoeutectic region defining the tie-line for the Lever rule at the freezing alloy C_o temperature T_f^* with a partition coefficient or segregation coefficient $k_o < 1$. This mean that k_o is associated with solute compositions in the solid and liquid phases.

Similarly, Fig. 7.3c exhibits the tie-line for developing the Lever rule at T_f^* with $k_o > 1$, where k_o is mathematically defined below. Moreover, the term microsegregation stands for non-uniform distribution of the solute in the solidifying melt induced by solute rejection at the smooth, flat, and uniform planar L-S interface or throughout a mushy layer of solid dendrites during solidification.

Microsegregation, in general, causes differences in local chemical composition, leading to a metallurgical problem called coring, which in turn, is caused by the rejection of solutes from a solidified melt into the liquid phase due to differences in solute solubility in these phases at the equilibrium temperature.

7.3.1 Lever Rule

Mass Fraction Method Fundamentally, the Lever rule requires a horizontal tie-line drawn horizontally at T_f^* intercepting the liquidus and solidus lines as shown in Fig. 7.3b, c. For an alloy mass $m_o = m_s + m_l$ with $m_s = m_\alpha$ or $m_s = m_\beta$, the mass balance is written as

$$C_s + C_l = C_o \quad \text{at } T = T_f^* \tag{7.4a}$$

$$m_s C_s + m_l C_l = m_o C_o \tag{7.4b}$$

$$\frac{m_s}{m_o} C_s + \frac{m_l}{m_o} C_l = C_o \tag{7.4c}$$

where C_s, C_l denote the solid and liquid solute compositions, m_s, m_l denote the mass of solid and liquid phases.

Let $f_s = m_s / m_o$ and $f_l = m_l / m_o$ be the solid and liquid fractions with $f_s + f_l = 1$ so that Eq. (7.4c) becomes

$$f_s C_s + f_l C_l = C_o \tag{7.5a}$$

$$f_s C_s + (1 - f_s) C_l = C_o \tag{7.5b}$$

from which

$$f_s = \frac{C_l - C_o}{C_l - C_s} \quad \& \quad f_l = \frac{C_o - C_s}{C_l - C_s} \tag{7.5c}$$

According to Fig. 7.3b or 7.3c, the first solid to form will have the composition $k_o C_o$, whereas the liquid concentration is $C_l \simeq C_o$ at $T = T_l$.

The Lever rule is the most common method in any material science course dealing with the determination of the weight fraction of phases using an equilibrium phase diagram. Thus, for a given binary composition C_o and a freezing temperature range $T_s < T_f^* < T_l$, both f_s and f_l are determined as per Eq. (7.5c). Moreover, Fig. 7.3c can also be used to carry out a similar procedure for deriving f_s and f_l equations for a hypereutectic binary alloy.

During solidification at $t > 0$, the solid-to-liquid concentration ratio C_s/C_l defines the partition coefficient k_o at L-S interface temperature (isothermal) T_f^* and pressure (isobaric) P. This ratio is metallurgically an important parameter that determines how a solute B or impurity is distributed between molten and solidified A-B binary alloy at a particular temperature. Hence,

$$C_s = k_o C_l \qquad \text{at } T_s < T_f^* < T_l \tag{7.6a}$$

$$k_o = \frac{C_s}{C_l} < 1 \quad \text{(hypoeutectic, Fig. 7.3b)} \tag{7.6b}$$

$$k_o = \frac{C_s}{C_l} > 1 \quad \text{(hypereutectic, Fig. 7.3c)} \tag{7.6c}$$

Combining Eqs. (7.5b) and (7.6a) yields the solid fraction for a hypoeutectic binary alloy written as

$$f_s = \frac{1}{1 - k_o} \left(1 - \frac{C_o}{C_l} \right) \tag{7.7}$$

which depends on two variables; k_o and C_l.

Combining Eqs. (7.5b) and (7.6a) yields

$$\frac{C_o}{C_l} = 1 - (1 - k_o) f_s \tag{7.8a}$$

$$f_s (C_o - C_s) = (1 - f_s)(C_l - C_o) \tag{7.8b}$$

$$f_s (C_o - C_s) \, dm = (1 - f_s)(C_l - C_o) \, dm \quad \text{(mass balance)} \tag{7.8c}$$

where dm is intentionally added to Eq. (7.8b) and it denotes the increment solute mass. For an ideal L-S interface (solidification front), the mass balance described by Eq. (7.8c) physically means that $f_s (C_o - C_s)$ is the amount of solute rejected by the solid phase and $(1 - f_s)(C_l - C_o)$ is the fraction transferred to the liquid phase. Hence, solute segregation ahead of the L-S interface prevails.

Using the idealized portion of the phase diagram in Fig. 7.3b, the mathematical modeling of the solidification paths of the liquidus and solidus as straight lines along with the T_A-C_l-T_f^* and T_A-C_o-T_l triangles yields the temperature equations with $T_f = T_A$, the slope β_l and the C_l/C_o ratio written as

$$T_l = T_f + \beta_l C_l \quad \& \quad T_s = T_f + \beta_s C_s \tag{7.9a}$$

$$\beta_l = \frac{T_f - T}{0 - C_l} \quad \& \quad \beta_l = \frac{T_f - T_1}{0 - C_o} \tag{7.9b}$$

$$\frac{C_l}{C_o} = \frac{T_f - T}{T_f - T_1} \tag{7.9c}$$

$$T_s = T_f + \beta_s C_s \tag{7.9d}$$

$$\beta_s = \frac{T_f - T_1}{0 - C_s} \quad \& \quad \beta_s = \frac{T_f - T_2}{0 - C_o} \tag{7.9e}$$

Equating Eqs. (7.8a) and (7.9c) yields the f_s equation and subsequently, differentiate f_s with respect to $T = T_f^*$. Thus,

$$f_s = \frac{1}{1 - k_o} \frac{T - T_1}{T - T_f} \quad \text{(Lever rule)} \tag{7.10a}$$

$$\frac{df_s}{dT} = \left(\frac{1}{k_o - 1} \right) \left[\frac{1}{T - T_f} - \frac{T - T_l}{(T - T_f)^2} \right] \tag{7.10b}$$

Volume Fraction Method In addition, the volume fraction with $V_s = V_\alpha$ and $\rho_s = \rho_\alpha$ can also be used to determine the phase fractions. If $V_s + V_l = 1$ and $m = V/\rho$, then the volume fraction of the solid and liquid phases becomes $f_s = V_s/V$ and $f_l = V_l/V$, where $V = m_s/\rho_s + m_l/\rho_l$.

So far the above method is quite elementary, easy to understand, and widely used to determine the phase fractions of binary alloys. A similar procedure can be used for a hypereutectic alloy (Fig. 7.3c).

For the sake of clarity, the aforementioned solidification process being characterized at a macroscale using the Lever rule (and Gulliver–Scheil model given below) occurs in a heterogeneous domain before solidification is complete. This is so because of the two-phase volume of material undergoing phase transformation, $L \rightarrow S$, at T_f or $T < T_f$ for an undercooled metal liquid. Moreover, the solid may be homogeneous if the solidified melt is entirely a single solid phase (α, β, etc.) or heterogeneous for a two-phase solid product (α-β).

7.3.2 Gulliver–Scheil Model

This model considers the mass balance between the amount of solute B in the solid phase and the amount of B that diffuses into the liquid. Thus, the L-S interface advances by an increment df_s and the liquid composition increases by dC_l. This solidification problem is implicitly illustrated in Fig. 7.4 by the hatched areas.

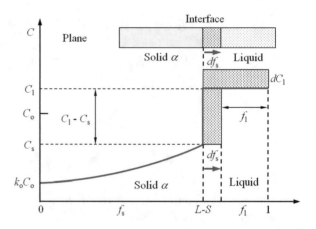

Fig. 7.4 Schematic distribution of the solute B concentration during alloy solidification

Thus, the mass balance with $f_s + f_l = 1$ may be written as the solute conservation of mass equation

$$(C_l - C_s)\, df_s = f_l dC_l = (1 - f_s)\, dC_l \quad \text{(mass balance)} \tag{7.11a}$$

$$\frac{df_s}{1 - f_s} = \frac{dC_l}{(1 - k_o)\, C_l} \tag{7.11b}$$

Notice that the rate of change of C_l depends on the rate of change of the solid fraction f_s induced by the evolved latent heat of fusion over a temperature range $T_2 < T_f^* < T_2$ for the binary A-B alloy with composition C_o.

Integrating Eq. (7.11b) yields

$$\int_0^{f_s} \frac{df_s}{1 - f_s} = \int_{C_o}^{C_l} \frac{dC_l}{(1 - k_o)\, C_l} \quad [\text{for } k_o = k_o\,(C_l)\,] \tag{7.12a}$$

$$\int_0^{f_s} \frac{df_s}{1 - f_s} = \frac{1}{(1 - k_o)} \int_{C_o}^{C_l} \frac{dC_l}{C_l} \quad (\text{for } k_o = \text{Constant}) \tag{7.12b}$$

$$-\ln\,(1 - f_s) = \frac{1}{(1 - k_o)} \ln\left(\frac{C_l}{C_o}\right) \tag{7.12c}$$

Reorganizing Eq. (7.12c) yields the *non-equilibrium lever rule* known as the *Gulliver–Scheil equation* for the concentration of solute B in the liquid and solid phases [2, p. 186]

$$C_l = C_o\,(1 - f_s)^{k_o - 1} = C_o\,(f_l)^{k_o - 1} \tag{7.13a}$$

$$C_s = k_o C_o\,(1 - f_s)^{k_o - 1} \tag{7.13b}$$

Rearranging Eq. (7.13a) describes the non-linear lever rule for the solid volume fraction as a concentration-dependent parameter

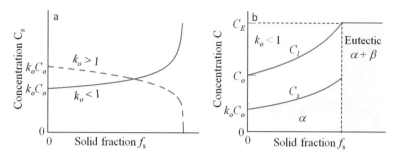

Fig. 7.5 Schematic concentration profiles. (**a**) Effect of partition coefficient k_o on C_s and (**b**) Scheil equations for the solid and liquid regions after solidification. The last liquid solidifies as a eutectic structure

$$f_l = \left(\frac{C_l}{C_o}\right)^{1/(k_o-1)} \tag{7.14a}$$

$$f_s = 1 - \left(\frac{C_l}{C_o}\right)^{1/(k_o-1)} \tag{7.14b}$$

$$f_s = 1 - \left(\frac{C_s}{k_o C_o}\right)^{1/(k_o-1)} \tag{7.14c}$$

These expressions represent the solute redistribution during solidification of an alloy with nominal composition C_o and without diffusion in the solid. Furthermore, Fig. 7.5a illustrates the behavior of the solid concentration C_s defined by Eq. (7.14b) for an A-B binary alloy with composition C_o and constant k_o values. A similar graphical representation of solute concentration can be found elsewhere for a Ni-based superalloy [3].

For solute partition coefficient $k_o < 1$, the liquidus and solidus lines have negative slopes, the alloy composition is in the range $0 < C \leq C_e$ and it is called a *hypoeutectic alloy* (left-hand side of the eutectic point E in Fig. 7.3b). Notice that C_s increases with increasing f_s and as expected, $C_s \rightarrow C_o$ very rapidly as $f_s \rightarrow 1$ for complete solidification with some eutectic composition. Moreover, a negative solid slope and $k_o < 1$ condition suggest that the excess of solute B at the L-S mushy zone is rejected into the liquid phase. Consequently, the solute in the liquid increases as solidification proceeds until the last liquid content reaches the eutectic composition at the eutectic temperature T_E.

For solute partition coefficient $k_o > 1$, the liquidus and solidus lines have positive slopes, the alloy composition is in the range $C \geq C_e$ and it is called a *hypereutectic alloy* (right-hand side of the eutectic point E in Fig. 7.3c), and C_s decreases with increasing f_s. Obviously, $C_s \rightarrow 0$ as $f_s \rightarrow 1$ for complete solidification. Hence, plotting Eq. (7.8) reveals increasing C_l and C_s concentrations with increasing f_s as shown in Fig. 7.5b.

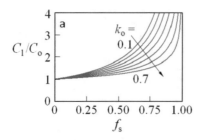 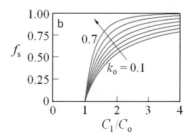

Fig. 7.6 Non-linear Lever rule showing trends for (**a**) $C_l/C_o = g(f_s)$ and (**b**) $f_s = g(C_l/C_o)$ functions for some values of $k_o < 1$

Notice that the C_l curve becomes asymptotic by acquiring the eutectic composition C_e at T_E (Fig. 7.3b). During solidification of a hypereutectic alloy, solid accommodates just the right amount of solute B and consequently, the liquid and solute contents continue to decrease until $T = T_E$, at which the last amount of liquid solidifies eutectically.

Furthermore, plotting Eq. (7.14b) as a particular example, the concentration ratio and solid fraction function, $C_1/C_o = g(f_s)$ and $f_s = g(C_1/C_o)$, for some values of $k_o < 1$ reveal additional non-linear trends as illustrated in Fig. 7.6.

Eutectic Solid Fraction For an alloy with $C_o = C_1$ (Fig. 7.3b), the first solid fraction is f_α. As solidification is near completion, the remaining liquid reaches the eutectic composition $C_l = C_e$ and subsequently, Eq. (7.6c) provides the eutectic solid $(\alpha + \beta)$ fraction f_E written as

$$\ln\left(\frac{C_e}{C_o}\right) = (k_o - 1)\ln(f_E) \quad \text{for } C_o < C_e \tag{7.15}$$

from which

$$f_E = \exp\left(\frac{\ln(C_e/C_o)}{k_o - 1}\right) \tag{7.16}$$

For consistency, the total solid fraction consists of two solid phases, namely $f_s = f_\alpha + f_{\alpha+\beta}$. This physically means that the resultant microstructure is composed of α and $(\alpha + \beta)$ grains.

For the sake of clarity, the freezing temperature T_f^* for an A-B binary alloy with nominal concentration C_o shown in Fig. 7.7a (T-C diagram) is directly related to the latent heat of fusion ΔH_f^* as depicted in Fig. 7.7b (H-T diagram).

Essentially, the enthalpy range $\Delta H = H_l - H_s$ for solidification (Fig. 7.7b) encloses the $L + S$ region at a temperature range, $\Delta T = H_l - T_s$, which is also depicted in Fig. 7.7a.

Accordingly, these diagrams play significant roles in the field of solidification of binary alloys.

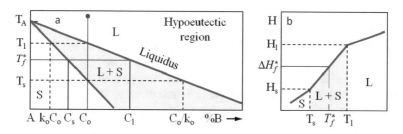

Fig. 7.7 Schematic mushy zones ($L+S$ phases). (**a**) Partial hypoeutectic diagram and (**b**) enthalpy diagram

Nonetheless, substituting Eq. (7.9c) into (7.14a) with $T = T_f^*$ and $k_o < 1$ yields the Gulliver–Scheil equation, which is compared to Lever rule. Thus,

$$f_l = \left(\frac{T - T_f}{T_1 - T_f}\right)^{1/(k_o-1)} \qquad \text{(non-linear Lever rule)} \qquad (7.17a)$$

$$f_s = 1 - \left(\frac{T - T_f}{T_1 - T_f}\right)^{1/(k_o-1)} \qquad \text{(Gulliver–Scheil equation)} \qquad (7.17b)$$

$$f_s = \frac{1}{1 - k_o}\frac{T - T_1}{T - T_f} \qquad \text{(linear Lever rule)} \qquad (7.17c)$$

Differentiating Eq. (7.17b) with respect to T and using the chain rule of differentiation yields the rate of formation of the solid fraction being dependent on the solidification rate (dT/dt)

$$\frac{df_s}{dT} = -\frac{1}{(k_o-1)(T_1-T_f)}\left(\frac{T-T_f}{T_1-T_f}\right)^{(2-k_o)/(k_o-1)} \qquad (7.18a)$$

$$\frac{df_s}{dt} = \frac{df_s}{dT}\frac{dT}{dt} \quad \text{(chain rule)} \qquad (7.18b)$$

$$\frac{\partial f_s}{\partial t} = \frac{1}{(k_o-1)(T_l-T_s)}\left(\frac{T_f^*-T}{T_l-T_s}\right)^{(2-k_o)/(k_o-1)}\frac{\partial T}{\partial t} \qquad (7.18c)$$

Note that Eqs. (7.18a), (7.18c) are for a solute partition coefficient $k_o < 1$. A similar procedure may follow for the case of a different solute partition coefficient $k_o > 1$. Hence, $\partial f_s/\partial t = f(\partial T/\partial t)$ at T.

Explicitly, the rate of formation the solid fraction $(\partial f_s/\partial t)$, as defined by Eq. (7.18c), characterizes (describes) the rate of solidification at a certain slow solidification velocity, which in turn, is analytically defined in a later section or appropriate chapter.

Example 7.1 Consider a 80*Cu*-20*Ag* (weight percent) alloy plate initially at 1100 °C and a 80*Sn*-20*Bi* (weight percent) alloy plate initially at 250 °C. The partial phase diagrams are given below. The former solidus and the latter liquidus curved phase boundaries are approximated by dashed straight lines. During an ideal solidification process for the 80*Cu*-20*Ag* alloy system, the melt is partially solidified at 900 °C. Determine the phase fractions using (a) the Lever rule with the actual liquidus and solidus phase boundaries, (b) the Lever rule with straight solidus phase boundary (dashed line), (c) the Gulliver–Scheil equation, and (d) plot T_f^* as a function of solid fraction f_s. Assume that the results from part (a) are the most accurate. Compute the percent difference (Δf_α) and the error percentage for the solid α-phase between these methods. (e) Repeat the required calculations for the 80*Sn*-20*Bi* alloy system at $T_f^* = 175$ °C.

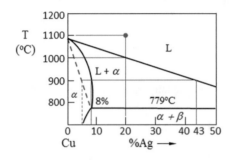
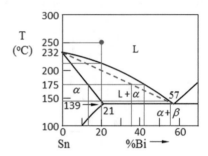

Solution

(a) **80*Cu*-20*Ag* alloy:** Use the solidus curved path at 900 °C so that

$$C_s \simeq 8\% Ag, \quad C_o = 20\% Ag, \quad C_l = 43\% Ag$$

$$f_l^{(a)} = \frac{C_o - C_s}{C_l - C_s} = \frac{20 - 8}{43 - 8} = 0.34 \quad \text{(curved solidus path)}$$

$$f_\alpha^{(a)} = 1 - f_l^{(a)} = 0.66$$

(b) Use the tie-line at 900 °C and read-off the *Ag* concentrations so that

$$C_s \simeq 6\% Ag, \quad C_o = 20\% Ag, \quad C_l = 43\% Ag$$

$$f_l^{(b)} = \frac{C_o - C_s}{C_l - C_s} = \frac{20 - 6}{43 - 6} = 0.38 \quad \text{(straight solidus path)}$$

$$f_\alpha^{(b)} = 1 - f_l^{(b)} = 0.62$$

Then,

$$\Delta f_\alpha = \frac{f_\alpha^{(a)} - f_\alpha^{(b)}}{\left(f_\alpha^{(a)} + f_\alpha^{(b)}\right)/2} \times 100\% = \frac{0.67 - 0.62}{(0.67 + 0.62)/2} \times 100\% \simeq 7.75\%$$

$$\text{error} = \frac{f_\alpha^{(a)} - f_\alpha^{(b)}}{f_\alpha^{(a)}} \times 100\% = \frac{0.66 - 0.62}{0.66} \times 100\% \simeq 6.06\%$$

(c) From Eq. (7.8a) along with solidus straight line

$$C_s \simeq 6\% Ag, \quad C_o = 20\% Ag, \quad C_l = 43\% Ag$$

$$k_o = \frac{C_s}{C_l} = \frac{6}{43} = 0.14$$

$$f_l^{(c)} = \left(\frac{C_l}{C_o}\right)^{1/(k_o-1)} = \left(\frac{43}{20}\right)^{1/(0.14-1)} = 0.41$$

$$f_\alpha^{(c)} = 1 - f_l^{(c)} = 0.59$$

From Eq. (7.8a) along with solidus curved path

$$C_s \simeq 8\% Ag, \quad C_o = 20\% Ag, \quad C_l = 43\% Ag$$

$$k_o = \frac{C_s}{C_l} = \frac{8}{43} = 0.19$$

$$f_l^{(c')} = \left(\frac{C_l}{C_o}\right)^{1/(k_o-1)} = \left(\frac{43}{20}\right)^{1/(0.19-1)} = 0.39 \quad \text{(Gulliver–Scheil equation)}$$

$$f_\alpha^{(c')} = 1 - f_l^{(c')} = 0.61$$

Comparing results from part (a) and (b) yields for the approximated straight solidus line

$$\Delta f_\alpha = \frac{f_\alpha^{(a)} - f_\alpha^{(c)}}{\left(f_\alpha^{(a)} + f_\alpha^{(c)}\right)/2} \times 100\% = \frac{0.66 - 0.59}{(0.66 + 0.59)/2} * 100\% = 11.20\%$$

$$\text{error} = \frac{f_\alpha^{(a)} - f_\alpha^{(c)}}{f_\alpha^{(a)}} \times 100\% = \frac{0.66 - 0.59}{0.66} \times 100\% = 10.61\%$$

and for the curved solidus path

$$\Delta f_\alpha = \frac{f_\alpha^{(a)} - f_\alpha^{(c')}}{\left(f_\alpha^{(a)} + f_\alpha^{(c')}\right)/2} \times 100\% = \frac{0.66 - 0.61}{(0.66 + 0.61)/2} * 100\% = 7.87\%$$

$$\text{error} = \frac{f_\alpha^{(a)} - f_\alpha^{(c')}}{f_\alpha^{(a)}} \times 100\% = \frac{0.66 - 0.61}{0.65} \times 100\% = 7.69\%$$

Therefore, approximating the solidus curve as a straight line induces a significant percentage error as shown in this example.

(d) From Eq. (7.17c) with $T_f = T_A$ and $T_f^* = T$, the freezing temperature equation with $k_o = 0.19$ is

$$T_f^* = \frac{T_1 - f_s\,(1 - k_o)\,T_f}{1 - f_s\,(1 - k_o)} = \frac{1000 - (1 - 0.19)\,(1083)\,f_s}{1 - (1 - 0.19)\,f_s}$$

$$T_f^* = \frac{1000 - 877.23\,f_s}{1 - 0.81\,f_s} \qquad \text{(Lever rule)}$$

and from (7.17b),

$$T_f^* = T_f + (1 - f_s)^{k_o-1}\,(T_1 - T_f) = 1083 + (1 - f_s)^{k_o-1}\,(1000 - 1083)$$

$$T_f^* = 1083 - 83\,(1 - f_s)^{-0.81} \quad \text{(Scheil equation)}$$

Plotting the Lever and Scheil equations, $T_f^* = g\,(f_s)$, in the figure below clearly shows a non-linear behavior. According to both models, T_f^* exhibits a non-linear behavior and as expected, it decreases very rapidly as $f_s \longrightarrow 1$.

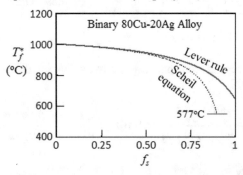

(e) **80Sn-20Bi alloy:** Carrying out similar calculations for this alloy is expected to draw different results and similar conclusions. Performing the required work is left out for the reader as problem to solve. Nevertheless, the solute Ag concentrations at $175\,^\circ C$ are

$$C_s = 13\%\,Bi, \quad C_o = 20\%\,Bi, \quad C_l = 36\%\,Bi \text{ (line)}, \quad C_l = 42\%\,Bi \text{ (curve)}$$

and

$$C_s \simeq 8\%Ag, \quad C_o = 20\%Ag, \quad C_l = 43\%Ag$$

$$f_l^{(a)} = \frac{C_o - C_s}{C_l - C_s} = \frac{20 - 13}{42 - 13} = 0.24 \quad \text{(curved solidus path)}$$

$$f_\alpha^{(a)} = 1 - f_l^{(a)} = 0.76$$

Remaining calculations for this alloy are left as an exercise for the reader.

7.4 Enthalpies of Binary Alloy Solidification

The fundamental law of conduction heat transfer during solidification dictates that the latent heat of fusion ΔH_f^* evolves from the liquid phase only at the freezing temperature T_f^*. Recall that $T = T_f^*$ is a constant temperature for a pure metal solidification (see Chap. 5) and $T_s < T = T_f^* < T_l$ for an alloy solidification.

In common foundry practices, the melt is chemically adjusted at the pouring temperature $T_p > T_f^*$, which induces a superheating effect on the liquid phase by developing a temperature gradient $\partial T_l / \partial x$, which affects the local solidification time t_s. This is the reason for using the *latent heat of solidification* $\Delta H_s > \Delta H_f^*$.

Thermodynamically, the specific enthalpy, the rate of specific enthalpy, and the specific heat at constant pressure are defined, respectively, in a general form by

$$H = c_p T \tag{7.19a}$$

$$\frac{\partial H}{\partial t} = c_p \frac{\partial T}{\partial t} \tag{7.19b}$$

$$c_p(T) = \frac{\partial H(T)}{\partial T} \tag{7.19c}$$

According to Poirier-Nandapurhar [4], the mushy zone enthalpy (H_m) can be defined as a solute concentration-dependent thermal energy at $T_s < T_f^* < T_l$. Thus, the enthalpies of an A-B binary alloy phase region during solidification can be generalized as

$$H_m = f_s H_s + (1 - f_s) H_l \tag{7.20a}$$

$$H_s = C_s H_{sA} + (1 - C_s) H_{sB} \tag{7.20b}$$

$$H_l = C_l H_{sl} + (1 - C_l) H_{lB} \tag{7.20c}$$

Here, C_s, C_l denote the usual fractions of concentration of phases and are (1) extracted directly from the equilibrium A-B binary phase diagram and f_s is determined from a suitable equation given above. In particular, the Gulliver–Scheil equation is used very extensively in the literature because it assumes local equilibrium at the L-S interface and complete solute mixing in the liquid, and no diffusion in the solid phase. Hence, the Gulliver–Scheil equation is an expression that depends on the segregation of the solute B or the redistribution of the solute B during solidification.

In addition, the mixing of mass densities in the mushy zone (mixture of phases) of A-B binary alloys can also be defined as [5]

$$\rho_m = f_s \rho_s + (1 - f_s) \rho_l \tag{7.21a}$$

$$\rho_s = C_s \rho_{sA} + (1 - C_s) \rho_{sB} \tag{7.21b}$$

$$\rho_l = C_l \rho_{s,l} + (1 - C_l) \rho_{l,B} \tag{7.21c}$$

For the thermal conductivities,

$$k_m = f_s k_s + (1 - f_s) k_l \tag{7.22a}$$

$$k_s = C_s k_{sA} + (1 - C_s) k_{sB} \tag{7.22b}$$

$$k_l = C_l k_{lA} + (1 - C_l) k_{lB} \tag{7.22c}$$

and for the specific heats

$$c_m = f_s c_s + (1 - f_s) c_l \tag{7.23a}$$

$$c_s = C_s c_{sA} + (1 - C_s) c_{sB} \tag{7.23b}$$

$$c_l = C_l c_{lA} + (1 - C_l) c_{lB} \tag{7.23c}$$

From Chap. 5, the thermal diffusivity α and the thermal inertia (thermal effusivity) γ equations are recast here for convenience. Thus,

$$\alpha_l = \frac{k_l}{\rho_l c_l} \quad \text{(liquid)} \tag{7.24a}$$

$$\alpha_s = \frac{k_s}{\rho_s c_s} \quad \text{(solid)} \tag{7.24b}$$

$$\alpha = \frac{k}{\rho c_p} \quad \text{(general)} \tag{7.24c}$$

$$\gamma_m = k_m / \sqrt{\alpha_m} = \sqrt{k_m \rho_m c_m} \tag{7.24d}$$

$$\gamma_s = k_s / \sqrt{\alpha_s} = \sqrt{k_s \rho_s c_s} \tag{7.24e}$$

For constant thermophysical constants, the governing heat and enthalpy equations are now described as

$$\frac{\partial T}{\partial t} = \alpha \frac{\partial^2 T}{\partial x^2} + \frac{dQ}{dt} = \alpha \frac{\partial^2 T}{\partial x^2} + \rho \Delta H_s \frac{\partial f_s}{\partial t} \tag{7.25a}$$

$$\rho \frac{\partial H}{\partial t} = k \frac{\partial^2 T}{\partial x^2} + \frac{dQ}{dt} = k \frac{\partial^2 T}{\partial x^2} + \rho \Delta H_s \frac{\partial f_s}{\partial t} \tag{7.25b}$$

$$\frac{dQ}{dt} = \rho \Delta H_s \frac{\partial f_s}{\partial t} \tag{7.25c}$$

Here, dQ/dt is the rate of heat flow during solidification or supply during melting. Notice that the latent heat of solidification ΔH_s replaces the latent heat of fusion ΔH_f in these expressions in order to account for the effect of superheating the melt.

In order to superheat the melt an additional amount of heat is added to the melting furnace or smelter to reach the desire temperature $T > T_f$ for a pure metal and $T > T_f^*$. Whichever the case may be, the superheat temperature is denoted as the pouring temperature $T = T_p$ prior to filling the mold cavity. This suggests that T_p must be upper bound of the superheated state of the melt; otherwise, severe liquid vibrational instability would induce casting problems. Nonetheless, T_p depends on the type of material and no attempt is made to include any theoretical approach to derive it in this section.

Coming back to the analytical issue at hand, substituting Eq. (7.18c) into (7.25a, 7.25b) yields

$$\frac{\partial T}{\partial t} = \alpha \frac{\partial^2 T}{\partial x^2} + \frac{\rho \Delta H_s}{(k_o - 1)\left(T_f^* - T_l\right)} \left(\frac{T_f^* - T}{T_l - T_s}\right)^{(2-k_o)/(k_o-1)} \frac{\partial T}{\partial t} \tag{7.26a}$$

$$\rho \frac{\partial H}{\partial t} = k \frac{\partial^2 T}{\partial x^2} + \frac{\rho \Delta H_s}{(k_o - 1)\left(T_f^* - T_l\right)} \left(\frac{T_f^* - T}{T_l - T_s}\right)^{(2-k_o)/(k_o-1)} \frac{\partial T}{\partial t} \tag{7.26b}$$

These expressions, Eq. (7.26), can be solved using the separation of variables method found in Chap. 5.

7.4.1 Latent Heat of Solidification of Binary Alloys

For convenience, Fig. 4.12 is recast as Fig. 7.8 for further clarification of relevant definitions of enthalpy terms related to pure metal and alloy solidification domains.

For an isothermal phase transition shown in Fig. 7.8a, the group of enthalpy terms are defined as

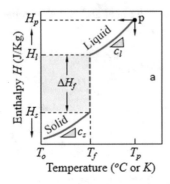
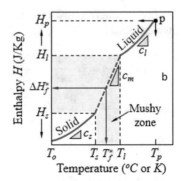

Fig. 7.8 Enthalpy-temperature diagrams (*H-T* diagrams) for (**a**) a pure metal and (**b**) an *A-B* binary alloy

$$H_s = c_s T \quad \text{at } T \leq T_s \tag{7.27a}$$

$$H = \Delta H_f^* + c_l T_f^* \quad \text{at } T \leq T_f^* \tag{7.27b}$$

$$\Delta H_f^* = c_s T_f \quad (H\text{-}T \text{ diagram "a")} \tag{7.27c}$$

Notice that the specific heats, c_s and c_l, must be constant properties under the current thermal process.

For a non-isothermal process (Fig. 7.8b),

$$H = c_s T \quad \text{for } T \leq T_s \tag{7.28a}$$

$$H = c_m T + \left(\frac{T - T_s}{T_l - T_s} \right) \Delta H_f^* \quad \text{for } T_s < T = T_f^* \leq T_l \tag{7.28b}$$

$$H = c_l T + \Delta H_f^* + c_m (T_l - T_s) \quad \text{for } T \geq T_l \tag{7.28c}$$

$$\Delta H_f^* = c_m (T_l - T_s) \quad (H\text{-}T \text{ diagram "b")} \tag{7.28d}$$

These expressions, Eqs. (7.27a), (7.27b) and (7.28a), (7.28b), (7.28c), can be used as controlling constraints in numerical analysis.

The latent heat of fusion depends on the transformation temperature T_f^*. However, one can do a literature survey and find different analytical forms of ΔH_f^* based on particular assumptions. A review on this topic is left out for the reader, but only one specific approach is undertaken in this section. Assume that the enthalpy H for the solid or liquid phase in a *A-B* binary alloy depends on the solute concentration B and temperature T_f^*. Thus,

$$\Delta H_f^* = \Delta H_{fA} + C_l \left| \Delta H_{fB} - \Delta H_{fA} \right| \tag{7.29}$$

Here, H_{fA} and H_{fB} denote the latent heat of fusion of solvent "A" and solute "B," respectively. In particular, $\Delta H_{fB} > \Delta H_{fA}$ since the solute B is assumed to be the most refractory element in a A-B binary alloy.

For complete *quasi-static solidification*, assume that the molten alloy is initially at $T_p > T_l$ and that the latent heat of solidification is defined as

$$\Delta H_s = \Delta H_f^* + \int_{T_s}^{T_p} c_p\,(C_s, T)\,dT \tag{7.30}$$

where ΔH_s accounts for superheating effects when the melt is at the pouring temperature $T_p > T_l$. If the specific heat at constant pressure (isobaric process) is $c_p = c_s$, then Eq. (7.30) along with (7.29) becomes

$$\Delta H_s = \Delta H_{fA} + C_l \left|\Delta H_{fB} - \Delta H_{fA}\right| + c_s\,(T_p - T_s) \tag{7.31}$$

Nevertheless, the solute liquid and solid fractions can be defined as enthalpy ratio entities [6, p. 213]

$$f_l = \frac{H_l}{\Delta H_s} \qquad \text{for } 0 < H_l < \Delta H_s \tag{7.32a}$$

$$f_s = 1 - f_l = 1 - \frac{H_l}{\Delta H_s} \tag{7.32b}$$

Substituting Eqs. (7.9c) and (7.13a) into (7.32b) along with $0 < f_s < 1$ and $T_E < T_f^* < T_l$ yields the latent heat of solidification as

$$\Delta H_s = H_{fA} + C_o\,(1 - f_s)^{k_o - 1}\,(H_{fB} - H_{fA}) + c_s\left(T_f^* - T_s\right) \tag{7.33a}$$

$$\Delta H_s = H_{fA} + C_o\left(\frac{T_A - T_f^*}{T_A - T_1}\right)(H_{fB} - H_{fA}) + c_s\left(T_f^* - T_s\right) \tag{7.33b}$$

Specifically, if $T_f^* = T_1 = T_E$, then Eq. (7.33b) becomes

$$\Delta H_s = H_{f,A} + C_o\,(H_{f,B} - H_{f,A}) + c_s\,(T_E - T_s) \tag{7.34}$$

For the melt initially at $T_p > T_l$ and the mold at $T_o < T_s$, Eq. (7.31) becomes

$$\Delta H_s = \Delta H_{fA} + C_l\left|\Delta H_{fB} - \Delta H_{fA}\right| - \int_{T_p}^{T_o} c_p\,(C_s, T)\,dT \tag{7.35a}$$

$$\Delta H_s = H_{f,A} + C_o\left|H_{f,B} - H_{f,A}\right| + c_s\,(T_p - T_o) \tag{7.35b}$$

where T_o is the mold or substrate initial temperature. In principle, Eq. (7.35b) is the same as Eq. (21) found in Xu et al. paper [7].

Additionally, the aforementioned solidification process accounts for a degree of superheating. Hence, at the equilibrium point

$$\Delta H_f(\text{freezing}) = -\Delta H_f(\text{melting}) \quad \text{at } T_f \text{ for pure metals}$$
$$\Delta H_f^*(\text{freezing}) = -\Delta H_f^*(\text{melting}) \quad \text{at } T_f^* \text{ for alloys}$$

and at the pouring temperature,

$$\Delta H_s > \Delta H_f \quad T_p > T_f \quad \text{for pure metals}$$
$$\Delta H_s > \Delta H_f^* \quad T_p > T_f^* \quad \text{for alloys}$$

Recall that T_p is treated henceforth as the uppermost pouring temperature prior to fill a sand or metallic mold cavity.

7.5   Single-Phase Solidification Models

Mathematical modeling of *solidification* is mainly controlled by conductive heat transfer and related rate of evolved latent heat of fusion ΔH_f^* at the *L-S* interface, which advances into the liquid at a particular velocity $ds(t)/dt$. Hence, $ds(t)/dt \to 0$ for slow (conventional) and $ds(t)/dt \to \infty$ for rapid solidification. The former is treated as a closed system containing an encapsulated cavity and the latter as an open system, where convective heat transfer may contribute significantly the solidification process.

The goal now is to find analytical solutions to the temperature field equations for (1) pure heat conduction and (2) coupled heat-mass transport models under controlled environments. The casting geometry considered in this chapter has a rectangular shape, such as a slab. As for pure metals characterized in Chap. 5, the one-dimensional heat transfer approach henceforth also considers a half-space $x > 0$ in a semi-infinite domain $0 < x < \infty$.

7.5.1   Ideal Mold–Solid Contact Model

For a one-dimensional heat transfer by conduction, the governing equations, continuity of temperatures and the heat flux balance at interfaces are

$$\frac{\partial T_m}{\partial t} = \alpha_m \frac{\partial^2 T_m}{\partial x^2}; \quad \frac{\partial T_s}{\partial t} = \alpha_s \frac{\partial^2 T_s}{\partial x^2}; \quad \frac{\partial T_l}{\partial t} = \alpha_l \frac{\partial^2 T_l}{\partial x^2} \tag{7.36a}$$

$$T_m(x,t) = T_s(x,t) = T_i \quad \text{at } x = 0, t > 0 \tag{7.36b}$$

$$k_m \frac{\partial T_m}{\partial x} = k_s \frac{\partial T_s}{\partial x} \quad \text{at } x = s\,(t)\,, t > 0 \ \ (m\text{-}s \text{ interface})\tag{7.36c}$$

$$T_s\,(x, t) = T_l\,(x, t) = T_f^* \quad \text{at } x = s\,(t)\,, t > 0 \tag{7.36d}$$

$$k_s \frac{\partial T_s}{\partial x} - k_l \frac{\partial T_l}{\partial x} = \rho_s \Delta H_s \frac{ds\,(t)}{dt} \quad \text{at } x = s\,(t) \ \ (L\text{-}S \text{ interface})\tag{7.36e}$$

where the subscript "m" denotes the mold or substrate. The heat fluxes are

$$q_m = -k_m \frac{\partial T_m}{\partial x}; \quad q_l = -k_l \frac{\partial T_l}{\partial x}; \quad q_s = -k_s \frac{\partial T_s}{\partial x}; \quad q_f = -\rho_l \Delta H_s \frac{ds\,(t)}{dt} \tag{7.37}$$

Consider a semi-infinite solidification system ($0 < x < \infty$) with an ideal mold–solid contact and negligible liquid temperature gradient ($\partial T_l / \partial x \to 0$). For such a system, it is appropriate to divide it into two subsystems.

The mold containing a cavity can be treated as a solid phase or subsystem that absorbs heat from the cooling solid phase. For one-dimensional analysis, the temperature field equation and the corresponding temperature gradient are

$$T_m = T_o + (T_i - T_o)\, \mathrm{erfc} \left(\frac{-x}{2\sqrt{\alpha_m t}} \right) \quad \text{(mold)} \tag{7.38a}$$

$$\frac{\partial T_m}{\partial x} = \frac{T_i - T_o}{\sqrt{\pi \alpha_m t}} \exp \left(\frac{-x}{2\sqrt{\alpha_m t}} \right) \tag{7.38b}$$

$$\left(\frac{\partial T_m}{\partial x} \right)_{x=0} = \frac{T_i - T_o}{\sqrt{\pi \alpha_m t}} \tag{7.38c}$$

Notice that Eq. (7.38a) is the solution of the governing expression defined by Eq. (7.36a) group.

For the solid phase, the temperature field equation and associated derivatives become

$$T_s = T_i + \frac{(T_p - T_i)}{\mathrm{erf}\,(\lambda)}\, \mathrm{erf} \left(\frac{x}{2\sqrt{\alpha_s t}} \right) \quad \text{(solid)} \tag{7.39a}$$

$$\frac{\partial T_s}{\partial t} = -\frac{(T_p - T_i)\,x}{2\,\mathrm{erf}\,(\lambda)\,t\sqrt{\pi \alpha_s t}} \exp \left(-\frac{x^2}{4\alpha_s t} \right) \tag{7.39b}$$

$$\frac{\partial T_s}{\partial x} = \frac{(T_p - T_i)}{\mathrm{erf}\,(\lambda)\,\sqrt{\pi \alpha_s t}} \exp \left(-\frac{x^2}{4\alpha_s t} \right) \tag{7.39c}$$

Evaluating the solid temperature gradient at $x = 0$ and $x = s\,(t)$ gives

$$\left(\frac{\partial T_s}{\partial x} \right)_{x=0} = \frac{(T_p - T_i)}{\mathrm{erf}\,(\lambda)\,\sqrt{\pi \alpha_s t}} \tag{7.40a}$$

$$\left(\frac{\partial T_s}{\partial x} \right)_{s(t)} = \frac{(T_p - T_i)\exp\left(-\lambda^2\right)}{\mathrm{erf}\,(\lambda)\,\sqrt{\pi \alpha_s t}} \tag{7.40b}$$

If $T_s = T_p = T_f^*$ at the L-S interface $x = s\,(t)$, then Eq. (7.39a) yields

$$\text{erf}\,(\lambda) = \text{erf}\left[\frac{s\,(t)}{2\sqrt{\alpha_s t}}\right] = \text{Constant} \tag{7.41}$$

From this expression, the characteristic length $s\,(t)$, the solution characteristic λ, the solidification time t, and the solidification velocity $ds\,(t)/dt$ are

$$s\,(t) = 2\lambda\sqrt{\alpha_s t}; \quad \lambda = \frac{s\,(t)}{2\sqrt{\alpha_s t}}; \quad t = \frac{s\,(t)^2}{4\lambda^2 \alpha_s}; \quad \frac{ds\,(t)}{dt} = \lambda\sqrt{\frac{\alpha_s}{t}} \tag{7.42}$$

For the heat flux balance at the mold–solid interface ($x = 0$), $q_m = q_s$ and

$$q_m = -k_m\left(\frac{\partial T_m}{\partial x}\right)_{x=0} \quad \& \quad q_s = -k_s\left(\frac{\partial T_s}{\partial x}\right)_{x=0} \tag{7.43a}$$

$$k_m\left(\frac{\partial T_m}{\partial x}\right)_{x=0} = k_s\left(\frac{\partial T_s}{\partial x}\right)_{x=0} \tag{7.43b}$$

$$\frac{k_m\,(T_i - T_o)}{\sqrt{\pi\alpha_m t}} = \frac{k_s\,(T_p - T_i)}{\text{erf}\,(\lambda)\,\sqrt{\pi\alpha_s t}} \tag{7.43c}$$

Combining Eqs. (7.24b) and (7.43c) yields the mold–solid interface temperature, $T_o < T_i < T_p$, as

$$T_i = \frac{\gamma_s T_p + \gamma_m\,\text{erf}\,(\lambda)\,T_o}{\gamma_s + \gamma_m\,\text{erf}\,(\lambda)} \tag{7.44}$$

Inserting Eq. (7.39c) at $x = s\,(t)$ into (7.36c) with $\partial T_l/\partial x = 0$ yields the transcendental equation given by

$$\sqrt{\pi}\lambda\exp\left(\lambda^2\right)\text{erf}\,(\lambda) = \frac{c_s\,(T_p - T_i)}{\Delta H_s} \tag{7.45}$$

which is similar to Eq. (5.41a).

Combining Eqs. (7.44) and (7.45) gives the classical transcendental equation written as

$$\lambda\exp\left(\lambda^2\right)\left[\text{erf}\,(\lambda) + \frac{\gamma_s}{\gamma_m}\right] = \frac{c_s\,(T_p - T_o)}{\sqrt{\pi}\,\Delta H_s} = \frac{S_t}{\sqrt{\pi}} \tag{7.46a}$$

$$\lambda\exp\left(\lambda^2\right)\left[\text{erf}\,(\lambda) + \sqrt{\frac{k_s\rho_s c_s}{k_m\rho_m c_m}}\right] = \frac{c_s\,(T_p - T_o)}{\sqrt{\pi}\,\Delta H_s} \tag{7.46b}$$

where S_t denotes the Stefan number for the solid phase. Moreover, the heat flux q_x crossing the mold–solid interface and the rate of thermal energy dQ/dt through the mold are, respectively,

$$q_x = -k_m \left(\frac{\partial T}{\partial x} \right)_{x=0} \tag{7.47a}$$

$$\frac{dQ}{dt} = -Ak_m \left(\frac{\partial T}{\partial x} \right)_{x=0} = \frac{Ak_m \left(T_p - T_o \right)}{\sqrt{\pi \alpha_m t}} \tag{7.47b}$$

Integrating Eq. (7.47b) yields the amount of thermal energy through the mold in the negative x-direction

$$Q_x = \frac{Ak_m \left(T_p - T_o \right)}{\sqrt{\pi \alpha_m}} \int_0^{t_s} \frac{dt}{\sqrt{t}} \tag{7.48a}$$

$$Q_x = \frac{2Ak_m \left(T_p - T_o \right)}{\sqrt{\pi \alpha_m}} \sqrt{t} \tag{7.48b}$$

and the amount of thermal energy due to the release of latent heat of solidification ΔH_s is written as

$$Q_s = m \Delta H_s = \rho_s V \Delta H_s = \rho_s As \left(t \right) \Delta H_s \tag{7.49}$$

where $V = As(t)$ is the volume of the casting geometry. This is the volume of material that is divided into two regions, liquid and solid, separated by the L-S interface during the solidification process in a semi-infinite domain $0 < x < \infty$.

Now, equating Eqs. (7.48b) and (7.49) yields the total solidification time t based on the mold properties and the temperature difference $\Delta T = T_p - T_o$

$$t = \frac{\pi}{4} \left[\frac{\sqrt{\alpha_m} \rho_s \Delta H_s}{k_m \left(T_p - T_o \right)} \right]^2 s \left(t \right)^2 = \frac{\pi}{4} \left[\frac{\rho_s \Delta H_s}{\gamma_m \left(T_p - T_o \right)} \right]^2 s \left(t \right)^2 \tag{7.50}$$

This particular equation implies that a long solidification time contributes to a reduction of casting defects attributable to atomic diffusion since atoms would have more time to accommodate themselves at their preferred lattice sites. However, gas bubbles ahead of the solidification front can rise to the casting surface or may be trapped, to an extent, causing microporosity.

Furthermore, a long solidification time allows a riser to feed liquid metal to the solidifying melt and fortunately, shrinkage porosity is avoided or reduced to an extent. In the end, the morphology of the microstructure depends on the type of material being solidified and it has a direct impact on properties. Most castings are heat treated after solidification is complete in order to homogenize the microstructure so that properties are enhanced uniformly.

7.5.2   Fourier's Law of Heat Conduction Solution

In this subsection, the Fourier's law of conductive heat transfer is used to derive the temperature field equations. Assume that the temperature equations are [7]

$$T_m = T_o + (T_i - T_o)\,\text{erf}\left(\frac{-x}{2\sqrt{\alpha_m t}}\right) \quad \text{(mold)} \tag{7.51a}$$

$$T_s = T_i + \left(T_p - T_i\right)\text{erf}\left(\frac{x}{2\sqrt{\alpha_s t}}\right) \quad \text{(solid)} \tag{7.51b}$$

From Eq. (7.43c), the mold–solid intermediate temperature becomes

$$T_i = \frac{\gamma_s T_p + \gamma_m T_o}{\gamma_s + \gamma_m} \tag{7.52}$$

Note that T_i strongly depends on the initial melt and mold temperatures. Additionally, using Eq. (7.51a) and Fourier's law yields the amount of thermal energy crossing the ideal mold–solid (m-S) interface $x = 0$

$$Q_m = A\int\left(\frac{\partial T_m}{\partial x}\right)_{x=0} dt = \frac{2 A k_m\,(T_i - T_o)\,\sqrt{t}}{\sqrt{\pi \alpha_m}} \tag{7.53a}$$

$$Q_m = \frac{2 A \gamma_m}{\sqrt{\pi}}\,(T_i - T_o)\,\sqrt{t} \tag{7.53b}$$

At the planar L-S interface position $x = s(t)$, the rate and the amount of thermal energy being released during solidification are, respectively, defined as

$$\frac{Q_s}{dt} = A\rho_s \Delta H_s \frac{ds(t)}{dt} \tag{7.54a}$$

$$Q_s = A\rho_s \Delta H_s s(t) \tag{7.54b}$$

Equating $Q_m = Q_s$ yields the total solidification time t using mold properties and $\Delta T = T_i - T_o$

$$t = \frac{\pi}{4}\left[\frac{\rho_s \Delta H_s}{\gamma_m \Delta T}\right]^2 s(t)^2 \tag{7.55a}$$

$$t = \frac{\pi}{4}\left[\frac{\rho_s \Delta H_s}{\gamma_m\,(T_i - T_o)}\right]^2 s(t)^2 \tag{7.55b}$$

In summary, knowing the total solidification time t leads to calculations of variables such as $ds(t)/dt$, Q_m, Q_s, $\partial T/\partial x$, and $\partial T/\partial t$.

According to Eqs. (7.50) and (7.55), the solidification time strongly depends on the latent heat of solidification so that $t \propto (\Delta H_s)^2$. However, ΔH_s is treated as thermophysical property for a certain material and it, then, has a fixed value as well as γ_m and ΔT. Consequently, the solidification time is proportional to the square casting thickness. Hence, $t \propto s(t)^2 = x_c^2$.

Recall that $s(t)$ represents the position of the solidification L-S interface (solidification front) at t, x_c defines the location of the casting centerline and that solidification is complete when $x = s(t) = x_c$. In fact, this has been explained in Chap. 5 to an extent. Nonetheless, Eqs. (7.50) and (7.55) are the outcome of the current analytical procedure for deriving the theoretical solidification time, which is subjected to comparison with experimental data.

Example 7A.2 (Substrate and Solid Thermal Resistances) This example particularly illustrates the use of the transcendental equation and the Fourier's law of conduction for an *Al-20Si* (see given phase diagram) binary alloy sample. Use the published data from Xu et al. paper [7].

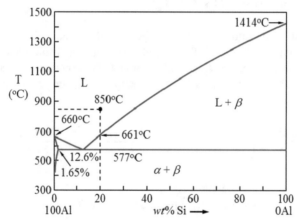

Assume that a 40-μm thick *Al-20Si* ribbon is to be produced on a rotating copper roller (substrate). Calculate (a) the solidification time t, (b) the solidification velocity dx_s/dt, (c) the solidification rate, and (d) the heat flux at the substrate–solid interface $x = 0$. The *Al-20Si* binary alloy is initially at its pouring temperature $T_p = 850\,°C$, while the copper mold is maintained at $T_o = 20\,°C$. The freezing temperatures of pure aluminum and pure silicon are $T_{Al} = 660\,°C$ and $T_{Si} = 1414\,°C$. The thermophysical properties and the *Al-20Si* equilibrium phase diagram are given below.

Material	k (W/m K)	ρ (kg/m³)	c_s (J/kg K)	ΔH_f (kJ/kg)
Al-20Si alloy	161	2627	854	$\Delta H_{Al} = 397$
Copper substrate	398	8930	386	$\Delta H_{Si} = 1790$

Solution

(a) **Similarity solution approach:**

$$\alpha_s = \frac{k_s}{\rho_s c_s} = 7.1764 \times 10^{-5}\,\text{m}^2\text{s}; \quad \alpha_m = \frac{k_m}{\rho_m c_m} = 1.1546 \times 10^{-4}\,\text{m}^2\text{s}$$

$$\gamma_m = \frac{k_m}{\sqrt{\alpha_m}} = \sqrt{k_m \rho_m c_m} = 37{,}040\,\frac{\text{W}\sqrt{\text{sec}}}{\text{m}^2\,\text{K}} = 37{,}040\,\frac{\text{J}}{\text{m}^2\,\text{K}\,\text{s}^{1/2}}$$

$$\gamma_s = \frac{k_s}{\sqrt{\alpha_s}} = \sqrt{k_s \rho_s c_s} = 19{,}005\,\frac{\text{W}\sqrt{\text{sec}}}{\text{m}^2\,\text{K}} = 19{,}005\,\frac{\text{J}}{\text{m}^2\,\text{K}\,\text{s}^{1/2}}$$

Using Eqs. (7.35b) with $C_o = 0.20$ and (7.46a) yields

$$\Delta H_s = \Delta H_{f,Al} + C_o \left| \Delta H_{f,Si} - \Delta H_{f,Al} \right| + c_s \left(T_p - T_E \right)$$

$$\Delta H_s = 9.0874 \times 10^5\,\text{J/kg}$$

This result takes into account the effect of superheating the liquid up to the pouring temperate $T_p > T_f$. Recall that the latent heat of solidification is greater than that for freezing, $\Delta H_s > \Delta H_f$, is now the thermophysical property being released at the solidification front at a certain velocity v. The transcendental equation gives

$$\lambda \exp\left(\lambda^2\right)\left[\text{erf}\left(\lambda\right) + \frac{\gamma_s}{\gamma_m}\right] = \frac{c_s\left(T_p - T_o\right)}{\sqrt{\pi}\,\Delta H_s} = \frac{(854)\,(850 - 20)}{\sqrt{\pi}\,(9.0874 \times 10^5)}$$

$$0 = \lambda \exp\left(\lambda^2\right)\left[\text{erf}\left(\lambda\right) + 0.51309\right] - 0.44007$$

$$\lambda = 0.39902$$

and from Eq. (7.44), the mold–solid interface (intermediate) temperature is calculated as

$$T_{i,1} = T_{i,2} = \frac{\gamma_s T_p + \gamma_m\,\text{erf}\left(\lambda\right) T_o}{\gamma_s + \gamma_m\,\text{erf}\left(\lambda\right)} = 472.79\,^\circ\text{C}$$

Now, the solidification time t can be calculated applying two different approaches. Using Eq. (7.42) with $\lambda = 0.39909$ and (7.55b) with $T_i = 472.79\,^\circ\text{C}$ yields, respectively,

$$t_1 = \frac{s\,(t)^2}{4\lambda^2 \alpha_s} = \frac{\left(40 \times 10^{-6}\,\text{m}\right)^2}{4\,(0.39908)^2\,(7.1764 \times 10^{-5})} = 3.50 \times 10^{-5}\,\text{s}$$

$$t_2 = \frac{\pi}{4} \left[\frac{\rho_s \Delta H_s}{\gamma_m \left(T_{i,2} - T_o \right)} \right]^2 s\,(t)^2 = 2.55 \times 10^{-5}\,\text{s}$$

Fourier's Law Solution Using Eqs. (7.52) and (7.55b) gives somewhat different intermediate temperature and solidification time. Thus,

$$T_{i,3} = \frac{\gamma_s T_p + \gamma_m T_o}{\gamma_s + \gamma_m} = 301.46\,^\circ\text{C}$$

$$t_3 = \frac{\pi}{4} \left[\frac{\rho_s \Delta H_s}{\gamma_m \left(T_{i,3} - T_o \right)} \right]^2 s\,(t)^2 = 6.59 \times 10^{-5}\,\text{s}$$

Therefore, the results for the solidification times are within reported values found in the literature.

(b) **Solidification velocity:**

$$\left[\frac{ds\,(t)}{dt} \right]_1 = \lambda \sqrt{\frac{\alpha_s}{t_1}} = 0.57137\,\text{m/s} = 571.37\,\text{mm/s}$$

From Eq. (7.39c),

$$\left(\frac{\partial T_s}{\partial x} \right)_{1,s(t)} = \frac{\left(T_p - T_{i,1} \right)}{\text{erf}\,(\lambda)\,\sqrt{\pi \alpha_s t_1}} \exp \left[-\frac{s\,(t)^2}{4\alpha_s t_1} \right] = 1.2320 \times 10^7\,^\circ\text{C/m}$$

$$\left(\frac{\partial T_s}{\partial x} \right)_{2,s(t)} = \frac{\left(T_p - T_{i,2} \right)}{\text{erf}\,(\lambda)\,\sqrt{\pi \alpha_s t_2}} \exp \left[-\frac{s\,(t)^2}{4\alpha_s t_2} \right] = 1.3602 \times 10^7\,^\circ\text{C/m}$$

and from Eq. (7.36e) with $\partial T_l / \partial x = 0$,

$$k_s \left(\frac{\partial T_s}{\partial x} \right)_{2,s(t)} = \rho_s \Delta H_s \left[\frac{ds\,(t)}{dt} \right]_2$$

$$\left[\frac{ds\,(t)}{dt} \right]_2 = \frac{k_s}{\rho_s \Delta H_s} \left(\frac{\partial T_s}{\partial x} \right)_{2,s(t)} = 0.91734\,\text{m/s} = 918.34\,\text{mm/s}$$

From Eqs. (7.51b) and (7.39c), respectively

$$T_s = T_{i,3} + \left(T_p - T_i \right) \text{erf} \left(\frac{x}{2\sqrt{\alpha_s t}} \right)$$

$$\left(\frac{\partial T_s}{\partial x} \right)_{3,s(t)} = \frac{\left(T_p - T_{i,3} \right)}{\sqrt{\pi \alpha_s t_3}} \exp \left[-\frac{s\,(t)^2}{4\alpha_s t_3} \right] = 4.1353 \times 10^6\,^\circ\text{C/m}$$

From Eq. (7.36e) with negligible liquid temperature gradient, the solidification rate at the solidification front is predicted to be

$$k_s \left(\frac{\partial T_s}{\partial x} \right)_{3,s(t)} = \rho_s \Delta H_s \left[\frac{ds\,(t)}{dt} \right]_3$$

$$\left[\frac{ds\,(t)}{dt} \right]_3 = \frac{k_s}{\rho_s \Delta H_s} \left(\frac{\partial T_s}{\partial x} \right)_{3,s(t)} = 0.27889 \,\text{m/s} = 278.89 \,\text{mm/s}$$

(c) **Solidification rate:** Using Eq. (7.39b) with $x = s\,(t)$ yields

$$\left[\frac{\partial T_s}{\partial t} \right]_1 = -\frac{\left(T_p - T_{i,1} \right) s\,(t)}{2 \,\text{erf}\,(\lambda)\, t_1 \sqrt{\pi \alpha_s t_1}} \exp \left[-\frac{s\,(t)^2}{4 \alpha_s t_1} \right] = -7.04 \times 10^6 \,{}^{\circ}\text{C/s}$$

$$\left[\frac{\partial T_s}{\partial t} \right]_2 = -\frac{\left(T_p - T_{i,2} \right) s\,(t)}{2 \,\text{erf}\,(\lambda)\, t_2 \sqrt{\pi \alpha_s t_2}} \exp \left[-\frac{s\,(t)^2}{4 \alpha_s t_2} \right] = -10.67 \times 10^6 \,{}^{\circ}\text{C/s}$$

From Eq. (7.51b) with $T_{i,3} = 301.46\,{}^{\circ}\text{C}$,

$$T_s = T_{i,3} + \left(T_p - T_i \right) \text{erf} \left[\frac{s\,(t)}{2\sqrt{\alpha_s t}} \right]$$

$$\left[\frac{\partial T_s}{\partial t} \right]_3 = -\frac{\left(T_p - T_{i,3} \right) s\,(t)}{2 t_3 \sqrt{\pi \alpha_s t_3}} \exp \left[-\frac{s\,(t)^2}{4 \alpha_s t_3} \right] = -1.26 \times 10^6 \,{}^{\circ}\text{C/s}$$

(d)   The heat flux, Eq. (7.37), at $x = 0$ is

$$q_{s,1} = -k_s \left(\frac{\partial T_s}{\partial x} \right)_{1,x=0} = \frac{k_s \left(T_p - T_{i,1} \right)}{\text{erf}\,(\lambda)\, \sqrt{\pi \alpha_s t_1}} = 1.60 \,\text{GW/} \left(\text{m}^2\, \text{s} \right)$$

$$q_{s,2} = -k_s \left(\frac{\partial T_s}{\partial x} \right)_{2,x=0} = \frac{k_s \left(T_p - T_{i,2} \right)}{\text{erf}\,(\lambda)\, \sqrt{\pi \alpha_s t_2}} = 1.87 \,\text{GW/} \left(\text{m}^2\, \text{s} \right)$$

$$q_{s,3} = -k_s \left(\frac{\partial T_s}{\partial x} \right)_{3,x=0} = \frac{k_s \left(T_p - T_{i,3} \right)}{\text{erf}\,(\lambda)\, \sqrt{\pi \alpha_s t_3}} = 1.70 \,\text{GW/} \left(\text{m}^2\, \text{s} \right)$$

Therefore, it can be assumed that all results are within reasonable limits.

7.6   Coupled Heat and Mass Transport Model

As for pure metals, the Rubinstein solidification model [8] takes into account conductive heat transfer and mass transport, and assumes a similarity solution at the solidification front in A-B binary alloys. The corresponding analytical solution for binary alloy solidification is an extension of the classical Stefan problem. Chronologically, the Rubinstein model published in 1971 [8] is analyzed or discussed along with valuable cited bibliographies by Crank in 1984 [9, chapters 1 and 3], Alexiades-Solomon in 1993 [6, pp. 105–107], Voller in 1997 [10, pp. 2869–2877], Voller in 2006 [11, pp. 1981–1955], Voller 2008 [12, pp. 696–706], and Tarzia in 2011 [13, pp. 463–465].

Additionally, in order to be consistent with the theoretical approach used in Chap. 5, the Rubinstein model [8] is analyzed in a half-space region $x > 0$ within a semi-infinite domain $0 < x < \infty$. This model considers (1) thermal resistances in the solid and liquid phases along with solute diffusion, and (2) excludes any mold–solid interface interaction at $x = 0$ by assuming perfect thermal contact and the effect of thermal gradient in the mold material. Moreover, the solidification problems for casting alloys are based on the general temperature function $T = f(n)$ as the *similarity solution* to a partial differential equation (PDE) with $n = x/\sqrt{4\alpha t}$ as the *similarity variable*. Fundamentally, a similarity solution strongly depends on position x and time t as indicated by $x/\sqrt{t}$. Recall that the analytical procedure for deriving the similarity solutions is given in Chap. 5.

Two-Phase Problem For a semi-infinite slab of a particular A-B binary alloy with solute B composition C_o, the solid phase is in the half-space $0 < x < s(t)$, the liquid is in the region $s(t) < x < \infty$ and the boundary conditions are based on a smooth and flat (or sharp interface) planar L-S interface boundary located at a distance defined by the characteristic length $x = s(t)$.

For one-dimensional heat transfer analysis with an ideal mold–solid contact surface area, the governing heat equations for the two-phase solidification problem, continuity of temperature at the planar L-S interface, the Fourier heat flux balance $(q_s - q_l = q_f)$, and the diffusion flux balance for solidifying an A-B binary alloy slab are, respectively, [8, pp. 54–55]

$$\frac{\partial T_m}{\partial t} = \alpha_m \frac{\partial^2 T_m}{\partial x^2} \quad \text{for } -\infty < x \leq 0 \tag{7.56a}$$

$$\frac{\partial T_s}{\partial t} = \alpha_s \frac{\partial^2 T_s}{\partial x^2}; \quad \frac{\partial C_s}{\partial t} = D_s \frac{\partial^2 T_s}{\partial x^2} \tag{7.56b}$$

$$\frac{\partial T_l}{\partial t} = \alpha_l \frac{\partial^2 T_l}{\partial x^2}; \quad \frac{\partial C_l}{\partial t} = D_l \frac{\partial^2 T_l}{\partial x^2} \tag{7.56c}$$

$$T_s\left[C_s\left(T_f^*\right)\right] = T_l\left[C_l\left(T_f^*\right)\right] = T_f^* \text{ at } x = s(t) \tag{7.56d}$$

$$k_s \frac{\partial T_s}{\partial x} - k_l \frac{\partial T_l}{\partial x} = \rho_s \Delta H_s \frac{ds(t)}{dt} \text{ at } x = s(t), t > 0 \tag{7.56e}$$

$$\left[C_l\left(T_f^*\right) - C_s\left(T_f^*\right)\right] \frac{ds(t)}{dt} = D_s \frac{\partial C_s}{\partial x} - D_l \frac{\partial C_l}{\partial x} \text{ at } x = s(t) \tag{7.56f}$$

Note that Eqs. (7.56a)–(7.56f) clearly indicate that the Rubinstein model for alloy solidification involves the heat diffusion and atomic diffusion fields. Thus, heat transfer and mass transport are combined to solve binary alloy solidification problems. Moreover, as for pure metals, a superheating temperature $T_p \rightarrow T_\infty$ can be imposed on the melt so that the latent heat of solidification becomes $\Delta H_s > \Delta H_f^*$.

Although Eq. (7.56e) represents the heat flux balance, it is also the heat flux boundary condition at the interface $x = s(t)$. This is attributed to the heat flux arising from the latent heat associated with the phase transformation and liquid diffusion, where $\rho_s dx(t)/dt$ represents the mass rate of phase transition per unit area of interface and $\rho_s \Delta H_s dx(t)/dt$ represents the total heat release per unit area.

Furthermore, ΔH_f^* has been replaced by ΔH_s in Eq. (7.56e) since ΔH_s accounts for the removal of sensible and latent heats when the alloy is initially superheated at $T_p > T_f^*$.

7.6.1   Dimensional Analysis

Next, a sequence of groups of suitable equations are given below for deriving the transcendental equations involved in binary alloy solidification. Hence, λ and T_f^* are the critical unknown variables. For the sake of convenience, the characterization of solidification using the coupled heat and mass transport approach for a slab is carried out using the dimensional analysis and the dimensionless analysis methods. The former method requires known slab thickness or characteristic length $s(t)$ and the latter is an analytical procedure for scaling the governing equations without the characteristic length.

Temperature Field Equations   The one-dimensional analytical solutions of the governing heat equations defined by Eqs. (7.56a)–(7.56c) for semi-infinite regions with $0 < x < \infty$ are adopted henceforth. These analytical solutions are appropriate for characterizing the solidification process in pure metals at a constant freezing temperature T_f. Interestingly, the resultant temperature field equations defined by Eqs. (5.15a), (5.17a), (5.20a), and (5.21e) are recast here for the solidification of binary alloys, provided that T_f is replaced by T_f^*. Accordingly, following the analytical procedure from Chap. 5, the similarity solutions for the actual temperature field equations due to heat flux flow are

$$T_m = T_o + (T_i - T_o)\,\mathrm{erfc}\left(\frac{-x}{2\sqrt{\alpha_m t}}\right) \qquad \text{(mold)} \qquad (7.57a)$$

$$T_s = T_i + \frac{\left(T_f^* - T_i\right)}{\mathrm{erf}(\lambda)}\,\mathrm{erf}\left(\frac{x}{2\sqrt{\alpha_s t}}\right) \qquad \text{(solid)} \qquad (7.57b)$$

$$T_l = T_p - \frac{\left(T_p - T_f^*\right)}{\mathrm{erfc}\left(\lambda\sqrt{\alpha_s/\alpha_l}\right)}\,\mathrm{erfc}\left(\frac{x}{2\sqrt{\alpha_l t}}\right) \quad \text{(liquid)} \qquad (7.57c)$$

$$T_i = \frac{\gamma_s T_f + \gamma_m T_o\,\mathrm{erf}\,(\lambda)}{\gamma_s + \gamma_m\,\mathrm{erf}\,(\lambda)} \qquad \text{(mold-solid)} \qquad (7.57d)$$

Furthermore, the freezing temperature T_f for a pure metal, as defined in Chap. 5, is a time-dependent variable $T_f = T_f(t)$, while $T_f^* = T_f^*\left(C_l^*\right)$ for an alloy is a concentration-dependent variable, which is known as the critical (intermediate) temperature. Moreover, T_i in Eq. (7.57d) is the temperature between the mold–solid interface and T_o is the initial mold temperature. Both T_i and T_o are assumed constant during the solidification process.

Solute Distribution Equations Assume that the analytical solution of the diffusion equation in the liquid state at $T_s < T_f^* < T_l$ admits a similarity solution. The general equation for the liquid concentration C_l and boundary conditions are

$$C_l = A_1 + A_2\,\mathrm{erfc}\left(\frac{x}{2\sqrt{D_l t}}\right) \quad \text{for } s\,(t) < x, t > 0 \qquad (7.58a)$$

$$\text{BC1 } C = C\,(\infty, t) = C_o \qquad (7.58b)$$

$$C_o = A_1 + A_2\,\mathrm{erfc}\,(\infty) \quad \rightarrow \quad A_1 = C_o \qquad (7.58c)$$

$$\text{BC2 } C_l = C\,[s\,(t), t] = C_l^* \qquad (7.58d)$$

$$C_l^* = C_o + A_2\,\mathrm{erfc}\left(\frac{s\,(t)}{2\sqrt{D_l t}}\right) \qquad (7.58e)$$

$$C_l^* = C_o + A_2\,\mathrm{erfc}\left(\frac{s\,(t)}{2\sqrt{D_l t}}\frac{\sqrt{\alpha_s}}{\sqrt{\alpha_s}}\right) \qquad (7.58f)$$

$$C_l^* = C_o + A_2\,\mathrm{erfc}\left(\lambda\sqrt{\frac{\alpha_s}{D_l}}\right) \qquad (7.58g)$$

$$A_2 = \frac{C_l^* - C_o}{\mathrm{erfc}\left(\lambda\sqrt{\alpha_s/D_l}\right)} \qquad (7.58h)$$

Assuming that there is no solute diffusion in the solid and substituting A_1, A_2 into Eq. (7.58a) yields

$$C_s = 0 \qquad (7.59a)$$

$$C_l = C_o + \frac{C_l^* - C_o}{\mathrm{erfc}\left(\lambda\sqrt{L_e}\right)}\,\mathrm{erfc}\left(\frac{x}{2\sqrt{D_l t}}\right) \qquad (7.59b)$$

where $L_e = \alpha_s/D_l$ denotes the Lewis number. The distribution of the solute B in the solid is taken as $C_s = 0$ since the Rubinstein solidification model does not considered solute diffusion in the solid state due to a stable solidified melt forming a specific crystalline structure during solidification.

Solute Concentrations Assume that a specific amount of solute atoms accommodate in the growing crystal structure and an excess solute B at the planar L-S interface diffuses into the liquid phase. Consequently, the solute concentration in the liquid and solid phases can be generalized as temperature-dependent concentrations for solidifying a binary alloy, where $x = s(t)$ and $T = T_f^*$. This implies that $C_l^* = C_l^*\left(T_f^*\right)$ and $C_s^* = C_s^*\left(T_f^*\right)$. Thus, the concentrations of the solute B in the liquid and solid phase at $T_s < T_f^* < T_l$ can be determined from a binary phase diagram, provided that the freezing temperature T_f^* and the alloy nominal composition C_o are known entities.

At this moment, T_f^* is not known and the solute concentrations are conveniently recast from Eqs. (7.6a) and (7.9a) as, respectively,

$$C_s^* = k_o C_l^* \tag{7.60a}$$

$$C_l^* = -\frac{T_A - T_f^*}{\beta_l} \tag{7.60b}$$

Notice that the partition coefficient k_o and the slope β_l of the liquidus line of a binary equilibrium phase diagram are fundamentally important variables that must be known a priori.

Solidification Front Position At any solidification time $t > 0$, the position of the L-S is referred to as the solidification front and the corresponding rate of the evolving planar L-S interface are defined as

$$s(t) = 2\lambda\sqrt{\alpha_s t} \tag{7.61a}$$

$$\frac{ds(t)}{dt} = \lambda\sqrt{\frac{\alpha_s}{t}} \tag{7.61b}$$

Temperature and Concentration Gradients Consider heat transfer by conduction at a smooth and flat planar L-S interface. In this case, differentiate Eqs. (7.57b), (7.57c) with respect to x to get the temperature gradients written as

$$\frac{\partial T_s}{\partial x} = \frac{\left(T_f^* - T_o\right)}{\sqrt{\pi \alpha_s t}\,\text{erf}(\lambda)}\exp\left(-\frac{x^2}{4\alpha_s t}\right) \tag{7.62a}$$

$$\frac{\partial T_l}{\partial x} = -\frac{T_f^* - T_p}{\sqrt{\pi \alpha_l t}\,\text{erfc}\left(\lambda\sqrt{\alpha_s/\alpha_l}\right)}\exp\left(-\frac{x^2}{4\alpha_l t}\right) \tag{7.62b}$$

and from (7.59) with constant $C_s = C_s\left(T_f^*\right)$, the concentration gradients written as

$$\frac{\partial C_s}{\partial x} = 0 \quad \text{at } x < x_s\,(t) \tag{7.63a}$$

$$\frac{\partial C_l}{\partial x} = -\frac{C_l^* - C_o}{\sqrt{\pi D_l t}\, \text{erfc}\left(\lambda\sqrt{\alpha_s/D_l}\right)} \exp\left(-\frac{x^2}{4D_l t}\right) \tag{7.63b}$$

Recall that Eq. (7.56e) defines the heat flux balance at the planar L-S interface and for convenience, the corresponding heat flux terms are individually written as

$$q_s = -k_s \frac{\partial T_s}{\partial x} \qquad \text{at } x = s\,(t)\,, t > 0 \tag{7.64a}$$

$$q_l = -\frac{\partial T_l}{\partial x} \qquad \text{at } x = s\,(t)\,, t > 0 \tag{7.64b}$$

$$q_f = -\rho_s \Delta H_s \frac{ds\,(t)}{dt} \quad \text{at } x = s\,(t)\,, t > 0 \tag{7.64c}$$

Transcendental Equation Due to Heat Diffusion Substituting Eqs. (7.62) into (7.56e) yields the first coupled transcendental equations for T_f^* due to heat conduction and λ due to diffusion of the solute B from the planar L-S interface into the liquid phase

$$\frac{\left(T_f^* - T_o\right)}{\lambda \exp\left(\lambda^2\right) \text{erf}\,(\lambda)} + \frac{\gamma_l\left(T_f^* - T_p\right) \exp\left(-\lambda^2 \alpha_s/\alpha_l\right)}{\gamma_s \lambda\, \text{erfc}\left(\lambda\sqrt{\alpha_s/\alpha_l}\right)} = \frac{\sqrt{\pi}\,\Delta H_s}{C_s} \tag{7.65}$$

Solving Eq. (7.65) for the unknown T_f^* yields

$$T_f^* = \frac{\sqrt{\pi}\,\rho_s \Delta H_s + T_o B_1\,(\lambda) + T_p B_2\,(\lambda)}{B_1\,(\lambda) + B_2\,(\lambda)} \tag{7.66}$$

with constants $B_1\,(\lambda)$ and $B_2\,(\lambda)$ being written as

$$B_1\,(\lambda) = \frac{k_s}{\alpha_s \lambda \exp\left(\lambda^2 \alpha_s/\alpha_l\right) \text{erf}\left(\lambda\sqrt{\alpha_s/\alpha_l}\right)} \tag{7.67a}$$

$$B_2\,(\lambda) = \frac{k_l}{\sqrt{\alpha_s \alpha_l}\,\lambda \exp\left(\lambda^2 \alpha_s/\alpha_l\right) \text{erfc}\left(\lambda\sqrt{\alpha_s/\alpha_l}\right)} \tag{7.67b}$$

Transcendental Equation Due to Solute Diffusion Substituting Eq. (7.63) into (7.56f) yields a second transcendental equation for λ

$$\sqrt{\pi}\lambda\sqrt{\frac{\alpha_s}{D_l}}\exp\left(\lambda^2\alpha_s/D_l\right)\text{erfc}\left(\lambda\sqrt{\alpha_s/D_l}\right) = \frac{C_l^* - C_o}{C_l^* - C_s^*} \tag{7.68a}$$

$$\sqrt{\pi L_e}\lambda\exp\left(\lambda^2 L_e\right)\text{erfc}\left(\lambda\sqrt{L_e}\right) = \frac{C_l^* - C_o}{C_l^* - C_s^*} \tag{7.68b}$$

Again, a convergent iterative process must be used for evaluating λ and C_f^*. Combining Eqs. (7.60) and (7.68b) yields the transcendental equation for the unknown T_f^*

$$T_f^* = T_A + \frac{\beta_l C_o}{1 + (k_o - 1) B_3(\lambda)} \tag{7.69a}$$

$$B_3(\lambda) = \sqrt{\pi L_e}\lambda\exp\left(\lambda^2 L_e\right)\text{erfc}\left(\lambda\sqrt{L_e}\right) \tag{7.69b}$$

Now, equating Eqs. (7.66) and (7.69a) yields the transcendental equation for λ only

$$\frac{\sqrt{\pi}\rho_s\Delta H_s + T_o B_1(\lambda) + T_p B_2(\lambda)}{B_1(\lambda) + B_2(\lambda)} = T_A + \frac{\beta_l C_o}{1 + (k_o - 1) B_3(\lambda)} \tag{7.70}$$

Once λ is evaluated using Eq. (7.70), one can proceed to calculate the remaining unknown variables T_f^* from (7.69a) and C_l^*, C_s^* from (7.60).

7.6.2 Dimensionless Analysis

The following analytical approach can be found in Dantzig-Rappaz book [1, pp. 173–180]. The purpose of this approach is to give the reader insights on the analytical procedure for scaling the governing equations using normalized variables and the Lewis number L_e. Using the casting half-thickness x_c and the characteristic length $s(t)$, the normalized variables (dimensionless variables) are

$$\delta = \frac{x}{x_c}; \quad \delta(\tau) = \frac{s(t)}{x_c}; \quad \tau = \frac{\alpha t}{x_c^2} \tag{7.71a}$$

$$\theta_s = \frac{T_s - T_i}{T_\infty - T_i} \quad \text{(solid)} \quad \& \quad \theta_l = \frac{T_l - T_i}{T_\infty - T_i} \quad \text{(liquid)} \tag{7.71b}$$

$$\theta_f = \frac{T_f - T_i}{T_\infty - T_i} \quad \text{(pure metal)} \tag{7.71c}$$

$$\theta_f^* = \frac{T_f^* - T_i}{T_\infty - T_i} \quad \text{(binary alloy)} \tag{7.71d}$$

The main reason for normalizing and scaling is to simplify equations and minimize round-off error. For instance, $0 < \delta < 1$ or $0 < \delta^* < 1$ allows minimization of round-off error as compared to $0 < s(t) \le x_c$.

7.6.3 Temperature Scaling

Combining Eqs. (7.56b), (7.56c), and (7.71b) yields the new dimensionless governing equations along with some conditions

$$\frac{\partial \theta_s}{\partial \tau} = \frac{\partial^2 \theta_s}{\partial \delta^2} \quad \text{for } 0 \le \delta \le \delta(\tau) \tag{7.72a}$$

$$\frac{\partial \theta_l}{\partial \tau} = \frac{\partial^2 \theta_l}{\partial \delta^2} \quad \text{for } \delta(\tau) \le \delta \le \infty \tag{7.72b}$$

$$\frac{\partial \theta_s}{\partial \delta} - \frac{\partial \theta_l}{\partial \delta} = \frac{1}{S_t} \frac{d\delta(\tau)}{d\tau} \quad \text{for } \delta = \delta(\tau) \tag{7.72c}$$

$$\delta(\tau) = 0 \quad \text{at } \tau = 0 \tag{7.72d}$$

$$\theta_s = 1 \quad \text{at } \delta = 0 \tag{7.72e}$$

$$\theta_s = \theta_f^* \quad \text{at } \delta = \delta(\tau) \tag{7.72f}$$

$$\theta_l = 1 \quad \text{at } \delta \to \infty \tag{7.72g}$$

$$\theta_l = 1 \quad \text{at } \tau = 0 \tag{7.72h}$$

where the Stefan number S_t and the dimensionless similarity variable $\delta(\tau)$ are defined by

$$S_t = \frac{c_s (T_p - T_o)}{\Delta H_s} \tag{7.73a}$$

$$\delta(\tau) = 2\lambda\sqrt{\tau} \tag{7.73b}$$

The similarity solutions for the dimensionless solid and liquid phases can be derived using the method of separation of variables cited in Appendix 5B. The resultant dimensionless temperature expressions are

$$\theta_s = \frac{\theta_f^*}{\text{erf}(\lambda)} \text{erf}\left(\frac{\delta}{2\sqrt{\tau}}\right) = \frac{\theta_f^*}{\text{erf}(\lambda)} \text{erf}\left(\frac{\lambda\delta}{\delta(\tau)}\right) \tag{7.74a}$$

$$\theta_l = 1 - \frac{1 - \theta_f^*}{\text{erfc}(\lambda)} \text{erfc}\left(\frac{\delta}{2\sqrt{\tau}}\right) = 1 - \frac{1 - \theta_f^*}{\text{erfc}(\lambda)} \text{erfc}\left(\frac{\lambda\delta}{\delta(\tau)}\right) \tag{7.74b}$$

Combining Eqs. (7.72c) and (7.74) yields the first transcendental function $\theta_f^* = f(\lambda)$, from which $d\theta_f^*/d\lambda = 0$ for finding the λ equation

$$\theta_f^* = \operatorname{erf}(\lambda)\left[1 + \frac{\sqrt{\pi}}{S_t}\lambda\exp\left(\lambda^2\right)\right] \tag{7.75a}$$

$$\theta_f^* = \operatorname{erf}(\lambda) + \frac{\sqrt{\pi}}{S_t}\lambda\operatorname{erf}(\lambda)\exp\left(\lambda^2\right) \tag{7.75b}$$

$$\frac{d\theta_f^*}{d\lambda} = \frac{2}{\sqrt{\pi}}\exp\left(-\lambda^2\right) + \frac{\sqrt{\pi}}{S_t}\left[\operatorname{erf}(\lambda)\exp\left(\lambda^2\right) + 2\lambda^2\operatorname{erf}(\lambda)\exp\left(\lambda^2\right) + \frac{2\lambda}{\sqrt{\pi}}\right] \tag{7.75c}$$

$$\frac{d\theta_f^*}{d\lambda} = \frac{2S_t}{\pi}\exp\left(-\lambda^2\right) + \operatorname{erf}(\lambda)\exp\left(\lambda^2\right) + 2\lambda^2\operatorname{erf}(\lambda)\exp\left(\lambda^2\right) + \frac{2\lambda}{\sqrt{\pi}} = 0 \tag{7.75d}$$

$$0 = \frac{2S_t}{\pi}\exp\left(-\lambda^2\right) + \operatorname{erf}(\lambda)\exp\left(\lambda^2\right) + 2\lambda^2\operatorname{erf}(\lambda)\exp\left(\lambda^2\right) + \frac{2\lambda}{\sqrt{\pi}} \tag{7.75e}$$

7.6.4 Solute Concentration Scaling

Now, applying the scaling given in Eq. (7.71a)–(7.71d) to (7.56b), (7.56c) for the solute diffusion counterparts gives the set of equations and conditions

$$\frac{\partial C_s}{\partial\tau} = \frac{D_s}{\alpha_s}\frac{\partial^2 C_s}{\partial\delta^2} \quad \text{for } 0 \leq \delta \leq \delta(\tau) \tag{7.76a}$$

$$\frac{\partial C_l}{\partial\tau} = \frac{D_l}{\alpha_l}\frac{\partial^2 C_l}{\partial\delta^2} \quad \text{for } \delta(\tau) \leq \delta < \infty \tag{7.76b}$$

$$\frac{D_s}{\alpha_s}\frac{\partial C_s}{\partial\delta} - \frac{D_l}{\alpha_l}\frac{\partial C_l}{\partial\delta} = C_l^*(1 - k_o)\frac{d\delta(\tau)}{d\tau} \quad \text{for } \delta = \delta(\tau) \tag{7.76c}$$

$$\frac{\partial C_s}{\partial\delta} = 0 \qquad \text{for } \delta = 0 \tag{7.76d}$$

$$C_s^* = k_o C_l^* \qquad \text{for } \delta = \delta(\tau) \tag{7.76e}$$

$$C_l^* = C_o \qquad \text{for } \delta \to \infty \tag{7.76f}$$

$$C_l^* = C_o \qquad \text{for } \tau = 0 \tag{7.76g}$$

According to the condition given by Eq. (7.63a), $\partial C_s/\partial\delta = \partial C_s/\partial x = 0$ and consequently, (7.76b), (7.76c) along with $L_e = \alpha_l/D_l$ become

$$\frac{\partial C_l}{\partial \tau} = L_e^{-1} \frac{\partial^2 C_l}{\partial \delta^2} \qquad \text{for } \delta(\tau) \leq \delta < \infty \tag{7.77a}$$

$$-L_e^{-1} \frac{\partial C_l}{\partial \delta} = C_l^* (1 - k_o) \frac{d\delta(\tau)}{d\tau} \qquad \text{for } \delta = \delta(\tau) \tag{7.77b}$$

with

$$C_l^* = \frac{T_i - T_\infty}{\beta_l} \left(\theta_f - \theta_f^* \right) \tag{7.78}$$

Substituting $L_e = \alpha_l / D_l$ and Eq. (7.71a) into (7.59b) gives the solution

$$C_l = C_o + \frac{C_l^* - C_o}{\text{erfc}\left(\lambda\sqrt{L_e}\right)} \text{erfc}\left(\frac{x}{2\sqrt{D_l t}}\right) \tag{7.79a}$$

$$C_l = C_o + \frac{C_l^* - C_o}{\text{erfc}\left(\lambda\sqrt{L_e}\right)} \text{erfc}\left[\frac{\delta x_c}{2\sqrt{D_l \left(\tau x_c^2 / \alpha_l\right)}}\right] \tag{7.79b}$$

$$C_l = C_o + \frac{C_l^* - C_o}{\text{erfc}\left(\lambda\sqrt{L_e}\right)} \text{erfc}\left(\frac{\delta\sqrt{L_e}}{2\sqrt{\tau}}\right) \tag{7.79c}$$

Combining Eqs. (7.75a), (7.77b) and (7.78) yields

$$\left[\theta_f - B_4(\lambda)\right] B_5(\lambda) = \frac{\beta_l C_o}{T_i - T_\infty} \tag{7.80}$$

where

$$B_4(\lambda) = \text{erf}(\lambda)\left[1 + \frac{\sqrt{\pi}}{S_t}\lambda \exp\left(\lambda^2\right)\text{erfc}(\lambda)\right] \tag{7.81a}$$

$$B_5(\lambda) = \left[1 + \sqrt{\pi L_e}(k_o - 1)\lambda \exp\left(\lambda^2 L_e\right)\text{erfc}\left(\lambda\sqrt{L_e}\right)\right] \tag{7.81b}$$

$$f(\lambda) = \left[\theta_f - B_4(\lambda)\right] B_5(\lambda) \tag{7.81c}$$

$$f(\lambda) = \frac{\beta_l C_o}{T_i - T_\infty} \tag{7.81d}$$

Finally, evaluate λ from Eq. (7.80), θ_f^* from (7.75a) and subsequently, calculate other variables as desired. For instance, Fig. 7.9 shows the trend of the dimensionless temperature, $\theta_f = f(\lambda)$, described by Eq. (7.80) for fixed k_o, S_t, β_l, $T_i = 1480.60\,°C$, $T_p = 1540\,°C$ and some L_e values.

Example 7A.3 Consider pouring liquid Fe-0.08C steel at $T_p = 1540\,°C$ into a sand mold with a slab cavity. The freezing temperature of pure iron in accordance with the Fe-C phase diagram shown in Fig. 7.2b is $T_f = T_A = 1538\,°C$. Assume that

Fig. 7.9 Dimensionless
temperature $\theta_f = f(\lambda)$
profile for fixed k_o, S_t, β_l, T_i,
T_p and L_e

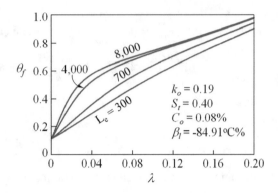

the diffusion coefficient of carbon into the liquid steel is $D_l = 2 \times 10^{-8}\,\mathrm{m^2/s}$, and
that the steel and the sand mold thermophysical properties given below represent
average values. Let $c_{p,s} = c_s$ and $c_{p,l} = c_l$.

Phase	T (°C)	k (W/m K)	ρ (kg/m³)	c_p (J/kg K)	ΔH_f^* (kJ/kg)
Solid steel		60	7872	678	
Liquid steel	1540	296	7000	680	276
Sand mold	25	0.40	1550	986	

Determine the temperature profiles at 5, 10, 20, and 35 s.

Solution From Fig. 7.2b with $T_1 = 1426\,°\mathrm{C}$ and $T_\mathrm{per} = 1493\,°\mathrm{C}$, and $T_f = T_{Fe} = 1538\,°\mathrm{C}$

$$\beta_l = \frac{T_f - T_\mathrm{per}}{0 - C_l} = \frac{1538 - 1493}{-0.53} = -84.91\,°\mathrm{C/mass\%}$$

$$\beta_s = \frac{T_f - T_\mathrm{per}}{0 - C_s} = \frac{1538 - 1493}{-0.09} = -500\,°\mathrm{C/mass\%}$$

$$k_o = \frac{\beta_l}{\beta_s} = \frac{(T_f - T_\mathrm{per})/C_l}{(T_f - T_\mathrm{per})/C_s} = \frac{0.09}{0.53} = 0.17$$

The thermal diffusivity terms are

$$\alpha_m = \frac{k_m}{\rho_m c_m} = 2.6173 \times 10^{-7}\,\mathrm{m^2/s}$$

$$\alpha_s = \frac{k_s}{\rho_s c_s} = 1.1242 \times 10^{-5}\,\mathrm{m^2/s}$$

$$\alpha_l = \frac{k_l}{\rho_l c_l} = 6.2185 \times 10^{-5}\,\mathrm{m^2/s}$$

The thermal inertia terms and the Lewis number are

$$\gamma_m = \frac{k_m}{\sqrt{\alpha_m}} = 781.87 \, \text{J} / \left(\text{m}^2 \, \text{K} \sqrt{s} \right)$$

$$\gamma_s = \frac{k_s}{\sqrt{\alpha_s}} = 17{,}895 \, \text{J} / \left(\text{m}^2 \, \text{K} \sqrt{s} \right)$$

$$\gamma_l = \frac{k_l}{\sqrt{\alpha_l}} = 37{,}536 \, \text{J} / \left(\text{m}^2 \, \text{K} \sqrt{s} \right)$$

$$L_e = \frac{\alpha_s}{D_l} = 562.10$$

Dimensional Analysis

Now the constants $B_i (\lambda)$ can be simplified for later use, where $i = 1, 2, 3$. From Eqs. (7.67) and (7.69b),

$$B_1 (\lambda) = \frac{k_s}{\alpha_s \lambda \exp \left(\lambda^2 \alpha_s / \alpha_l \right) \operatorname{erf} \left(\lambda \sqrt{\alpha_s / \alpha_l} \right)}$$

$$B_1 (\lambda) = \frac{5.3371 \times 10^6}{\lambda \left(1 - \operatorname{erfc} (0.42519\lambda) \right) \exp \left(0.18078\lambda^2 \right)}$$

$$B_2 (\lambda) = \frac{k_l}{\sqrt{\alpha_s \alpha_l} \lambda \exp \left(\lambda^2 \alpha_s / \alpha_l \right) \operatorname{erfc} \left(\lambda \sqrt{\alpha_s / \alpha_l} \right)}$$

$$B_2 (\lambda) = \frac{1.1195 \times 10^7}{\lambda \exp \left(0.18078\lambda^2 \right) \operatorname{erfc} (0.42519\lambda)}$$

$$B_3 (\lambda) = \sqrt{\pi L_e} \lambda \exp \left(\lambda^2 L_e \right) \operatorname{erfc} \left(\lambda \sqrt{L_e} \right)$$

$$B_3 (\lambda) = 42.022\lambda \exp \left(562.1\lambda^2 \right) \operatorname{erfc} (23.709\lambda)$$

Substituting these constants into Eq. (7.70) yields the root λ of the transcendental with $T_o = 25 \,^\circ\text{C}$ and $T_p = 1540 \,^\circ\text{C}$

$$0 = \frac{\sqrt{\pi} \rho_s \Delta H_s + T_o B_1 (\lambda) + T_p B_2 (\lambda)}{B_1 (\lambda) + B_2 (\lambda)} - T_A - \frac{\beta_l C_o}{1 + (k_o - 1) B_3 (\lambda)}$$

$$\lambda = 1.8096$$

Inserting λ into Eq. (7.69a) gives the steel freezing temperature T_f^* and in turn, Eq. (7.60) yields

$$T_f^* = T_f + \frac{\beta_l C_o}{1 + (k_o - 1) B_3 (\lambda)} = 1500.10 \,^\circ\text{C}$$

$$C_l^* = -\frac{T_f - T_f^*}{\beta_l} = -\frac{1538\,^\circ\text{C} - 1500.10\,^\circ\text{C}}{-84.91\,^\circ\text{C/mass\%}} = 0.45\%$$

$$C_s^* = k_o C_l^* = (0.17)\,(0.45\%) = 0.08\%$$

Now, the temperature at the mold–solid interface, Eq. (7.57d), is

$$T_i = \frac{\gamma_s T_f^* + \gamma_m T_o \, \text{erf}\,(\lambda)}{\gamma_s + \gamma_m \, \text{erf}\,(\lambda)} = 1480.60\,^\circ\text{C}$$

As expected, $T_i < T_f$ and $T_i < T_f^*$.

Dimensionless Analysis

This is based on normalized variables defined by Eq. (7.71). Using Eqs. (7.71c), (7.73a), and (7.81a), (7.81b) yields the dimensionless temperature parameter and the Stefan number. Calculations give

$$\theta_f = \frac{T_f - T_i}{T_p - T_i} = 0.96633 \quad \& \quad S_t = \frac{c_s\,(T_p - T_i)}{\Delta H_f^*} = 145.92$$

and the $B_4\,(\lambda)$ and $B_5\,(\lambda)$ constants are

$$B_4\,(\lambda) = \text{erf}\,(z)\left(1 + \frac{\sqrt{\pi}}{S_t} z \exp\left(z^2\right) \text{erfc}\,(z)\right)$$

$$B_4\,(\lambda) = \left[5.9082\lambda \, \text{erfc}\,(\lambda) \exp\left(\lambda^2\right) + 1\right][1 - \text{erfc}\,(\lambda)]$$

$$B_5\,(\lambda) = 1 + \sqrt{\pi L_e}\,(k_o - 1)\,\lambda \exp\left(\lambda^2 L_e\right) \text{erfc}\left(\lambda\sqrt{L_e}\right)$$

$$B_5\,(\lambda) = 1 - 34.879\lambda \exp\left(562.1\lambda^2\right) \text{erfc}\,(23.709\lambda)$$

Then, Eq. (7.80) gives

$$\left[\theta_f - B_4\,(\lambda)\right] B_5\,(\lambda) = \frac{\beta_l C_o}{T_i - T_p}$$

$$\lambda = 0.17737$$

and (7.75a) yields

$$\theta_f^* = \text{erf}\,(\lambda)\left[1 + \frac{\sqrt{\pi}}{S_t}\lambda \exp\left(\lambda^2\right) \text{erfc}\,(\lambda)\right] = 0.37$$

Note that $\lambda \simeq 1.80$ using both methods. Thus, the temperature equations, Eqs. (7.57a)–(7.57c), based on this value are

$$T_m = 25 + 1455.6\,\mathrm{erfc}\left(-977.33x/\sqrt{t}\right)$$

$$T_s = 1480.60 + 19.715\,\mathrm{erf}\left(149.12x/\sqrt{t}\right)$$

$$T_l = 1540 - 142.96\,\mathrm{erfc}\left(63.406x/\sqrt{t}\right)$$

The plots below show the temperature profiles at different solidification times.

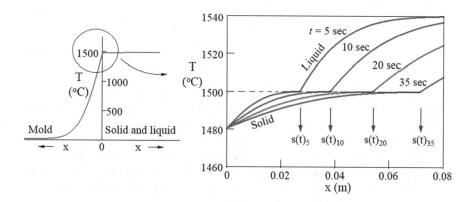

In particular, $T = f(x)$ exhibits non-linear behavior, while the freezing and mold–solid interface temperatures, $T_f^* = 1500.10\,^\circ\mathrm{C}$ and $T_i = 1480.60\,^\circ\mathrm{C}$, respectively, remain fixed as the L-S interface temperature advances.

7.7 Coupled Heat and Solid Fraction Model

This particular case is referred to as the Tien–Geiger binary-alloy solidification model [14], which considers (1) the one-dimensional unidirectional solidification process of a binary eutectic alloy in a semi-infinite region and (2) the release of the latent heat of fusion occurs at a temperature range $T_s < T_f^* < T_l$ within a mushy zone ahead of the planar L-S interface having a characteristic thickness $x_s(t) \le \delta(t) \le x_s(t)$.

Moreover, this model couples heat conduction and solid fraction at the planar L-S interface and it is analyzed in a half-space region $x > 0$ within a semi-infinite domain $0 < x < \infty$.

The governing equations along with $\rho = \rho_s = \rho_l$, boundary conditions and solid fraction function are [3, p.109–111]

$$\frac{\partial T_s}{\partial t} = \alpha_s \frac{\partial^2 T_s}{\partial x^2} \quad \text{at } 0 < x < x_s(t), \ t > 0 \tag{7.82a}$$

$$\frac{\partial T_m}{\partial t} = \alpha_m \frac{\partial^2 T_s}{\partial x^2} + \frac{\Delta H_s}{c_m} \frac{df_s}{dt} \quad \text{at } x_s(t) \leq x \leq x_l(t), \ t > 0 \tag{7.82b}$$

$$k_s \left(\frac{\partial T_s}{\partial x}\right)_{x_s(t)} - k_m \left(\frac{\partial T_m}{\partial x}\right)_{x_s(t)} = \rho \Delta H_s (1 - f_s) \frac{dx_s(t)}{dt} \tag{7.82c}$$

$$f_s = f_o \left[\frac{x_l(t) - x}{x_l(t) - x_s(t)}\right] \tag{7.82d}$$

where $(\Delta H_s/c_m)(df_s/dt)$ denotes the latent heat released by the mushy zone, df_s/dt is the rate of formation of the solid fraction, f_o defines a solid fraction function with $0 < f_s < 1$, and f_s denotes a solid fraction function.

Additionally, the subscript "m" denotes the mushy zone (two-phase free-boundary problem), which is an unstable complex phase constituted by layers of dendritic crystals that commonly form during solidification of binary or multi-phase alloys. Therefore, it is modeled as a layer instead of a simple smooth and flat interface boundary. This implies that a mushy zone is simply a thickening layer separating the solidified melt and the remaining melt during solidification.

This solidification model does not include any thermal interaction at the mold–solid interface $x = 0$ and the effect of thermal gradient in the mold material on the solidification time. Therefore, Tien–Geiger model [9] assumes thermal resistance in the solid phase and in the mushy zone only.

Further, the characteristic length and the solidification velocity equations are, respectively,

$$x_s(t) = 2\lambda_s \sqrt{\alpha_s t} \tag{7.83a}$$

$$\frac{dx_s(t)}{dt} = \lambda_s \sqrt{\frac{\alpha_s}{t}} \tag{7.83b}$$

$$x_l(t) = 2\lambda_l \sqrt{\alpha_l t} \tag{7.83c}$$

$$\frac{dx_l(t)}{dt} = \lambda_s \sqrt{\frac{\alpha_l}{t}} \tag{7.83d}$$

The actual temperature field equation for the solid phase with $T_o < T_2$ (solidus temperature in Fig. 7.3b) is given by

$$T_s = A_1 + A_2 \operatorname{erf}\left(\frac{x}{2\sqrt{\alpha_s t}}\right) \quad \text{for } 0 < x < s(t), t > 0 \tag{7.84a}$$

BC1 $T_s = T_s(0, t) = T_o$ (7.84b)

$\quad T_o = A_1 + A_2 \, \mathrm{erf}\,(0)$ (7.84c)

$\quad A_1 = T_o$ (7.84d)

BC2 $T_s = T_s[x_s(t), t] = T_2$ (7.84e)

$$T_f^* = T_o + A_2 \, \mathrm{erf}\left(\frac{x_s(t)}{2\sqrt{\alpha_s t}}\right) = T_o + A_2 \, \mathrm{erf}\,(\lambda_s) \tag{7.84f}$$

$$A_2 = \frac{(T_2 - T_o)}{\mathrm{erf}\,(\lambda_s)} \tag{7.84g}$$

$$T_s = T_o + \frac{(T_2 - T_o)}{\mathrm{erf}\,(\lambda_s)} \, \mathrm{erf}\left(\frac{x}{2\sqrt{\alpha_s t}}\right) \quad \text{(solid)} \tag{7.84h}$$

and that in the mushy zone is [3, p. 111]

$$T_m = T_1 + \frac{f_s \Delta H_s}{c_m(\lambda_l - \lambda_s)}\left(\lambda_l - \frac{x}{2\sqrt{\alpha_s t}}\right) \quad \text{(mushy)} \tag{7.85}$$

$$- \left(T_1 - T_2 + \frac{f_s \Delta H_s}{c_m}\right) \frac{\mathrm{erf}\left(x/2\sqrt{\alpha_m t}\right) - \mathrm{erf}\left(\lambda_l \sqrt{\alpha_s/\alpha_m}\right)}{\mathrm{erf}\left(\lambda_s \sqrt{\alpha_s/\alpha_m}\right) - \mathrm{erf}\left(\lambda_l \sqrt{\alpha_s/\alpha_m}\right)}$$

Hence, Eq. (7.82c) gives

$$\sqrt{\pi} f_s \exp\left(\lambda_l^2 \mu^2\right)\left[\mathrm{erf}\,(\lambda_l \mu) - \mathrm{erf}\,(\lambda_s \mu)\right] = \mu(\lambda_l - \lambda_s)\left(S_{t,m} - f_s\right) \tag{7.86a}$$

$$\frac{S_{t,m}}{\sqrt{\pi}\lambda_s \exp\left(\lambda_s^2\right)\mathrm{erf}\,(\lambda_s)} = 1 + f_s + f_s \frac{\exp\left[\mu^2\left(\lambda_l^2 - \lambda_s^2\right)\right] - 1/2}{\mu^2 \lambda_s(\lambda_l - \lambda_s)} \tag{7.86b}$$

$$\frac{S_{t,m}}{\sqrt{\pi}\lambda_s \exp\left(\lambda_s^2\right)\mathrm{erf}\,(\lambda_s)} - f_s\left(1 + \frac{\exp\left[\mu^2\left(\lambda_l^2 - \lambda_s^2\right)\right] - 1/2}{\mu^2 \lambda_s(\lambda_l - \lambda_s)}\right) = 1 \tag{7.86c}$$

with the Stefan numbers for the mushy zone and solid phases, respectively, and the thermal diffusivity ratio being written as

$$S_{t,m} = \frac{c_m(T_1 - T_2)}{\Delta H_s} \tag{7.87a}$$

$$S_{t,s} = \frac{c_s(T_2 - T_o)}{\Delta H_s} \tag{7.87b}$$

$$\mu = \sqrt{\frac{\alpha_s}{\alpha_m}} \tag{7.87c}$$

Additionally, $S_{t,m}$ and $S_{t,s}$ are Stefan numbers for the mushy zone and solid regions in a semi-infinite $(0 < x < \infty)$ solidification domain. These constants are strongly dependent on the temperature differences and the released latent heat of solidification. Actually, $S_{t,m}$ and $S_{t,s}$ take into account the effect of superheating the liquid metal.

7.8   Alloy Constitutional Supercooling

Consider a one-dimensional solidification heat transfer of a binary-alloy melt undergoing constitutional supercooling mechanism ahead of the L-S interface at time $t \geq 0$, where the onset of solidification starts as a uniform thin layer at the mold–liquid interface. In general, constitutional supercooling refers to compositional solid changes adjacent to the moving L-S interface with solute segregation buildup caused by solute rejection by the solid into the liquid.

Firstly, the sketches shown in Fig. 7.10 generalize the graphical representation of the binary alloy constitutional supercooling problem [1, p. 339]. For an alloy with nominal composition C_o, the actual melt temperature $T_l(x)_a$ in Fig. 7.10a acquires a curvature below the liquidus temperature, the liquid freezing temperature is $T_f^* = f(C_l)$, and the constitutional supercooling is caused by the undercooling $\Delta T = T_l(C_l) - T_l(x)_a$.

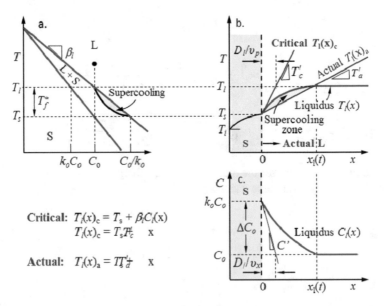

Fig. 7.10 Illustration of the constitutional supercooling phenomenon. (**a**) Partial equilibrium hypoeutectic phase diagram, (**b**) supercooling zone ahead of the L-S interface and (**c**) concentration profile. After ref. [1, p. 339]

Secondly, Fig. 7.10b depicts the actual temperature gradient $T'_a = (dT_l/dx)_a$ and the critical temperature gradient $T'_c = (dT_l/dx)_c$, which are the variables needed for deducing a simple morphological stability theory based on linear temperature equations cited in Fig. 7.10b for $T_l(x)_c$ and $T_l(x)_a$.

Thirdly, Fig. 7.10c illustrates a typical solute concentration distribution in the liquid phase ahead of the L-S interface and the extent of the diffusion characteristic length or thickness of the boundary layer $\delta_c = D_l/v_x$, where D_l is the diffusion coefficient of the solute in the liquid phase and v_x is the velocity of the interface.

For the sake of clarity, the magnitude of v_x significantly depends on the evolved latent heat of fusion ΔH_f for pure metals and ΔH_f^* for alloys. If the solute composition ahead of the L-S boundary changes during the solidification process, then the liquid is under a constitutional supercooling condition.

7.9   Plane-Front Stability Criterion

The actual supercooled freezing temperature $T_l(x)_a$ is obviously lower than the one predicted by the equilibrium phase diagram as $T_l(C_l)_e$. Hence, $T_l(x)_a$ is a function of distance x from the interface, $T_l(C_l)_e$ is a function of solute composition and as expected, the corresponding T-C and T-x diagrams are essential for characterizing the actual solidification problem.

The stability physical problem is illustrated in Fig. 7.11a as an ideal case for a stable planar L-S interface, where the equilibrium and the actual temperature gradients (slopes) are equal; $T'_a = T'_c$. This is a graphical representation of the advancing solidification front at a steady-state velocity v_x induced by the rate of release of the latent heat of solidification. Ultimately, the liquid solidifies as equiaxed grains in the absence of protrusions.

On the other hand, Fig. 7.11b schematically shows the interface instability as a constitutional supercooling zone having a limit $0 \le x \le x_l(t)$ above the actual melt temperature T_a ahead of the advancing L-S interface in the form of protrusion when

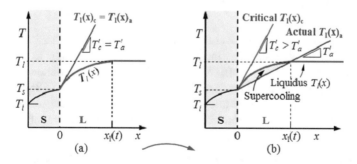

Fig. 7.11 (a) Stable planar front for equiaxed grain growth and (b) unstable planar front due to constitutional supercooling related to cellular or dendritic structure ahead of the interface

$T_a' < T_c'$. This implies that the initially planar solidification front in the form of a thin solid layer at the mold–solid interface breaks down as a result of constitutional supercooling.

Actually, the thermal condition given in Fig. 7.11b suggests that supercooling occurs when $T_a' < T_c'$ and consequently, a cellular or dendritic grain morphology develops ahead the L-S interface. Conversely, if $T_a' \geq T_c'$, then equiaxed grains may form under a stable planar L-S interface.

Furthermore, for the hypoeutectic binary alloy having a linear liquidus path (Fig. 7.10a), the linear stability, known as the Tiller's stability criterion [15], is based on conductive heat transfer and diffusion in the liquid phase. The Tiller's stability criterion is conveniently called plane-front stability criterion since it sets the rule for plane-front solidification, where solute atoms in excess for forming the solid are rejected into the liquid phase.

In reality, Tiller's stability criterion indicates or defines the thermal condition for interface stability using concepts related to temperature gradients. Therefore, this criterion does not predict the instance the interface becomes unstable. it only provides the condition for such an event.

This criterion can be deduced by defining the origin of the solidification front at $x = 0$ and by letting the actual and critical temperature gradients (Fig. 7.11) ahead of the planar L-S interface be related as shown below.

Normally, the concept of morphological instability of the liquid–solid interface refers to perturbations in morphology. Henceforth, the planar-front criterion in instability analysis is mathematically denoted in terms of temperature gradients for determining if the solidifying melt is under constitutional supercooling related to the morphology of the microstructure ahead of the moving interface. Thus,

$$\left(\frac{dT_l(x)_a}{dx} \right)_{x=0} \geq \left(\frac{dT_l(x)_c}{dx} \right)_{x=0} = \frac{dT_l}{dC_l} \left(\frac{dC_l}{dx} \right)_{x=0} \quad \text{(stability)} \qquad (7.88a)$$

$$\left(\frac{dT_l(x)_a}{dx} \right)_{x=0} \geq \beta_l \left(\frac{dC_l}{dx} \right)_{x=0} \quad \text{or} \quad T_a' \geq \beta_l C' \quad \text{(stability)} \qquad (7.88b)$$

$$\left(\frac{dT_l(x)_a}{dx} \right)_{x=0} < \beta_l \left(\frac{dC_l}{dx} \right)_{z=0} \quad \text{or} \quad T_a' < \beta_l C' \quad \text{(instability)} \qquad (7.88c)$$

Mathematically, the changeover from stability, Eq. (7.88b), to instability, (7.88c), is carried out by reversing the inequality. Physically, according to the thermal condition defined by Eqs. (7.88a), (7.88b), the planar-front growth should be smooth and flat for the formation of equiaxed crystals. Conversely, Eq. (7.88c) defines the instability condition for a liquid alloy undergoing constitutional supercooling ahead of the planar L-S interface before solidification proceeds in the form of protrusions, such as cellular or dendritic crystals.

Directional Solidification Process Assume that one-dimensional quasi-static solidification is coupled to conduction-driven and liquid diffusion-driven solidification at the moving L-S interface. This implies that the mold stationary

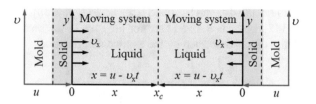

Fig. 7.12 Transformation of stationary coordinate (u,v) system to moving L-S interface coordinate system (x,y). The interface is ideally defined as a smooth and flat vertical plane moving in the x-direction with a velocity v_x

coordinate system (u, v) must be transformed to the moving interface coordinate system (x, y) as schematically shown in Fig. 7.12 for a smooth and flat plane (planar).

The relationship between these coordinate systems is defined as the Galilean transformation $x = u - v_x t$, where x denotes the interface position at time $t > 0$, u denotes the casting dimension, and $dx/dt = -v_x$ denotes the normal interface velocity [1, pp. 181 and 338], and [16, pp. 140–141].

Upon using the selected mold coordinate system, the one-dimensional diffusion is mathematically described by Fick's second law of diffusion as the concentration rate of the solute B in the liquid. Thus,

$$\frac{\partial C_l}{\partial t} = D_l \frac{\partial^2 C_l}{\partial u^2} \tag{7.89}$$

which can be transformed to moving coordinates.

Consider the following boundary conditions (BC's)

$$dx/du = 1 \tag{7.90a}$$

$$\frac{\partial C_l}{\partial u} = \frac{\partial C_l}{\partial x} \frac{\partial x}{\partial u} = \frac{\partial C_l}{\partial x} \tag{7.90b}$$

$$\frac{\partial^2 C_l}{\partial u^2} = \frac{\partial^2 C_l}{\partial x^2} \tag{7.90c}$$

Now, the set of equations to be solved are

$$\frac{\partial C_l}{\partial t} = D_l \frac{\partial^2 C_l}{\partial x^2} \qquad \text{(Fick's second law)} \tag{7.91a}$$

$$\frac{\partial C_l}{\partial t} + \frac{\partial x}{\partial t} \frac{\partial C_l}{\partial x} = D_l \frac{\partial^2 C_l}{\partial x^2} \qquad \text{(chain rule)} \tag{7.91b}$$

$$\frac{\partial C_l}{\partial t} - v_x \frac{\partial C_l}{\partial x} = D_l \frac{\partial^2 C_l}{\partial x^2} \qquad \text{(transformed)} \tag{7.91c}$$

$$0 = \frac{\partial^2 C_l}{\partial x^2} + \frac{v_x}{D_l} \frac{\partial C_l}{\partial x} \qquad \text{(steady-state)} \tag{7.91d}$$

The solution to Eq. (7.91d) can be found elsewhere [1, pp. 338], [16, p. 141], [17, p. 163], [18, p. 148], but it is derived here for the sake of the reader. Assume that the liquid concentration is described by the exponential function

$$C_l = C_l(x) = A_1 + A_2 \exp\left(-\frac{x}{D_l/v_x}\right) \tag{7.92a}$$

$$\text{BC1}\ \ C_l = C_l(0) = C_o/k_o \ \ \& \ \ \exp(0) = 1 \tag{7.92b}$$

$$C_o/k_o = A_1 + A_2 \tag{7.92c}$$

$$\text{BC2}\ \ C_l = C_l(-\infty) = C_o \ \ \& \ \ \exp(-\infty) = 0 \tag{7.92d}$$

$$C_o = A_1 \tag{7.92e}$$

$$A_2 = C_o/k_o - A_1 = C_o/k_o - C_o \tag{7.92f}$$

Substituting A_1, A_2 into Eq. (7.92a) yields

$$C_l = C_o + \left(\frac{C_o}{k_o} - C_o\right) \exp\left(-\frac{x}{D_l/v_x}\right) \tag{7.93a}$$

$$C_l = C_o + \Delta C_o \exp\left(-\frac{x}{D_l/v_x}\right) \tag{7.93b}$$

from which the unknown concentration gradient dC_l/dx at $x = 0$ becomes inversely proportional to the diffusion characteristic length (D_l/v_x)

$$C' = \left(\frac{dC_l}{dx}\right)_{x=0} = -\frac{(1 - k_o)\,v_p C_o}{k_o D_l} \tag{7.94}$$

The liquidus temperature (defined here as a critical entity) related to the solute concentration at any point x ahead of the interface is defined from slope of the liquidus line on the partial equilibrium phase diagram (Fig. 7.10a). Hence,

$$T_l(x)_c = T_f + \beta_l C_l(x) \quad \text{(liquidus)} \tag{7.95a}$$

$$T_l(x)_c = T_f + \beta_l C_o \left[1 + \frac{1 - k_o}{k_o} \exp\left(-\frac{x}{D_l/v_x}\right)\right] \tag{7.95b}$$

$$\frac{T_l(x)_c}{dx} = -\frac{\beta_l C_o(1 - k_o)}{k_o D_l/v_x} \exp\left(-\frac{x}{D_l/v_x}\right) \tag{7.95c}$$

where $\beta_l < 0$ and the equilibrium freezing temperature and composition ranges (Fig. 7.9a) are written as

$$\Delta T_o = -\beta_l \Delta C_o = T_l - T_s \tag{7.96a}$$

$$\Delta C_o = \frac{C_o(1 - k_o)}{k_o} = \frac{C_o}{k_o} - C_o \tag{7.96b}$$

Critical Temperature Gradient   Now, using the chain rule on the critical liquidus temperature gradient yields

$$\left(\frac{T_l(x)_c}{dx}\right)_{x=0} = \frac{dT_l}{dC_l}\left(\frac{dC_l}{dx}\right)_{z=0} \quad \text{(critical)} \tag{7.97a}$$

$$T'_c = \left[\frac{T_l(x)_c}{dx}\right]_{x=0} = -\frac{\beta_l C_o(1 - k_o)}{k_o D_l/v_x}\left(\frac{dC_l}{dx}\right)_{x=0} \tag{7.97b}$$

$$T'_c = \frac{\Delta T_o v_x}{D_l} \quad \text{(critical)} \tag{7.97c}$$

Note that T'_c is inversely proportional to the diffusion characteristic length D_l/v_x. Physically, $T'_c \to 0$ as $v_x \to 0$ for a stationary L-S interface.

Actual Temperature Gradient   The actual liquid temperature ahead of the planar L-S interface is intentionally defined as a linear relationship in accord with Figs. 7.10b and 7.11b

$$T_l(x)_a = T_s + \left[\frac{T_l(x)_a}{dx}\right]x \tag{7.98}$$

In this case, the temperature gradient is simply the slope of Eq. (7.98) and it determines the steepness of the liquid temperature ahead of the L-S interface.

Stability Criterion   Combining Eqs. (7.88a) and (7.95c) gives the stability criterion written as

$$\left[T_l(x)_a/dx\right]_{x=0} \geq -\frac{(1 - k_o)\beta_l C_o v_x}{k_o D_l} = \frac{\Delta T_o v_x}{D_l} \quad \text{(stability)} \tag{7.99a}$$

$$\left[T_l(x)_a/dx\right]_{x=0} < -\frac{(1 - k_o)\beta_l C_o}{k_o D_l} = \frac{\Delta T_o v_x}{D_l} \quad \text{(instability)} \tag{7.99b}$$

Equating Eq. (7.99a) yields the limit of the constitutional supercooling and the L-S interface velocity (v_x)

$$\frac{\left[T_l(x)_a/dx\right]_{x=0}}{v_x} = \frac{\Delta T_o}{D_l} \quad \text{(limit)} \tag{7.100a}$$

$$v_x = \frac{D_l}{\Delta T_o}[dT_l(x)_a/dx]_{x=0} \quad \text{(critical)} \tag{7.100b}$$

Accord with the nomenclature used in the literature, the actual liquid temperature gradient (G_a) in Eqs. (7.99) and (7.100c) is redefined as [15]

$$\frac{G_a}{\upsilon_x} \geq \frac{\Delta T_o}{D_l} \qquad \text{(stability)} \qquad\qquad (7.101\text{a})$$

$$\frac{G_a}{\upsilon_x} < \frac{\Delta T_o}{D_l} \qquad \text{(instability)} \qquad\qquad (7.101\text{b})$$

$$\upsilon_x = \frac{D_l G_a}{\Delta T_o} \qquad \text{(critical)} \qquad\qquad (7.101\text{c})$$

Theoretically, the stability criterion suppresses the formation of cellular and dendritic structures ahead of the L-S interface and as a result, the liquid alloy solidifies as equiaxed grains with homogeneous solute distribution in the solid state. This microstructural feature is desirable for producing castings with uniform or isotropic mechanical properties.

On the other hand, the instability criterion occurs by the formation of protrusions (cellular or dendritic grains), and consequently, post-solidification heat treatment follows for homogenizing the microstructure morphology and solute distribution.

Furthermore, the classic stability criterion, Eq. (7.99a), for a planar L-S interface during the solidification of binary alloys strongly depends on the growth velocity (υ_x) of the interface, which in turn, depends on the rate of release of the latent heat of fusion ΔH_f^*.

Finally, the position of the planar L-S interface is always at the coordinate origin $x = 0$ and the stability criterion is directly related to the conservation law of the heat fluxes at the interface with an assumed constant phase density $\rho = \rho_s = \rho_l$

$$k_s \left(\frac{\partial T_s}{\partial x}\right)_{x_s(t)} - k_l \left(\frac{\partial T_l}{\partial x}\right)_{x_s(t)} = \rho \Delta H_s \frac{dx_s(t)}{dt} \qquad\qquad (7.102)$$

This partial differential equation (PDE) governs the evolution of solid morphology at a particular solidification velocity and indirectly influences the solid properties, and it is deduced from elementary conservation laws of physics. It is also classified as a conservation equation due to the evolution of latent heat of solidification ΔH_s, provided that $x_s(t) > 0$. If $\Delta H_s = 0$, then $x_s(t) = 0$ and solidification has not yet stated.

Example 7A.4 Consider the solidification of a hypothetical binary alloy with nominal solute composition $C_o = 0.10\,\text{wt}\%$, equilibrium partition coefficient $k_o = 0.02$ and liquidus line slope liquid $\beta_l = -40\,\text{K/wt}\%$. If the solute liquid diffusion coefficient and the actual liquid temperature gradient ahead of the planar L-S interface are $D_l = 10^{-8}\,\text{m}^2/\text{s}$ and $T_a' = 10^5\,\text{K/m}$, respectively, calculate (a) the L-S interface velocity υ_x, (b) the L-S interface characteristic length δ_c, (c) the time it takes the L-S interface to reach a length equals to its characteristic length, and (d) What does the stability criterion predict?

Solution

(a) From Eq. (7.95c), the L-S interface velocity is

$$T_c' = \left[\frac{T_l(x)_c}{dx}\right]_{x=0} = \frac{\beta_l C_o (1-k_o)}{k_o D_l / v_x}$$

$$v_x = \frac{k_o D_l T_c'}{\beta_l C_o (1-k_o)} = \frac{(0.02)\left(10^{-8}\,\text{m}^2/\text{s}\right)\left(10^5\,\text{K/m}\right)}{(40\,K/\%)\,(0.10\%)\,(1-0.02)} = 5.10 \times 10^{-6}\,\text{m/s}$$

(b) The L-S interface supercooling length is

$$\delta_c = \frac{D_l}{v_x} = \frac{10^{-8}\,\text{m}^2/\text{s}}{5.10 \times 10^{-6}\,\text{m/s}} = 1.96 \times 10^{-3}\,\text{m} = 1.96\,\text{mm}$$

(c) The time for length δ_c is

$$t = \frac{\delta_c}{v_x} = \frac{1.96\,\text{mm}}{5.10 \times 10^{-3}\,\text{mm/s}} = 384.31\,\text{s} = 6.41\,\text{min}$$

(d) From Eq. (7.96), the concentration and temperature ranges are

$$\Delta C_o = \frac{C_o (1-k_o)}{k_o} = \frac{(0.1\%)\,(1-0.02)}{0.02} = 4.9\,\%$$

$$\Delta T_o = -\beta_l \Delta C_o = -(-40\,K/\%)\,(4.9\%) = 196\,K$$

and (7.97c) gives

$$T_c' = \frac{T_l(x)_c}{dx} = \frac{\Delta T_o v_x}{D_l} = \frac{(196)\left(5.10 \times 10^{-6}\right)}{10^{-8}} = 10^5\,\text{K/m}$$

Note that $T_a' = 10^5\ K/m$ (given) and T_c' compare

$$T_a' = T_c' = 10^5 \frac{K}{m} \quad \text{stability!}$$

Example 7A.5 Consider the solidification of a Fe-$0.08C$ steel with nominal solute composition $C_o = 0.10\,\text{wt\%}$, equilibrium partition coefficient $k_o = 0.17$ and liquidus line slope liquid $\beta_l = -84.91\,\text{K/wt\%}$. If the solute liquid diffusion coefficient ahead of the planar L-S interface and the L-S interface velocity are $D_l = 2 \times 10^{-8}\,\text{m}^2/\text{s}$ and $v_x = 10^5\,\text{m/s}$, respectively, calculate (a) the L-S interface characteristic length δ_c, (b) the time it takes the L-S interface to reach a length equals to its characteristic length, and (c) What does the stability criterion predict?

Solution

(a) From Eq. (7.95c),

$$\delta_c = \frac{D_l}{\upsilon_x} = \frac{2 \times 10^{-8}}{10^{-5}} = 0.002\,\text{m}$$

(b) The time for the advancing planar L-S interface to reach a length equals to its characteristic length is

$$t = \frac{\delta_c}{\upsilon_x} = \frac{0.002}{10^{-5}} = 200\,\text{s}$$

(c) Stability criterion prediction:

$$T_c' = \left[\frac{T_l\,(x)_c}{dx}\right]_{x=0} = \frac{\beta_l C_o\,(1 - k_o)}{k_o D_l/\upsilon_x} = -\frac{(-84.91)\,(0.08)\,(1 - 0.17)}{(0.17)\,\left(2 \times 10^{-8}\right)/\left(8 \times 10^{-6}\right)}$$

$$T_c' = 13{,}266\,\text{K/m}$$

and from Example 6.3, the actual liquid temperature is

$$T_l\,(x)_a = 1540 - 142.96\,\text{erfc}\left(63.406x/\sqrt{200}\right)$$

$$T_l\,(x)_a = 1540 - 142.96\,\text{erfc}\,(4.4835x)$$

from which

$$\left[\frac{dT_l\,(x)_a}{dx}\right] = -\frac{2\,(4.4835)\,(-142.96)}{\sqrt{\pi}}\exp\left(-\,(4.4835x)^2\right)$$

$$\left[\frac{dT_l\,(x)_a}{dx}\right] = 723.25\exp\left(-20.102x^2\right)$$

$$T_a' = \left[\frac{dT_l\,(x)_a}{dx}\right]_{x=0} = 723.25\,\text{K/m}$$

Therefore, $T_a' < T_c'$ instability! The stability criterion predicts an unstable plane-front solidification because $T_a' < T_c'$. This means that the planar L-S interface breaks down and cellular or dendritic solidification will take place.

7.10 Microstructural Design

During solidification processing (SP), the freezing or cooling rate is defined as per the chain rule

$$\frac{dT}{dt} = \frac{dx}{dt}\frac{dT}{dx} \tag{7.103}$$

where dx/dt denotes the solidification velocity or simply the solidification front velocity and dT/dx denotes the temperature gradient along the x-direction. This expression indicates that during rapid solidification processing (RSP), $dT/dt \rightarrow \infty$ as $dx/dt \rightarrow \infty$ at $dT/dx > 0$ and as a result, the final casting product may consist of fine grains or amorphous structure. Conversely, during slow solidification (SS), $dT/dt \rightarrow 0$ as $dx/dt \rightarrow 0$ at $dT/dx > 0$.

Theoretically, imposing a freezing rate $dT/dt > 0$ on the solidification process causes evolution of latent heat of solidification (ΔH_s), which is extracted as thermal energy (Q), which in turn, dissipates through the solidification environment. Therefore, ΔH_s and Q are directly linked as the foundation for heat transfer theory applied to nucleation and growth of the solid phase from the melt (liquid metal at $T \geq T_m$).

Equation (7.103) can also be analyzed in a different manner. Thus,

- If $dT/dt \rightarrow \infty$ is imposed on the solidification of a small casting dimension, such as a strip, then rapid solidification prevails as the phase transformation process (PTP), where the melt is thermodynamically unstable and the final microstructure can be entirely amorphous since there is not enough time for a phase partition to develop. Hence, rapid solidification lacks microsegregation.
- If $0 < dT/dt \ll \infty$, then the casting dimensions can be very large as in the case of ingot production. In this case, the solidification process can be thermodynamically stable and the final microstructure may be equiaxed, dendritic, or a combination of these microstructural morphologies. In this case, a phase partition develops as described in a previous chapter.

Nonetheless, manipulating or controlling the temperature gradient dT/dt during solidification leads to grain refinement associated with heterogeneous nucleation and minimum segregation effects.

7.11 Summary

It has been shown that the phase fractions can be determined using the mass fraction, volume fraction, concentration ratio, temperature ratio, enthalpy ratio, and the Gulliver–Scheil equation, which is considered as a non-linear Lever rule that strongly depends on the partition coefficient k_o. Also useful in this chapter is the definition of k_o since it is related to hypoeutectic binary alloys ($k_o < 1$) and hypereutectic ($k_o > 1$) binary alloys. Thus, the eutectic point E represents the phase boundary for defining the type of an alloy under solidification.

Furthermore, a simple analytical procedure is introduced in order to derive the latent heat of fusion that evolves from the liquid phase at the freezing temperature T_f^* within the temperature range $T_s < T_f^* < T_l$ and solute concentration range $C_s < C_l^* < C_l$. Hence, $T_f^* = T\left(C_l^*\right)$ and $C_l^* = C\left(T_f^*\right)$.

The analytical procedures for modeling solidification of binary alloys are mainly due to heat conduction and mass transport. For the sake of simplicity, convection and radiation effects are not considered in this chapter. In particular, the Rubinstein [8] and Tien–Geiger [14] models use a direct conductive heat transfer approach to determine the unknown variables $T(x, t)$ and $s(t)$ and the mass transport with constant diffusion coefficients. Despite that these models are based on coupled heat-mass transfer, they exclude any analytical approach for explaining supercooling of alloys.

In addition, the binary alloy constitutional supercooling problem is assessed in a simple manner so that the stability criterion reveals the importance of controlling the solidification process in eutectic systems. This led to the determination and comparison of the actual and critical temperature gradients related to the resultant morphology of the solidified microstructure. Another aspect of eutectic solidification is the degree of undercooling for the formation of the α-β lamellar structure.

7.12 Problems

7.1 Consider a quasi-static solidification of a $Sn\text{-}15Bi$ alloy with $C_o = 15\%$ (by weight) from the liquid state initially at $240\,^\circ\text{C}$. For a very slow solidification process, follow the equilibrium phase diagram and determine the partition coefficient k_o at the onset of solidification ($\simeq 216\,^\circ\text{C}$), $200\,^\circ\text{C}$ and at the end on solidification ($\simeq 165\,^\circ\text{C}$). Is $k_o = f(T)$ a linear relationship? What can you conclude from these results?

7.2 Consider the solidification of $Pb\text{-}10Sn$ alloy into a small slab shape. Assume that the diffusion flux J_x (Fick's first law) of solute Sn away from the moving $L\text{-}S$ interface must equal the rate of solute Sn rejection $v_p\Delta C$ into the liquid phase, and that Fick's first law is applicable at the $L\text{-}S$ interface at $x = 0$. For a constant $L\text{-}S$ interface velocity of 10^{-4} m/s, the actual liquid temperature gradient ahead of the interface is $dT_l/dx = 10^4$ K/m and Sn diffusion coefficient in the liquid is $D_l = 5 \times 10^{-9}$ m^2/s. The liquidus slope β_l and the partition coefficient k_o must be determined from the equilibrium $Pb\text{-}Sn$ diagram (not given). Based on this information, determine (a) the critical liquid temperature gradient and (b) the condition for an unstable interface under steady-state condition. What type of microstructure will be produced upon completion of solidification?

7.3 Assume that the diffusion flux J_x (Fick's first law) of solute Sn away from the moving L-S interface must equal the rate of solute Sn rejection $v_p \Delta C$ into the liquid phase during solidification of a Pb-$10Sn$ alloy into a small slab shape and that Fick's first law is applicable at the L-S interface at $x = 0$. For a constant L-S interface velocity of 1.5×10^{-4} m/s, the actual liquid temperature gradient ahead of the interface is $dT_l/dx = 10^4$ K/m and Sn diffusion coefficient in the liquid is $D_l = 5 \times 10^{-9}$ m^2/s. The liquidus slope β_l and the partition coefficient k_o must be determined from the equilibrium Pb-Sn diagram (not given). Based on this information, determine (a) the critical liquid temperature gradient and (b) the condition for an unstable interface under steady-state condition. What type of microstructure will be produced upon completion of solidification?

7.4 For the diffusion flux J_x (Fick's first law) of solute Sn away from the moving L-S interface must equal the rate of solute Sn rejection $v_p \Delta C$ into the liquid phase during solidification of a Pb-$10Sn$ alloy into a small slab shape and that Fick's first law is applicable at the L-S interface at $x = 0$. For a constant L-S interface velocity of 2×10^{-4} m/s, the actual liquid temperature gradient ahead of the interface is $dT_l/dx = 10^4$ K/m and Sn diffusion coefficient in the liquid is $D_l = 5 \times 10^{-9}$ m^2/s. The liquidus slope β_l and the partition coefficient k_o must be determined from the equilibrium Pb-Sn diagram (not given). Based on this information, determine (a) the critical liquid temperature gradient and (b) the condition for an unstable interface under steady-state condition. What type of microstructure will be produced upon completion of solidification?

7.5 For directional solidification of Al-$3Cu$ and Al-$5Cu$ at $v_x = 0.10$ m/s, use $k_o = 0.14$, $\beta_l = -2.6 °C/wt\%$ and $D_l = 5 \times 10^{-5}$ m^2/s to determine the value of the critical temperature gradient for stability of the solidifying melt.

7.6 Consider the directional solidification of Al-$3Cu$ and Al-$5Cu$ at $v_x = 0.20$ m/s. If $k_o = 0.14$, $\beta_l = -2.6 °C/wt\%$ and $D_l = 5 \times 10^{-5}$ m^2/s, then determine the value of the critical temperature gradient $\left[T_l (x)_c /dx \right]_{x=0}$ for stability of the solidifying melt. Which alloy requires a higher $\left[T_l (x)_c /dx \right]_{x=0}$.

7.7 Assume that a 40-μm thick Al-$20Si$ ribbon is produced on a rotating copper roller (substrate). Calculate (a) the solidification time t, (b) the solidification velocity dx_s/dt, (c) the solidification rate, and (d) the heat flux at the substrate–solid interface $x = 0$. The Al-$20Si$ binary alloy is initially at its pouring temperature $T_p = 850 °C$, while the copper mold is maintained at $T_o = 25 °C$. The freezing temperatures of pure aluminum and pure silicon are $T_{Al} = 660 °C$ and $T_{Si} = 1414 °C$. The thermophysical properties and the Al-Si equilibrium phase diagram are given in Example 7A.2.

7.8 Suppose that you are asked to produce a 3-cm thick Al-$20Si$ flat plate in a stationary copper mold (enclosed substrate). The Al-$20Si$ binary alloy is initially at its pouring temperature $T_p = 850 °C$, while the copper mold is maintained at $T_o = 25 °C$. The freezing temperatures of pure aluminum and pure silicon are $T_{Al} = 660 °C$ and $T_{Si} = 1414 °C$. The thermophysical properties and the

Al-Si equilibrium phase diagram are given in Example 7A.2. Calculate (a) the solidification time t, (b) the solidification velocity dx_s/dt, (c) the solidification rate, and (d) the heat flux at the substrate–solid interface $x = 0$.

7.9 Suppose that you are asked to produce a 2-cm thick *Al-20Si* flat plate in a stationary copper mold (enclosed substrate). The *Al-20Si* binary alloy is initially at its pouring temperature $T_p = 850\,°C$, while the copper mold is maintained at $T_o = 25\,°C$. The freezing temperatures of pure aluminum and pure silicon are $T_{Al} = 660\,°C$ and $T_{Si} = 1414\,°C$. The thermophysical properties and the *Al-Si* equilibrium phase diagram are given in Example 7A.2. Calculate (a) the solidification time t, (b) the solidification velocity dx_s/dt, (c) the solidification rate, and (d) the heat flux at the substrate–solid interface $x = 0$.

7.10 Consider the solidification of a 2-cm thick flat plate made out of an equiatomic *Ni-Ti* alloy ($T_f = 1310\,°C$) in a stationary copper mold (enclosed substrate at $T_o = 20\,°C$). Let the pouring temperature be $T_p = 1380\,°C$ and use the data given below to calculate (a) the solidification time t, (b) the solidification velocity dx_s/dt, (c) the solidification rate, and (d) the heat flux at the substrate–solid interface $x = 0$.

Material	k (W/m K)	ρ (kg/m^3)	c_s (J/kg K)	ΔH_f (kJ/kg)
Ni-Ti alloy	18	6450	480	$\Delta H_{Ti} = 322.75$
Steel substrate	35	7845	500	$\Delta H_{Ni} = 297.67$

References

1. J.A. Dantzig, M. Rappaz, *Solidification*. Revised & Expanded, 2nd edn. (EPFL Press, Lausanne, 2016)
2. H. Fredriksson, U. Akerlind, *Materials Processing During Casting* (Wiley, Chichester, 2006)
3. S.M. Seo, J.H. Lee, Y.S. Yoo, C.Y. Jo, H. Miyahara, K. Ogi, Solute redistribution during planar and dendritic growth of directionally solidified Ni-base superalloy CMSX-10, in *11th International Symposium on Superalloys 2008, TMS (The Minerals, Metals & Materials Society)*, ed. by R.C. Reed, K.A. Green, P. Caron, T.P. Gabb, M.G. Fahrmann, E.S. Huron, S.A. Woodard. Pennsylvania TMS 2008 (Wiley, Champion, 2008), pp. 277–286
4. D.R. Poirier, P. Nandapurkar, Enthalpies of a binary alloy during solidification. Metall. Trans. A **19A**, 3057–3061 (1988)
5. J.P.J. Cherng, The effects of deposit thermal history on microstructure produced by uniform droplet spray forming, Dissertation, Massachusetts Institute of Technology, Boston, 2002
6. V. Alexiades, A.D. Solomon, *Mathematical Modeling of Melting and Freezing Processes* (Hemisphere Publishing Corporation, Washington, 1993)
7. C.L. Xu, H.Y. Wang, F. Qiu, Y.F. Yang, Q.C. Jiang, Cooling rate and microstructure of rapidly solidified Al-20wt.%Si alloy. Mater. Sci. Eng. A **417**, 275–280 (2006)
8. L.I. Rubinstein, *The Stefan Problem*. Translations of the American Monographs, vol. 27 (American Mathematical Society, Providence, 1971)
9. J. Crank, *Free and Moving Boundary Problems* (Oxford University Press, New York, 1984)

10. V.R. Voller, A numerical method for the Rubinstein binary-alloy problem in the presence of an under-cooled liquid. Int. J. Heat Mass Trans. **51**, 696–706 (2008)
11. V.R. Voller, A similarity solution for solidification of an under-cooled binary alloy. Int. J. Heat Mass Trans. **49**(11), 1981–1985 (2006)
12. V.R. Voller, A similarity solution for the solidification of a multicomponent alloy. Int. J. Heat Mass Trans. **40**(12), 2869–2877 (1997)
13. D.A. Tarzia, Explicit and approximated solutions for heat and mass transfer problems with a moving interface, in *Advanced Topics in Mass Transfer*, ed. by M. El-Amin, ch. 20 (InTech, Rijeka, 2011), pp. 439–484. intechweb.org
14. R.H. Tien, G.E. Geiger, A heat-transfer analysis of the solidification of a binary eutectic system. J. Heat Trans. **89**(3), 230–233 (1967)
15. W.A. Tiller, K.A. Jackson, J.W. Rutter, B. Chalmers, The redistribution of solute atoms during the solidification of metals. Acta Metall. **1**(4), 428–437 (1953)
16. R. Asthana, A. Kumar, N.B. Dahotre, *Materials Processing and Manufacturing Science* (Elsevier Science & Technology, New York, 2005)
17. W. Kurz, D.J. Fisher, *Fundamentals of Solidification*, 3rd edn. (Trans Tech Publications, Kapellweg, 1992)
18. M.E. Glicksman, *Principles of Solidification: An Introduction to Modern Casting and Crystal Growth Concepts* (Springer Science+Business Media, LLC, New York, 2011)

Chapter 8
Alloy Solidification II

8.1 Introduction

This chapter is devoted to solidification of binary A-B alloys of nominal solute composition C_o. The interpretation of a binary phase diagram depends on the composition of an alloy ($C_o < C_e$, $C_o = C_e$ and $C_o > C_e$) and its freezing temperature range $T_s \leq T_f^* \leq T_l$, the solidification velocity. Assuming that the alloy is at equilibrium at a low solidification velocity, one has to determine the type and morphology of the solid phases resulting from solidification, compositions of the liquid and solid phases at a particular freezing temperature, the mass fractions of the phases, and the degree of any undercooling ΔT.

The objective in this chapter is to introduce analytical procedures to derive analytical solutions to eutectic solidification problems related to regular α-β lamellar phase morphology induced by the eutectic reaction $L \rightarrow \alpha + \beta$ at a rate of evolved latent heat of fusion. This reaction is the reason for using the phrase "eutectic solidification" which means that a binary A-B alloy may have a nominal solute B composition $C_\alpha \leq C_o \leq C_\beta$, where C_α, C_β denote the minimum and maximum, respectively, solute compositions at the eutectic temperature T_E. Moreover, the analytical analysis is carried out under the assumption of a small eutectic undercooling (or undercooling) ΔT and a steady solidification at low α-β lamellar interface advancement, which is strongly influenced by the local coupled diffusion mechanisms in the xz solidification plane; lateral diffusion (x-axis) and axial diffusion (z-axis). The former diffusion mechanism is considered an oscillatory process associated with a contoured α-β lamellar interface.

© Springer Nature Switzerland AG 2020
N. Perez, *Phase Transformation in Metals*,
https://doi.org/10.1007/978-3-030-49168-0_8

8.2 Morphology of Eutectic Solidification

The evolution and morphology of unidirectional solidification microstructures containing the coupled α-β solid phases (Fig. 8.1, [1, p. 184]) are important aspects in the crystal growth field since the L-α-β region is the most active place during solidification. Notice the microstructural features (grains) in pure Al, off-eutectic Al-Si alloys, and pure Si. Consequently, the diversity of mechanical properties of Al-Si alloys is attributed to these microstructural features and the mathematical modeling of solidification is very difficult and complicated.

A few microstructural images are also included in order for the reader to understand and easily interpret the connection between equilibrium phase diagrams and the corresponding equilibrium solidification process as per the eutectic reaction $L \rightarrow \alpha + \beta$ at the eutectic temperature T_E. In this particular case, Fig. 8.1 shows the eutectic composition $C_e = 12.6\% Si$ in the aluminum (Al) matrix at $T_E = 577\,^\circ C$ [1, p. 184].

Noticeably, an Al-Si composition (Fig. 8.1) induces a particular characteristic black-white microstructural morphology. Therefore, one can deduce that both melting and solidification processes are related to the characteristics of the phase diagrams, and that there is a strong correlation between microstructure and mechanical properties. However, large β particles in hypereutectic Al-20 Si and Al-50Si alloys limit their industrial applications.

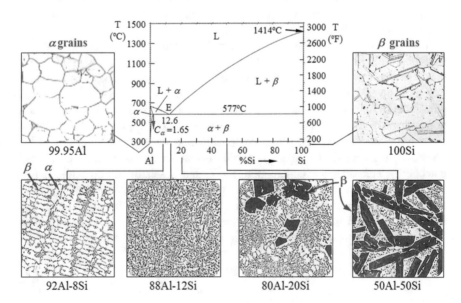

Fig. 8.1 Al-Si equilibrium phase diagram and morphology of microstructure of pure Al, pure Si, and some Al-Si alloys [1, p. 184]

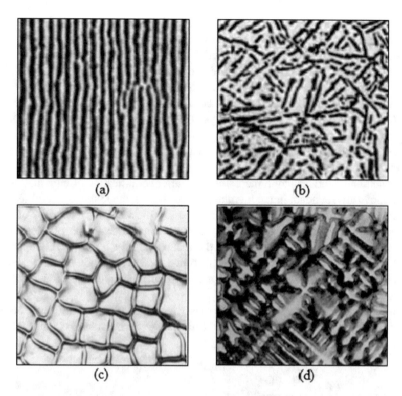

Fig. 8.2 Eutectic microstructures (**a**) Regular *Al-Si* lamellar, (**b**) irregular eutectic structure in *Al-Si* [2, p. 57], (**c**) cellular eutectic in dilute *Sn-Pb* alloy [2, p. 16], and (**d**) dendrites [3, p. 9]

Microscopically, the *Si*-rich β phase solidifies as plate-like crystals, *Al*-rich α phase is a face-centered cubic (FCC) and the Si-rich β phase has a diamond-like cubic structure, and *Al-Si* alloys form a substitutional phase because these elements are comparable in size; $r_{Al} = 118$ pm and $r_{Si} = 111$ pm.

In fact, the *Al-Si* phase diagram has only one single-solid phase region identified as α-phase and the corresponding hypoeutectic *Al-Si* alloys are limited to a narrow compositional range $C_\alpha = 1.65\% Si \leq C \leq C_e = 12.6\% Si$ by weight.

8.2.1 Classification of Eutectic Structures

The variety of geometrical arrangements of the α and β solid phases can be used to classify the eutectic microstructures. Regardless of the type of alloy and magnification, Fig. 8.2 shows typical regular and irregular α-β lamellar (plate-like) eutectic structures in *Al-Si* alloys [2], Fig. 8.2c exhibits a cellular (hexagonal-like) α-β structure in a *Pb-Sn* alloy, and Fig. 8.2d depicts a dendritic structure in some alloy [3, p. 9]. The idea here is to show the reader the diversity in microstructural analysis one has to consider as a result of the complex solidification process.

Furthermore, in Fig. 8.2a, both α and β solid phases are assumed to be fundamentally comparable dimensionally and geometrically.

These individual phases are crystallographically different because they may have dissimilar mass densities and different crystal structures, such as BCC and FCC, BCC and tetragonal, or any other crystal combination as determined by X-ray diffraction. Thus, both type of atomic structures and the reflected incident light influenced by the surface roughness are reasons for observing different discoloration on the surface of microstructures under the microscope.

Nonetheless, the resultant microstructures after solidification of pure metals and their alloys are abundant in the literature. Nevertheless, the microstructural features are responsible for mechanical properties, to say the least, and behavior of solid materials in service. This means that one has to understand the physics of the problem in order to derive or develop the mathematics as a significant support to predict the material performance related to phase transformation or solidification, which is the main focus of this chapter. Thus, further analysis of the eutectic solidification process is given next.

In addition, Fig. 8.3 shows snapshots of α-β lamellar structures during directional solidification in carbon tetrabromide and 0.21 wt% hexachloroethane (CBr_4-

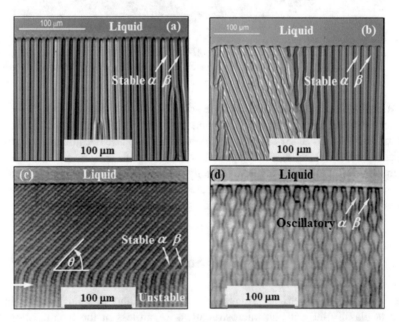

Fig. 8.3 Lamellar eutectic in directional solidification. For NPG-45.3DC samples [4], (**a**) and (**b**) regular lamellar showing different crystallographic orientations. For CBr_4-0.21%C_2Cl_6 samples ([5], (**c**) regular transition (see arrow and angle θ) from $v_z = 0.85\,\mu$m/s to stable tilted growth to $v_z = 3.30\,\mu$m/s) at $G_a = 80$ K/cm and (**d**) oscillatory lamellar structure at $v_z = 2\,\mu$m/s and $G_a = 80$ K/cm

$0.21C_2Cl_6$) [4] and neopentylglycol-(D)camphor (NPG-45.3DC) transparent organic alloy samples (Fig. 8.3c, d) [5].

These images are ideal for research and academic purposes, and for clarity, the α (light gray) and β (dark gray) notations are assigned to the phases for identifying these eutectic structures. Some additional information on the conditions for eutectic solidification is given in the figure caption.

Interestingly, the well aligned α-β solid phases shown in Fig. 8.3a are ideal for determining the lamellar spacing λ as a function of growth velocity υ_z and temperature gradient $G_l = dT_l/dz$. Specifically, the lamellar spacing $\lambda = f(\upsilon_z)$ and the undercooling $\Delta T = f(\lambda)$ functions are derived in a later section.

Additional details on these α-β eutectic structures can be found in the cited references [6–8] and in the literature. Only the regular α-β eutectic structure (Fig. 8.3a) is considered in a later section to introduce the procedures for deriving well-known analytical solutions to this type of eutectic solidification problem. Actually, the mathematical modeling of solidification is a very difficult task, but one can use analytical solutions available in the literature.

For instance, growth of eutectic structure as rod and lamellar morphology appears to be the most studied analytically for characterizing the evolution of two-phase solidification microstructures. The solidification rate seems to have a fundamental or practical importance in producing eutectic or eutectic-containing microstructures. However, the amount and morphology of the eutectic structure in a specific alloy dictate magnitude of properties. This is one of the reasons for characterizing the regular or irregular lamellar spacing λ under the influence of a temperature gradient, such as $G_l = dT_l/dz$, and related alloy liquid (melt) undercooling $\Delta T = f(\lambda)$.

8.3 Dendritic Growth

The onset of solidification starts at the mold–melt interface as a thin layer with inevitable defects, which are the potential sites for subsequent interface growth. For alloys, the planar interface morphology is commonly destabilized due to temperature gradients, solute segregation, and crystallographic anisotropy of the evolving microstructure. Consequently, dendrite growth dominates the evolution of the as-cast microstructure in most industrial cases.

Figure 8.4a shows an X-ray tomography dendritic image in a 1-mm diameter sample cooled at 2 °C/min [9]. Note that the shown vertical dendrite is composed of a predominantly solid cylindrical (stalk or stem) with secondary and tertiary dendrite arms with spherical caps. For clarity, the blue and red colored areas are located at the shown distances from the dendrite tip as per the original authors and the main dendrite growth is assumed to occur in the z-direction at velocity υ_z (primary stalk or stem) along with growing arms at υ_x and υ_{xz} into the liquid phase. For instance, the secondary arms grow at υ_x along the <100> crystallographic direction.

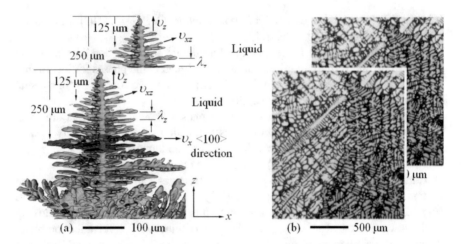

Fig. 8.4 Dendrite morphology. (**a**) Al-24Cu dendrites obtained by using X-ray tomographic technique on 1-mm diameter sample cooled at $2\,^{\circ}C/\min$ [9] and (**b**) as-cast dendritic structure in some metallic material [10]

Moreover, Fig. 8.4b depicts a particular as-cast dendritic microstructure in an unidentified Cu-based eutectic alloy taken from the 2005 Metals Conservation Summer Institute presentation [10].

In addition, NASA Isothermal Dendritic Growth Experiment (IDGE) provides insights on the effect of microgravity on dendrite formation in order to understand how heat and mass transfer affects materials processing. It is clearly evident in Fig. 8.4a that the morphology of dendrites consists of tree-like branched crystals with solidification velocity $v_z > v_x, v_{xz}$ into the liquid phase. As solidification progresses, the solidification velocity v_z and heat flow q_z occur in the opposite directions ($v_z \rightarrow$ and $q_z \leftarrow$) and fundamentally, a dendrite forms from an unstable nucleus, treated as particle with a critical radius r_c.

In practice, solidified industrial components or as-cast components with dendritic structures are usually heat treated to obtain equiaxed microstructures with uniform properties. This suggests that the microstructure-property coupling is an undercooling-dependent process. Thus, the resultant microstructural evolution is strongly affected by the solidification velocity to either produce equiaxed or dendritic microstructure.

8.3.1 Extended Metalstable Phase Fields

First of all, consider the schematic binary equilibrium phase diagram shown in Fig. 8.5a with stable field boundaries (red lines) and a symmetric coupled zone (orange area) having extended metalstable boundary fields (dashed liquidus lines) at $T < T_E$. Thus, cooperative growth of α and β phases occurs as a diffusion couple

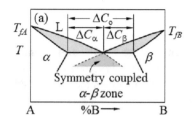
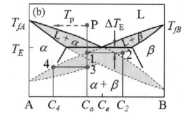

Fig. 8.5 (**a**) Schematic eutectic phase diagram showing extensions of the phase-field boundaries and (**b**) lamellar eutectic growth model. Borrowed phase diagram from Ref. [16, p. 398]

having L-α-β triple junctions during eutectic solidification. Conversely, a divorced growth is just the opposite of cooperative growth.

Moreover, Fig. 8.5b illustrates completely and nearly symmetric metalstable extension of the metalstable phase fields (gray areas) at $T < T_E$ and related phase-field boundaries (dashed lines). For simplicity, $(L + \alpha)_{area} \simeq (L + \beta)_{area}$ in Fig. 8.5, but most common equilibrium phase diagrams exhibit unsymmetrical phase fields since $(L + \alpha)_{area} > (L + \beta)_{area}$ or vice versa.

Another interesting information one can extract from Fig. 8.5b is the solute compositions using the extended gray phase fields

- Draw a vertical straight line from point "P" down to the x-axis.
- Draw a horizontal 1-2 line intercepting the extended dashed solidus and liquidus curves below T_E for the α-phase and a vertical line from point 2 down to the x-axis. Thus, $C_1 = C_o$ and $C_2 = C_o / k_o^{\alpha}$.
- Similarly, draw a horizontal 4-3 line connecting the hatched solidus and liquidus curves for the β-phase and a vertical line from point "4" down to the x-axis. Hence, $C_3 = C_o$ and $C_4 = 1 - (1 - C_o) / k_o^{\beta}$.

Also included in Fig. 8.5a are elementary definitions of some solute compositions that are useful for kinetic eutectic solidification. For a superheated hypoeutectic binary alloy with nominal composition C_o initially at the pouring temperature T_p (point "P" in Fig. 8.5b), a slow solidification (equilibrium solidification) produces α and α-β grains as the alloy reaches a condition where the temperature is $T < T_E$. As the solidification proceeds, the last amount of liquid transform according to the eutectic invariant reaction $L \rightarrow \alpha + \beta$, where β nucleates and grow concurrently with α. Thus, the eutectic solidification involves an exchange of solute by diffusion (mass transport) in the liquid adjacent to liquid–solid interface at some undercooling ΔT.

In contrast, a rapid solidification (non-equilibrium solidification) of a metallic liquid (melt) is fundamentally a rapid quenching process, where the final microstructure may be extremely fine or amorphous. Recall that these solidification processes have been described in a previous chapter.

8.3.2   Invariant Eutectic Reaction

Upon slow solidification above the eutectic temperature T_E, the $L + \alpha$ and $L + \beta$ gray areas in Fig. 8.5 represent the equilibrium phase-field regions and $L \rightarrow \alpha + \beta$ at $T = T_E$ is the eutectic invariant reaction for the formation of the eutectic structure with a eutectic composition C_e. From Fig. 8.5a,

$$\Delta C_o = \Delta C_\alpha + \Delta C_\beta \tag{8.1a}$$

$$\Delta C_\alpha = \left(1 - k_o^\alpha\right) C_e = C_e - C_\alpha \tag{8.1b}$$

$$\Delta C_\beta = 1 - k_o^\beta \left(1 - C_e\right) = C_\beta - C_e \tag{8.1c}$$

Below T_E, the gray areas represent the extended metalstable phase-field regions for $L \rightarrow \alpha + \beta$ at a wide range of alloy nominal compositions. This is achievable if a liquid alloy C_o (point "P" in Fig. 8.5b) solidifies at $T < T_E$ as a eutectic dual-phase solid. From Fig. 8.5b,

$$C_\alpha = k_o^\alpha C_e \tag{8.2a}$$

$$C_\beta = 1 - k_o^\beta \left(1 - C_e\right) \tag{8.2b}$$

Furthermore, the reactions representing the solidification of alloys having a compositional range $C_\alpha \leq C_o \leq C_\beta$ are

$$L_p \rightarrow L + \alpha \rightarrow \alpha + (\alpha + \beta) \quad \text{(hypoeutectic)} \tag{8.3a}$$

$$L_p \rightarrow (\alpha + \beta) \qquad\qquad \text{(eutectic)} \tag{8.3b}$$

$$L_p \rightarrow L + \beta \rightarrow \beta + (\alpha + \beta) \quad \text{(hypereutectic)} \tag{8.3c}$$

where L_p denotes the initial liquid alloy phase at the pouring temperature T_p, which is only shown for a hypoeutectic alloy with $C_o < C_e$ in Fig. 8.5b.

In a hypoeutectic alloy shown in Fig. 8.5b, the gray α-phase field, the single α-phase denotes the primary solid in the form of grains that precipitates out of the liquid and the two-$(\alpha + \beta)$ couple denotes the eutectic structure (grains) that represents the last liquid to solidify at just below the eutectic temperature T_E. For instance, this particular hypoeutectic solidification is evident in Fig. 8.1 for the 92Al-8Si alloy, in which α is white phase and β is dark.

For the solidification of a hypereutectic alloy with $C_o > C_e$, Fig. 8.5a or b predicts that the reaction defined by Eq. (8.3c) forms β-grains as the primary solidified solid phase and subsequently, as solidification proceeds under the influence of an undercooling ΔT, the last liquid at just below T_E solidifies as the eutectic structure. As a result, the microstructure consists of β and $(\alpha + \beta)$ grains as indicated in Fig. 8.1 for the 80Al-20 Si and 50Al-50 Si alloys.

So far solidification has been explained under the assumption of equilibrium solidification, which implies that the general phase transformation $L \rightarrow S$ occurs very slowly. However, the industrial production of alloys is basically a process that deviates from the equilibrium conditions predicted by phase diagrams. This implies that the solidification process variables like are altered or deviated from their equilibrium state. In principle, the solidification process depends on variables such as undercooling, alloy constitution, temperature gradient, growth rate, and rate of evolution of the latent heat of solidification.

In practice, application of alloy systems, such as *Al-Si* alloys in Fig. 8.1, is subjected to mechanical and corrosion behavior, and properties such as electric and thermal conductivity, mechanical properties, and so forth.

8.4 Lamellar Directional Solidification

The morphology of microstructures strongly depends on the solidification velocity; relatively slow as in industrial casting technology and extremely fast as in the rapid solidification technology (*RST*). The latter can be related to the production of metallic glasses, which are fundamentally frozen metallic melts in the amorphous state.

In general, precise understanding of the solidification as a whole and accurate control of related parameters, such as temperature gradients, solidification velocity, and the like, are essential in producing the right casting with appropriate microstructural morphology and suitable material properties for structural integrity and component functionality. All this implies that phase-field models are used to develop reliable analytical or numerical solutions to directional or unidirectional solidification problems.

The general overview of the directional solidification (phase transformation) process is illustrated in Fig. 8.6.

By definition, *directional solidification* starts at one end towards the extreme end of the mold in one direction only at a constant imposed velocity or induced fluctuating velocity υ_z and actual thermal gradient $G_a = (dT/dz)_a$.

On the other hand, *progressive solidification* starts from the mold walls towards the casting geometry centerline, where solidification is complete. The former is characterized below as per Tiller's [6] and Jackson–Hunt [7] eutectic solidification models, which are considered classic solidification models subjected to modifications in order to enhance the theoretical background for a better understanding of eutectic solidification. Recall that directional solidification of pure metals is described in Chap. 6 for producing single crystals using the Czochralski (Cz) process.

In this section, the theory of directional solidification is presented as per Tiller's model published in 1958 [6] and later defined by Jackson–Hunt model in 1966 [7] for characterizing alloy eutectic solidification. The former considers

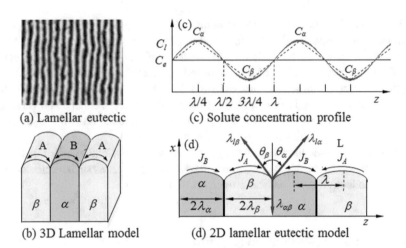

Fig. 8.6 Lamellar eutectic structure. (**a**) Real regular eutectic morphology, (**b**) 3D eutectic model, (**c**) z-periodicity and linearized (dashed lines) concentration profiles due to diffusion of A and B ahead of the lamellar solid phases, and (**d**) 2D model for lamellar eutectic

one-dimensional and the latter assumes a two-dimensional diffusion under the assumption of steady-state eutectic solidification.

The general assumptions and conditions for explaining this type of phase transformation at steady solidification velocity are

- Among different types of eutectic structures induced by the invariant eutectic reaction $L \rightarrow \alpha + \beta$, emphasis is given to modeling of the regularly spaced or symmetric α and β lamellae structure depicted in Fig. 8.6a. Note that the solid phases form in adjacent layers or lamellae. This implies that α and β grow continuously from the melt.
- The model assumes a plate-like lamellar structure (Fig. 8.6b), where the advancing α-β interface (solidification front) is the most active region the solidification process due to the coupled axial-lateral solute diffusion in the xz plane.
- In particular, the solute concentration profile $C_l = C_l(z)$ is also assumed to have a lateral periodicity along the z-axis (Fig. 8.6c).
- The molar fluxes (J_A and J_B) for A and B atoms in a binary alloy represent the excess solute in front of the growing α-β interface and the L-α-β triple junction defines the origin of the surface tension (surface energy) terms as shown in Fig. 8.6d.

In general, the eutectic structure is an mixture of A and B atoms by simultaneous cooperative growth mechanism and as a result, α-β faceted, lamellar, cellular, or dendritic morphology forms on surface defects (steps), while simultaneous solute rejection from one phase to another progresses.

In particular, the lamellar eutectic structure consisting of a coupled α-β is modeled as a plate-like structure having a symmetrical interface contour (Fig. 8.6b, d), which advances at a steady speed v_z in front of the supercooled liquid. The arrows in the sketches (Fig. 8.6b, d) indicate that A-atoms and B-atoms are rejected by the α-phase and β-phase, respectively. Thus, there exists lateral diffusion (z-axis) as solidification proceeds perpendicularly along the z-axis at a slow rate.

Especially, the excess solute B-atoms ahead of the lamellar α-phase diffuses laterally to the lamellar β-phase and conversely, excess solvent A-atoms ahead of the lamellar β-phase is rejected towards the lamellar α-phase. Moreover, the amount of excess atoms in front of the solid phases is the molar or mass flux J_A or J_B (Fig. 8.6d) as per Fick's first law of steady-state diffusion characterized analytically in Chap. 3.

The morphology of the eutectic structure may be in the form of lamellar (plate-like), rods, flake, faceted, or any other form. The former is adopted henceforth as shown in Fig. 8.6b. Once the coupled α-β lamellar structure initially forms as a thin layer, both α-interface and β-interface grow symmetrically due to lateral diffusion as indicated by the arrows in Fig. 8.6b or d.

As previously stated, the nucleation and growth of the α-interface and β-interface are subjected to mass transport (diffusion) in the liquid under the influence of a certain liquid undercooling ΔT, which in turn, it depends on the type of alloy composition and the freezing rate associated with the rate of latent heat of solidification $\Delta H_s > \Delta H_f^*$. Recall that an alloy latent heat of fusion depends on a temperature range based on the alloy nominal composition C_o. That is, $\Delta H_f^*(T_s) \leq \Delta H_f^*(T_f^*) \leq \Delta H_f^*(T_l)$.

8.4.1 Lamellar Size

Here, the lamellar thickness is defined as $2\lambda_\alpha$ and $2\lambda_\beta$, and $\lambda_\alpha = \lambda_\beta = \lambda/2$ for symmetry. Also, $\gamma_{l\alpha}, \gamma_{l\beta}, \gamma_{\alpha\beta}$ are the local interface surface energies at the triple junction. Note that the oscillatory concentration in Fig. 8.6c implies that the local liquid solute concentration in the melt is raised above the eutectic composition ($C_{l\alpha} > C_e$) ahead of the α-phase at a position $\lambda/4$, depressed ($C_{l\beta} < C_e$) ahead of the β-phase at $3\lambda/4$ and $C_e - C_{l\alpha} = -\left(C_e - C_{l\beta}\right)$ at the α-β interface. Moreover, the resultant liquid solute oscillatory concentration profile can be conveniently linearized (dashed lines) as per Glicksman [11, pp. 407–408] in order to simplify the analytical solution to the steady-state diffusion mechanism (Fick's first law of diffusion) in the liquid. Additional theoretical and mathematical details on this model can be found elsewhere [4, 11–17].

8.4.2 Conservation of Mass

Consider a melt undergoing phase change in an xz plane under the assumption of a small eutectic undercooling ΔT and a steady growth velocity v_z. Now, assume that

- The eutectic reaction $L \rightarrow \alpha + \beta$ occurs in the form of thermal stable plate-like called lamellar structure (Fig. 8.5b) having preferred crystallographic orientations, which are known as habits.
- The melt steadily and directionally solidifies as α-β lamellar crystalline structure under the assumption of a small eutectic undercooling ΔT and a relatively low growth velocity v_z.
- The coupled diffusion mechanism occurs at the α-β lamellar interface exhibiting a curvature and the solute diffusion coefficient D_l into the liquid phase is constant laterally and axially in the xz plane.
- The triple junction L-α-β exhibits low interfacial (surface) energies as schematically shown in Fig. 8.6d.
- The concentration field, denoted as $C_l = C_l(x, z)$, exhibits periodicity along the lateral $\pm x$-axis.

The mass balance for the formation of a lamellar eutectic structure can be carried out using a hypoeutectic or hypereutectic alloy with nominal composition C_o. In order to accomplish this, it is assumed that a planar α-β interface advances at a steady speed v_z, while latent heat of fusion (ΔH_f^*) evolves at a steady rate $ds(t)/dx$. Among the variables that are defined below, the main goal is to derive the eutectic undercooling (ΔT) equation for $L \rightarrow \alpha + \beta$ at $T < T_E$. Moreover, it is assumed that α and β have similar mass densities $(\rho_\alpha \simeq \rho_\beta)$ and different crystal structures.

Now, the mathematical modeling of directional solidification is introduced as compilations of Tiller [6] and Jackson–Hunt [7] models. The main goal is now based on the analytical approach these authors originally described their solidification models and include the relevant analytical procedures. It is not intended henceforth to critically compare these models. Actually, each model has its own merits based on principles of solidification.

8.5   Tiller's Eutectic Solidification Model

Consider Tiller's solidification theory for lamellar eutectic [6]. For a hypoeutectic alloy C_o and partition coefficient $k_o^\alpha = C_s^\alpha / C_l^\alpha < 0$, a one-dimensional steady-state mass transport of solute B-atoms is considered relevant in order to define the molar flux J_B due to lateral diffusion (advection) and J_B due to axial diffusion (Fick's first law) [11, pp. 407–408]. Thus,

$$J_B = v_z \left(C_l^\alpha - C_s^\alpha \right) = v_z \left(1 - k_o^\alpha \right) C_l^\alpha \quad \text{(lateral diffusion)} \qquad (8.4a)$$

$$J_B = -\phi D_l \frac{\partial C}{\partial x} \simeq \pm \phi D_l \frac{C_l^\alpha - C_l^\beta}{\lambda/2} \quad \text{(axial diffusion)} \qquad (8.4b)$$

where C_l^α, C_s^α, and C_l^β denote the solute concentrations in the α and β phases, respectively. Also, ϕ denotes an unknown geometric correction factor ($\phi \simeq 1$) and λ denotes the lamellar spacing.

Equating Eqs. (8.4a) and (8.4b) with $C_l^\alpha = C_e$ in (8.4a) yields the concentration difference in the liquid between α and β phases

$$v_z \left(1 - k_o^\alpha\right) C_e = \phi D_l \frac{C_l^\alpha - C_l^\beta}{\lambda/2} \tag{8.5a}$$

$$C_l^\alpha - C_l^\beta = \frac{\lambda}{2\phi} \frac{v_z}{D_l} \left(1 - k_o^\alpha\right) C_e \tag{8.5b}$$

Assume that the equilibrium liquidus field boundaries are linear in the equilibrium phase diagram. Thus, the linear temperature equations with slopes $m_{l\alpha} < 0$ and $m_{l\beta} > 0$, and the equilibrium temperature range ΔT expressions are defined by

$$T_l^\alpha = T_E + m_{l\alpha}(C_e - C_l^\alpha) \tag{8.6a}$$

$$T_l^\beta = T_E + m_{l\beta}\left(C_e - C_l^\beta\right) \tag{8.6b}$$

$$\Delta T_d = T_l^\alpha - T_E = m_{l\alpha}(C_e - C_l^\alpha) \tag{8.6c}$$

$$\Delta T_d = T_l^\beta - T_E = m_{l\beta}\left(C_e - C_l^\beta\right) \tag{8.6d}$$

where ΔT_d is now the driving force for lateral diffusion [6]. For convenience, rearrange Eqs. (8.6c) and (8.6d) to get

$$\frac{\Delta T_d}{m_{l\alpha}} = C_e - C_l^\alpha \tag{8.7a}$$

$$\frac{\Delta T_d}{m_{l\beta}} = C_e - C_l^\beta \tag{8.7b}$$

Subtracting these equations gives

$$\Delta T_C = \frac{m_{l\alpha} m_{l\beta}}{m_{l\alpha} - m_{l\beta}} \left(C_l^\alpha - C_l^\beta\right) \tag{8.8}$$

Combining Eqs. (8.5b) and (8.8) yields

$$\Delta T_C = \frac{m_{l\alpha} m_{l\beta}}{m_{l\alpha} - m_{l\beta}} \frac{\lambda}{2\phi} \frac{v_z}{D_l} \left(1 - k_o^\alpha\right) C_e \tag{8.9a}$$

$$\frac{\Delta T_C}{v_z} = \frac{m_{l\alpha} m_{l\beta}}{m_{l\alpha} - m_{l\beta}} \frac{\lambda}{2\phi D_l} \left(1 - k_o^\alpha\right) C_e = \text{constant} \tag{8.9b}$$

For symmetry, assume that $m_{l\alpha} = -m_{l\beta}$ so that Eq. (8.9a) becomes the sought undercooling term for the isothermal α-β interface

$$\Delta T_C = -\frac{\lambda m_{l\alpha}}{4\phi} \frac{v_z}{D_l} \left(1 - k_o^\alpha\right) C_e \tag{8.10a}$$

$$\Delta T_C = +\frac{\lambda m_{l\beta}}{4\phi} \frac{v_z}{D_l} \left(1 - k_o^\alpha\right) C_e \tag{8.10b}$$

Furthermore, the rate of free energy for the formation of the coupled α-β lamellar eutectic and for interfacial supercooling, as defined in Tiller's model [6] and Glicksman's book [11, p. 411], are written, respectively, as

$$\frac{d\left(\Delta G_{\alpha\beta}\right)}{dt} = \frac{2v_z}{\lambda}\gamma_{\alpha\beta} \tag{8.11a}$$

$$\frac{d\left(\Delta G_{\alpha\beta}\right)}{dt} = \Delta T_{\alpha\beta}\rho_s \Delta S_f^* v_z \tag{8.11b}$$

Equating Eqs. (8.11) and using (4.6) with $T = T_E$ yields the undercooling term due to any interface curvature

$$\Delta T_R = \frac{2\gamma_{\alpha\beta}}{\lambda\rho_s \Delta S_f^*} = \frac{2\gamma_{\alpha\beta} T_E}{\lambda\rho_s \Delta H_f^*} \tag{8.12}$$

Now that ΔT_C, ΔT_R have been defined, it is convenient to follow Jackson–Hunt analogy [7] in order to define the total undercooling, $\Delta T = (T_E - T) > 0$, for a two-phase alloy with $C_\alpha \leq C_o \leq C_\beta$. Thus,

$$\Delta T = \Delta T_C + \Delta T_R + \Delta T_K \tag{8.13}$$

where ΔT_C is due to departure from local eutectic composition, ΔT_R is due to effect of interface curvature, and ΔT_K is the non-equilibrium kinetic term.

Essentially, $\Delta T_K \ll \Delta T_C$, ΔT_R and subsequently, Eq. (8.13) can be simplified to [7]

$$\Delta T = \Delta T_C + \Delta T_R \tag{8.14a}$$

$$\Delta T = \frac{m_{l\alpha} m_{l\beta}}{m_{l\alpha} - m_{l\beta}} \frac{\lambda}{2\phi} \frac{v_z}{D_l} \left(1 - k_o^\alpha\right) C_e + \frac{2\gamma_{\alpha\beta} T_E}{\lambda\rho_s \Delta H_f^*} \tag{8.14b}$$

$$\Delta T = K_C v_z \lambda + \frac{K_R}{\lambda} \tag{8.14c}$$

where the constants K_C, K_R are Tiller's constant defined by

$$K_C = \left(\frac{m_{l\alpha} m_{l\beta}}{m_{l\alpha} - m_{l\beta}} \right) \frac{\left(1 - k_o^\alpha\right) C_e}{2\phi D_l} \tag{8.14d}$$

$$K_R = \frac{2\gamma_{\alpha\beta} T_E}{\rho_s \Delta H_f^*} = \frac{2\gamma_{\alpha\beta}}{\rho_s \Delta S_f^*} \tag{8.14e}$$

From Eq. (8.14c), $|d\left(\Delta T\right)/d\lambda|_{\lambda=\lambda_c} = K_C v_z - K_R/\lambda_c^2 = 0$ yields the critical (minimum) value of the interface spacing for a minimum undercooling

$$\lambda_c = \sqrt{\frac{K_R}{K_C} \frac{1}{v_z}} \tag{8.15a}$$

$$\Delta T_c = 2\sqrt{K_C K_R v_z} \tag{8.15b}$$

Rearrange Eq. (8.15) to get common expressions found in the literature as

$$\lambda_c^2 v_z = \frac{K_R}{K_C} = \text{constant} \tag{8.16a}$$

$$\frac{\Delta T_c^2}{v_z} = 4K_C K_R = \text{constant} \tag{8.16b}$$

Substituting Eqs. (8.14d) and (8.14e) into (8.16) gives

$$\lambda_c^2 v_z = \frac{4\phi D_l}{\left(1 - k_o^\alpha\right) C_e} \frac{\gamma_{\alpha\beta} \Gamma_{\alpha\beta}}{\Delta S_f^*} \left(\frac{m_{l\alpha} - m_{l\beta}}{m_{l\alpha} m_{l\beta}} \right) = \text{constant} \tag{8.17a}$$

$$\frac{\Delta T_c^2}{v_z} = \frac{4\left(1 - k_o^\alpha\right) C_e}{\phi D_l} \frac{\gamma_{\alpha\beta}}{\rho_s \Delta S_f^*} \left(\frac{m_{l\alpha} m_{l\beta}}{m_{l\alpha} - m_{l\beta}} \right) = \text{constant} \tag{8.17b}$$

For $m_{l\alpha} = -m_{l\beta}$ or $m_{l\beta} = -m_{l\alpha}$, Eq. (8.17) becomes Tiller's equations for eutectic solidification

$$\lambda_c^2 v_z = -\frac{8\phi D_l}{m_{l\alpha} \left(1 - k_o^\alpha\right) C_e} \frac{\gamma_{\alpha\beta}}{\rho_s \Delta S_f^*} = \text{constant} \tag{8.18a}$$

$$\frac{\Delta T_c^2}{v_z} = -\frac{2m_{l\alpha} \left(1 - k_o^\alpha\right) C_e}{\phi D_l} \frac{\gamma_{\alpha\beta}}{\rho_s \Delta S_f^*} = \text{constant} \tag{8.18b}$$

Essentially, Eqs. (8.16) through (8.18) can be treated a Tiller's equations for quantifying eutectic solidification, where $\lambda_c = f\left(v_z\right)$ and $T_c = f\left(v_z\right)$ or simply stated these entities are inversely proportional to the square root of the interface velocity; $\lambda_c \propto v_z^{-1/2}$ and $\Delta T_c \propto v_z^{-1/2}$. Actually, these are the conditions for planar interface instability as per Tiller's theory of eutectic solidification [6].

Accordingly, Glicksman [11, p. 414] cites relevant authors who successfully verified experimentally Tiller's constants and informs (confirms) that the interfacial undercooling for typical metallic eutectic solidification lies in the range $1\,\mathrm{K} \leq \Delta T \leq 10\,\mathrm{K}$ in the order of 1 mm/s solidification velocity.

Additionally, Tiller's constants can be used as scaling laws between the critical lamellar spacing λ_c and the solidification velocity v_z, and the undercooling ΔT and the solidification velocity v_z. In other words, the scaling law can be defined graphically using $\lambda_c = f(v_z)$ and $\Delta T = f(v_z)$.

With respect to the interfacial undercooling rage $1\,\mathrm{K} \leq \Delta T \leq 10\,\mathrm{K}$, it is clear that the eutectic solidification is assumed to occur close to the equilibrium conditions imposed on an equilibrium phase diagram. Thus, Tiller's assumption related to the interfacial undercooling (L-S interface) is based on a small ΔT, while the moving L-S interface is coupled through solute diffusion in the adjacent liquid phase.

Fundamentally, Tiller's theory of eutectic solidification considers steady-state conditions for near-equilibrium solidification, diffusion in the liquid phase and ignores solid-state diffusion. Moreover, Tiller's theory provides an excellent basic background for characterizing eutectic solidification. For instance, Tiller also discussed the lamellar and rod-like morphologies of the eutectic structure, lamellar to rod, and rod to globules transitions at relatively high growth rate v_z and large undercooling ΔT_c [6]. This, then, leads to lamellar eutectic, rod-like eutectic and globular eutectic structures. Only the lamellar eutectic has been treated in this chapter due to lack of space.

8.6   Jackson–Hunt Lamellar Solidification Model

The Jackson–Hunt (JH) solidification theory followed the work of Hillert [18] to establish the starting point for determining the scaling parameter in directional solidification at steady solidification velocity. For time-independent two-dimensional diffusion, the governing partial differential equation (PDE) is

$$\frac{\partial^2 C_l}{\partial x^2} + \frac{\partial^2 C_l}{\partial z^2} + \frac{v_z}{D_l}\frac{\partial C_l}{\partial z} = 0 \tag{8.19}$$

The diffusion molar flux (J_z) as per Fick's first law of steady-state diffusion for the liquid concentration $C_l = C_l(x, z)$ at the α-β interface acting as boundary conditions is

$$J_z = -D_l \frac{\partial C_l}{\partial x} = 0 \quad \text{for } x = 0, \lambda/2 \tag{8.20a}$$

$$D_l \frac{\partial C_l}{\partial z} = -v_z \left(1 - k_o^\alpha\right) C_l \quad \text{for } z = 0, 0 \leq x \leq \frac{f_\alpha \lambda}{2} \tag{8.20b}$$

$$= v_z \Delta C_\alpha \quad \text{for } z = 0, 0 \leq x \leq \lambda_\alpha$$

$$D_l \frac{\partial C_l}{\partial z} = v_z \left(1 - k_o^\beta\right) (1 - C_l) \quad \text{for } z = 0, \frac{f_\alpha \lambda}{2} \leq x \leq \frac{\lambda}{2} \qquad (8.20c)$$

$$= -v_z \Delta C_\beta \quad \text{for } z = 0, \lambda_\alpha \leq x \leq \lambda_\alpha + \lambda_\beta$$

$$C_l = C_o \quad \text{for } z \to \infty \qquad (8.20d)$$

In principle, this diffusion mechanism can occur on a planar $(-)$ or contoured (convex $\smile$ and concave $\frown$) L-S interface located at $z = 0$.

The analytical solution to the governing diffusion equation, Eq. (8.19), is derived in Appendix 8A as [8, p. 394]

$$C_l(x, z) = C_o + \sum_{n=0}^{\infty} F_n \cos\left(\frac{n\pi}{\lambda} x\right) \exp\left(-d_n z\right) \quad \text{for } C_o \neq C_e \qquad (8.21a)$$

$$C_l(x, z) = C_e + \sum_{n=0}^{\infty} F_n \cos\left(\frac{n\pi}{\lambda} x\right) \exp\left(-d_n z\right) \quad \text{for } C_o = C_e \qquad (8.21b)$$

For consistency, $d_o = v_z/D_l$ at $n = 0$ and Eq. (8.21) reduces to

$$C_l(x, z) = C_o + F_o \exp\left(-\frac{v_z}{D_l} z\right) + \sum_{n=1}^{\infty} F_n \cos\left(\frac{2n\pi x}{\lambda}\right) \exp\left(-d_n z\right) \qquad (8.22a)$$

$$C_l(x, z) = C_e + F_o \exp\left(-\frac{v_z}{D_l} z\right) + \sum_{n=1}^{\infty} F_n \cos\left(\frac{2n\pi x}{\lambda}\right) \exp\left(-d_n z\right) \qquad (8.22b)$$

where C_o denotes the alloy nominal concentration, F_n denote the Fourier coefficients known as amplitudes of the solute fluxes, F_o denotes the coefficient related to concentration change (defined below) in the boundary layer, and D_l/v_z denotes the usual characteristic thickness of the diffusion layer in the z-axis.

According to Jackson–Hunt [7], the eutectic composition C_e with respect to α and β phases is the only near-equilibrium entity in the aforementioned equations at the eutectic temperature T_E and the diffusion length (characteristic thickness) is greater than the lamellar spacing; $D_l/v_z > \lambda$.

8.6.1 Decay Coefficient

In addition, the decay coefficient d_n can easily be derived by taking the appropriate derivatives of $C_l(x, z)$ as per Eq. (8.19). The resultant expression is a quadratic equation defined by

$$-\sum_{n=0}^{\infty}\left(\frac{2n\pi}{\lambda}\right)^2 + \sum_{n=0}^{\infty} d_n^2 - \frac{v_z}{D_l}\sum_{n=0}^{\infty} d_n = 0 \qquad (8.23a)$$

$$-\left(\frac{2n\pi}{\lambda}\right)^2 + d_n^2 - \frac{v_z}{D_l}d_n = 0 \qquad (8.23b)$$

$$d_n^2 - \frac{v_z}{D_l}d_n - \left(\frac{2n\pi}{\lambda}\right)^2 = 0 \qquad (8.23c)$$

Solving this quadratic expression, Eq. (8.23c), yields the decay coefficient

$$d_n = \frac{v_z}{2D_l} + \sqrt{\left(\frac{v_z}{2D_l}\right)^2 + \left(\frac{2n\pi}{\lambda}\right)^2} \quad \text{for } n \geq 0 \qquad (8.24a)$$

$$d_o = \frac{v_z}{D_l} \quad \text{for } n = 0, \text{ planar interface} \qquad (8.24b)$$

$$d_n = \frac{2n\pi}{\lambda} \quad \text{if } \frac{v_z}{2D_l} << \frac{2n\pi}{\lambda} \ \& \ n \geq 1 \qquad (8.24c)$$

which follows Jackson–Hunt assumptions [7].

8.6.2   Fourier Coefficients

Consider the extension of the boundary fields shown in Fig. 8.5 and assume that $C_l \rightarrow C_e$ is a reasonable approximation for finding F_o and F_n as defined by Eqs. (8A.29b) and (8A.29c), Appendix 8A. In summary [4, p. 395],

$$F_o = \Delta C_\alpha f_\alpha - \Delta C_\beta f_\beta \qquad (8.25a)$$

$$= \left(1 - k_o^\alpha\right) f_\alpha C_e - \left(1 - k_o^\beta\right)(1 - C_e) f_\beta$$

$$F_n = \frac{v_z \lambda \Delta C_o}{(n\pi)^2 D_l} \sin(n\pi f_\alpha) = \frac{v_z \lambda \left(\Delta C_\alpha + \Delta C_\beta\right)}{(n\pi)^2 D_l} \sin(n\pi f_\alpha) \qquad (8.25b)$$

Note that $\Delta C_o = \Delta C_\alpha + \Delta C_\beta$. Hence, Eq. (8.22) becomes the analytical solution to (8.19)

$$C_l(x, z) = C_o + \left(\Delta C_\alpha f_\alpha - \Delta C_\beta f_\beta\right) \exp\left(-\frac{v_z}{D_l} z\right) \qquad (8.26)$$

$$+ \frac{v_z \lambda \Delta C_o}{D_l} \sum_{n=1}^{\infty} \frac{\sin(n\pi f_\alpha)}{(n\pi)^2} \cos\left(\frac{2n\pi x}{\lambda}\right) \exp\left(-\frac{n\pi z}{\lambda}\right)$$

If $n = 0$ and $C_o = k_o^\alpha C_e$ in Eq. (8.26), then (7.93b) is recovered

$$C_l(x, z) = C_o + \Delta C_\alpha \exp\left(-\frac{z}{D_l/v_x}\right) \tag{8.27a}$$

$$C_l(x, z) = C_o + \left(1 - k_o^\alpha\right) C_e \exp\left(-\frac{z}{D_l/v_x}\right) \tag{8.27b}$$

$$C_l(x, z) = C_o + \left(\frac{1 - k_o^\alpha}{k_o^\alpha}\right) C_o \exp\left(-\frac{z}{D_l/v_x}\right) \tag{8.27c}$$

In addition, evaluating Eq. (8.27c) along the z-direction (axial diffusion) leads to the following simplifications

$$C_l(x, z) = C_o + \Delta C_\alpha = C_o + \left(\frac{1 - k_o^\alpha}{k_o^\alpha}\right) C_o \quad \text{at } z = 0 \tag{8.28a}$$

$$C_l(x, 0) = C_o/k_o^\alpha \quad \text{at } z = 0 \tag{8.28b}$$

$$C_l(x, \infty) = C_o \quad \text{at } z = \infty \tag{8.28c}$$

Note that Eq. (8.28c) is the anticipated condition defined by (8.20d). Therefore, Eq. (8.26) is a reliable expression under the assumption of small undercooling for steady directional solidification. Recall that the change of solute compositions ΔC_α, ΔC_β, ΔC_o are extracted from an equilibrium phase diagram (Fig. 8.5a).

8.6.3 Regular Lamellar Supercooling

The analytical procedure for determining the degree of undercooling ΔT equation for any arbitrary binary alloy with a nominal composition C_o (off-eutectic composition $C_o \neq C_e$ or eutectic composition $C_o = C_e$) is described below [7, 8].

Assume that the liquidus and solidus lines in an equilibrium phase diagram are linear, the α-β interface growth is isothermal (constant $T_{\alpha\beta}$), the surface attachment kinetics (related to crystal growth along crystallographic planes) is negligible. Thus, the average quantities needed for deriving ΔT are

$$\overline{C}_l^\alpha = \frac{2}{f_\alpha \lambda} \int_0^{f_\alpha \lambda/2} C_l(x, 0)\, dx = C_o + \left(f_\alpha \Delta C_\alpha - f_\beta \Delta C_\beta\right) + \frac{v_z \lambda \Delta C_o}{f_\alpha} \overline{F}_n \tag{8.29a}$$

$$\overline{C}_l^\beta = \frac{2}{f_\beta \lambda} \int_{f_\alpha \lambda/2}^{\lambda/2} C_l(x, 0)\, dx = C_o + \left(f_\alpha \Delta C_\alpha - f_\beta \Delta C_\beta\right) - \frac{v_z \lambda \Delta C_o}{f_\beta} \overline{F}_n \tag{8.29b}$$

$$\overline{F}_n = \sum_{n=1}^{\infty} \frac{\sin(n\pi f_\alpha)}{(n\pi)^3} \tag{8.29c}$$

The accurate magnitude of $\overline{F}_n$ in Eq. (8.29c) depends on the number of terms in the series. Moreover, the average solute undercooling for α and β phases are mathematically defined by

$$\overline{\Delta T}_C^\alpha = |m_{l\alpha}| \left(\overline{C}_l^\alpha - C_e \right) \tag{8.29d}$$

$$= |m_{l\alpha}| \left(C_o - C_e + f_\alpha \Delta C_\alpha - f_\beta \Delta C_\beta \right) + \frac{P_e |m_{l\alpha}| \Delta C_o}{f_\alpha} \overline{F}_n$$

$$\overline{\Delta T}_C^\beta = |m_{l\beta}| \left(C_e - \overline{C}_l^\beta \right) \tag{8.29e}$$

$$= |m_{l\beta}| \left(C_e - C_o - f_\alpha \Delta C_\alpha + f_\beta \Delta C_\beta \right) + \frac{P_e |m_{l\beta}| \Delta C_o}{f_\beta} \overline{F}_n$$

In reality, $\overline{\Delta T}_C^\alpha, \overline{\Delta T}_C^\beta$ denote the average solute supercooling for each lamellae and P_e denotes the Peclet number defined by

$$P_e = \frac{\upsilon_z \lambda}{D_l} \tag{8.30}$$

Now, the average curvature undercooling along with the interface curvature $k(z)$ of an arc in the form $z = f(x)$ for each phase is

$$\overline{\Delta T}_R^\alpha = \frac{2\Gamma_{l\alpha}}{f_\alpha \lambda} \int_0^{f_\alpha \lambda/2} k_\alpha(z)\, dx \tag{8.31a}$$

$$\overline{\Delta T}_R^\beta = \frac{2\Gamma_{l\beta}}{f_\beta \lambda} \int_{f_\alpha \lambda/2}^{\lambda/2} k_\beta(z)\, dx \tag{8.31b}$$

Here, $k(z)$ is unknown, but it can be derived using elementary calculus as illustrated below.

8.6.4   Symmetric Boundary Groove

The curvature function $z(x)$ can represent the temperature gradient $\pm dT/dx$ along α or β curvature as schematically shown in Fig. 8.7.

By definition, θ is the dihedral angle between two intersecting α and β planes. The quantitative measurement of a dihedral angle θ can be made at high scanning

Fig. 8.7 (**a**) Triple junction between α and β convex $z(x)$ curves and (**b**) coordinate system with tangential slope $z'(x)$

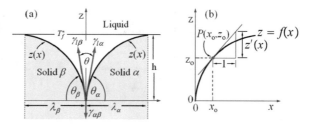

electron microscope (*SEM*) magnification. According to Fig. 8.7a, the relevant interfacial surface energies take the form

$$\gamma_{\alpha\beta} = \gamma_{l\alpha} \cos(\theta_\alpha) + \gamma_{l\beta} \cos(\theta_\beta) \tag{8.32}$$

$$\gamma_{l\alpha} \sin(\theta_\alpha) = \gamma_{l\beta} \sin(\theta_\beta) \tag{8.33}$$

These expressions arise due to surface tension between α and β phases. This is attributed to the cohesive forces that keep these phases parallel to each other.

Following Protter-Morrey, Jr [19, p. 399], the local interface curvature $k(z)$ of an arc $z = f(x)$ (Fig. 8.7b) is the rate of change of the dihedral angle θ (Fig. 8.7a) with respect to the arc length s. For an arc of the form $z = f(x)$, the chain rule conveniently gives

$$k(z) = \frac{d\theta}{ds} = \frac{d\theta}{dx}\frac{dx}{ds} \tag{8.34}$$

where

$$\frac{dz}{dx} = f'(x) = \tan(\theta) \tag{8.35a}$$

$$\frac{d\theta}{dx} = \frac{d}{dx}\left[\arctan\left(\frac{dz}{dx}\right)\right] = \frac{d}{dx}\left[\arctan f'(x)\right] = \frac{f''(x)}{1 + [f'(x)]^2} \tag{8.35b}$$

with $f''(x) = d^2z/dx^2$. Now, the arc length s and the corresponding derivative with rest to the x-direction are written as

$$s = \int \sqrt{1 + [f'(x)]^2}\,dx \tag{8.36a}$$

$$ds = \sqrt{1 + [f'(x)]^2}\,dx \tag{8.36b}$$

$$\frac{dx}{ds} = \frac{1}{\sqrt{1 + [f'(x)]^2}} \tag{8.36c}$$

Substituting Eqs. (8.35b) and (8.36c) into (8.34) yields the general and fundamental equation of curvature $k(z)$ of an arc $z = f(x)$ (see Appendix 8B)

$$k(z) = \frac{f''(x)}{1 + [f'(x)]^2} \frac{1}{\sqrt{1 + [f'(x)]^2}} = \frac{f''(x)}{\left[1 + [f'(x)]^2\right]^{3/2}} \qquad (8.37a)$$

from which,

$$k(z) = \frac{d^2 z(x)/dx^2}{\left[1 + [dz(x)/dx]^2\right]^{3/2}} = \frac{d[\tan(\theta)]/dx}{\sqrt{\left[1 + \tan^2(\theta)\right]^3}} = \frac{\left[1 + \tan^2(\theta)\right] d\theta/dx}{\sqrt{\left[1 + \tan^2(\theta)\right]^3}} \qquad (8.37b)$$

$$k(z) = \frac{1}{\sqrt{1 + \tan^2(\theta)}} \frac{d\theta}{dx} = \frac{1}{\sqrt{1 + \sin^2(\theta)/\cos^2(\theta)}} \frac{d\theta}{dx} = \cos(\theta) \frac{d\theta}{dx} \qquad (8.37c)$$

Inserting Eq. (8.37c) into (8.31) and defining $k(z)$ for both α and β lamellae yields [20, p. 267]

$$\overline{\Delta T}^{\alpha}_{R} = \frac{2\Gamma_{l\alpha}}{f_{\alpha}\lambda} \int_0^{\theta_{\alpha}} \frac{d^2 z(x)/dx^2}{\left[1 + (dz(x)/dx)^2\right]^{3/2}} dx \qquad (8.38a)$$

$$= \frac{2\Gamma_{l\alpha}}{f_{\alpha}\lambda} \int_0^{\theta_{\alpha}} \cos(\theta) \frac{d\theta}{dx} dx = \frac{2\Gamma_{l\alpha}\sin\theta_{\alpha}}{f_{\alpha}\lambda}$$

$$\overline{\Delta T}^{\beta}_{R} = \frac{2\Gamma_{l\beta}}{f_{\beta}\lambda} \int_0^{\theta_{\beta}} \frac{d^2 z(x)/dx^2}{\left[1 + (dz(x)/dx)^2\right]^{3/2}} dx \qquad (8.38b)$$

$$= \frac{2\Gamma_{l\beta}}{f_{\beta}\lambda} \int_0^{\theta_{\beta}} \cos(\theta) \frac{d\theta}{dx} dx = \frac{2\Gamma_{l\beta}\sin\theta_{\beta}}{f_{\beta}\lambda}$$

For an isothermal α-β interface, $\overline{\Delta T}^{\alpha} = \overline{\Delta T}^{\beta}$ and the average eutectic undercooling with positive weighted factors, $|m_{l\alpha}|$ and $|m_{l\beta}|$, is written as ([7, p. 398], [16, p. 268] and [20, p. 266])

$$\Delta T = \frac{\overline{\Delta T}^{\alpha} |m_{l\alpha}|^{-1} + \overline{\Delta T}^{\beta} |m_{l\beta}|^{-1}}{|m_{l\alpha}|^{-1} + |m_{l\beta}|^{-1}} \qquad (8.39)$$

Combining Eqs. (8.38) and (8.39) yields the classic Jackson–Hunt total supercooling equation with positive liquidus slopes $m_{l\alpha}$ and $m_{l\beta}$ [7]. For convenience, the resultant ΔT expression is written as per Dantzig-Rappaz [8]

$$\Delta T = \frac{\lambda v_z}{D_l} \frac{\Delta C_o}{f_\alpha f_\beta} \frac{|m_{l\alpha}|\,|m_{l\beta}|}{|m_{l\alpha}| + |m_{l\beta}|} \sum_{n=1}^{\infty} \frac{\sin^2(n\pi f_\alpha)}{(n\pi)^3} \tag{8.40a}$$

$$+ \frac{1}{\lambda} \frac{|m_{l\alpha}|\,|m_{l\beta}|}{|m_{l\alpha}| + |m_{l\beta}|} \left[\frac{2\Gamma_{l\alpha}\sin(\theta_\alpha)}{|m_{l\alpha}| f_\alpha} + \frac{2\Gamma_{l\beta}\sin(\theta_\beta)}{|m_{l\beta}| f_\beta} \right]$$

In fact, the amount of undercooling ΔT in Eq. (8.40a) is defined as per (8.14a) and it is recast here for convenience

$$\Delta T = K_C v_z \lambda + \frac{K_R}{\lambda} \tag{8.41}$$

with

$$K_C = \frac{\Delta C_o}{f_\alpha f_\beta D_l} \frac{|m_{l\alpha}|\,|m_{l\beta}|}{|m_{l\alpha}| + |m_{l\beta}|} \sum_{n=1}^{\infty} \frac{\sin(n\pi f_\alpha)}{(n\pi)^3} \tag{8.42a}$$

$$K_R = \frac{|m_{l\alpha}|\,|m_{l\beta}|}{|m_{l\alpha}| + |m_{l\beta}|} \left[\frac{2\Gamma_{l\alpha}\sin(\theta_\alpha)}{|m_{l\alpha}| f_\alpha} + \frac{2\Gamma_{l\beta}\sin(\theta_\beta)}{|m_{l\beta}| f_\beta} \right] \tag{8.42b}$$

where $\Gamma_{l\alpha}$, $\Gamma_{l\beta}$ denote the Gibbs–Thomson coefficients, $|m_{l\alpha}|$, $|m_{l\beta}|$ denote the absolute values of the liquidus slopes (Fig. 8.5), and K_C, K_R have the same physical meaning in Tiller's model [6] and Jackson–Hunt model [7], but different mathematical definitions.

8.6.5 Gibbs–Thomson Coefficients

The Gibbs–Thomson coefficients, $\Gamma_{l\alpha}$ and $\Gamma_{l\beta}$, for the L-α and L-β interfaces are material properties related to interfacial or surface energy due to surface tension and thermodynamic entities, such as entropy change of fusion ΔS_f or enthalpy change of fusion ΔH_f (latent heat of fusion). Thus, Gibbs–Thomson coefficients are written as [20, p. 7]

$$\Gamma_{l\alpha} = \frac{\gamma_{l\alpha}}{\rho_s^\alpha \Delta S_f^\alpha} = \left(\frac{\gamma_{l\alpha}}{\rho_s^\alpha \Delta H_f^\alpha} \right) T_E \tag{8.43a}$$

$$\Gamma_{l\beta} = \frac{\gamma_{l\beta}}{\rho_s^\beta \, \Delta S_f^\beta} = \left(\frac{\gamma_{l\beta}}{\rho_s^\beta \, \Delta H_f^\beta}\right) T_E \tag{8.43b}$$

where the entropy change of fusion defined by Eq. (4.4b) is recast as a general expression

$$\Delta S_f = \frac{\Delta H_f}{T_E} \tag{8.44}$$

With regard to Eq. (8.43), ρ_s^α, ρ_s^β denote the mass densities (kg/m^3), ΔS_f^α, ΔS_f^β denote the specific entropy change of fusion (J/kg K), $\gamma_{l\alpha}$, $\gamma_{l\beta}$ denote the surface energies (Fig. 8.5d), and ΔH_f^α, ΔH_f^β denote the enthalpy change per volume (J/m^3) for α and β solid phases. Moreover, $\gamma_{l\alpha}$ and $\gamma_{l\beta}$ can be defined as the work required to create a unit area of interface, and subsequently, $\Gamma_{l\alpha}$ and $\Gamma_{l\beta}$ may be treated as work-related coefficients.

8.6.6   Extremum Condition

In general, "extremum" is a common name for maximum and minimum value of a function evaluated at some particular independent variable. Fundamentally, Eq. (8.41) gives $|d(\Delta T)/d\lambda|_{\lambda=\lambda_c} = 0$ and $d^2(\Delta T)/d\lambda^2 > 0$ for a critical (minimum) ΔT at a critical λ_c. Thus,

$$\lambda_c^2 v_z = \frac{K_R}{K_C} = K_z = \text{constant} \tag{8.45a}$$

$$\lambda_c = \sqrt{\frac{K_R}{K_C} \frac{1}{v_z}} \tag{8.45b}$$

Substituting Eq. (8.45b) into (8.41) yields the minimum ΔT_c and a material constant $\Delta T_c / \sqrt{v_z}$ under the JH extremum condition

$$\Delta T_c = \left(2\sqrt{K_C K_R}\right)\sqrt{v_z} = K_e \sqrt{v_z} \tag{8.46a}$$

$$\frac{\Delta T_c}{\sqrt{v_z}} = 2\sqrt{K_C K_R} = K_e = \text{Constant} \tag{8.46b}$$

It is worth noting that $\Delta T_c = f(v_z)$ is a non-linear model with a material constant $K_e = 2\sqrt{K_C K_R}$.

With reference to Eqs. (8.45a) and (8.46b), Table 8.1 lists a compilation of experimental values for some binary alloys that solidified as regular lamellar eutectic structure under suitable conditions [21, p. 53].

Table 8.1 Regular lamellar values under the JH extremum condition [21]

Alloy	T_E (°C)	C_e (%)	$\lambda_c^2 v_z$ ($\mu m^3/s$)	$\Delta T_c / \sqrt{v_z}$ [K$(\mu m/s)^{-1/2}$]
Ag-Pb	328	95.5	120	0.14
Cu-Ag	779	71.9	80	0.18
Al-Zn	382	95.0	64	0.05
Bi-Zn	255	2.7	69	0.04
Cd-Pb	248	82.5	21	0.08
Cd-Sn	176	66.6	72	0.05
Pb-Sn	183	61.9	33	0.13

Notice that the extremum condition indicates that λ_c, v_z, and ΔT_c are common solidification parameters independent of the liquid thermal gradient dT_l/dz ahead of the α-β lamellar eutectic interface. According to Dantzig-Rappaz [8], these parameters, as defined by Eqs. (8.45a) and (8.46b), can be applied to equiaxed solidification at constant local undercooling ΔT_c.

8.7 Free-Energy Solidification Model

This section is devoted to solidification of eutectic binary alloys using the concept of free energy. Recall that alloy solidification is mainly controlled by solute diffusion, whereas pure metals are controlled by heat conduction.

Assumed that during solidification both α-β phases grow simultaneously due to lateral diffusion along the x-axis and axial diffusion into the liquid z-axis, the α-β interface growth is isothermal and that the α-β interface has a lamellae morphology as indicated in Fig. 8.5b, d.

Consider the solidification of a rectangular hypoeutectic alloy with nominal composition $C_\alpha \leq C_o \leq C_e$ (Fig. 8.8a) and the corresponding Gibbs energy changes (Fig. 8.8b) as per Porter et al. [22, p. 223]. From Eq. (4.10) with a rectangular volume $xyz = \lambda yz$ and cross-sectional area yz, the general Gibbs energy change can be written as

$$\Delta G = -(\lambda yz) \frac{\rho_s \Delta H_f^* \Delta T_R}{T_E} + (2yz)\gamma_{\alpha\beta} \qquad (8.47)$$

where ΔT_R is the undercooling contribution for α-β interface formation. This partial supercooling is required to overcome the interfacial curvature effects. The subscript "R" in ΔT_R stands for the radius of curvature. Hence, ΔT_R is the undercooling due to curvature effects and $\Delta G_{z > 0}$ in Fig. 8.8b denotes the infinite free-energy change ΔG_∞ as normally found in the literature.

If $\Delta G = 0$ at equilibrium, then the undercooling (ΔT_R) and the minimum lamellar $\alpha\beta$ spacing (λ) become

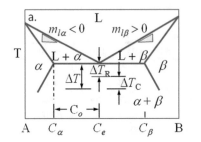
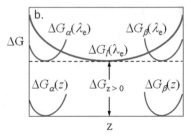

Fig. 8.8 Schematic (**a**) free-energy diagram after Ref. [22, p. 223] and (**b**) phase diagram showing the level of the eutectic supercooling

$$\Delta T_R = \frac{2\gamma_{\alpha\beta} T_E}{\rho_s \Delta H_f^*} \frac{1}{\lambda} \tag{8.48a}$$

$$\lambda = \frac{2\gamma_{\alpha\beta}}{\rho_s \Delta H_f^*} \frac{T_E}{\Delta T_R} \tag{8.48b}$$

It is worth mentioning that ΔT_R and λ are simple expressions as compared with Tiller's and Jackson–Hunt models.

For one-dimensional steady-state diffusion, the molar flux along the axial diffusional dimension, z-direction, is conveniently written as

$$J_z = -D_l \frac{\partial C_l}{\partial z} \simeq D_l \frac{\Delta C_l}{\lambda/2} \tag{8.49a}$$

$$J_z = \left(\frac{C_e - C_{l\alpha}}{2}\right) \frac{dz}{dt} = \left(\frac{C_e - C_{l\alpha}}{2}\right) v_z \tag{8.49b}$$

where $\Delta C_l = \Delta T_C / m_{l\alpha}$ is the slope of the hypoeutectic liquidus line (Fig. 8.8b). The subscript "C" in ΔT_C stands for concentration or solute composition. The term ΔT_C is simply a mathematical temperature difference.

Combining Eq. (8.49) yields the solidification velocity for a hypoeutectic alloy with $C_o = C_{l\alpha}$ as

$$v_z = \frac{4D_l \Delta C_l}{\lambda (C_e - C_o)} = \frac{4D_l \Delta T_C}{\lambda m_{l\alpha} (C_e - C_o)} \tag{8.50a}$$

$$\Delta T_C = \left[\frac{m_{l\alpha} (C_e - C_o)}{4D_l}\right] v_z \lambda > 0 \tag{8.50b}$$

Now, ΔT_C becomes the undercooling required for atomic steady-state diffusion related to the molar flux (J_z) and a concentration gradient as per Fick's first law of diffusion. Here, $m_{l\alpha} < 0$ and $C_e < C_o$.

For high atom mobility across the α-β interface, the total undercooling is defined by

$$\Delta T = \Delta T_C + \Delta T_R \tag{8.51a}$$

$$\Delta T = \left[\frac{m_{l\alpha}\,(C_e - C_o)}{4D_l}\right] v_z \lambda + \frac{2\gamma_{\alpha\beta}\,T_E}{\rho_s\,\Delta H_f^*}\frac{1}{\lambda} \tag{8.51b}$$

$$\Delta T = K_C v_z \lambda + \frac{K_R}{\lambda} \tag{8.51c}$$

where

$$K_C = \frac{m_{l\alpha}\,(C_e - C_o)}{4D_l} \tag{8.52a}$$

$$K_R = \frac{2\gamma_{\alpha\beta}\,T_E}{\rho_s\,\Delta H_f^*} \tag{8.52b}$$

Notably, Eq. (8.51c) has the same definition as (8.13) and (8.41), but with different K_C and K_R expressions. In addition, Eq. (8.52) are very simple mathematical expressions that do not take into account diffusion in the solid phase during solidification. Specifically, K_C strongly depends on the diffusion coefficient D_l and K_R depends on the thermophysical properties of the material.

For $\Delta T = T_E - T$, the supercooled liquid temperature is at

$$T = T_E - \Delta T \tag{8.53a}$$

$$T = T_E + \left[\frac{(C_e - C_o)}{4D_l}\right] v_z \lambda + \frac{2\gamma_{\alpha\beta}\,T_E}{\rho_s\,\Delta H_f^*}\frac{1}{\lambda} \tag{8.53b}$$

and when $C_o = C_e$, Eqs. (8.51b) and (8.53b) reduce to

$$\Delta T = \frac{2\gamma_{\alpha\beta}\,T_E}{\rho_s\,\Delta H_f^*}\frac{1}{\lambda} \tag{8.54a}$$

$$T = T_E + \frac{2\gamma_{\alpha\beta}\,T_E}{\rho_s\,\Delta H_f^*}\frac{1}{\lambda} \tag{8.54b}$$

Remarkably, the above treatment is a simple analytical solution to solidification problem related to a two-phase microstructure without the effect of the interface curvature across the α-β interface.

Example 8A.1 Consider a hypothetical binary A-B alloy with $C_o = 8\%$ and $C_e = 72\%$ undergoing solidification at $v_z = 200, 400, 600\,\mu m/s$ under the assumption of small eutectic undercooling. The available data set is

Characterize $\Delta T = f(\lambda)$ at v_z.

$m_{l\alpha} = -m_{l\beta} = -4.22\,°\text{C}/\%$	$D_l = 10^{-9}\,\text{m}^2/\text{s}$
$\phi = 1$	$\Delta C_o = 64\%$
$x = s\,(t) = 0.001\,\text{m}$	$\Gamma_{l\alpha} = \Gamma_{l\beta} = 1.70 \times 10^{-7}\,\text{K m}$
$f_\alpha = f_\beta = 0.5$	$\theta_\alpha = \theta_\beta = \pi/3$
$\gamma_{\alpha\beta} = 0.059\,\text{J/m}^2$	$k_o^\alpha = 0.2$
$\Delta H_f^* = \Delta H_s = 104.14\,\text{kJ/kg}$	$c_s = 231\,\text{J/(kg K)}$
$\rho_s = 8650\,\text{kg/m}^3$	$T_E = 449\,\text{K}$

Solution *Tiller's Model [6]* From Eqs. (8.14d), (8.14e) and (8.14c), respectively,

$$K_C = \left(\frac{m_{l\alpha} m_{l\beta}}{m_{l\alpha} - m_{l\beta}} \right) \frac{\left(1 - k_o^\alpha \right) C_e}{2\phi D_l} = 6.0768 \times 10^{10} \frac{\text{K s}}{\text{m}^2} = 6.0768 \times 10^{-2} \frac{\text{K s}}{\mu\text{m}^2}$$

$$K_R = \frac{2\gamma_{\alpha\beta} T_E}{\rho_s \Delta H_f^*} = \frac{2\gamma_{\alpha\beta}}{\Delta S_f^*} = 5.8816 \times 10^{-7}\,\text{K m} = 0.58816\,\text{K}\,\mu\text{m}$$

$$\Delta T = K_C \upsilon_z \lambda + \frac{K_R}{\lambda} = \left(6.0768 \times 10^{-2} \frac{\text{K s}}{\mu\text{m}^2} \right) \upsilon_z \lambda + \frac{0.58816\,\text{K}\,\mu\text{m}}{\lambda}$$

For $\upsilon_z = 200,\ 400$ and $600\,\mu\text{m/s}$,

$$\Delta T = (6.0768 \times 10^{-2})\,(200)\,\lambda + \frac{0.58816}{\lambda} = 12.154\lambda + \frac{0.58816}{\lambda}$$

$$\Delta T = (6.0768 \times 10^{-2})\,(400)\,\lambda + \frac{0.58816}{\lambda} = 24.307\lambda + \frac{0.58816}{\lambda}$$

$$\Delta T = (6.0768 \times 10^{-2})\,(600)\,\lambda + \frac{0.58816}{\lambda} = 36.461\lambda + \frac{0.58816}{\lambda}$$

From Eq. (8.15) with $\lambda = \lambda_c$ and $\Delta T = \Delta T_c$,

$$\lambda_c = \sqrt{\frac{K_R}{K_C \upsilon_z}} = (3.1111)\sqrt{\frac{1}{200}} = 0.21999\,\mu\text{m}$$

$$\lambda_c = \sqrt{\frac{K_R}{K_C \upsilon_z}} = (3.1111)\sqrt{\frac{1}{400}} = 0.15556\,\mu\text{m}$$

$$\lambda_c = \sqrt{\frac{K_R}{K_C \upsilon_z}} = (3.1111)\sqrt{\frac{1}{600}} = 0.12701\,\mu\text{m}$$

and

$$\Delta T_c = 2\sqrt{K_C K_R \upsilon_z} = (0.3781)\sqrt{200} = 5.3471\,\text{K}$$

$$\Delta T_c = 2\sqrt{K_C K_R \upsilon_z} = (0.3781)\sqrt{400} = 7.5620\,\text{K}$$

$$\Delta T_c = 2\sqrt{K_C K_R \upsilon_z} = (0.3781)\sqrt{600} = 9.2615\,\text{K}$$

Finally, the constants related to λ_c and ΔT_c are verified and given as

$$\lambda_c^2 \upsilon_z = 9.6788\,\mu\text{m}^3/\text{s}$$

$$\frac{\Delta T_c}{\sqrt{\upsilon_z}} = 0.3781\,\text{K}\,(\mu\text{m/s})^{-1/2}$$

Curve fitting or non-linear least squares (regression) yields the critical or minimum undercooling $\Delta T_c = f(\lambda_c)$ as

$$\Delta T_c = 1.1763/\lambda_c \quad (minimum)$$

The minimum λ_c values for $\Delta T_c = f(\lambda_c)$ represent stable lamellar growth to perturbations. For convenience, $\Delta T = f(\lambda)$ and $\Delta T_c = f(\lambda_c)$ are plotted as solid red curves towards the end of this example.

Jackson–Hunt Model (JH) [7] From Eq. (8.42) with 1000 terms in the series, the K_C and K_R constants are

$$K_C = \frac{\Delta C_o}{f_\alpha f_\beta D_l} \frac{|m_{l\alpha}|\,|m_{l\beta}|}{|m_{l\alpha}| + |m_{l\beta}|} \sum_{n=1}^{1000} \frac{\sin(n\pi f_\alpha)}{(n\pi)^3} = 1.8323 \times 10^{-2}\,\frac{\text{K}\,\text{s}}{\mu\text{m}^2}$$

$$K_R = \frac{|m_{l\alpha}|\,|m_{l\beta}|}{|m_{l\alpha}| + |m_{l\beta}|} \left[\frac{2\Gamma_{l\alpha}\sin(\theta_\alpha)}{|m_{l\alpha}|\,f_\alpha} + \frac{2\Gamma_{l\beta}\sin(\theta_\beta)}{|m_{l\beta}|\,f_\beta} \right] = 0.34\,\text{K}\,\mu\text{m}$$

Substituting these constants into Eq. (8.41) for the given solidification velocities yields the ΔT equations as

$$\Delta T = \left(3.6646\,\frac{\text{K}}{\mu\text{m}}\right)\lambda + \frac{0.34\,\text{K}\,\mu\text{m}}{\lambda} \quad \text{at } \upsilon_z = 200\frac{\mu\text{m}}{\text{s}}$$

$$\Delta T = \left(7.3292\,\frac{\text{K}}{\mu\text{m}}\right)\lambda + \frac{0.34\,\text{K}\,\mu\text{m}}{\lambda} \quad \text{at } \upsilon_z = 400\frac{\mu\text{m}}{\text{s}}$$

$$\Delta T = \left(10.9940\,\frac{\text{K}}{\mu\text{m}}\right)\lambda + \frac{0.34\,\text{K}\,\mu\text{m}}{\lambda} \quad \text{at } \upsilon_z = 600\frac{\mu\text{m}}{\text{s}}$$

The λ_c values for ΔT_c are

$$\lambda_c = \sqrt{\frac{K_R}{K_C}\frac{1}{\upsilon_z}} = \left(4.3077\,\sqrt{\mu\text{m}^3/\text{s}}\right)\sqrt{\frac{1}{200\,\mu\text{m/s}}} = 0.3046\,\mu\text{m}$$

$$\lambda_c = \left(4.3077\sqrt{\mu m^3/s}\right)\sqrt{\frac{1}{400\,\mu m/s}} = 0.21539\,\mu m$$

$$\lambda_c = \left(4.3077\sqrt{\mu m^3/s}\right)\sqrt{\frac{1}{600\,\mu m/s}} = 0.17586\,\mu m$$

where

$$\Delta T_c = \left(2\sqrt{K_C K_R}\right)\sqrt{v_z} = \left(0.15786\,K\sqrt{s/\mu m}\right)\sqrt{v_z}$$

$$\Delta T_c = \left(0.15786\,K\sqrt{s/\mu m}\right)\sqrt{200} = 2.2325\,K$$

$$\Delta T_c = \left(0.15786\,K\sqrt{s/\mu m}\right)\sqrt{400} = 3.1572\,K$$

$$\Delta T_c = \left(0.15786\,K\sqrt{s/\mu m}\right)\sqrt{600} = 3.8668\,K$$

Then,

$$\lambda_c^2 v_z = 18.556\,\mu m^3/s$$

$$\frac{\Delta T_c}{\sqrt{v_z}} = 0.15786\,K\,(\mu m/s)^{-1/2}$$

Curve fitting yields the minimum undercooling $\Delta T_c = f(\lambda_c)$ as

$$\Delta T_c = 0.68/\lambda_c \;(minimum)$$

The JH corresponding plots for $\Delta T = f(\lambda)$ and $\Delta T_c = f(\lambda_c)$ at v_z are

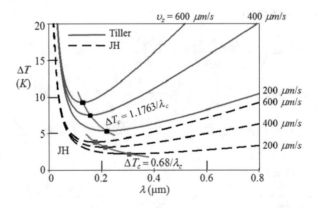

Free-Energy Approach   From Eq. (8.52), the constants K_C and K_R are

$$K_C = \frac{m_{l\alpha}(C_e - C_o)}{4D_l} = 7.6804 \times 10^{-3} \, \text{K s/}\mu\text{m}^2$$

$$K_R = \frac{2\gamma_{\alpha\beta}T_E}{\rho_s \Delta H_f^*} = 0.59 \, \text{K}\,\mu\text{m}$$

Thus, the undercooling $\Delta T = f(\lambda)$ becomes

$$\Delta T = \left(7.6804 \times 10^{-3}\right) v_z\lambda + \frac{0.59}{\lambda}$$

Plot $\Delta T = K_C v_z\lambda + K_R/\lambda$ equation at $v_z = 200\,\mu\text{m/s}$ for all models.

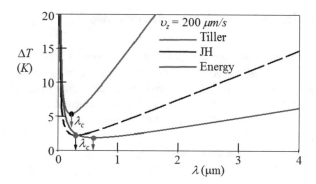

where $d\Delta T/d\lambda = 0$ leads to $\lambda = \lambda_c$.

8.8   Multicomponent Alloy Solidification

In general, most commercial casting alloys are multicomponent systems, which require sophisticated theoretical work for characterizing these materials [23–32].Among many publications on phase transformation, Catalina et al. [23] in 2015 and by Senninger-Voorhees [24] in 2016 followed Jackson–Hunt model [7] to provide additional insights on eutectic solidification of multicomponent alloys.

The two-dimensional partial differential equation (PDE) for the liquid concentration field $C_{l,i}(x, z)$ variable is written as

$$\frac{\partial^2 C_{l,i}}{\partial x^2} + \frac{\partial^2 C_{l,i}}{\partial z^2} + \frac{v_z}{D_{l,i}}\frac{\partial C_{l,i}}{\partial z} = 0 \tag{8.55}$$

where $D_{l,i}$ denotes the diffusion coefficient for elements $i = 1, 2, \ldots N$ and v_z denotes the rate of growth of the two-phase (α-β) eutectic structure.

The analytical solution to Eq. (8.55) for multicomponent alloy solidification is similar to Eq. (8.22) [23–25]. Thus,

$$C_{l,i}\,(x, z) = C_{o,i} + F_o \exp\left(-\frac{v_z}{D_{l,i}}z\right) + \sum_{n=1}^{\infty} F_{n,i} \cos\left(\frac{2n\pi x}{\lambda}\right) \exp\left(-\frac{n\pi z}{\lambda}\right)$$

$$(8.56)$$

Here, $C_{o,i}$ is the alloy initial nominal composition of any element i and $C_{l,i}\,(x, z)$ is subjected to coordinate conditions ahead of the eutectic interface during steady solidification at v_z.

8.8.1   Multicomponent Eutectic Growth Model

For multicomponent alloys, the work of Catalina, Voorhees, Huff, and Genau [23], named here as the CVHG eutectic growth model, is adopted in this section to show the reader one particular modified Jackson–Hunt eutectic growth model. Catalina et al. [23] documented comparable theoretical and experimental results for the Fe-C and Al-Si based alloys with nominal carbon compositions near the eutectic point and relatively low solidification velocities.

The goal now is to derive the undercooling (ΔT) equation similar to Eq. (8.14c) or (8.41) using Catalina et al. [23] proposed eutectic growth model, which introduces

- The precipitation of faceted phases with fixed chemical composition and the change of volume fraction of the eutectic phases.
- The concept of isothermal α-β interface with only the β phase having a faceted morphology.

For one-dimensional analysis, Fick's first law of diffusion for a multicomponent (i) eutectic growth the CVHG model gives the steady-state molar flux $J_{x,i}$ written as

$$J_{x,i} = -D_{l,i}\frac{\partial C_{l,i}}{\partial x} \quad \text{for } i = 1, 2, 3, \ldots N \qquad (8.57)$$

where $j = \alpha, \beta$ denote the eutectic phases.

The steady-state conditions are

$$\frac{D_{l,i}}{v_z}\left(\frac{\partial C_{l,i}}{\partial x}\right)_{z=0} = C_{s,i} - \eta_\alpha k_j^\alpha C_{l,j}^\alpha \quad \text{for } 0 \le x \le f_\alpha \lambda/2 \qquad (8.58a)$$

$$\frac{D_{l,i}}{v_z}\left(\frac{\partial C_{l,i}}{\partial x}\right)_{z=0} = C_{s,i} - \eta_\beta C_{l,j}^\beta \quad \text{for } f_\alpha \lambda/2 \le x \le \lambda/2 \qquad (8.58b)$$

with the density ratio $\eta_j = \rho_j/\rho_l$ $(j = \alpha, \beta)$ and

$$C_{l,j}^\alpha = \frac{C_{o,j} - (1 - f_\alpha)\,\eta_\beta C_{s,i}^\beta}{\eta_\alpha k_j^\alpha f_\alpha} \tag{8.59a}$$

$$C_{l,i}^\beta = \frac{\left(\eta_\alpha k_j^\alpha - 1\right) C_{o,j} + (1 - f_\alpha)\,\eta_\beta C_{s,j}^\beta + k_j^\alpha F_{o,j}}{(1 - f_\alpha)\,\eta_\alpha k_j^\alpha f_\alpha} \tag{8.59b}$$

$$F_{o,j} = \frac{\left(1 - \eta_\alpha k_j^\alpha f_\alpha\right) C_{o,j} - (1 - f_\alpha)\,\eta_\beta C_{s,j}^\beta}{\eta_\alpha k_j^\alpha f_\alpha} - P_{e,i} A_i S_\alpha / f_\alpha^2 \tag{8.59c}$$

where the Peclet number $P_{e,j}$ and constants A_i, S_α are written as

$$P_{e,j} = \frac{\lambda v_z}{D_{l,i}} \tag{8.60a}$$

$$A_i = \eta_\beta C_{s,i}^\beta - C_{o,i} \tag{8.60b}$$

$$S_\alpha = \sum_{n=1}^\infty \frac{\sin^2 (n\pi f_\alpha)}{(n\pi)^3} \tag{8.60c}$$

The average undercooling in front of the $j = \alpha, \beta$ solid phases $(f_\alpha + f_\beta = 1)$ is defined by Catalina et al. [23] as

$$\overline{\Delta T}_\alpha = m_{l\alpha}\left(C_e - \overline{C}_l^\alpha\right) + \sum_{i=2}^N m_{l\alpha,i}\left(C_{o,i} - \overline{C}_{l,i}^\alpha\right) + \frac{2\Gamma_{l\alpha}\cos(\theta_\alpha)}{\lambda f_\alpha} \tag{8.61a}$$

$$\overline{\Delta T}_\beta = m_{l\beta}\left(C_e - \overline{C}_l^\beta\right) + \sum_{i=2}^N m_{l\beta,i}\left(C_{o,i} - \overline{C}_{l,i}^\beta\right) + \frac{2\Gamma_{l\beta}\cos(\theta_\beta)}{\lambda f_\beta} \tag{8.61b}$$

where C_e denotes the concentration of the main element in the multicomponent alloy at the eutectic composition, as in Fe-4C-0.5 Si steel with $C_e = 4\%C$.

Let $K_C = \overline{\Delta T}_\alpha$ and $K_R = \overline{\Delta T}_\beta$ so that Eq. (8.41) for a multicomponent alloy becomes

$$\Delta T = \overline{\Delta T}_\alpha v_z \lambda + \frac{\overline{\Delta T}_\beta}{\lambda} \tag{8.62}$$

Letting $|d\,(\Delta T)/d\lambda|_{\lambda=\lambda_c} = 0$ yields $\lambda = \lambda_c$ for a minimum ΔT, provided that $d^2\,(\Delta T)/d\lambda^2 > 0$

$$\left.\frac{d\,\Delta T}{d\lambda}\right|_{\lambda=\lambda_c} = \overline{\Delta T}_\alpha v_z - \frac{\overline{\Delta T}_\beta}{\lambda_c^2} = 0 \tag{8.63a}$$

$$\lambda_c^2 v_z = \frac{\overline{\Delta T}_\beta}{\overline{\Delta T}_\alpha} = \text{constant} \tag{8.63b}$$

which is similar to Tiller's and Jackson–Hunt expressions defined by Eqs. (8.16a) and (8.45a), respectively.

8.9 Summary

It has been shown that the amount of undercooling is just a temperature difference ΔT parameter necessary for supercooling the melt at $T < T_E$. The three solidification models introduced in this chapter predict $\Delta T = f(\lambda)$ and $\Delta T \to \Delta T_c$ as $\lambda \to \lambda_c$, where ΔT_c and λ_c are the critical solidification parameters treated in the literature as minimum quantities.

Only the curved lamellar α-β interface morphology is characterized by the two-phase solidification models for binary and multicomponent alloys using a common expression defined by Eqs. (8.14c), (8.41), (8.51c), and (8.61) for steady-state directional solidification. These equations are renumbered for convenience. Nonetheless, the undercooling is defined by

$$\Delta T = K_C v_z \lambda + \frac{K_R}{\lambda} \tag{8.64}$$

from which $\Delta T_c = f(\lambda_c)$ if $|d(\Delta T)/d\lambda|_{\lambda=\lambda_c} = 0$ and $d^2(\Delta T)/d\lambda^2 > 0$. Moreover, all directional solidification models above and many published experimental data indicate that

$$\lambda_c^2 v_z = \frac{K_R}{K_C} = \text{constant} \tag{8.65a}$$

$$\frac{\Delta T_c}{\sqrt{v_z}} = 2\sqrt{K_C K_R} = \text{constant} \tag{8.65b}$$

Since v_z is fixed during an experiment, the most elegant and specific or appropriate representation of these equations is via

$$\lambda_c = \sqrt{\frac{K_R}{K_C}\frac{1}{v_z}} = \text{constant} \tag{8.66a}$$

$$\Delta T_c = 2\sqrt{(K_C K_R)\,v_z} = \text{constant} \tag{8.66b}$$

Note that λ_c and ΔT_c are constants that depend on other constants; K_C, K_R, and υ_z. Mathematically,

$$\lambda_c \to \infty, \ \Delta T_c \to 0 \text{ as } \upsilon_z \to 0 \tag{8.67a}$$

$$\lambda_c \to 0, \ \Delta T_c \to \infty \text{ as } \upsilon_z \to \infty \tag{8.67b}$$

Experimentally, the solidification velocity is within the range $0 < \upsilon_z < \infty$. For slow solidification, $\upsilon_z \to 0$ and for rapid solidification $\upsilon_z \to \infty$.

Furthermore, one can find in the literature that the Jackson–Hunt solidification model [7] is the most complete analytical solution to two-dimensional diffusion under steady-state phase transition in binary alloys. Appendix 8C contains some data sets for Pb-Sn alloys available in the literature [33].

Lastly, solidification models for multicomponent alloys are still a challenging analytical stage. For clarity, only the Catalina et al. [23] model was briefly introduced in this chapter.

8.10   Appendix 8A: Separation of Variables

The method of separation of variables is suitable for finding the solutions of partial differential equations (PDE) [2, p. 201]. Below is an illustration of the method for solving the two-dimensional diffusion problems. The governing equation for the liquid ahead of the two-phase (α-β) solid interface is described by Eq. (8.11) as

$$\frac{\partial^2 W}{\partial x^2} + \frac{\partial^2 W}{\partial z^2} + \frac{\upsilon_z}{D_l} \frac{\partial W}{\partial z} = 0 \tag{8A.1}$$

Let $W(x, z)$ be a product of $P(x)$ and $Q(z)$ so that

$$W(x, z) = P(x) Q(z) \tag{8A.2a}$$

$$\frac{\partial W}{\partial x} = Q \frac{\partial P}{\partial x} \quad \& \quad \frac{\partial W}{\partial z} = P \frac{\partial Q}{\partial z} \tag{8A.2b}$$

$$\frac{\partial W}{\partial x} = \frac{W}{P} \frac{\partial P}{\partial x} \quad \& \quad \frac{\partial W}{\partial z} = \frac{W}{Q} \frac{\partial Q}{\partial z} \tag{8A.2c}$$

$$\frac{\partial^2 W}{\partial x^2} = \frac{W}{P} \frac{\partial^2 P}{\partial x^2} = Q \frac{\partial P}{\partial x} \quad \& \quad \frac{\partial^2 W}{\partial z^2} = \frac{W}{Q} \frac{\partial^2 Q}{\partial z^2} \tag{8A.2d}$$

Insert Eqs. (8A.2c) and (8A.2d) into (8A.1) to get

$$\frac{W}{P} \frac{\partial^2 P}{\partial x^2} + \frac{W}{Q} \frac{\partial^2 Q}{\partial z^2} + \frac{\upsilon_z}{D_l} \frac{W}{Q} \frac{\partial Q}{\partial z} = 0 \tag{8A.3a}$$

$$\frac{1}{P}\frac{\partial^2 P}{\partial x^2} + \frac{1}{Q}\frac{\partial^2 Q}{\partial z^2} + \frac{v_z}{D_l}\frac{1}{Q}\frac{\partial Q}{\partial z} = 0 \tag{8A.3b}$$

$$\frac{1}{Q}\frac{\partial^2 Q}{\partial z^2} + \frac{v_z}{D_l}\frac{1}{Q}\frac{\partial Q}{\partial z} = -\frac{1}{P}\frac{\partial^2 P}{\partial x^2} = \gamma_s \tag{8A.3c}$$

where γ_s denotes the separation variable. Grouping Eq. (8A.3c) in the form of homogeneous partial differential equations (PDE) is a convenient approach for finding the corresponding solutions.

First Equation For the axial diffusion ahead of the growing α-β interface at steady-state condition, the second-order partial derivative equation (PDE) for characterizing diffusional changes or differences with a steady mass transfer treatment is written as a homogeneous expression

$$\frac{1}{Q}\frac{\partial^2 Q}{\partial z^2} + \frac{v_z}{D_l}\frac{1}{Q}\frac{\partial Q}{\partial z} = \gamma_s \tag{8A.4a}$$

$$\frac{\partial^2 Q}{\partial z^2} + \frac{v_z}{D_l}\frac{\partial Q}{\partial z} - Q\gamma_s = 0 \tag{8A.4b}$$

Assume a general analytical solution and its respective partial derivatives of the form

$$Q = C_1 \exp(-C_2 z) \tag{8A.5a}$$

$$\frac{\partial Q}{\partial z} = -C_1 C_2 \exp(-C_2 z) \tag{8A.5b}$$

$$\frac{\partial^2 Q}{\partial z^2} = -C_1 C_2^2 \exp(-C_2 z) \tag{8A.5c}$$

Substituting Eqs. (8A.5a)–(8A.5c) into (8A.4b) yields an expression that can be solved for C_2

$$0 = -C_1 C_2^2 \exp(-C_2 z) - \frac{v_z}{D_l}C_1 C_2 \exp(-C_2 z) \tag{8A.6a}$$

$$\qquad - C_1\gamma_s \exp(-C_2 z)$$

$$0 = C_2^2 - \frac{v_z}{D_l}C_2 - \gamma_s \tag{8A.6b}$$

$$C_2 = -\left(\frac{v_z}{2D_l}\right) \pm \sqrt{\left(\frac{v_z}{2D_l}\right)^2 + \gamma_s} \quad \text{for } \gamma_s > 0 \tag{8A.6c}$$

Letting $\gamma_s = 0$ in this quadratic expression, Eq. (8A.6c), one obtains

$$C_2 = -\left(\frac{v_z}{2D_l}\right) \pm \left(\frac{v_z}{2D_l}\right) \qquad \text{for } \gamma_s = 0 \qquad (8A.7a)$$

$$C_2 = -\left(\frac{v_z}{2D_l}\right) - \left(\frac{v_z}{2D_l}\right) = -\frac{v_z}{D_l} \neq 0 \qquad (8A.7b)$$

$$C_2 = -\left(\frac{v_z}{2D_l}\right) + \left(\frac{v_z}{2D_l}\right) = 0 \qquad \text{(disregard)} \qquad (8A.7c)$$

$$C_2 = -\left(\frac{v_z}{2D_l}\right) - \sqrt{\left(\frac{v_z}{2D_l}\right)^2 + \gamma_s} \qquad \text{(choice)} \qquad (8A.7d)$$

Substituting Eq. (8A.7d) into (8A.5a) yields

$$Q = C_1 \exp\left\{\left[-\left(\frac{v_z}{2D_l}\right) - \sqrt{\left(\frac{v_z}{2D_l}\right)^2 + \gamma_s}\right] z\right\} \qquad (8A.8)$$

Second Equation For the lateral diffusion along the curvature of the α-β interface in Fig. 8.7b, Eq. (8A.3c) gives

$$\frac{1}{P}\frac{\partial^2 P}{\partial x^2} + \gamma_s = 0 \qquad (8A.9a)$$

$$\frac{\partial^2 P}{\partial x^2} + P\gamma_s = 0 \qquad (8A.9b)$$

Assume a general solution and its derivatives of the form

$$P = C_3 \cos\left(C_4 x\right) + C_5 \sin\left(C_6 x\right) \qquad (8A.10a)$$

$$\frac{\partial P}{\partial x} = -C_3 C_4 \sin\left(C_4 x\right) + C_5 C_6 \cos\left(C_6 x\right) \qquad (8A.10b)$$

$$\frac{\partial^2 P}{\partial x^2} = -C_3 C_4^2 \cos\left(C_4 x\right) - C_5 C_6^2 \sin\left(C_6 x\right) \qquad (8A.10c)$$

Then, Eq. (8A.9b) becomes

$$-C_3 C_4^2 \cos\left(C_4 x\right) - C_5 C_6^2 \sin\left(C_6 x\right) + C_3 \gamma_s \cos\left(C_4 x\right) + C_5 \gamma_s \sin\left(C_6 x\right) = 0$$

from which, the cosine and sine groups are

$$-C_3 C_4^2 \cos\left(C_4 x\right) + C_3 \gamma_s \cos\left(C_4 x\right) = 0 \qquad (8A.11a)$$

$$C_4 = \sqrt{\gamma_s} \qquad (8A.11b)$$

$$-C_5 C_6^2 \sin\left(C_6 x\right) + C_5 \gamma_s \sin\left(C_6 x\right) = 0 \qquad (8A.11c)$$

$$C_6 = \sqrt{\gamma_s} \tag{8A.11d}$$

Substitute $C_4 = C_6 = \sqrt{\gamma_s}$ into Eq. (8A.10a) to get

$$P = C_3 \cos\left(\sqrt{\gamma_s}x\right) + C_5 \sin\left(\sqrt{\gamma_s}x\right) \tag{8A.12a}$$

$$\frac{\partial P}{\partial x} = -C_3\sqrt{\gamma_s} \sin\left(\sqrt{\gamma_s}x\right) + C_5\sqrt{\gamma_s} \cos\left(\sqrt{\gamma_s}x\right) \tag{8A.12b}$$

Evaluating $\partial P/\partial x = 0$ at $x = 0$ yields

$$\left(\frac{\partial P}{\partial x}\right)_{x=0} = C_5\sqrt{\gamma_s} = 0 \tag{8A.13a}$$

$$C_5 = 0 \tag{8A.13b}$$

and $\partial P/\partial x = 0$ at $x = \lambda = \lambda_\alpha + \lambda_\beta$ gives

$$\left(\frac{\partial P}{\partial x}\right)_{x=\lambda} = -C_3\sqrt{\gamma_s} \sin\left(\sqrt{\gamma_s}\lambda\right) + C_5\sqrt{\gamma_s} \cos\left(\sqrt{\gamma_s}\lambda\right) = 0 \tag{8A.14a}$$

$$\left(\frac{\partial P}{\partial x}\right)_{x=\lambda} = -C_3\sqrt{\gamma_s} \sin\left(\sqrt{\gamma_s}\lambda\right) = 0 \tag{8A.14b}$$

$$C_3 \sin\left(\sqrt{\gamma_s}\lambda\right) = 0 \tag{8A.14c}$$

Now, manipulating this simple expression, Eq. (8A.14c), one defined the separation variable γ_s under the assumption that $C_3 \neq 0$.

Thus, $C_3 \sin\left(\sqrt{\gamma_s}\lambda\right) = C_n \sin(n\pi)$ for $n \geq 0$ and $\sqrt{\gamma_s}\lambda = n\pi$ so that

$$\sin\left(\sqrt{\gamma_s}\lambda\right) = \sin(n\pi) \quad \text{for } n \geq 0 \tag{8A.15a}$$

$$\sqrt{\gamma_s}\lambda = n\pi \tag{8A.15b}$$

$$\gamma_s = \left(\frac{n\pi}{\lambda}\right)^2 \tag{8A.15c}$$

Then, Eqs. (8A.12a) and (8A.8) become eigenfunctions

$$P = C_3 \cos\left(\frac{n\pi}{\lambda}x\right) \tag{8A.16a}$$

$$Q = C_1 \exp(-d_n z) \tag{8A.16b}$$

with the decay coefficient being defined by

$$d_n = \left(\frac{v_z}{2D_l}\right) + \sqrt{\left(\frac{v_z}{2D_l}\right)^2 + \left(\frac{n\pi}{\lambda}\right)^2} \tag{8A.17}$$

Substituting Eq. (8A.16) into (8A.2a) yields $W = W(x, z)$ under the assumption that $C_o \neq C_e$ (arbitrary C_o) and $v_z / 2D_l << n\pi/\lambda/2$ yields

$$W(x, z) = P(x) Q(z) \tag{8A.18a}$$

$$W(x, z) = C_3 \cos\left(\frac{n\pi}{\lambda}x\right) . C_1 \exp(-d_n z) \tag{8A.18b}$$

Let $F_n = C_n = C_1 C_3$ be the Fourier coefficient for $n = 0, 1, 2, 3, \ldots, \infty$ and for convenient, express Eq. (8A.18) as a Fourier cosine series for a short-range diffusion at the scale of lamellar spacing λ [4, p. 395]

$$W(x, z) = \sum_{n=0}^{\infty} F_n \cos\left(\frac{2n\pi}{\lambda}x\right) \exp(-d_n z) \quad \text{for } C_o \neq C_e \tag{8A.19}$$

Here, $\Delta C = W(x, z) = C_l(x, y) - C_o$ represents the local change in solute concentration during diffusion at the α-β lamellar interface in the xz plane. Thus, the concentration for the steady advection-diffusion process with two particular conditions takes the form [8, p. 394]

$$C_l(x, z) = C_o + \sum_{n=0}^{\infty} F_n \cos\left(\frac{2n\pi}{\lambda}x\right) \exp(-d_n z) \text{ at } C_o \neq C_e \tag{8A.20a}$$

$$C_l(x, z) = C_e + \sum_{n=0}^{\infty} F_n \cos\left(\frac{2n\pi}{\lambda}x\right) \exp(-d_n z) \text{ at } C_o = C_e \tag{8A.20b}$$

At this moment, Eq. (8A.20a) defines the liquid concentration field equation generalized as $C_l(x, z)$ for an element or component in any A-B system as an off-eutectic alloy.

Note that Eq. (8A.20) must comply with the following boundary conditions

- BC1: $C_l(x, z) = C_o$ and $C_l(x, z) = C_e$ at $z \to \infty$ since $\exp(-\infty) = 0$, respectively.
- BC2: $\partial C_l(x, z) / \partial x = 0$ at $x = 0$ and $x = \lambda/2$ since

$$\partial C_l(x, z) / \partial x = -\sum_{n=0}^{\infty} F_n \sin\left(\frac{2n\pi}{\lambda}x\right) \exp(-d_n z) = 0 \tag{8A.21}$$

and $\sin(0) = \sin(n\pi) = 0$.

The liquid concentration field $C_l(x, z)$ of any element undergoing eutectic growth in a two-phase (α-β) multicomponent alloys can be expressed as

$$C_{l,i}(x, z) = C_o + \sum_{n=0}^{\infty} F_{n,i} \cos\left(\frac{2n\pi}{\lambda}x\right) \exp(-d_n z) \text{ at } C_o \neq C_e \qquad (8A.22a)$$

$$C_{l,i}(x, z) = C_e + \sum_{n=0}^{\infty} F_{n,i} \cos\left(\frac{2n\pi}{\lambda}x\right) \exp(-d_n z) \text{ at } C_o = C_e \qquad (8A.22b)$$

where $i = 1, 2, \ldots, N$ denotes the number of elements in the alloy.

8.10.1   Defining the Fourier Coefficients

Evaluate F_o at $n = 0$ and F_n at $n > 0$ using the concentration field expression $C_l(x, z)$, Eq. (8A.22b), and its derivative $\partial C_l(x, z)/\partial z$

$$C_l(x, z) = C_o + F_o \exp\left(-\frac{v_z}{D_l}z\right) \qquad (8A.23a)$$

$$+ \sum_{n=1}^{\infty} F_n \cos\left(\frac{2n\pi}{\lambda}x\right) \exp\left(-\frac{2n\pi}{\lambda}z\right)$$

$$\left(\frac{\partial C_l}{\partial z}\right)_{z>0} = -\frac{v_z}{D_l}F_o - \sum_{n=1}^{\infty} \frac{2n\pi}{\lambda} F_n \cos\left(\frac{2n\pi}{\lambda}x\right) \exp\left(-\frac{2n\pi}{\lambda}z\right) \qquad (8A.23b)$$

$$\left(\frac{\partial C_l}{\partial z}\right)_{z=0} = -\frac{v_z}{D_l}F_o - \sum_{n=1}^{\infty} \frac{2n\pi}{\lambda} F_n \cos\left(\frac{2n\pi}{\lambda}x\right) \qquad (8A.23c)$$

The diffusion flux conditions defined by (8.20b) and (8.20c) are

$$\left(\frac{\partial C_l}{\partial z}\right)_{z=0} = -\frac{v_z \Delta C_\alpha}{D_l} \quad \text{for } z = 0, 0 \leq x \leq \lambda_\alpha \qquad (8A.24a)$$

$$\left(\frac{\partial C_l}{\partial z}\right)_{z=0} = \frac{v_z \Delta C_\beta}{D_l} \quad \text{for } z = 0, \lambda_\alpha \leq x \leq \lambda_\alpha + \lambda_\beta \qquad (8A.24b)$$

Combining Eqs. (8A.23c) and (8A.24) yields

$$v_z \Delta C_\alpha = v_z F_o + D_l \sum_{n=1}^{\infty} \frac{2n\pi}{\lambda} F_n \cos\left(\frac{2n\pi}{\lambda}x\right) \quad \text{for } 0 \leq x \leq \lambda_\alpha \qquad (8A.25a)$$

$$v_z \Delta C_\beta = v_z F_o + D_l \sum_{n=1}^{\infty} \frac{2n\pi}{\lambda} F_n \cos\left(\frac{2n\pi}{\lambda} x\right) \quad \text{for } \lambda_\alpha \leq x \leq \lambda/2 \qquad (8A.25b)$$

which become

$$(F_o - \Delta C_\alpha) v_z = -D_l \sum_{n=1}^{\infty} \frac{2n\pi}{\lambda} F_n \cos\left(\frac{2n\pi}{\lambda} x\right) \quad \text{for } 0 \leq x \leq \lambda_\alpha \qquad (8A.26a)$$

$$\left(F_o - \Delta C_\beta\right) v_z = -D_l \sum_{n=1}^{\infty} \frac{2n\pi}{\lambda} F_n \cos\left(\frac{2n\pi}{\lambda} x\right) \quad \text{for } \lambda_\alpha \leq x \leq \lambda/2$$

$$(8A.26b)$$

In classical Fourier analysis, use of orthogonality integrals is a standard procedure for finding each of the Fourier coefficients (amplitudes) F_n. Following Glicksman's approach [11, p. 486], the orthogonality integrals are

$$\frac{2\pi}{\lambda} \int_0^{\lambda/2} \cos\left(\frac{2m\pi x}{\lambda}\right) \cos\left(\frac{2n\pi x}{\lambda}\right) dx = 0 \quad \text{for } m \neq 0, n > 0 \qquad (8A.27a)$$

$$\frac{2\pi}{\lambda} \int_0^{\lambda/2} \cos\left(\frac{2m\pi x}{\lambda}\right) \cos\left(\frac{2n\pi x}{\lambda}\right) dx = \frac{\pi}{2} \quad \text{for } m = n = 1 \qquad (8A.27b)$$

$$\frac{2\pi}{\lambda} \int_0^{\lambda/2} \cos\left(\frac{2n\pi x}{\lambda}\right) \cos\left(\frac{2n\pi x}{\lambda}\right) dx = \pi \quad \text{for } m = n \neq 1 \qquad (8A.27c)$$

Now, multiply Eq. (8A.26) by a weight function, say, cosine function $\cos(2n\pi x/\lambda)$ and integrate the resultant expression with appropriate limits of integration illustrates the application of the orthogonality integral approach.

For one periodic unit of the eutectic,

$$\frac{v_z}{D_l} \int_0^{\lambda_\alpha} (F_o - \Delta C_\alpha) \cos\left(\frac{2n\pi x}{\lambda}\right) dx + \frac{v_z}{D_l} \int_{\lambda_\alpha}^{\lambda/2} \left(F_o - \Delta C_\beta\right) \cos\left(\frac{2n\pi x}{\lambda}\right) dx$$

$$= \int_{\lambda_\alpha}^{\lambda/2} \sum_{n=1}^{\infty} \frac{2n\pi}{\lambda} F_n \cos\left(\frac{2m\pi}{\lambda} x\right) \cos\left(\frac{2n\pi}{\lambda} x\right) dx \qquad (8A.28)$$

Integrating Eq. (8A.28) with $m = n > 0$ and rearranging the resultant expression yields the Fourier coefficients

$$-\frac{n\pi}{2} F_n = \frac{\lambda}{2n\pi} \left(F_o - \Delta C_\alpha\right) \sin \left(\frac{2n\pi x}{\lambda}\right)_0^{\lambda_\alpha} \tag{8A.29a}$$

$$+ \frac{\lambda}{2n\pi} \left(F_o - \Delta C_\beta\right) \sin \left(\frac{2n\pi x}{\lambda}\right)_{\lambda_\alpha}^{\lambda/2}$$

$$F_n = \frac{v_z \, \lambda \Delta C_o}{D_l \, (n\pi)^2} \sin \left(\frac{2n\pi \lambda_\alpha}{\lambda}\right) \tag{8A.29b}$$

$$F_o = \frac{\lambda_\alpha \Delta C_\alpha - \lambda_\beta \Delta C_\beta}{\lambda} \tag{8A.29c}$$

where F_o is included here for comparison and $\Delta C_o = \Delta C_\alpha + \Delta C_\beta$. Note that $F_n \propto 1/n^2$ in Eq. (8A.29b) and it becomes a singularity defined as $F_n \to \infty$ when $n = 0$.

8.11   Appendix 8B: Shape of Interfaces

This Appendix includes a thermodynamic approach to the general theory of curved interfaces characterized by the scalar dilation (or dilatation), shear tensions, bending, and torsion moments. A dilation is a type of transformation that occurs when a figure or interface is enlarged or reduced and correspondingly, it requires a center point and a scale factor.

In phase transformation, the shape of the lamellae interface and related triple junction is important for defining the surface tension on a plane, say $q_{ls} = F_{ls}/x$, which is converted to surface energy per unit area $\gamma_{ls} = F_{ls}/A_{ls}$. Here, F_{ls} is the force acting on a unit length of the perimeter of the L-S interface area.

Consider the curved surfaces or interfaces shown in Fig. 8B.1, where P is a point on the $z(x)$ curve in Fig. 8B.1. Draw a tangent line at P to each s segment of the osculating circles. Thus, the radius of the circle of curvature is called the radius of curvature denoted by $k(z)$.

Let point P on the $z(x)$ curve be connected to a tangent line and a perpendicular line to the radius of an inscribed (osculating) circle. Actually, the sign of $k(z)$ indicates the convexity and concavity of the curve in the neighborhood of point P. Note that both curves are convex with $k(z) < 0$ in Fig. 8B.1a and $k(z) > 0$ in Fig. 8B.1b.

Using the Circle Approach The general radius of curvature $k(z)$ of an arc $z = f(x)$ function is the reciprocal of the always positive radius of a circle. Accordingly, the radius of curvature $k(z)$, the equation of a circle, and the $\tan(\theta)$ function are, respectively, defined by

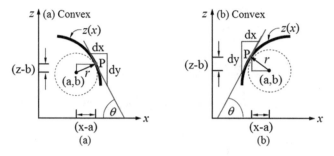

Fig. 8B.1 Shape of the convex interfaces with positive radius of curvature $k(z)$. (**a**) Negative slope dz/dx and (**b**) positive slope dz/dx

$$k\left(z\right) = \frac{1}{r} \tag{8B.1a}$$

$$\left(x - a\right)^2 + \left(z - b\right)^2 = r^2 \tag{8B.1b}$$

$$\tan\left(\theta\right) = -\frac{dz}{dx} \tag{8B.1c}$$

These equations are interrelated and they can be manipulated to derive expressions related to convex curves shown in Fig. 8B.1. For the sake of clarity, these curves represent the ideal or theoretical curved edges of the eutectic α-β interface moving in the z-direction of an ideal planar solidification front. As a matter of fact, Fig. 8.1 is enormously helpful in setting up the model for the theoretical eutectic growth.

Solving Eq. (8B.1b) for z and taking dz/dx yields

$$z = b + \sqrt{r^2 - \left(x - a\right)^2} \tag{8B.2a}$$

$$\frac{dz}{dx} = -\frac{\left(x - a\right)}{\sqrt{r^2 - \left(x - a\right)^2}} = -\frac{\left(x - a\right)}{\sqrt{\left(x - a\right)^2 + \left(z - b\right)^2 - \left(x - a\right)^2}} \tag{8B.2b}$$

$$\frac{dz}{dx} = -\frac{\left(x - a\right)}{\left(z - b\right)} = -\tan\left(\theta\right) \tag{8B.2c}$$

Combining Eqs. (8B.1c) and (8B.2c) gives

$$\tan\left(\theta\right) = \frac{\left(x - a\right)}{\left(z - b\right)} = \frac{\sin\left(\theta\right)}{\cos\left(\theta\right)} = -\frac{dz}{dx} \tag{8B.3a}$$

$$\cos\left(\theta\right) = \frac{\left(z - b\right)}{\left(x - a\right)}\sin\left(\theta\right) = \frac{\left(z - b\right)}{r} \tag{8B.3b}$$

$$\frac{d\cos\left(\theta\right)}{dz} = \frac{1}{r} \tag{8B.3c}$$

The arc length ds/dx of a small segment of the AB curve can be derived using the definition of $\cos(\theta)$; that is,

$$\cos(\theta) = \frac{1}{\sec(\theta)} = \frac{dx}{ds} \tag{8B.4a}$$

$$\frac{ds}{dx} = \frac{1}{\cos(\theta)} = \sec(\theta) = \sqrt{1 + \tan^2(\theta)} \tag{8B.4b}$$

$$\frac{ds}{dx} = \sqrt{1 + \left(-\frac{dz}{dx}\right)^2} = \left[1 + \left(\frac{dz}{dx}\right)^2\right]^{1/2} \tag{8B.4c}$$

Differentiating Eq. (8B.2c) with respect to (wrt) x yields the sought second-order ordinary differential equation

$$\frac{d^2z}{dx^2} = -\sec^2(\theta)\frac{d\theta}{dx} = \left[1 + \tan^2(\theta)\right]^{1/2}\frac{d\theta}{dx} \tag{8B.5a}$$

$$\frac{d^2z}{dx^2} = \left[1 + \tan^2(\theta)\right]^{1/2}\frac{d\theta}{ds}\frac{ds}{dx} \quad \text{(positive only)} \tag{8B.5b}$$

Combining Eqs. (8B.2c) and (8B.5b) gives

$$\frac{d^2z}{dx^2} = \pm\frac{1}{r}\left[1 + \left(\frac{dz}{dx}\right)^2\right]^{3/2} \tag{8B.6a}$$

$$k(z) = \frac{1}{r} = \frac{d\cos(\theta)}{dz} = \pm\frac{d^2z/dx^2}{\left[1 + (dz/dx)^2\right]^{3/2}} \tag{8B.6b}$$

For the convex $k(z) > 0$ (Fig. 8B.1) and the concave $k(z) < 0$ (not shown) AB curves, the radius of curvature becomes

$$k(z) = +\frac{d^2z/dx^2}{\left[1 + (dz/dx)^2\right]^{3/2}} \quad \text{(convex)} \tag{8B.7a}$$

$$k(z) = -\frac{d^2z/dx^2}{\left[1 + (dz/dx)^2\right]^{3/2}} \quad \text{(concave)} \tag{8B.7b}$$

Imposed Triangle and Pythagorean Theorem Approach   This is a classic analytical approach for defining small increments of the arc length s described by the curvature function $z(x)$.

Accordingly, for a small segment ds, the fundamental Pythagorean theorem that deals with the lengths of the sides of a right triangle gives

$$(ds)^2 = (dx)^2 + (dz)^2 \tag{8B.8a}$$

$$\frac{ds}{dx} = \pm\sqrt{1 + \left(\frac{dz}{dx}\right)^2} \tag{8B.8b}$$

and for a curvature function $z = f(x)$,

$$\frac{dz}{dx} = \frac{df(x)}{dx} = \tan\theta \tag{8B.9a}$$

$$\frac{dz}{dx} = \tan\theta = \frac{\sin(\theta)}{\cos(\theta)} = \frac{dz/ds}{dx/ds} \tag{8B.9b}$$

$$\frac{d^2z}{dx^2} = \sec^2(\theta)\frac{d\theta}{dx} = \left[1 + \tan^2(\theta)\right]\frac{d\theta}{dx} \tag{8B.9c}$$

Inserting Eq. (8B.9a) into (8B.9c) yields

$$\frac{d^2z}{dx^2} = \left[1 + \left(\frac{dz}{dx}\right)^2\right]\frac{d\theta}{dx} \tag{8B.10a}$$

$$\frac{d\theta}{dx} = \frac{d^2z/d^2x}{1 + (dz/dx)^2} \tag{8B.10b}$$

The general definition of the radius of curvature is written as

$$k(z) = \frac{d\theta}{ds} = \frac{d\theta/dx}{ds/dx} \tag{8B.11}$$

Substituting Eqs. (8B.8b) and (8B.10b) into (8B.11) gives, again, the usual convex $k(z)$ equation, which is compared with its concave counterpart.

$$k(z) = +\frac{d^2z/d^2x}{\left[1 + (dz/dx)^2\right]^{3/2}} \quad \text{(convex)} \tag{8B.12a}$$

$$k(z) = -\frac{d^2z/d^2x}{\left[1 + (dz/dx)^2\right]^{3/2}} \quad \text{(concave)} \tag{8B.12b}$$

This concludes the analytical procedure for deriving the equation of the radius of curvature as $k(z) > 0$.

Fig. 8C.1 *Pb-Sn* equilibrium phase diagram

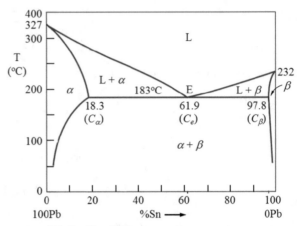

Table 8C.1 Thermophysical properties of Pb-Sn alloys [33]

Properties	Units	Pb	Sn	Pb-2.2Sn	Pb-2.5Sn
k_s	W/m K	33.0	23.8	32.8	32.8
k_l	W/m K	29.7	19.7	29.5	29.4
c_s	J/kg K	129.8	209.0	131.5	131.8
c_l	J/kg K	138.2	239.0	140.4	140.7
ρ_s	kg/m^3	11,340	6680	11,238	11,224
ρ_l	kg/m^3	10,678	6483	10,586	10,573
ΔH_f	J/kg	26,205	163,000	29,215	29,625
D_l	m^2/s			3.00×10^{-9}	3.00×10^{-9}
Γ	m K			9.89×10^{-9}	9.89×10^{-9}
T_l	°C			313.0	312.0
T_s	°C			289.0	287.0
T_E	°C			183	183
m_l	°C/%			6.696	6.696
k_o				0.3125	0.3125

Properties	Units	Pb-3Sn	Pb-6.6Sn	Pb-11.2Sn
k_s	W/m K	32.7	32.4	32.0
k_l	W/m K	29.4	29.0	28.6
c_s	J/kg K	132.2	135.1	138.7
c_l	J/kg K	141.2	144.9	149.5
ρ_s	kg/m^3	11,200	11,032	10,818
ρ_l	kg/m^3	10,552	10,401	10,208
ΔH_f	J/kg	30,309	35,234	41,526
D_l	m^2/s	3.00×10^{-9}	3.00×10^{-9}	3.00×10^{-9}
Γ	m K	9.89×10^{-9}	9.89×10^{-9}	9.89×10^{-9}
T_l	°C	306.50	289.00	251.7
T_s	°C	266.20	251.70	251.7
T_E	°C	183.00	183.00	183.00
m_l	°C/%	-6.696	-6.696	-6.696
k_o^α		0.3125	0.3125	0.3125

8.12 Appendix 8C: Thermophysical Data

For convenience, the Pb-Sn equilibrium phase diagram is shown in Fig. 8C.1 and relevant thermophysical properties of Pb-Sn alloys [33] are given in Table 8C.1, respectively.

8.13 Problems

8.1 It is known that the solidification processing is an important materials fabrication technique for producing microstructural features related to mechanical properties and mechanical behavior of alloys. For directional solidification of a hypereutectic Al-$20Si$ alloy in the form of a flat plate sample with suitable dimensions, use the Tiller's theory of lamellar eutectic solidification to calculate (a) the latent heat of solidification ΔH_s, (b) the partition coefficients k_o^α and k_o^β, (c) the linear liquidus slopes, $m_{l\alpha}$ and $m_{l\beta}$, (d) the constants K_C and K_R with $\phi = 1$, (e) the parameters $\lambda = \lambda_c$ and $\Delta T = \Delta T_c$ using the solidification velocities $v_z = 100$, 200, 400, 600, and 800 μm/s. Explain the results. Determine (f) the constants for $\lambda_c^2 v_z$ and $\Delta T_c / \sqrt{v_z}$, (g) find the best fit equation using the calculated $\lambda_c, \Delta T_c$ data and plot the resultant function $\Delta T_c = f(\lambda_c)$, and (h) superimpose the functions $\Delta T = f(\lambda)$ and $\Delta T_c = f(\lambda_c)$ on the graph as per Example 8A.1. (i) What happens to the solidification process if $\Delta T = 0$? Use the pouring temperature $T_p = 850\,^\circ$C, one $x = s(t) = 0.01$ m-thick flat sample, equal Gibbs–Thomson coefficients $\Gamma_{l\alpha} = \Gamma_{l\beta}$, equal solid fractions $f_\alpha = f_\beta = 0.5$, equal lamellar curvature angles $\theta_\alpha = \theta_\beta = \pi/3$ and the given data along with the Al-Si equilibrium phase diagram from Example 7A.2.

$k = 161$ W/m K	$c_s = 854$ J/kg K	$\rho = 2627$ kg/m^3
$\Gamma_{l\alpha} = 2 \times 10^{-7}$ K m	$D_l = 2.94 \times 10^{-12}$ m^2/s	$\gamma_{\alpha\beta} = 0.12$ J/m^2
$\Delta H_{f,Al} = 397$ kJ/kg	$\Delta H_{f,Si} = 1790$ kJ/kg	$\Delta H_s = \Delta H_f^* = ?$

8.2 For directional solidification of a hypereutectic Al-$20Si$ alloy in the form of a flat plate sample with suitable dimensions, calculate (a) the latent heat of solidification ΔH_s, (b) the partition coefficients k_o^α and k_o^β, (c) the linear liquidus slopes, $m_{l\alpha}$ and $m_{l\beta}$, (d) the constants K_C and K_R for the Jackson–Hunt (JH) eutectic solidification model, (e) the critical parameters $\lambda = \lambda_c$ and $\Delta T = \Delta T_c$ using the solidification velocities $v_z = 100$, 200, 400, 600 and 800 μm/s. Explain the results. Determine (f) the constants for $\lambda_c^2 v_z$ and $\Delta T_c / \sqrt{v_z}$, (g) find the best fit equation using the calculated $\lambda_c, \Delta T_c$ data and plot the resultant function $\Delta T_c = f(\lambda_c)$, and (h) superimpose the functions $\Delta T = f(\lambda)$ and $\Delta T_c = f(\lambda_c)$ on the graph as per Example 8A.1.(i) What happens to the solidification process if $\Delta T = 0$? Use the pouring temperature $T_p = 850\,^\circ$C, one $x = s(t) = 0.01$ m-thick flat sample, equal Gibbs–Thomson coefficients $\Gamma_{l\alpha} = \Gamma_{l\beta}$, equal solid fractions $f_\alpha = f_\beta = 0.5$, equal lamellar curvature angles $\theta_\alpha = \theta_\beta = \pi/3$ and the given data in problem 8.1 along with the *Al-Si* equilibrium phase diagram.

8.3 Use the free-energy solidification model with the solidification velocities $v_z = 100, 200, 400, 600$, and $800\,\mu\text{m/s}$ to determine the undercooling profiles $\Delta T = f(\lambda)$ and $\Delta T_c = f(\lambda_c)$ for a directional solidification process to produce an *Al-20Si* alloy. Use a pouring temperature $T_p = 850\,°\text{C}$, a $x = s(t) = 0.01$ m-thick flat sample and the given data in Problem 8.1 along with the *Al-Si* equilibrium phase diagram.

8.4 Apparently, Tiller's theory of lamellar eutectic solidification stands out because it is completely general. For a constant moving velocity of the solid–liquid interface during directional solidification of a hypoeutectic *Al-6Cu* alloy, the solute *Cu* content at the interface changes and causes a change in the interfacial temperature. Calculate (a) the partition coefficients k_o^α and k_o^β, (b) the liquidus and solidus $m_{l\alpha}, m_{s\alpha}$ and $m_{l\beta}, m_{s\beta}$ slopes, (c) the latent heat of solidification ΔH_s and the corresponding entropy change ΔS_f for the alloy, (d) the undercooling using the free-energy solidification model with the solidification velocities $v_z = 100\,\mu\text{m/s}$, and (e) the critical temperature gradient G_{λ_c}. Assume a planar *L-S* interface in one $x = s(t) = 0.01$ m-thick flat sample and a pouring temperature $T_p = 660\,°\text{C}$, and use the following data

$k = 190\,\text{W/m K}$	$c_s = 900\,\text{J/kg K}$	$\rho = 2600\,\text{kg/m}^3$
$\Gamma_{l\alpha} = 1.2 \times 10^{-7}\,\text{K m}$	$D_l = 1.50 \times 10^{-9}\,\text{m}^2/\text{s}$	$\gamma_{\alpha\beta} = 0.60\,\text{J/m}^2$
$\Delta H_{f,Al} = 397\text{kJ/kg}$	$\Delta H_{f,Cu} = 205\,\text{kJ/kg}$	$\Delta H_s = \Delta H_f^* = ?$

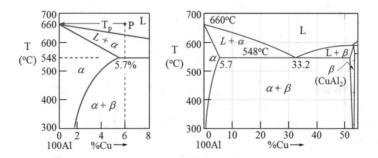

8.5 Use the free-energy solidification model to analyze the *Al-20Si* melt, initially at the pouring temperature $T_p = 850\,°\text{C}$. For one directionally solidified $x = s(t) = 0.01$ m-thick flat plate at $v_z = 100\,\mu\text{m/s}$, use the Gibbs–Thomson coefficients $\Gamma_{l\alpha} = \Gamma_{l\beta}$, solid fractions $f_\alpha = f_\beta = 0.5$, and the given data in Problem 8.1

along with the *Al-Si* equilibrium phase diagram (Example 7A.2) to calculate (a) the Fourier's coefficients F_o and F_n at $n = 1$, and (b) plot $C_l(x, z)$ at $0 \le z \le 0.2\,\mu\text{m}$ and $\lambda = \lambda_c = 20.43 \times 10^{-3}\,\mu\text{m}$. Interpret the resultant $C_l(x, z)$ non-linear trend; determine a possible valid interval for z and approximate the thickness of an apparent mushy zone.

8.6 Assume that an amount of *Al-5Ni* melt initially at $T_p = 660\,°\text{C}$ and subsequently, directionally solidified at $v_z = 100\,\mu\text{m/s}$ as a $x = s(t) = 0.01\,\text{m}$-thick flat plate. Use the given data and the free-energy solidification model to calculate (a) the partition coefficient k_o^α, (b) the liquidus and solidus slopes $m_{l\alpha}$ and $m_{s\alpha}$, (c) the latent heat of solidification ΔH_s and the entropy change of fusion ΔS_f, (d) the critical undercooling ΔT_c and the corresponding lamellar spacing λ_c, and (e) the temperature gradient G_{λ_c} at λ_c.

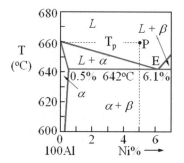
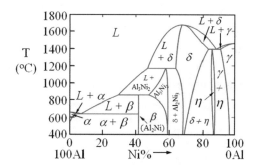

$k_l = 70\,\text{W/m K}$	$c_s = 1200\,\text{J/kg K}$	$\rho = 2500\,\text{kg/m}^3$
$\Gamma_{l\alpha} = 2 \times 10^{-7}\,\text{K m}$	$D_l = 5 \times 10^{-9}\,\text{m}^2/\text{s}$	$\gamma_{\alpha\beta} = 0.80\,\text{J/m}^2$
$\Delta H_{f,Al} = 397\,\text{kJ/kg}$	$\Delta H_{f,Ni} = 92\,\text{kJ/kg}$	$\Delta H_s = ?\,\text{kJ/kg}$

8.7 Consider an amount of *50Pb-50Sn* melt initially at $T_p = 250\,°\text{C}$ and subsequently, directionally solidified at $v_z = 100\,\mu\text{m/s}$ as one $x = s(t) = 0.01\,\text{m}$-thick flat plate. Use the free-energy solidification model and the data given below to calculate (a) the partition coefficient k_o^α, (b) the liquidus and solidus linear slopes $m_{l\alpha}$ and $m_{s\alpha}$, (c) the latent heat of solidification ΔH_s and the entropy change of fusion ΔS_f, (d) the critical undercooling ΔT_c and the corresponding lamellar spacing λ_c, and (e) the temperature gradient at λ_c.

$k_l = 0.50\,\text{W/m K}$	$c_s = 28\,\text{J/kg K}$	$\rho = 8870\,\text{kg/m}^3$
$\Gamma_{l\alpha} = 8 \times 10^{-7}\,\text{K m}$	$D_l = 3 \times 10^{-9}\,\text{m}^2/\text{s}$	$\gamma_{\alpha\beta} = 0.13\,\text{J/m}^2$
$\Delta H_{f,Pb} = 23\,\text{kJ/kg}$	$\Delta H_{f,Sn} = 60\,\text{kJ/kg}$	$\Delta H_s = ?\,\text{kJ/kg}$

8.8 Consider an amount of $Ni\text{-}30Sb$ eutectic alloy melt initially at $T_p = 1300\,°C$ and subsequently, directionally solidified at υ_z as one $x = s(t) = 0.01\,\text{m}$-thick flat plate. Use the free-energy solidification model to determine (a) The latent heat of solidification ΔH_s, (b) the linear hypoeutectic liquidus $m_{l\alpha}$ and $m_{l\beta}$ slopes, (c) the partition coefficients k_o^α and k_o^β, (d) the constants K_C and K_R, (e) the total undercooling function $\Delta T = f(\lambda)$ when $\upsilon_z = 100\,\mu\text{m/s}$, (f) the critical undercooling ΔT_c, the critical lamellar spacing λ_c and the corresponding temperature gradient $[d\Delta T/d\lambda]_{\lambda=\lambda_c}$, and (g) plot $\Delta T = f(\lambda)$ and the point $(\lambda_c, \Delta T_c)$.

$k_l = 0.50\,\text{W/m K}$	$c_s = 28\,\text{J/kg K}$	$\rho = 8870\,\text{kg/m}^3$
$\Gamma_{l\alpha} = 2 \times 10^{-7}\,\text{K m}$	$D_l = 3 \times 10^{-9}\,\text{m}^2/\text{s}$	$\gamma_{\alpha\beta} = 0.13\,\text{J/m}^2$
$\Delta H_{f,Ni} = 92\,\text{kJ/kg}$	$\Delta H_{f,Sb} = 163\,\text{kJ/kg}$	$\Delta H_s = ?\,\text{kJ/kg}$

References

1. A.G. Guy, J.J. Hren, *Elements of Physical Metallurgy*, 3rd edn. (Addison-Wesley Publishing Company, Reading, 1959)
2. R. Elliott, *Eutectic Solidification Processing Crystalline and Glassy Alloys* (Butterworths & Co (Publishers) Ltd., New York, 1983)
3. A.R. Bailey, *The Role of Microstructure in Metals*, 2nd edn. (Metallurgical Services Betchworth, Surrey, 1972)
4. V.T. Witusiewicz, L. Sturz, U. Hecht, S. Rex, Lamellar coupled growth in the neopentylglycol-(D)camphor eutectic. J. Cryst. Growth **386**, 69–75 (2014)
5. A. Karma, A. Sarkissian, Morphological instabilities of lamellar eutectics. Metall. Mater. Trans. **27A**(3), 635–656 (1996)
6. W.A. Tiller, Polyphase solidification, in *Liquid Metals and Solidification: A Seminar Held During the Thirty-Ninth National Metal Congress and Exposition*, Chicago 1957, vol. 276 (ASM, Cleveland, 1958), pp. 279–318
7. K.A. Jackson, J.D. Hunt, Lamellar and rod eutectic growth. Trans. Metal. Soc. AIME **236**(1129), 129–142 (1966)
8. J.A. Dantzig, M. Rappaz, *Solidification*, 2nd edn. Revised & Expanded (EPFL Press, Lausanne, 2016)
9. J.W. Gibbs et al., The three-dimensional morphology of growing dendrites. Sci. Rep. **5** (2015). Article No. 11824. https://doi.org/10.1038/srep11824
10. R.E. Napolita, *The Structure of Cast Metals* (01 June 2005). PowerPoint presentation found at https://web.wpi.edu/Images/CMS/MCSI/2006napolitano1.pdf
11. M.E. Glicksman, *Principles of Solidification: An Introduction to Modern Casting and Crystal Growth Concepts* (Springer Science+Business Media, LLC, New York, 2011)
12. H. Fredriksson, U. Akerlind, *Materials Processing During Casting* (Wiley, Chichester, 2006)
13. H. Fredriksson, U. Akerlind, *Solidification and Crystallization Processing in Metals and Alloys* (Wiley, New York, 2012)
14. D.M. Stefanescu, *Science and Engineering of Casting Solidification*, 2nd edn. (Springer Science+Business Media, LLC, New York, 2009)

15. D.M. Herlach, P. Galenko, D. Holland-Moritz, in *Metastable Solids from Undercooled Melts*, ed. by R. Cahn (Elsevier B.V., Amsterdam, 2007)
16. W. Kurz, D.J. Fisher, *Fundamentals of Solidification*, vol. 1, 3rd edn. (Trans Tech Publications Ltd, Zurich, 1986)
17. R. Caram, M. Banan, W.R. Wilcox, Directional solidification of Pb-Sn eutectics with vibration. J. Cryst. Growth **114**(1–2), 249–254 (1991)
18. M. Hillert, Role of interfacial energy during solid-state phase transformations. Jernkontorets Annaler **141**, 757–789 (1957)
19. M.H. Protter, C.B. Morrey Jr., *College Calculus with Analytic Geometry*, 2nd edn. (Addison-Wesley Publishing Company, Reading, 1970)
20. A. Choudhury, M. Plapp, B. Nestler, Theoretical and numerical study of lamellar eutectic three-phase growth in ternary alloys. Phys. Rev. E **83**(5), 051608 (2011)
21. M. Serefoglu, Pattern selection dynamics in rod eutectics. Graduate Theses and Dissertations #10642, Iowa State University City of Ames, Iowa (2009)
22. D.A. Porter, K. Easterling, M.Y. Sherif, *Phase Transformations in Metals and Alloys*, 3rd edn. (CRC Press, Taylor & Francis Group, LLC, Boca Raton, 2009)
23. A.V. Catalina, P.W. Voorhees, R.K. Huff, A.L. Genau, A model for eutectic growth in multicomponent alloys. IOP Conf. Ser. Mater. Sci. Eng. **84**(1), 012085 (2015)
24. O. Senninger, P.W. Voorhees, Eutectic growth in two-phase multicomponent alloys. Acta Mater. **116**, 308–320 (2016)
25. W.J. Boettinger, The solidification of multicomponent alloys. J. Phase Equilib. Diffus. **37**(1), 4–18 (2016)
26. J. De Wilde, L. Froyen, V. Witusiewicz, U. Hecht, Two-phase planar and regular lamellar coupled growth along the univariant eutectic reaction in ternary alloys: an analytical approach and application to the Al-Cu-Ag system. J. Appl. Phys. **97**(11), 113515 (2005)
27. A. Ludwig, S. Leibbrandt, Generalised Jackson-Hunt model for eutectic solidification at low and large Peclet numbers and any binary eutectic phase diagram. Mater. Sci. Eng. A **375**, 540–546 (2004)
28. D. McCartney, J. Hunt, R. Jordan, The structures expected in a simple ternary eutectic system: Part 1. Theory. Metall. Trans. A **11** (8), 1243–1249 (1980)
29. J. Fridberg, M. Hillert, Ortho-pearlite in silicon steels. Acta Metall. **18**(12), 1253–1260 (1970)
30. M.C. Gao, J.W. Yeh, P.K. Liaw, Y. Zhang (eds.), *High-Entropy Alloys: Fundamentals and Applications* (Springer International Publishing, Cham, 2016)
31. J.W. Yeh, S.K. Chen, S.J. Lin, J.Y. Gan, T.S. Chin, T.T. Shun, Nanostructured high entropy alloys with multiple principal elements: novel alloy design concepts and outcomes. Adv. Eng. Mater. **6**(5), 299–303 (2004)
32. F. Otto, Y. Yang, H. Bei, E.P. George, Relative effects of enthalpy and entropy on the phase stability of equiatomic high entropy alloys. Acta Mater. **61**(7), 2628–2638 (2013)
33. D.M. Rosa, J.E. Spinelli, I.L. Ferreira, A. Garcia, Cellular growth during transient directional solidification of Pb-Sb alloys. J. Alloys Compd. **422**(1–2), 227–238 (2006)

Chapter 9
Solid-State Phase Change

9.1 Introduction

This chapter initially introduces crystallography as an important tool employed to study crystal structures and their corresponding transformation from one type of solid atomic arrangement induced thermally or mechanically to another with different morphology. In essence, crystallography and phase transformation modeling are specifically coupled in order to explain the formation of new phases in the solid state.

Also included in this chapter is a compilation of analytical solutions to one-dimensional phase transformation problems in the solid state with moving grain boundary ()migration. Moreover, the solid-state phase transformation is controlled by heating samples at temperatures T below the solidus temperatures T_s, followed by slow or rapid cooling in thermal conductive media.

The corresponding phase transformation mechanism is either diffusion driven by a concentration gradient ($\partial C/\partial x$) or by a diffusionless (displacive) process in the solid-phase domain. The former is a slow atomic process that induces nucleation or precipitation of a new crystalline phase from the parent (old) phase. The latter requires rapid cooling in order to induce phase transformation by a quenching and as a result, an amorphous of glassy structure is obtained with excellent properties.

In practice, the extreme cases for phase transformation in the solid state are based on diffusional (annealing) and diffusionless (quenching) processes. The former may be defined as a random walk by diffusion mechanism during slow cooling and the latter is a displacive (abrupt) mechanism during rapid cooling.

© Springer Nature Switzerland AG 2020
N. Perez, *Phase Transformation in Metals*,
https://doi.org/10.1007/978-3-030-49168-0_9

9.2 Thermally-Induced Phase Change

Crystal structural changes in thermally-induced phase transformation (transition) of a solid phase can be determined using conventional metallography and a diffraction technique, such as X-ray diffraction. The former reveals the microstructural features of a crystalline material and the latter provides a diffraction pattern related to crystallographic (hkl) planes.

In general, assume that a "parent" or "old" α-phase is the matrix to be transformed to β-phase at a particular critical temperature T_c upon slow cooling. Regardless of the physical conditions of the matrix, the phase-change process occurs by diffusional transformation in the solid state associated with some type of crystallographic adjoining at the growing α-β interface, while latent heat of transformation evolves at a critical temperature T_c.

The phrase "solid-state" used in this chapter, henceforth, refers to phase changes from solid-to-solid denoted by invariant reactions such as

Ferrite to austenite	$= \delta\,(BCC) \rightarrow \gamma\,(FCC) = A$	for steels
Austenite to ferrite	$= \gamma\,(FCC) \rightarrow \alpha\,(BCC) = F$	for steels
Austenite to pearlite	$= \gamma\,(FCC) \rightarrow \alpha\,(BCC) + Fe_3C$	for steels
Austenite to martensite	$= \gamma\,(FCC) \rightarrow \alpha'\,(BCT) = M$	for steels
Austenite to martensite	$= \gamma\,(FCC) \rightarrow \alpha'\,(BCT) = M$	for Nitinol
Alpha to beta phase	$= \alpha\,(HCP) \rightarrow \beta\,(BCC)$	for Ti-alloys
Beta to alpha	$= \beta\,(BCT) \rightarrow \alpha\,(cubic)$	for Tin (Sn)

Crystal structural changes in thermally-induced phase transformation (transition) of a solid phase can be determined using conventional metallography and a diffraction technique, such as X-ray diffraction. The former reveals the microstructural features of a crystalline material and the latter provides a diffraction pattern related to crystallographic (hkl) planes.

9.2.1 Phase-Transformation Crystallography

This section describes the effects of the crystallographic orientation relationship (OR) on the phase transformation of a metal or alloy lattice structure. Among the seven (7) types of crystal structures, the face-centered orthorhombic (FCO) or face-centered cubic (FCC) unit cell is a suitable atomic arrangement for showing how to deduce the crystallography of the transformed phase at a particular critical temperature (T_c) and pressure. This is shown in Fig. 9.1 for a pure orthorhombic or cubic metal.

The general crystallography between two different crystal lattices (say, α and β) is defined in terms of parallel or near parallel planes and directions in each lattice. The corresponding notation and condition are denoted as

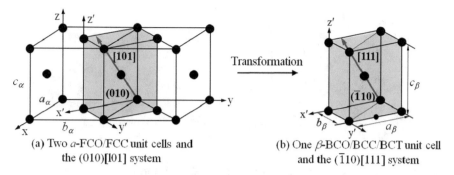

(a) Two a-FCO/FCC unit cells and the $(010)[\bar{1}01]$ system

(b) One β-BCO/BCC/BCT unit cell and the $(\bar{1}10)[111]$ system

Fig. 9.1 Solid-state phase transformation of a face-centered orthorhombic (FCO) or face-centered cubic (FCC) into a BCO, BCC, or BCT unit cell. (**a**) Two α-FCO/FCC unit cells and the $(010)[\bar{1}01]$ system. (**b**) One β-BCO/BCC/BCT unit cell and the $(\bar{1}10)[111]$ system

$$(hkl)_\alpha \parallel (hkl)_\beta \ \& \ [uvw]_\alpha \parallel [uvw]_\beta \qquad (9.1a)$$

$$hu + kv + lw = 0 \quad \text{for } (hkl) \parallel [uvw] \qquad (9.1b)$$

This means that $(hkl)_\alpha$ plane is parallel ($\parallel$) to $(hkl)_\beta$ plane and $[uvw]_\alpha$ direction is parallel to $[uvw]_\beta$ direction. However, $[uvw]_\alpha$ must lie in the $(hkl)_\alpha$ plane and similarly, $[uvw]_\beta$ is in the $(hkl)_\beta$ plane. It can be deduced from Fig. 9.1 that $(010)_\alpha \parallel (\bar{1}10)_\beta$ and $[101]_\alpha \parallel [111]_\beta$.

Note that the selected unit cell in gray color is assumed to transform from the FCO unit cell with $a_\alpha \neq b_\alpha \neq c_\alpha$ or FCC unit cell with $a_\alpha = b_\alpha = c_\alpha$ as a BCO or FCC unit cell (Fig. 9.1b), respectively.

In particular, the Miller indices for the crystallographic planes shown in Fig. 9.1 and the general notation given by Eq. (9.1a) indicate that

$$(hkl)_\alpha = (010)_\alpha \quad \text{and} \quad [uvw]_\alpha = [101]_\alpha$$

$$(hkl)_\beta = (\bar{1}10)_\beta \quad \text{and} \quad [uvw]_\beta = [111]_\beta$$

where $[101]_\alpha$ is in the $(010)_\alpha$ plane and $[111]_\beta$ is in the $(110)_\beta$. Thus, the crystallographic orientation relationship between FCO and BCO is clearly illustrated in Fig. 9.1. Moreover, using the $(010)_\alpha [101]_\alpha$ and $(\bar{1}10)_\beta [111]_\beta$ systems on Eq. (9.1b) verifies that these directions are parallel to their respective planes. A similar analysis can be carried out for any other crystal structure.

Furthermore, using the closest packed planes in the α_{FCC} and β_{BCC} structures one can deduce that there are 4 {111} close-packed planes in the FCC unit cell and 3 ⟨110⟩ close-packed directions per each {111} plane. On the other hand, there are 6 {110} planes and 2 ⟨111⟩ close-packed directions per each {110} plane in the

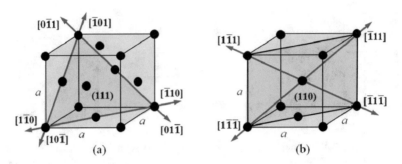

Fig. 9.2 Close-packed planes and directions in (**a**) FCC and (**b**) BCC unit cells

BCC unit cell. For illustration purposes, Fig. 9.2 shows the (111)-{110} and (110)-{111} systems for FCC and BCC unit cells, respectively.

The above crystallographic approach for determining the orientation relationships (OR's) between planes and directions during solid-phase transformation is essentially ideal or simply hypothetical. However, slight misorientation in the parallelism of crystallographic directions is possible. Moreover, texture and azimuthal orientations are also used for parallelism between planes and directions, respectively. The azimuthal orientation refers to an angle between projected and reference vectors and texture orientation is related to surface shape.

In crystallography, the crystal orientation is fundamentally defined by a $(hkl)[uvw]$ slip system and depicted in Fig. 9.2. For instance, one can select the (111) and (110) planes and the shown $[uvw]$ directions to define several crystal orientations, such as

Fig. 9.2a   Fig. 9.2b

$(111)[01\bar{1}]$ $(111)[\bar{1}10]$ $(110)[\bar{1}1\bar{1}]$ $(110)[\bar{1}11]$

$(111)[\bar{1}01]$ $(111)[0\bar{1}1]$ $(110)[1\bar{1}1]$ $(110)[1\bar{1}\bar{1}]$

$(111)[1\bar{1}0]$ $(111)[10\bar{1}]$

These are simple and essential notations necessary to understand the crystallography of atomic structure transformation associated with the concept of phase transformation. In particular, the orientation relationship (OR) at the interface between two distinct (dissimilar) crystal structures. In other words, this type of crystallographic notation imparts the relation between specific planes and directions of two crystals on either side of boundary, as in FCC and BCC unit cells shown in Fig. 9.2. For the sake of clarity, the above selected pairs of directions ($[uvw]_\alpha$, $[uvw]_\beta$), and pairs of planes and directions, $(hkl)[uvw]$, are commonly determined experimentally using X-ray diffraction.

Nevertheless, the lattice misorientation associated with interfaces, such as grain boundaries and crystals, is an important aspect of microstructures. For instance, if a substrate–monolayer interface does not have a matching atomic arrangement, then a crystallographic orientation relationship (OR) and related orientation angle are defined as useful aspects of atomic deposition by adsorption or diffusion.

9.2.2 Interface Coherency

This section includes phenomenological models that describe the logical interconnections or interface boundaries between two crystals having similar or different atomic arrangements. The general features of interface boundaries can be understood with the aid of Fig. 9.3a–f, suggesting that the interface boundaries (ib) between α and β grains (Fig. 9.3) and between β-particles within α-phase (Fig. 9.3d,e,f) are viable (applicable) options to understand the solid-state phase transformation process.

These models are classic geometrical sketches used for visualizing the atomic arrangement of α and β phases and related type of interface boundary (ib) coherency ([1, pp. 146–149] and [2, p. 422]). One can assume that the solid surfaces of two crystals are smooth and flat atomic planes having partial atomic bonding and that their surface planes repeat themselves underneath such surfaces. Moreover, during transformation in the solid state, a portion of the source solid (α-phase) completely or partially changes to a new solid (β-phase) with a critical volume $V_{c,\beta}$ and a critical radius r_c, preferably, along grain boundaries (gb). Consequently, a α-β

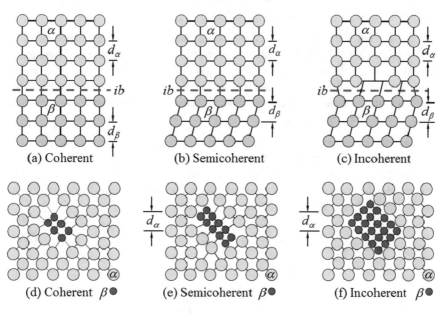

(a) Coherent   (b) Semicoherent   (c) Incoherent

(d) Coherent $\beta\bullet$ (e) Semicoherent $\beta\bullet$ (f) Incoherent $\beta\bullet$

Fig. 9.3 Phase boundary models for (**a,b,c**) α-β-grains [1, p. 147] and (**d,e,f**) β-particles within α-grains [2, p. 422]

interface with a total surface area A forms and moves (migrates), while the volume of the β-phase increases at the expense of the original volume of the α-phase.

Normally, the interplanar spacings (d_α and d_β) or the lattice parameters (a_α and a_β) and the surface energy $\gamma_{\alpha\beta}$ are the relevant macroscopic entities one can used to characterize the results of phase transformation. The interplanar spacing, d_α-spacing or d_β-spacing, defines the distance between atomic planes.

Coherent Interface The coherent interface (Fig. 9.3a,d) arises when there is a perfect atomic match between the two adjoining α and β metallic phases with equal lattice parameters are continuous across the interface. However, a small lattice mismatch induces a small elastic strain in the adjacent crystals. This implies that the lattice parameters (a_α and a_β) are similar to an extent and the surface energy due to straining is $\gamma_{strain} = 0$.

Further, this type of interface boundary is found in some *Cu-Si* alloys as indicated by the crystallographic orientation relationship (OR) between α-FCC and $\kappa = \beta$-HCP phases. The notation for the α-β orientation relationship is based on the parallelism between FCC and HCP structures [1, p. 148]. Thus,

$$(111)_\alpha \parallel (0001)_\beta \quad \& \quad \left[\bar{1}10\right]_\alpha \parallel \left[11\bar{2}0\right]_\kappa$$

$$\epsilon_{\alpha\beta} \simeq 0 \quad \text{since} \quad a_\alpha \simeq a_\beta$$

which indicate that the $(hkl)_\alpha$ plane of the α crystal lies parallel to the $(hkl)_\beta$ plane of the β crystal. This is a parallelism denoted by the symbol $\parallel$ or by the double-slant symbol $//$. The former prevails in this chapter.

Semi-coherent Interface The semi-coherent interface (Fig. 9.3b,e) occurs when dislocations (Fig. 9.5a) distort adjacent atomic planes, causing the rise of the surface energy, $\gamma_{strain} > 0$, of the system. This means that a slight lattice mismatch causes a periodic array of misfit dislocations, leading to a significant elastic strain. This implies that a sufficiently large atomic misfit causes the disregistry by dislocations. Hence, semi-coherent interface.

Incoherent Interface The incoherent interface (Fig. 9.3c,f) represents the most antisymmetric phase boundary with the highest lattice distortion (Fig. 9.5b), leading a high lattice misfit at the interface. Thus, the coherent surface energy $\gamma_{\alpha\beta}$ and related elastic strain $\epsilon_{\alpha\beta}$ due to elastic distortion caused by dimensional misfit of strain-free lattices at the α-β interface are given by

$$\gamma_{\alpha\beta} = \gamma_{chem} + \gamma_{strain} \tag{9.2a}$$

$$\epsilon_{\alpha\beta} = \frac{\lfloor d_\alpha - d_\beta \rfloor}{d_\alpha} \simeq \frac{\lfloor d_\alpha - d_\beta \rfloor}{d_\beta} \tag{9.2b}$$

$$\epsilon_{\alpha\beta} = \frac{\lfloor a_\alpha - a_\beta \rfloor}{a_\alpha} \simeq \frac{\lfloor a_\alpha - a_\beta \rfloor}{a_\beta} \tag{9.2c}$$

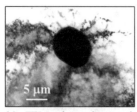 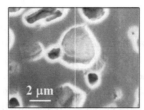 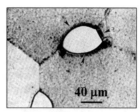

(a) TEM photomicrograph showing a $M_{23}C_6$ particle in stainless steel 304 (author)

(b) SEM photomicrograph showing boride particles in Devitrium 7025 alloy. (author)

(c) Optical photomicrograph showing one α precipitate in Cu-In alloy [1]

Fig. 9.4 Coherency of particles (precipitates) in different metal matrices. (**a**) Heat treated (HT) 304 stainless steel at 1000°C for 0.5 h and air cooled, (**b**) HT Devitrium 7025 alloy at 1100°C for 24 h and air cooled, and (**c**) HT Cu-In alloy [1]

Here, γ_{chem} denotes the chemical interfacial energy, γ_{strain} denotes the interfacial strain surface energy due to dislocations at the interface, and a_α denotes the reference lattice parameter. Moreover, Eqs. (9.2b,9.2c) represent the lattice misfit at the interface. This suggests that the higher $\epsilon_{\alpha\beta}$ the higher the array of dislocations and the higher the strain energy at the interface boundary.

Note that the d-spacings, d_α and d_β, depend on the lattice parameters determined by a diffraction technique, such as X-ray diffraction introduced in Chap. 1. Thus, knowledge of X-ray diffraction is an essential complement to the analysis of crystallography and related atomic structure of solid phases.

With respect to particle-matrix atomic coherency, one can deduce that coherent particles embedded in the metal matrix have matching atomic arrangement at the interfaces, semi-coherent particles have at least one straight edge, and completely incoherent particles are either spherical, semi-spherical, or elliptical in shape. This rough deduction is due to the shape of unit cells. Figure 9.4 shows distinguishable semi-coherent and incoherent particles (precipitates) in three different alloys.

Figure 9.4a depicts a near-circular carbide particle located at grain boundary triple point, Fig. 9.4b clearly shows boride particles embedded in an amorphous Ni, Mo-based Devitrium 7025 alloy matrix; $Ni_{53}Mo_{35}B_9Fe_2$ with 1% impurities. This is a non-distinguishable grain boundary material containing different boride crystals with dissimilar morphologies. This alloy was rapidly solidified, consolidated, and heat treated. Figure 9.4c clearly illustrates a particle at a grain boundary triple point (taken from [1, p. 159]).

Essentially, precipitated crystalline particles (Fig. 9.4b) embedded in an amorphous matrix form semi-coherent and incoherent interfaces as indicated by the models in Fig. 9.3b,c. Amazingly, the consolidated Devitrium 7025 alloy exhibited a strong resistance to crystallization after a prolong heating at 1100°C for 24 h. Therefore, the effect of annealing on the amorphous matrix was not observed on the amorphous matrix, instead stable boride particles precipitated, mainly as a hard intermetallic phase Mo_2NiB_2.

A single type of particle is considered a secondary phase embedded and randomly distributed within the host matrix. As a whole, the effect of many hard or hard enough particles is that they impede the easy motion of dislocations during mechanical deformation of specimens and as a result, the microstructure is strengthen and mechanical properties are significantly enhanced. Hence, a precipitation strengthening mechanism has a similar physical meaning as dispersion hardening (by adding a dispersed secondary phase into the host matrix). Moreover, it is clear now the fundamentally importance of crystallographic analysis and material characterization needed to understand the theory of particle coherency related to material strengthening. For precipitates in a matrix, the corresponding lattice parameters are accommodated by elastic strains as indicated by Eq. (9.2).

9.3   Energy of Straight Dislocations

This section focuses on the common analytical solutions to dislocation problems in elastically deformed crystals during solid-state phase transformation. The classic core-cylindrical ring models are used for deriving the energy of straight dislocations that generate during phase transformation. As a result, dislocations induce an elastic distortion on adjacent atomic planes. Thus, the stress field close to the dislocation core and the energy of dislocations can be predicted using the theory of elasticity.

A dislocation can be modeled as a straight or curved volume element. The former case is adopted henceforth for deriving the energy of dislocations. For instance, Fig. 9.5a shows a *bubble raft* model containing an array of dislocations (darker points) within a grain [3]. This microscale linear defect can be defined as a dislocation pile-up, which induces a pressure to the upper grain boundary.

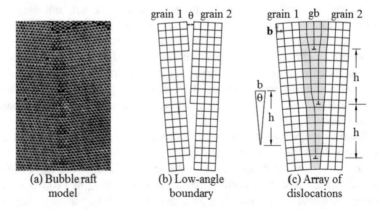

(a) Bubble raft   (b) Low-angle   (c) Array of
     model   boundary   dislocations

Fig. 9.5 Low-angle grain boundary model. (**a**) Bubble raft model [3], (**b**) two-grain morphology showing the low angle θ, and (**c**) an array of edge dislocations separated by a distance h in a hypothetical solid [4, p. 157]

Figure 9.5b exhibits the classic symmetrical low-angle tilt boundary (θ) between two crystals and Fig. 9.5c depicts the model for the array of edge dislocations with a common Burgers vector $\mathbf{b} = \overrightarrow{b}$ within a crystal [4, p. 157]. This array of dislocations represent the atomic misorientation of the lattice.

Accordingly, the number of dislocations per unit length of boundary (Fig. 9.5c) can be defined by

$$n = \frac{1}{h} = \frac{2\sin(\theta/2)}{b} \simeq \frac{\theta}{b} \tag{9.3}$$

where b is the magnitude of the Burgers vector and θ is the tilt angle in radians. For modeling purposes, the array of straight dislocations has a common Burgers vector b and spacing h (Fig. 9.5c). In addition, an array dislocations related to low-angle grain boundaries is normally used to derive equations for stress, strain, and displacement components, and elastic strain energy near dislocation cores as function of the misorientation angle θ.

Read–Shockley Analytical Procedure According to this dislocation model [5], the grain boundaries between α and β phases (or crystallites) must have a small orientation. This suggests that the model considers semi-coherent interface boundaries for deriving the grain boundary energy (GB) equation as a function of the angle of misfit (θ) and the orientation of the boundary. Actually, the Read–Shockley theory elucidates the fundamental importance for understanding the concept of grain boundary (GB) energy during elastic deformation of crystalline materials.

The mathematical modeling of a local atomic distortion of a crystalline solid induced by internal shear stresses leads (1) to an atomic-scale characterization of dislocations and (2) to a macroscale derivation of the elastic strain energy of dislocations.

For convenience, only the top view of a cylindrical ring with core radius $r_o \geq b$ and length $L(r)$ for an edge dislocation is shown in Fig. 9.6 [5, p. 357]. Notice the dislocation energy regions; U_I denotes the core energy region (Fig. 9.6a), U_{II} denotes the cylinder energy region (Fig. 9.6), U_{III} denotes the outside energy region, and dU_{II} denotes the infinitesimal dislocation energy change (Fig. 9.6b).

Let the shear stress component on the slip plane along the slip direction be defined as

$$\tau = \tau_o \left(\frac{b}{r}\right) \tag{9.4}$$

with

$$\tau_o = \frac{Gb^2}{2\pi} \qquad \text{(screw dislocation)} \tag{9.5a}$$

$$\tau_o = \frac{Gb^2}{2\pi(1-\upsilon)} \quad \text{(edge dislocation)} \tag{9.5b}$$

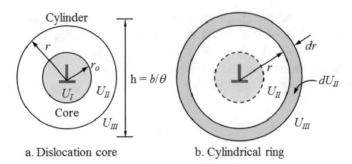

a. Dislocation core   b. Cylindrical ring

Fig. 9.6 Cylindrical ring model and dislocation core. (**a**) Dislocation core and related energy regions and (**b**) cylindrical ring due to changes in dislocation energy dU_{II} in region U_{II} [5, p. 357]

$$\frac{dr}{r} = \frac{dh}{h} = -\frac{d\theta}{\theta} \tag{9.5c}$$

where G denotes the shear modulus and υ denotes the Poisson's ratio.

Let E be the energy per unit area of the boundary and $U/b = U\theta/b$ be the energy per dislocation per unit length outside the dislocation core so that

$$\frac{d}{d\theta}\left(\frac{U\theta}{b}\right) = \frac{dU}{d\theta} \tag{9.6}$$

The general elastic energy stored in the volume element dV of a dislocation is

$$d\left(\frac{U}{\theta}\right) = \frac{1}{2}\tau b dr = \frac{1}{2}\tau_o b^2 \frac{dr}{r} = -\frac{1}{2}\tau_o b^2 \frac{d\theta}{\theta} \tag{9.7}$$

Integrating Eq. (9.7) yields the elastic energy U

$$\frac{U}{\theta} = \frac{\tau_o b^2}{2}[A - \ln(\theta)] = E_o[C - \ln(\theta)] \tag{9.8a}$$

$$U = U_o\theta[A - \ln(\theta)] \tag{9.8b}$$

$$U = U_o\theta\left[1 + \ln\left(\frac{b}{2\pi r_o}\right) - \ln(\theta)\right] \tag{9.8c}$$

where U_o is a constant defining the type of dislocation energy and A is the constant of integration.

These constants are defined by [6, p. 359]

$$U_o = \frac{\tau_o b^2}{2} \tag{9.9a}$$

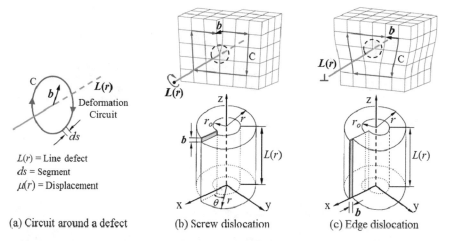

Fig. 9.7 Deformation circuits around dislocations. (**a**) The circuit around a dislocation line, (**b**) core of an edge dislocation, and (**c**) core of a screw dislocation. After [7, 8]

$$A = 1 + \ln\left(\frac{b}{2\pi r_o}\right) + \frac{2}{\tau_o b^2}(E_I + E_{II}) \simeq 1 + \ln\left(\frac{b}{2\pi r_o}\right) \qquad (9.9b)$$

Combining Eq. (9.5) and (9.8c) yields the elastic strain energy for edge and screw dislocations as functions of the angle of misfit (θ)

$$U_{screw} = \frac{Gb^2\theta}{2\pi}\left[1 + \ln\left(\frac{b}{2\pi r_o}\right) - \ln(\theta)\right] \qquad (9.10a)$$

$$U_{edge} = \frac{Gb^2\theta}{2\pi(1-v)}\left[1 + \ln\left(\frac{b}{2\pi r_o}\right) - \ln(\theta)\right] \qquad (9.10b)$$

$$U_{screw} = (1-v)\, E_{edge} \qquad (9.10c)$$

Most metals have a Poisson's ratio $v = 1/3$. In such a case, screw dislocation energy becomes $U_{screw} = (2/3)\, E_{edge} = 0.67 E_{edge}$. This simply implies that $U_{screw} < E_{edge}$. Moreover, if the dislocation core radius is $r_o > b$, which is usually the case, then $\ln(b/2\pi r_o) < 0$ and $[\ln(b/2\pi r_o) - \ln(\theta)] < 1$ so that the dislocation energy terms are always positive; $U_{screw} > 0$ and $U_{edge} > 0$.

Hull–Bacon Analytical Procedure Consider a closed clockwise curve known as the *Burgers circuit* around a line defect $L(r)$ depicted in Fig. 9.7 [7], and the dislocations modeled as cylindrical shells of radius r and thickness dr shown in Fig. 9.6b,c [8, pp. 67 & 70].

The local Burgers vector of the linear defect surrounded by an arbitrary contour "C" is given by the line integral [7] and its directions are based on vector algebra. For characterizing dislocations,

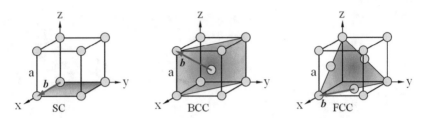

Fig. 9.8 Burgers vectors in cubic crystals

$$\vec{b} = \oint \frac{\partial \mu\,(r)}{\partial s} ds \tag{9.11a}$$

$$\vec{b} \times \vec{L}\,(r) = 0 \quad \text{(screw \circlearrowright, parallel)} \tag{9.11b}$$

$$\vec{b} \cdot \vec{L}\,(r) = 0 \quad \text{(edge \perp, perpendicular)} \tag{9.11c}$$

For a line defect called dislocation, the magnitude of the Burgers vector is non-zero $b \neq 0$ and the dislocation displacement $\mu\,(r)$ decays in the x-y plane at a distance r. The latter indicates that a dislocation ends at a free surface, grain boundary or another dislocation line. Moreover, the *Burgers vector* $\mathbf{b} = \vec{b}$ (Fig. 9.8) represents the magnitude and direction of the lattice distortion due to dislocation motion and symbolizes a step required to close the Burgers circuit.

The magnitude of the Burgers vector, $b = \lfloor \mathbf{b} \rfloor$, is the smallest interatomic spacing shown by the arrow between two atoms in a close-packed crystallographic direction and it has been assigned an arbitrary location and direction in cubic crystal structures (see Chap. 1) for illustration purposes.

The magnitude of the Burgers vector (b) in cubic crystals can be defined by

$$\vec{b} = a\,[uvw] \qquad \text{(simple cubic)} \tag{9.12a}$$

$$b = a\sqrt{u^2 + v^3 + w^2} \quad \text{(simple cubic)} \tag{9.12b}$$

$$\vec{b} = \frac{a}{2}\,[uvw] \qquad (FCC \text{ or } BCC) \tag{9.12c}$$

$$b = \frac{a}{2}\sqrt{u^2 + v^3 + w^2} \quad (FCC \text{ or } BCC) \tag{9.12d}$$

Among excellent resources on dislocation theory available in the literature, the subsequent analytical procedure for deriving the energy of dislocations is found in the Hull–Bacon book [8, chapter 4].

Cartesian Coordinates The mathematical modeling of a *screw dislocation* is based on the elastic strain field within a dislocation cylinder (Fig. 9.8b), where the non-zero displacement (μ_z) and elastic strains (ϵ_{ij}) on a radial slit are

$$\mu_z = \frac{b\theta}{2\pi} = \arctan\left(\frac{y}{x}\right) \tag{9.13a}$$

$$\epsilon_{xz} = \epsilon_{zx} = \frac{1}{2}\left(\frac{\partial\mu_z}{\partial x} + \frac{\partial\mu_x}{\partial z}\right) = -\frac{b}{4\pi}\frac{y}{x^2+y^2} = -\frac{b}{4\pi}\frac{\sin\theta}{r} \tag{9.13b}$$

$$\epsilon_{yz} = \epsilon_{zy} = \frac{1}{2}\left(\frac{\partial\mu_y}{\partial x} + \frac{\partial\mu_x}{\partial y}\right) = +\frac{b}{4\pi}\frac{x}{x^2+y^2} = +\frac{b}{4\pi}\frac{\cos\theta}{r} \tag{9.13c}$$

and according to the theory of elasticity, the relevant non-zero elastic shear stresses are defined by

$$\tau_{xz} = \tau_{zx} = 2G\epsilon_{xz} = -\frac{Gb}{4\pi}\frac{\sin\theta}{r} \tag{9.14a}$$

$$\tau_{yz} = \tau_{zy} = 2G\epsilon_{xy} = +\frac{Gb}{4\pi}\frac{\cos\theta}{r} \tag{9.14b}$$

Here, the axial (E) and shear (G) modulus of elasticity are material's properties and they are related as

$$E = 2G\left(1 + \upsilon\right) \tag{9.15}$$

The differential elastic energy for a screw dislocation (dU_s) and for an edge dislocation (dU_e) can be defined as

$$dU_s = \frac{1}{2}\tau_{yz}bdA = \frac{Gb^2}{4\pi}\frac{dr}{r} \tag{9.16a}$$

$$dU_e = \frac{1}{2}\tau_{xy}bdA = \frac{Gb^2}{4\pi\left(1-\upsilon\right)}\frac{dr}{r} \tag{9.16b}$$

In principle, the strained atomic bonds outside the dislocation core in an elastically strained material can be envisioned as the source of dislocation energy. Hence, dislocation strain energy. Mathematically, Eq. (9.16) provides a significant input on the magnitude of the dislocation energy. Notice that $dU_e > dU_s$ since $(1 - \upsilon) < 1$.

Thus, the dislocation elastic energy terms are

$$U_s = \frac{Gb^2}{4\pi}\int_{r_o}^{r}\frac{dr}{r} = \frac{Gb^2}{4\pi}\ln\left(\frac{r}{r_o}\right) \tag{9.17a}$$

$$U_e = \frac{Gb^2}{4\pi\left(1-\upsilon\right)}\int_{r_o}^{r}\frac{dr}{r} \tag{9.17b}$$

$$U_e = \frac{Gb^2}{4\pi\left(1-\upsilon\right)}\ln\left(\frac{r}{r_o}\right) \tag{9.17c}$$

$$U_s = (1 - \upsilon) U_e \tag{9.17d}$$

Polar Coordinates The non-zero shear stresses (τ_{ij}) and the shear strain (γ_{ij}) are

$$\tau_{rz} = +\tau_{xz} \cos\theta + \tau_{yz} \sin\theta \tag{9.18a}$$

$$\tau_{\theta z} = -\tau_{xz} \sin\theta + \tau_{yz} \cos\theta \tag{9.18b}$$

$$\gamma_{\theta z} = \gamma_{z\theta} = \frac{b}{4\pi r} \tag{9.18c}$$

For an element of volume dV, the elastic strain energy dU_ϵ in Cartesian and polar coordinates is defined by Hooke's law

$$dU_\epsilon = \frac{1}{2}dV \sum_{i=x,y,z} \sum_{j=x,y,z} \sigma_{ij}\epsilon_{ij} \tag{9.19a}$$

$$dU_\epsilon = \frac{1}{2}dV \sum_{i=r,\theta,z} \sum_{j=r,\theta,z} \sigma_{ij}\epsilon_{ij} \tag{9.19b}$$

where $\sigma_{ij}\epsilon_{ij} = \tau_{ij}\gamma_{ij}$ for $i \neq j$ and $\sigma_{ij}\epsilon_{ij} = \sigma_{ji}\epsilon_{ji}$ due to symmetry. Moreover, the screw dislocation energy change dU_s (Fig. 9.7b) is defined as the elastic energy (dU_ϵ) stored per unit dislocation length $L = L(r)$ (Fig. 9.7) within the volume dV.

For a cylindrical ring, the infinitely differential volume dV, the general shear strain γ_{ij} and strain energy U, the shear stress τ_{ij} and the strain-energy volume U_V are, respectively, written as

$$dV = 2\pi r L(r)\, dr \tag{9.20a}$$

$$\gamma_{ij} = \frac{b}{2\pi r} \tag{9.20b}$$

$$U = \frac{1}{2}\tau_{ij}\gamma_{ij} \tag{9.20c}$$

$$\tau_{ij} = G\gamma_{ij} \tag{9.20d}$$

$$U_V = \frac{1}{2}G\gamma_{ij}^2 = \frac{1}{2}\frac{Gb^2}{(2\pi r)^2} \tag{9.20e}$$

Thus,

$$dU = \left[\frac{1}{2}\frac{Gb^2}{(2\pi r)^2}\right] dV = \left[\frac{1}{2}\frac{Gb^2}{(2\pi r)^2}\right][2\pi r L(r)\, dr] = \frac{Gb^2 L(r)}{4\pi r}dr \tag{9.21}$$

For a straight screw dislocation with $L = L(r)$, $dU_s = dU/L(r)$ so that

$$dU_s = \frac{1}{2}\left(2\pi r dr\right)\left(\tau_{\theta z}\gamma_{\theta z} + \tau_{z\theta}\gamma_{z\theta}\right) = \left(4\pi r dr\right)G\gamma_{\theta z}^2 \tag{9.22a}$$

$$dU_s = 4\pi r dr G\left(\frac{b}{4\pi r}\right)^2 = \frac{Gb^2}{4\pi}\frac{dr}{r} \tag{9.22b}$$

$$U_s = \frac{Gb^2}{4\pi}\int_{r_o}^{r}\frac{dr}{r} = \frac{Gb^2}{4\pi}\ln\left(\frac{r}{r_o}\right) \tag{9.22c}$$

and for a straight edge dislocation with $dU_e = dU/L\,(r) = dU/L$,

$$dU_e = \frac{1}{2}2\pi r dr\left(\tau_{\theta z}\gamma_{\theta z} + \tau_{z\theta}\gamma_{z\theta}\right) = 4\pi r dr G\gamma_{\theta z}^2 \tag{9.23a}$$

$$dU_e = \frac{4\pi r dr G}{1-\upsilon}\left(\frac{b}{4\pi r}\right)^2 = \frac{Gb^2}{4\pi\left(1-\upsilon\right)}\frac{dr}{r} \tag{9.23b}$$

$$U_e = \frac{Gb^2}{4\pi\left(1-\upsilon\right)}\int_{r_o}^{r}\frac{dr}{r} = \frac{Gb^2}{4\pi\left(1-\upsilon\right)}\ln\left(\frac{r}{r_o}\right) \tag{9.23c}$$

So far the preceding analytical procedure is for deriving the strain energy of a dislocation in its purest state or conditions. In real crystals, dislocations are three-dimensional mixed defects that control the physical behavior of materials and consequently, mechanical properties and corrosion are affected by the presence or generation of dislocations.

Assuming the above statement is correct, dislocations in real materials are most commonly mixed in nature as schematically shown in Fig. 9.9, after Callister–Rethwisch [9, p. 117].

A mixed dislocation may have its Burgers vector defined in a general form as $b_{screw} \leq b_{mix} \leq b_{edge}$ or $b_{edge} \leq b_{mix} \leq b_{screw}$ and it is located at an intermediate angle with respect to the direction of the dislocation line. Hence, the strain energy of a mixed dislocation is $U_{mix} = U_s + U_e$.

For a mixed dislocation line, the strain energy is defined by

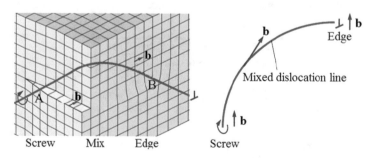

Fig. 9.9 Pure screw (A) → mixed (AB) → pure edge (B) dislocations with Burgers vector **b**. After [9, p. 117]

$$U_{mix} = U_s + U_e = \left(\frac{1+\upsilon}{\upsilon}\right)\frac{Gb^2}{4\pi}\ln\left(\frac{r}{r_o}\right) \tag{9.24}$$

Remarkably, the classical theory of elasticity breaks down at an atomic level due to the fundamental notion of a large length scale.

Setting the limits $r = 10^5 b$ and $r_o = 0.349b$ [10, p. 144] yields $\ln(r/r_o) = 4\pi$ and Eqs. (9.22c), (9.23c) and (9.24) become

$$U_s \simeq Gb^2 \tag{9.25a}$$

$$U_e \simeq \frac{Gb^2}{1-\upsilon} \tag{9.25b}$$

$$U_s \simeq (1-\upsilon)\,U_e \tag{9.25c}$$

$$U_{mix} = \left(\frac{1+\upsilon}{\upsilon}\right)Gb^2 \tag{9.25d}$$

If the dislocation length $L = \left|\vec{L}\,(r)\right|$ is known, then the dislocation strain energies are

$$U_s \simeq Gb^2 L \tag{9.26a}$$

$$U_e \simeq \frac{Gb^2 L}{1-\upsilon} \tag{9.26b}$$

$$U_s \simeq (1-\upsilon)\,U_e \tag{9.26c}$$

$$U_{mix} = U_s + U_e = \left(\frac{2-\upsilon}{1-\upsilon}\right)Gb^2 L \tag{9.26d}$$

Therefore, the mixed strain energy U_{mix} is significantly most realistic.

Example 9A.1 Consider edge, screw, and mixed dislocations in a diluted aluminum (Al) alloy. Use the given data below to calculate the dislocation strain energy and the general dislocation density with $L = 100$ nm.

$G = 28$ GPa	$b = 0.29$ nm	$v = 1/3$
$r = 10^5 b$	$r_o = 0.348b$	$L = 100$ nm

Solution From Eq. (9.23c),

$$U_e = \frac{Gb^2}{4\pi\,(1-\upsilon)}\ln\left(\frac{r}{r_o}\right)$$

$$U_e = \frac{\left(28 \times 10^9 \text{ N/m}^2\right)\left(0.29 \times 10^{-9} \text{ m}\right)^2}{4\pi\left(1 - 1/3\right)} \ln\left(\frac{5 \times 10^{-6} \text{ m}}{0.0174 \times 10^{-9} \text{ m}}\right)$$

$$U_e = 3.53 \times 10^{-9} \text{ N} = 3.53 \times 10^{-9} \text{ N.m/m} = 3.53 \times 10^{-9} \text{ J/m}$$

$$U_e = \left(\frac{3.53 \times 10^{-9} \text{ J/m}}{1.602 \times 10^{-19} \text{ J/eV}}\right) = 2.20 \times 10^{10} \text{ eV/m}$$

and from Eqs. (9.25c) and (9.26d),

$$U_s = (1 - v)\,U_e = (1 - 1/3)\left(2.20 \times 10^{10} \text{ eV/m}\right) = 1.47 \times 10^{10} \text{ eV/m}$$

$$U_{mix} = U_s + U_e = 2.20 \times 10^{10} \text{ eV/m} + 1.47 \times 10^{10} \text{ eV/m} = 3.67 \times 10^{10} \text{ eV/m}$$

The general dislocation density is

$$\rho = 1/L^2 = 1/\left(100 \times 10^{-9} \text{ m}\right)^2 = 10^{14} \text{ lines/m}^2$$

9.4   Transformation of Austenite in Steels

This section is devoted to solid-state phase transformation in steels upon slow cooling the austenite γ-phase from an austenitic temperature T_γ. The austenite region is the shaded field in the Fe-C equilibrium phase diagram (T-C diagram) shown in Fig. 9.10, adopted from Callister–Rethwisch book [9, p. 333] and found in Askeland's book [11, p. 495]. This phase diagram is useful in understanding phase transformation in steels being either annealed or normalized at a relatively slow transformation rate and low cooling rate dT_γ/dt.

For steels containing $0 < C < 2.15\ \%wt$, the austenitic γ-region is the FCC structure that must be homogenized (heat treated) prior to a slow cooling process in a suitable media, such as water, air, and so forth. The recommended heat treatment temperature for austenitizing steels are (Fig. 9.10)

$$T_\gamma = A_3 + 50°C \quad \text{(for hypoeutectoid steels)} \tag{9.27a}$$

$$T_\gamma = A_{cm} + 50°C \quad \text{(for hypereutectoid steels)} \tag{9.27b}$$

Obviously, the recommended T_γ, as defined by Eq. (9.27) can be altered, provided that T_γ is not too high in order to avoid high-temperature corrosion during heat treatment and other diffusion-related complications.

Phase Transformation Problem   Consider annealing one semi-infinite slab ($x \geq 0$) made out of a hypoeutectoid plain carbon steel with nominal composition $C_o < C_E = 0.76\%$ (Fig. 9.10). If the slab is initially at an austenitizing temperature T_γ

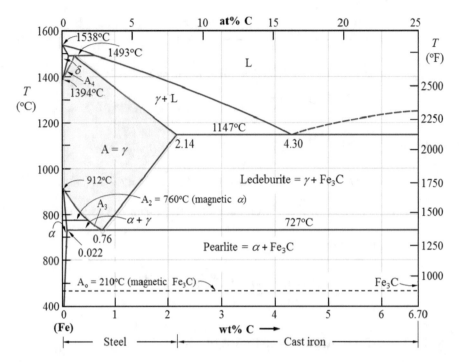

Fig. 9.10 The Fe-C phase diagram showing the austenitic region. Adapted from Callister–Rethwisch book [9, p. 333]

and it is slowly cooled in a furnace air, then phase transformation (decomposition) of austenite to ferrite occurs by substitutional diffusion mechanism, where the atoms from the FCC crystals rearrange themselves to form the BCC structure with a velocity defined as $v = M\Delta G$, where M denotes the atom mobility and ΔG denotes the Gibbs energy change.

Solution Undoubtedly, the solution to this generalized solid-state phase transformation problem involves the Fe-C phase diagram shown in Fig. 9.10 and a suitable nucleation model in order to explain the mechanism of phase changes represented by invariant reactions. Moreover, the following sections are essential for finding a solution to the annealing problem.

9.4.1 Slow Austenitic Transformation

In general, the phase transformation process adopted henceforth is based on cooling austenite at a particular cooling rate dT_γ/dt. For pure iron, the austenite-to-ferrite transformation, $\gamma_{FCC} \rightarrow \alpha_{BCC}$, occurs at $T < T_c = 1185\ K = 912°C$ (Fig. 9.10). A suitable microstructural representation of γ_{FCC} transformation is shown in

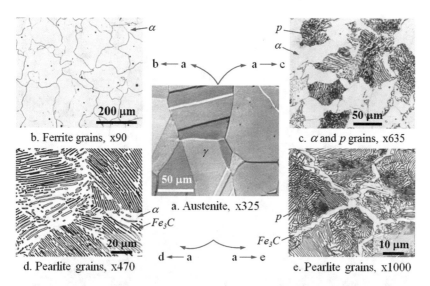

Fig. 9.11 Solid-state phase transformation of the austenite γ-phase upon slow cooling. (**a**) High-temperature austenite (FCC phase), (**b**) pure iron (*Fe*) at $T < T_c = 912°C$, (**c**) *Fe*-0.38C steel at $T < T_E = 727°C$, (**d**) *Fe*-0.80C steel at $T < T_E$, and (**e**) *Fe*-1.40C steel at $T < T_E$. All microstructures are taken from Callister–Rethwisch book [9, chapter 10]

Fig. 9.11a,b,c,d,e. These microstructures can be found in Callister–Rethwisch book [9, chapter 10].

According to the *Fe-C* diagram, the steel microstructure depends on the $\%wt\ C$ or simply $\%C$ and the cooling rate dT_γ/dt for austenite transformation. The reversible ($\rightleftarrows$) invariant reactions associated with the solid-state phase transformation of the γ-phase in pure iron (Fe) at $T < 912°C$ and plain carbon steels (*Fe-C*) at $T < T_E = 727°C$ are defined as

$$\gamma \rightleftarrows \alpha \quad \text{for } C = 0, T < 912°C \tag{9.28a}$$

$$\gamma \rightleftarrows \alpha + p \quad \text{for } 0 < C < 0.80\%, T < T_E = 727°C \tag{9.28b}$$

$$\gamma \rightleftarrows p \quad \text{for } C = 0.80\%, T < T_E = 727°C \tag{9.28c}$$

$$\gamma \rightleftarrows Fe_3C + p \quad \text{for } 0.80\% < C < 2.11\%, T < T_E = 727°C \tag{9.28d}$$

where the arrows represent cooling ($\rightarrow$) and heating ($\leftarrow$).

9.4.2 Solid-Phase Transformation Models

Modeling solid-phase transformation depends on the type of crystal symmetry subjected to a thermally-induced or strain-induced phase transitions at a particular

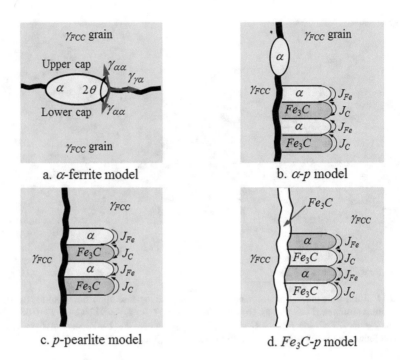

Fig. 9.12 Phase transformation models from austenite. (**a**) α-ferrite model. (**b**) α-p model. (**c**) p-pearlite model. (**d**) Fe_3C-p model

temperature in a suitable environment. For instance, Fig. 9.12 illustrates the ideal models for transforming austenite upon slow cooling (SC) particular specimens under controlled conditions, such as constant cooling rate dT/dt and cooling media, so that atomic diffusion becomes the dominant mechanism. As a result, one can obtain transformed microstructures as depicted in Fig. 9.1b,c,d,e.

Theoretically, it can be assumed that diffusional transformation, indicated by the molar fluxes J_C and J_{Fe}, is the mechanism controlling the atomic arrangement in the new solid phase, where the excess carbon in the α-phase and excess Fe in the Fe_3C-phase are rejected as shown by the arrows.

The morphology of a transformed phase is not as uniform as depicted in Fig. 9.11, but the above models are simple and provide a start point in understanding solid-phase transformation. Notice that only solid–solid-phase transformations in inorganic materials, such as carbon steels, have been considered. This physical phenomenon in steel alloys and other non-ferrous alloys are most difficult to model.

In particular, annealing involves a diffusion process upon slow cooling, while quenching is a diffusionless mechanism upon rapid cooling. It is convenient to start modeling solid-phase transformation using the former process in order to characterize the transformation of austenite as per Fig. 9.11. In practice, the annealing process is a common heat treatment of steels which requires a sufficiently long heat treatment time at an austenitic temperature T_γ for homogenizing the microstructure. Subsequently, a slow cooling rate (dT/dt) follows so that diffusion takes place in a natural manner.

9.4.2.1   Ferrite Model

The solid-state phase transformation from austenite (γ) to ferrite (α) is modeled as shown in Fig. 9.12a, represented by Fig. 9.11a $\rightarrow$ Fig. 9.11b and denoted by the invariant reaction $\gamma_{FCC}\text{-}Fe \rightarrow \alpha_{BCC}\text{-}Fe$ at $T < T_c = 912°C$. The resultant microstructure is composed of equiaxed ferrite grains in pure iron Fe (Fig. 9.11b).

Furthermore, the orientation relationship (OR) between the austenite and ferrite can be generalized as groups or family of planes and common directions with small misfit angles. Thus,

$$\{111\}_\gamma \parallel \{101\}_\alpha \quad \text{(planes)}$$

$$\langle 110 \rangle_\gamma \parallel \langle 110 \rangle_\alpha \quad \text{(directions)}$$

Detail information on OR's known as Kurdjumov–Sachs (KS) and Nishiyama–Wasserman can be found elsewhere [12, p. 22–24].

9.4.2.2   Ferrite–Pearlite Model

The transformation of austenite in hypoeutectoid steels with $0 < C < 0.8$ is modeled in Fig. 9.12b, represented by Fig. 9.11a $\rightarrow$ Fig. 9.11c, and denoted by invariant reactions $\gamma \rightarrow \alpha + \gamma_1 \rightarrow \alpha + p$, where α is the equiaxed *proeutectoid ferrite* grains (light) and γ_1 is the remaining austenite that transforms to pearlite $p = \alpha + Fe_3C$ with a lamellar pearlitic structure (light α and dark Fe_3C). The resultant microstructure is a combination of equiaxed *proeutectoid ferrite* α-grains and lamellar pearlite grains in a Fe-0.38C hypoeutectoid steel, where 0.38C is the carbon content in units of weight percentage $(wt\%)$, sometimes denoted as percentage (%).

9.4.2.3   Pearlite Model

This corresponds to the grain boundary eutectic invariant reaction $\gamma \rightarrow p = \alpha + Fe_3C$ modeled in Fig. 9.12c and represented by Fig. 9.11a $\rightarrow$ Fig. 9.11d The resultant microstructure is composed of pearlite grains (light α and dark Fe_3C) in a eutectoid Fe-0.80C steel. This invariant reaction ideally occurs at $T_E = 727°C$ as indicated in the Fe-C equilibrium phase diagram (Fig. 9.10). However, if an extra element X is added to a Fe-C steel, the eutectic point E (Fig. 9.10) is displaced and as result, the ferrous material is an alloy steel denoted as $Fe\text{-}C\text{-}X$, where $X = Cr, Ni, Mn$ or any other element.

9.4.2.4 Pearlite and Cementite Model

This particular model involves *proeutectoid cementite* Fe_3C phase as the light grain boundary surrounding the pearlite grains, denoted as pearlite colonies in Fig. 9.12d. Thus, Fig. 9.11a $\rightarrow$ Fig. 9.11e as per Fig. 9.12d, which represents a diffusional transformation. This corresponds to the invariant reactions

$$\gamma \rightarrow Fe_3C + \gamma_1 \rightarrow Fe_3C + p$$

$$\gamma_1 \rightarrow p = (\alpha + Fe_3C) \quad \text{(grains)}$$

where Fe_3C denotes the grain boundary and γ_1 denotes the remaining austenite that transforms to pearlite.

9.4.3 Martensite Formation

In particular, martensite, after Adolf Martens (1850-1914), derives from a soft face-centered cubic (γ-FCC) austenite as a hard metalstable phase in steels with a body-centered tetragonal (α'-BCT) lattice structure induced by a dimensionless mechanism at a high cooling rate. In fact, austenite can transform to martensite ($\gamma \rightarrow \alpha'$) upon quenching at a significant fraction of the speed of sound.

Following Cullity-Stock [13, p. 350], the martensitic transformation based on the classical *Bain Distortion Model* for steels is schematically shown in Fig. 9.13. Initially, the austenitic FCC unit cells are composed of iron (Fe) atoms (red) with carbon atoms (black) in interstitial positions or sites. This indicates that martensite may be derived from austenite by considering two FCC unit cells (two neighboring austenite cells in Fig. 9.13a) and the drawn BCT unit cell (Fig. 9.13a,b). This phase transformation can be denoted as a coordinate transformation with fixed z-axis (rotation about the z-axis) so that $(x, y, z)_\gamma \rightarrow (x', y', z')_{\alpha'}$ and $z = z'$.

As a result, this BCT unit cell (Fig. 9.13b) forms due to a fast dimensionless (displacive) transformation that suppresses the formation of ferrite or pearlite phases. In practice, a typical martensitic microstructure is composed of a randomly grown needle-like or plate shaped BCT structure embedded in the retained austenitic FCC structure as shown in Example 9A.3.

Notice that the drawn BCT unit cell (Fig. 9.13b) between the two γ-FCC unit cells and the $(011)_{BCT}$ plane (1-2-3 triangle) laying on the $(111)_\gamma$ (1-2'-3 triangle) are crystallographic geometries used here to present the martensitic transformation with its corresponding unit cell lattice parameters defined below.

Example 9A.2 Find the curve fitting equations for the data in Fig. 9.13c.

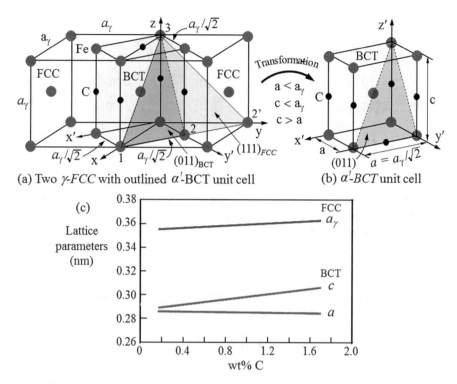

(a) Two γ-FCC with outlined α'-BCT unit cell

(b) α'-BCT unit cell

Fig. 9.13 Bain distortion model including (**a**) austenite unit cells showing the unit cell to be transferred as a martensitic structure and (**b**) martensite unit cell showing lattice sites for carbon atoms. (**c**) Austenite and martensite lattice parameters as functions of weight percent carbon (wt% C). After Cullity-Stock [13, p. 350]

$$a_\gamma = 0.35457 + \left(4.902 \times 10^{-3}\right)(\%C)$$

$$c = 0.28675 + (0.01125)(\%C)$$

$$a = 0.28637 - \left(1.3158 \times 10^{-3}\right)(\%C)$$

$$c/a = 1.0012 + \left(4.4267 \times 10^{-2}\right)(\%C)$$

These equations are valid for $0.20 \leq \%C \leq 1.70$ and are convenient to have when solving solid-state phase transformation problems related to plain carbon steels.

For the sake of clarity, according to the given FCC and drawn BCT (picked out of the FCC unit cell in Fig. 9.13a) illustrated in Figs. 9.12a and 9.13b, the lattice parameters ($a' < a_\gamma$ or simply $a' = 0.7071a_\gamma$), the principal crystallographic directions in the Cartesian coordinate system for these unit cells, and the Bain matrix (B_m) containing the principal strains are denoted, respectively, as

$$a' = \frac{a_\gamma}{\sqrt{2}} \tag{9.29a}$$

$$c = c_\gamma \tag{9.29b}$$

$$FCC = [100]_x, \quad [010]_y \ \& \ [001]_z \ \Rightarrow [100]_\gamma, \quad [010]_\gamma \ \& \ [001]_\gamma \tag{9.29c}$$

$$BCT = [100]_{x'}, \quad [010]_{y'} \ \& \ [001]_{z'} \ \Rightarrow [100]_{\alpha'}, \quad [010]_{\alpha'} \ \& \ [001]_{\alpha'} \tag{9.29d}$$

$$B_m = \begin{pmatrix} \epsilon_1 & 0 & 0 \\ 0 & \epsilon_2 & 0 \\ 0 & 0 & \epsilon_3 \end{pmatrix} \tag{9.29e}$$

Upon quenching from an austenitic temperature $T_\gamma > T_{A_3}$ or T_{cm} (Fig. 9.10), the martensitic transformation due to an atomic diffusionless or displacive mechanism is represented by the reaction

$$\gamma\text{-}BCT \rightarrow \alpha'\text{-}BCT \quad \text{with } c = c_{\alpha'} < c_\gamma = b_\gamma = a_\gamma \tag{9.30}$$

In general, the OR can be generalized using the $\{hkl\}$ closest packed planes and related parallelism of $\langle uvw \rangle$ crystallographic directions

$$\{hkl\}_\gamma \ \| \ \{hkl\}_{\alpha'} \tag{9.31a}$$

$$\langle uvw \rangle_\gamma \ \| \ \langle uvw \rangle_{\alpha'} \tag{9.31b}$$

where $\langle uvw \rangle_\gamma$ and $\langle uvw \rangle_{\alpha'}$ are the edge directions in the $\{hkl\}_\gamma$ and $\langle uvw \rangle_{\alpha'}$, respectively. Crystallographically, note that the $(011)_{\alpha'}$ plane outlined within the two austenite γ-FCC unit cells (Fig. 9.13a) is shown in the martensite BCT unit cell as illustrated in Fig. 9.13b.

Actually the lattice parameters of these crystal structures are the used as the relevant variables for determining the contraction and expansion of the unit cells in particular directions and planes, which are dependent on the carbon content in the steel matrix. Apparently, 20% contraction occurs along the crystallographic [001] direction and approximately 12% expansion takes place on the (001) plane.

Furthermore, the crystallographic lattice parameters can be determined by using X-ray diffraction, which requires precise measurements under controlled conditions. Thus, quantitative predictions of these parameters lead to the determination of the habit plane and the rotation angle θ, say, between the x and x' axes in Fig. 9.13a. In other words, the precise prediction of the orientation relationship (OR) is of fundamental importance in the theory of martensitic transformation.

In general, the martensitic transformation is characterized by the parallel crystallographic planes and directions written below as per most common cited work in the literature ([1, p. 393] and [12, pp. 22–24])

$$(111)_\gamma \ \| \ (011)_{\alpha'} \quad \text{(Bain \& other below)} \tag{9.32a}$$

$$\left[\bar{1}01\right]_\gamma \parallel \left[\bar{1}\bar{1}1\right]_{\alpha'} \quad \text{(Kurdjumov–Sachs relation)} \tag{9.32b}$$

$$\left[\bar{1}\bar{1}2\right]_\gamma \parallel \left[0\bar{1}1\right]_{\alpha'} \quad \text{(Nishiyama–Wassermann)} \tag{9.32c}$$

$$\left[\bar{1}01\right]_\gamma \cong\parallel \left[\bar{1}\bar{1}1\right]_{\alpha'} \quad \text{(Greninger–Troiano)} \tag{9.32d}$$

The macroscopic scale related to the martensitic transformation process is now focused on the lattice parameters for the γ-FCC and α'-BCT unit cells as depicted in Fig. 9.13c [13, p. 350]. It can be anticipated that the time-temperature-transformation (TTT) diagram introduced in a later section provides means to predict the particular phases.

Interestingly, the solid-phase transformation is based on nucleation and growth. With regard to martensitic transformation, the former is the suitable mechanism used to determine the kinetics of martensitic transformation since the latter is difficult to characterize at a fraction of the speed of sound. Based on this transformation problem, the TTT diagram becomes the most useful resource for kinetic studies. Moreover, during martensitic transformation some austenite is retained (untransformed), which depends mainly on chemical composition and cooling rate. Additional details on kinetics and TTT can be found in sections below and elsewhere ([14, p. 353] and [15]).

Example 9A.3 Consider the Bain distortion model illustrated in Fig. 9.13a for martensitic transformation in a 1040 carbon steel. A typical martensite microstructure is shown below.

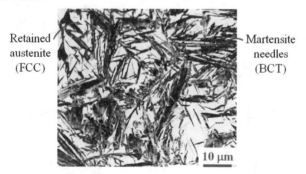

Retained austenite (FCC)

Martensite needles (BCT)

10 μm

Assume that the lattice parameters of austenite and ferrite are 0.356 nm and 0.291 nm, respectively, and that the actual lattice parameter ratio of martensite BCT crystal is $c/a < 1.633$ (theoretical value). Based on this information, calculate (a) the maximum displacement (μ_{1040} and μ_{1060}, respectively), (b) the Bain strains experienced by iron (Fe) atom during martensitic transformation $\gamma \to \alpha'$, and (c) deduce the orientation relationships corresponding to $[001]_\gamma$ and $[111]_\gamma$. Below is a suitable martensite microstructure with retained austenite [9, p. 377].

Solution

(a) From Fig. 9.13c or Example 9A.2 with 0.40%C in the carbon steel

$$a_\gamma = 0.356 \text{ nm}$$
$$a_{\alpha'} = 0.286 \text{ nm}$$
$$c_{\alpha'} = 0.291 \text{ nm}$$
$$(c/a)_{\alpha'} = 0.291/0.286 = 1.018 \text{ nm}$$

Then, the change in lattice parameters along with the maximum displacement is

$$\Delta c = a_\gamma - c_{\alpha'} = 0.356 - 0.291 = 0.065 \text{ nm}$$

$$\Delta a = a_{\alpha'} - \frac{a_\gamma}{\sqrt{2}} = 0.286 - \frac{0.356}{\sqrt{2}} = 0.034 \text{ nm}$$

$$\mu = \sqrt{\Delta a^2 + \Delta c^2} = \sqrt{(0.034)^2 + (0.065)^2} = 0.073 \text{ nm}$$

Consequently, these results indicate that the phase transformation induces atoms to move vertically $\Delta c = 0.065$ nm and horizontally $\Delta a = 0.034$ nm, and a maximum distance of $\mu = 0.073$ nm. Notice that the equation used to calculate μ is the resultant expression of vector addition. In other words, Δa and Δc are the edges (legs) of a right triangle whose hypotenuse (diagonal length) $\mu = 0.073$ nm at an angle $\theta = tan^{-1}(\Delta c/\Delta a) = 62.39°$.

(b) The lattice strains in tension (+) and compression (−) are

$$\epsilon_{([100]_\gamma} = \epsilon_{([010]_\gamma} = \frac{\sqrt{2}a_{\alpha'} - a_\gamma}{a_\gamma} = \frac{\sqrt{2}(0.286) - 0.356}{0.356} = 0.14$$

$$\epsilon_{[001]_\gamma} = \frac{c_{\alpha'} - a_\gamma}{a_\gamma} = \frac{0.291 - 0.356}{0.356} = -0.18 \quad \text{(compression)}$$

and the Bain matrix representing the Bain strains or distortion is

$$B_m = \begin{pmatrix} \epsilon_{([100]_\gamma} & 0 & 0 \\ 0 & \epsilon_{([010]_\gamma} & 0 \\ 0 & 0 & \epsilon_{[001]_\gamma} \end{pmatrix} = \begin{pmatrix} 0.14 & 0 & 0 \\ 0 & 0.14 & 0 \\ 0 & 0 & -0.18 \end{pmatrix}$$

(c) The orientation relationships are

$$[001]_\gamma \parallel [001]_{\alpha'}$$
$$[111]_\gamma \parallel [010]_{\alpha'}$$

Therefore, these results indicate that the austenite (FCC) γ unit cell contracts along the crystallographic c-axis since $c_{\alpha'} < a_\gamma$ and $\epsilon_{[001]_\gamma} < 0$.

9.5   Kinetics of Solid-Phase Transformation

This section aims at the kinetics of solid-state phase transformation under the influence of certain undercooling. Normally, the nucleation rate I_s and the time-temperature-transformation (TTT) diagram suffice the kinetic studies. The latter is conveniently introduced in the next section.

In general, one can root the characterization of phase transformation into thermodynamics, physical metallurgy and crystallography, kinetics and heat transfer theory. However, the analytical solutions to phase transformation problems are mostly based on heat transfer and thermodynamics.

Consider the homogeneous nucleation process. Conveniently, Eq. (4.51) is recast for predicting the number of stable nuclei as per Arrhenius-type equation, in which the Gibbs energy change ΔG is treated as the activation energy needed for the solid-state phase transformation process to take place isothermally. For comparison, the number of stable nuclei can be written as,

$$n_s = n_o \exp\left(-\frac{\Delta G_{hom}}{k_B T}\right) \tag{9.33a}$$

$$n_s = n_o \exp\left(-\frac{\Delta G_{het}}{k_B T}\right) \tag{9.33b}$$

where $\Delta G = \Delta G_{hom}$ or $\Delta G = \Delta G_{het}$. Recall that the Gibbs energy change for heterogeneous is $\Delta G_{het} < \Delta G_{hom}$.

Similarly, the nucleation rate described by Eq. (4.56) is also adopted henceforth. Accordingly,

$$I_s = \frac{n_o k_B T}{h} \exp\left[-\frac{16\pi \gamma_{\alpha\beta}^3}{3k_B (\Delta H_t)^2} \frac{1}{T} \left(\frac{T_c}{\Delta T}\right)^2 - \frac{Q_d}{k_B T}\right] \tag{9.34a}$$

$$I_s = I_o \exp\left[-\frac{Q_x}{k_B T}\right] \tag{9.34b}$$

with

$$Q_x = \frac{16\pi \gamma_{\alpha\beta}^3}{3 (\Delta H_t)^2} \left(\frac{T_c}{\Delta T}\right)^2 + Q_d \tag{9.35}$$

where ΔT denotes the undercooling, I_o denotes a proportionality constant, Q_x denotes the activation energy due to diffusion and evolved latent heat of transformation in the solid state, and the other variables have the usual definitions.

Notice that the I_o strongly depends on the temperature T, Q_x is governed by the interfacial energies (ΔH_t and $\gamma_{\alpha\beta}$), and the degree of undercooling.

Of special interest, for example, the theoretical rate of nucleation defined by Eq. (9.34a) implies that the solid-phase transformation is governed by the nucleation mechanism, which in turn, depends on the degree of undercooling and the local temperature $T = T_c - \Delta T$ at the moving interface.

All this is suggesting that solid-state phase transformation must occur explicitly and must be controlled by thermodynamic instability under steady-state condition. Thus, modeling the kinetics of the solid-state transformation has its fundamental importance during the isothermal and non-isothermal phase transformation. So far the former approach prevails in this section. Additional details on this subject matter can be found in Chap. 4.

9.6 TTT Diagram

First of all, consider an isokinetic analysis (constant velocity) for determining the volume fraction of the solid phase (f_i) isothermally using the Avrami equation as the transformation kinetics model [16, 17]. Thus, the Avrami equation is

$$f_i = 1 - \exp\left(-k_x t^n\right) \tag{9.36}$$

where k_x, n denote constants, t denotes the time at a constant temperature T, and i denotes the new solid phase. If experimental data sets do not obey Eq. (9.36), then the expressions given below can be used for curve fitting purposes

$$f_i = a_1 + a_2 \tanh\left(\frac{t - a_3}{a_4}\right) \tag{9.37a}$$

$$f_i = b_1 + b_2 \arctan\left(\frac{t - b_3}{b_4}\right) \tag{9.37b}$$

Here, a_i, b_i are coefficients to be determined numerically and $i = 1, 2.3, 4$. These expressions are of special interest for non-linear curve fitting. Plotting Eq. (9.36) with variable k_x, n yields the schematic sigmoidal transformation curves shown in Fig. 9.14. Many individual isothermal curves are used to construct a time-temperature-transformation (TTT) diagram as depicted in Fig. 9.15.

Figure 9.15a depicts two isothermal curves being plotted using the Avrami equation. Note that the rate of austenite-to-pearlite transformation ($\gamma \rightarrow p$) under isothermal conditions at T_1 and T_2 is $(df_i/dt)_1 > (df_i/dt)_2$. Only the start and finish data points of individual isothermal curves are used to construct the TTT diagram shown in Fig. 9.15b.

Of special interest in the current subject matter is the enthalpy data shown in Fig. 9.16a published by Kulawik's paper [18]. It provides a valuable graphical representation of the enthalpy function $\Delta H_t = f(T)$ for characterizing the solid-state phase transformation of iron (Fe) and medium carbon steels (Fe-C alloys) at a macroscale, where $\gamma \rightarrow \alpha, p, B$ or α'. Notice that $\Delta H_t = \Delta H_{\gamma \rightarrow \alpha'}$ for martensite formation from austenite exhibits a linear trend at a small temperature range due to

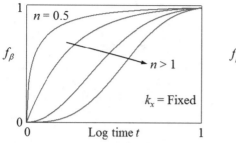 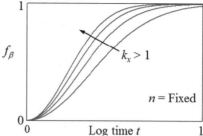

Fig. 9.14 Avrami equation. (**a**) Fixed k_x and (**b**) fixed n

Fig. 9.15 Solid-state transformation diagram. (**a**) Avrami-type and (**b**) TTT diagrams. Legend: SC = slow cooling and RC = rapid cooling, p = pearlite and B = bainite phases

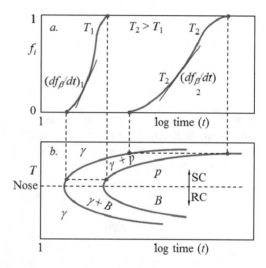

a fast displacive atomic rearrangement of the austenite crystal structure. Conversely, austenite transformation into a new phase (α, p or B) is non-linear, attributed to the coupled heat-diffusion process, which in turn, depends on the rate of evolved latent heat of transformation $d(\Delta H_t)/dt$.

In addition, Fig. 9.16b schematically shows the TTT diagram for a eutectoid steel ([9, p. 378] and [19, p. 16]). Note that the temperature marked with a horizontal dashed line is called the "nose" temperature, which divides the TTT diagram into a slow cooling (SC) region for pearlite formation and a rapid cooling (RC) region for bainite and martensite formation. Moreover, "s" and "f" stand for start and finish transformation lines.

Assume that three identical steel samples are austenitized at 900°C and subsequently, cooled at different rates as shown by the blue lines in Fig. 9.16b.

Line 1 Line 1 represents the displacive transformation at a fast cooling rate or quenching rate $(dT/dt)_1$ in a suitable medium. This line represents a continuous cooling transformation (CCT) process. Thus,

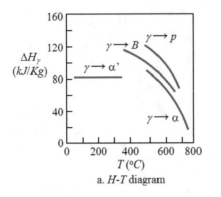

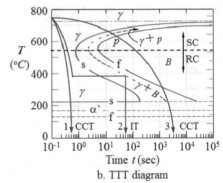

a. *H-T* diagram   b. TTT diagram

Fig. 9.16 Transformation diagrams. (**a**) Latent heat of transformation: *H-T* diagram [18] and (**b**) *TTT* diagram [9, p. 378] and [19, p. 16]

$$\gamma \rightarrow 100\% M \quad \text{at} \quad t = 0.5 \text{ s}$$

Line 2 Line 2 denotes an isothermal transformation (IT) process, which indicates that a sample is quenched at $(dT/dt)_2$ and held at $T < T_{nose}$ for some time $t > 0$ until $\gamma \rightarrow B$ transformation is complete and subsequently, quenched to room temperature. Hence,

$$\gamma \rightarrow 50\% B + 50\% M \quad \text{at} \quad t = 50 \text{ s}$$

Line 3 Line 3 is also a CCT process for $\gamma \rightarrow p = \alpha + Fe_3C$ at $(dT/dt)_3 << (dT/dt)_1$. Extracting data from Fig. 9.16b reveals a complete pearlitic trans-formation as the relevant eutectic structure, which is divided into fractions of ferrite (α) and cementite (Fe_3C). The computed results for a quenching rate $(dT/dt)_3$ are

$$\gamma \rightarrow 100\% p \quad \text{at} \quad t = 20 \text{ s}$$

$$f_\alpha = \frac{6.67 - 0.8}{6.67 - 0.0218} = 0.883 \text{ or } 88.30\%$$

$$f_{Fe_3C} = 100 - f_\alpha = 100 - 88.30 = 0.117 \text{ or } 11.70\%$$

Notice that pearlite is mostly composed of ductile ferrite containing embedded cementite. This means that pearlite is a two-phase metallic composite, which has a significant influence on mechanical behavior and mechanical properties of steels. Recall that the morphology of the cementite phase in pearlite may have a lamellar, rod-like or globular shape.

Moreover, Appendix 9A includes reference TTT diagrams for 1050, 1080, 9440, and 4340 steels. These TTT diagrams represent the kinetic changes in plain carbon and alloyed steels. Clearly, the kinetics of solid-state phase transformation is most suited through TTT diagrams, which are convenient phase maps of predictable solid phases. Moreover, metallography is the most prevalent technique because it is cost effective and simple.

9.7 Hardenability

Hardenability is a term used to describe the ability of steels to transform to martensite and on the other hand, hardness defines the resistance to indentation or deformation in compression. Hence, hardenability of steel determines the depth and distribution of hardness induced by austenite transformation upon a rapid cooling (quenching) process. Figure 9.17 depicts some hardenability data.

In general, hardenability is characterized by the Jominy End Quench Test as per the American Society for Testing Materials (ASTM A 255). Figure 9.17 shows hardenability diagram and a portion of the TTT diagram for a eutectoid steel. Figure 9.17a illustrates the Jominy bar and the Rockwell hardness scale-C (RHC) profile and some CCT curves superimposed on the 1080 steel TTT diagram ([9, pp. 444–445] and [19, p. 11]).

Figure 9.17b depicts $HRC = f(x)$ profiles for some Jominy bars made out carbon and steel alloys containing 0.40%C. As expected, the 4140, 5140, and 8140 steel alloys exhibit higher hardness than the 1040 steel due to the alloying elements in solid solution.

In practice, steel is austenitized and subsequently quenched at a specific rate to produce ferrite, pearlite, bainite, martensite, or a combination of these phases. If

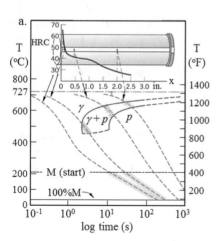
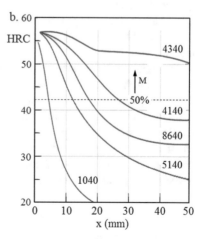

Fig. 9.17 Hardenability data of some steels. (**a**) Jominy test for hardness measurements and TTT diagram for 1080 steel and (**b**) hardness of some steels containing 0.40%C. Adapted from H.E. Boyer (editor) book (Callister–Rethwisch book [9, pp. 444–445] and [19, p. 11])

the final product is martensite, then a secondary heat treatment follows to obtain tempered martensite with a suitable toughness at the expense of some hardness. Nonetheless, Fig. 9.17a illustrates the correlation of the end-quench hardenability with the IT diagram along with CCT curves. This is the usual representation of hardenability data for hardened steels.

The Physics of Hardenability It is clear now that hardenability is a measure of martensite formation at certain depth into a specimen. A hardenability test (Jominy test) is made by quenching a round bar and subsequently, by measuring the hardness at different depths (cross section). This, then, establishes a criterion of depth for hardening steel; 50% martensite at the center of the bar. This criterion assures the usability of martensitic steels. However, a total economical factor dictates the type of martensitic steel to be produced. The final product undergoes shear deformation due to the creation of dislocations and acquires a high hardness.

Habit Plane This is fundamentally a flat or curved interface that defines a particular crystallographic plane between the martensitic and austenitic (parent or matrix) phases. In steels, the martensitic transformation is designated by a phase invariant reaction of the form $\gamma \, (FCC) \rightarrow \alpha' \, (BCT)$, which is a representation of a highly strained martensite platelet (lath) related to an interfacial and strain energies since.

It is evident nowadays that martensitic transformation also occurs in other materials, such as pure Ni, Ck-Al alloys, Ti-Mn alloys, and so forth. In general, the growth of martensite phase may occur at high velocity near or a fraction of the speed of sound due to diffusionless transformation at low temperatures.

In practice, the effects of quenching temperature and quenching rate on the hardenability of steel are normally characterized using the composition-dependent hardness and the corresponding microstructural features. Experimentally, the Jominy End Quench Test (JEQT) plays an important role in the material characterization field since it measures the ability of a steel to harden by transforming austenite into martensite.

Nonetheless, mechanical properties of steels or metallic alloys strongly depend on the morphology of the microstructure, which in turn, depends on the cooling rate in a particular medium (salt bath). Commonly, metallography and hardness measurements follow TTT experiments. However, characterization of materials may require use of other techniques such as X-ray diffraction, density measurements, electrical resistivity, acoustic emission, and so on. Moreover, TTT, CCT, and IT diagrams for general steels dominate the solid-state phase transformation field due to their importance in the steelmaking industry.

9.8 TTT Diagram for Glass Formation

In principle, conventional heterogeneous or slow solidification (SS) initiates at suspended particles and mold inner surfaces under the influence of small under-

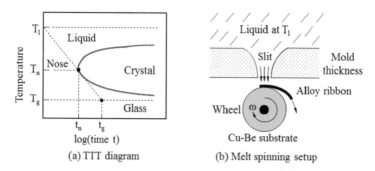

(a) TTT diagram   (b) Melt spinning setup

Fig. 9.18 Rapid solidification. (**a**) Schematic TTT diagram for glass formation and (**b**) an experimental setup

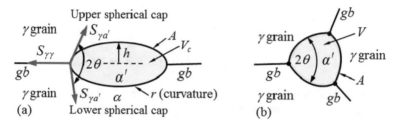

Fig. 9.19 Nucleation model for (**a**) a γ-nucleus growing along an γ-γ grain boundary (gb) with two adjoined spherical caps and (**b**) a α'-nucleus growing at an γ-grain triple junction with three adjoined γ-α' grain boundaries

cooling ΔT_{SS}. On the other hand, rapid solidification (RS) of alloys initiates on a relatively cooled substrate surface at $\Delta T_{RS} >> \Delta T_{SS}$, leading to glass (amorphous structure) formation. This type of casting is known as a rapidly solidified alloy (RSA). Nonetheless, RS is suitable for alloys due to $\Delta T_{RS} >> \Delta T_{SS}$. For the sake of clarity, the TTT diagram schematically shown in Fig. 9.19a along with the corresponding rapid solidification melt spinning process setup in Fig. 9.19b is an important graphical representation of the kinetics for glass formation. Notice that "C-shape" curve in Fig. 9.18 denotes the L-S transition at temperature T.

Initially the metallic solution (alloy melt) is the pouring temperature $T_p = T_l$ and subsequent rapid freezing at deep undercooling $\Delta T_{RS} >> \Delta T_{SS}$ the melt transforms into solid glass. Assuming that the freezing rate is linear, the T_l-T_g line intercepting the "nose" point at T_n can be used to define the respective critical or maximum cooling rate dT/dt for glass formation at T_g during rapid solidification.

Usually, thin ribbons can be produced using the RS technique by imposing a very large undercooling ΔT_{RS}, which in turn, suppresses atomic diffusion and enhances heat extraction at an enormous rate from the alloy melt on a common Cu-Be alloy wheel surface.

Example 9A.4 Assume that a hypothetical $A_x B_y C_z$ alloy melt is initially at $T_l = 1540°C$. Using the melt spinning technique (Fig. 9.18b) produces, say, amorphous

$A_x B_y C_z$ strips with uniform thickness. Also assume that the heat transfer coefficient between the copper-beryllium (Cu-Be) wheel substrate at $T_o = 20°C$ and the melt at $1540°C$ is $h = 18,000 \; W/(m^2.K)$. Determine (a) the solidification rate and (b) the solidification time for a ribbon thickness of $\delta = s(t) = 15 \; \mu m$. Given data:

$$\rho_l = \rho_s = 6800 \; kg/m^3 \quad \mid \quad c_p = 660 \; J/Kg.K$$

Recall that an amorphous structure is treated as a metallic glass that lacks atomic symmetry and periodicity compared to crystalline structures.

Solution

(a) First of all, the latent heat of solidification is approximated as

$$\Delta H_s = c_s dT \simeq c_s \Delta T = c_s \left(T_p - T_o\right)$$

$$\Delta H_s = (660 \; J/Kg.K)\,(1540 - 20) \; K = 10^6 \; J/Kg$$

From Eq. (6.3a), the amount of thermal energy due to convective heat transfer at the melt–substrate interface is

$$q_{x=0} = h\left(T_p - T_o\right) \quad \text{(Newton's law)}$$

$$\left[\frac{dQ}{dt}\right]_{x=0} = Ah\left(T_p - T_o\right)$$

$$dQ_{x=0} = Ah\left(T_p - T_o\right)dt$$

and on the other hand, the amount of thermal energy due to the release of latent heat of solidification ΔH_s is approximated as

$$Q_s = m\Delta H_s = V\rho_s \Delta H_s = As\,(t)\,\rho_s \Delta H_s$$

$$dQ_s = As\,(t)\,\rho_s c_s dT$$

For $dQ_{x=0} = dQ_s$, the thermal heat balance yields the solidification rate as

$$Ah\,(T_i - T_o)\,dt = \rho_s As\,(t)\,c_s dT$$

$$\frac{dT}{dt} = \frac{h\left(T_p - T_o\right)}{s\,(t)\,\rho_s c_s}$$

Using $h = 18,000 \; W/(m^2.K) = h = 18,000 \; J\,(m^2.K.s)$ and the remaining given data gives

$$\frac{dT}{dt} = \frac{\left[18,000 \text{ J/} \left(\text{m}^2.\text{K.s}\right)\right](1540 - 20) \text{ K}}{\left(15 \times 10^{-6} \text{ m}\right)\left(6800 \text{ kg/m}^3\right)(660 \text{ J/kg.K})} = 0.41 \times 10^6 \text{ K/s}$$

(b) Integrating the above dT/dt equation yields

$$\frac{dT}{dt} = \frac{h\left(T_p - T_o\right)}{s\left(t\right)\rho_s c_s}$$

$$\int_{T_o}^{T_p} dT = \frac{h\left(T_p - T_o\right)}{s\left(t\right)\rho_s c_s} \int_0^t dt$$

$$T_p - T_o = \frac{h\left(T_p - T_o\right)}{s\left(t\right)\rho_s c_s} t$$

from which the solidification time is approximated as

$$t = \frac{s\left(t\right)\rho_s c_s}{h} = \frac{\left(15 \times 10^{-6} \text{ m}\right)\left(6800 \text{ kg/m}^3\right)(660 \text{ J/kg.K})}{\left[18,000 \text{ J/}\left(\text{m}^2.\text{K.s}\right)\right]}$$

$$t = 3.74 \times 10^{-3} \text{ s}$$

Therefore, metallography or X-ray diffraction verifies the nature of the solid structure; amorphous, semi-crystalline, or crystalline.

9.9 Thermodynamics of Transformation

Normally, one can consider an infinite thermally conductive medium characterized by Fourier's law, in which a sub-domain can be treated as a small solid-phase, particle, or inclusion, subjected to a prescribed uniform thermal field that may or may not impose an insignificant or negligible temperature gradient within the growing new solid phase. However, thermodynamics simplifies the characterization of solid-phase transformation by considering the microscopic model shown in Fig. 9.19.

This model indicates that γ-grains transform into a α'-phase or α'-particle at γ grain boundaries. Notice that the surface energies $S_{\gamma\gamma}$, $S_{\gamma a'}$ arise as the δ-γ interface moves during the evolution of latent heat of transformation at a critical temperature T_c.

Accordingly, Fig. 9.19a shows a typical model for the precipitation or formation of a thin oblate spheroidal coherent martensitic phase having two adjoined (abutted) spherical caps, representing a spherical γ-nucleus. Notice that the pertinent surface energies related to martensitic transformation are purposely grown at one end of the oblate spheroidal phase, which can be treated as a particle for calculation purposes.

On the other hand, Fig. 9.19b depicts the model for an arbitrary γ-nucleus growing at a triple junction with three adjoined δ-γ curved grain boundaries. Moreover, A and V denote the surface area and volume, respectively.

Eventually, the morphology of the α' phase is small due to the growth constraints imposed at the γ grain boundaries and the corresponding nucleus growth direction is at random. Apparently, discontinuities in the γ phase are the potential sites for the initiation of martensitic transformation.

Adjacent lattice defects (such as high-energy grain boundaries, dislocations, and vacancies) are ideal for the onset of transformation, provided that there exists an evolving latent heat of transformation (ΔH_t), which acts as the activation energy at a temperature T slightly below T_c. Thus, a small undercooling, $\Delta T_t = (T_c - T) \to 0$, is needed for such a phase transformation to occur.

In the classical nucleation theory, it is assumed that martensite nuclei form along a path of constant composition and structure since there must exist an invariant crystallographic (hkl) plane related to an invariant reaction $\gamma \to \alpha'$ at a critical temperature T_c. Mathematically, the Gibbs energy for transforming austenite γ-FCC into martensite α'-BCT structure is described as $G_\gamma > G_{\alpha'}$ with $\Delta G = (G_{\alpha'} - G_\gamma) < 0$ as the driving force at $M_s \leq T_c \leq M_f$, where M_s and M_f are the start and finish temperatures for martensitic transformation.

Following Olson–Cohen model for solid-state nucleation theory ([20] and [21, p. yy]) found in Sinha's handbook [22, p. 8.40–8.43], the solid-state homogeneous and heterogeneous Gibbs energy changes ($\Delta G's$) for martensitic transformation are written as

$$\Delta G_{\text{hom}} = -V \Delta G_V + V E_s + A S_{\gamma \alpha'} \tag{9.38a}$$

$$\Delta G_{het} = -V \Delta G_V + V E_s + A S_{\gamma \alpha'} + \Delta G_d + \Delta G_i \tag{9.38b}$$

$$\Delta G_s = \Delta G_\varepsilon + \Delta G_p \tag{9.38c}$$

where V denotes the nucleus volume, A denotes the embryo surface area, E_s denotes the total strain energy due to the difference in volume of γ and α' phases, ΔG_ε denotes the elastic strain energy, ΔG_p denotes the elastic energy due to distortion of the habit plane, ΔG_v denotes the volumetric Gibbs energy change, ΔG_d denotes the release Gibbs energy term due to the destruction or annihilation of defects (grain boundaries, vacancies, dislocations, etc.), and ΔG_i denotes the nucleus-defect interaction energy.

From Olson–Cohen book [21, p. 303], E_s and K_ε can be defined as per Eshelby [23] elasticity approach for determining the elastic field of an ellipsoidal inclusion. The resultant mathematical equations suitable for the current thermodynamical approach are written as

$$E_s = \frac{K_\varepsilon h}{r} \tag{9.39a}$$

$$K_\varepsilon = \frac{\pi\,(2-v)}{8\,(1-v)}G_\varepsilon\gamma_\varepsilon^2 + \frac{\pi\,(1-v)}{4}G_\varepsilon\varepsilon_n^2 \tag{9.39b}$$

with radius of curvature $r \gg h$, aspect ratio h/r, strain-energy proportionality factor K_ε in the order of 1 GPa, Poisson's ratio v, isotropic shear modulus G_ε, total transformation shear strain γ_ε, and axial (normal) strain ε_n.

For homogeneous nucleation ($\Delta G_d = \Delta G_i = 0$) of an isolated thin oblate spheroidal coherent martensitic particle (Fig. 9.19a) with radius r, semi-thickness h, volume $V = (4/3)\,\pi r^2 h$, and area $A = 4\pi r^2$, Eqs. (9.38a) along with (9.38c) becomes the Kaufman–Cohen equation [20], which can be found in Olson–Cohen book [21, p. 356]

$$\Delta G_{\text{hom}} = \frac{4}{3}\pi r^2 h\left(\Delta G_p - \Delta G_V\right) + \frac{4}{3}\pi r h^2 K_\varepsilon + 4\pi r^2 S_{\gamma\alpha'} \tag{9.40}$$

Letting $|d\,(\Delta G_{\text{hom}})\,/dr|_{r=r_c} = |d\,(\Delta G_{\text{hom}})\,/dh|_{h=h_c} = 0$, solving the resultant expressions simultaneously for the critical radius $r = r_c$ of a stable α'-phase and the critical semi-thickness $h = h_c$, and evaluating Eq. (9.40) accordingly yields the critical Gibbs energy (nucleation energy barrier) ΔG_c. Thus,

$$r_c = \frac{4K_\varepsilon S_{\gamma\alpha'}}{\left(\Delta G_p - \Delta G_V\right)^2} \tag{9.41a}$$

$$h_c = -\frac{2S_{\gamma\alpha'}}{\Delta G_p - \Delta G_V} \tag{9.41b}$$

$$\Delta G_{c,\text{hom}} = \frac{32\pi K_\varepsilon^2 S_{\gamma\alpha'}^3}{3\left(\Delta G_p - \Delta G_V\right)^4} \tag{9.41c}$$

$$\Delta G_{c,het} < \Delta G_{c,\text{hom}} \text{ if } \Delta G_{c,\text{hom}} > k_B T \tag{9.41d}$$

Here, k_B is the Boltzmann constant and $k_B T$ in Eq. (9.41d) is the thermal activation energy used as a transition energy for defining the limit between homogeneous and heterogeneous nucleation of martensite. The inequality $\Delta G_{c,het} \le k_B T < \Delta G_{c,\text{hom}}$, theoretically, sets the limit for the type of martensitic transformation, starting at the critical temperature denoted as $T = T_c$.

It is evident that the magnitude of $\Delta G_{c,\text{hom}}$ is larger than the experimentally found barrier energy ($\Delta G_{c,\text{hom}} > \Delta G_{c,\text{exp}}$) for spontaneous martensitic transformation [22, p. 8.42]. Therefore, martensite nucleation must be a heterogeneous process with nucleation at a lattice defect or inhomogeneity. A comprehensive analysis of heterogeneous nucleation by defect dissociation can be found in Olson–Cohen work [21, p. 357].

In essence, $\Delta G_{c,het}$ defined by Eq. (9.41d) is the mathematical model that explicitly dictates that γ-phase growth occurs by lowering its Gibbs energy at a certain rate, leading to phase stability during γ-α' boundary motion. Mathematically, solid-state phase transformation, as described by the theory of nucleation and growth, is not strictly confined to thermodynamics since there is heat transfer that can be taken into account, but the inequality $\Delta G_v > \Delta G_s$ at $r \geq r_c$ is a significant analytical approach since $\Delta G_{c,het} \to 0$ upon completion of the atomic rearrangement process. This, then, controls the resultant microstructure and related macroscale entities such as mechanical properties.

Additionally, the rate of steady-state nucleation per unit volume (I_v) is defined by Eq. (9.35), which is redefined here as

$$I_v = \frac{n_o k_B T}{h} \exp\left(-\frac{\Delta G_c}{k_B T}\right)' \exp\left(-\frac{Q_d}{k_B T}\right) \tag{9.42a}$$

$$\Delta G_c = \Delta H_t - T_c \Delta S_t = \int_0^{T_c} c_p(T)\, dT - T_c \int_0^{T_c} \frac{c_p(T)\, dT}{T} \tag{9.42b}$$

where ΔH_t denotes the evolved latent heat of solid-phase transformation and ΔS_t denotes the entropy change for $\gamma \to \alpha'$ transformation.

In fact, Eqs. (9.42) are not restricted to martensitic transformation and subsequently, they can be generalized as analytical models that can be applied to any solid-state phase transformation. In general, a phase transformation initially evolves through nucleation of a new single or dual phase. This implies that original equilibrium phase destabilizes mainly due to temperature changes that causes heat transfer from a moving interface to the surroundings.

Example 9A.5 Use thermodynamics to characterize the quenching process of a small carbon steel plate in agitated water at 25°C. In order for martensitic transformation to occur, austenite destabilizes at a relatively low temperature. This is the main condition for nucleation of martensite. Now, assume a homogeneous nucleation for martensite formation without composition change. Based on this information and the given data below, calculate the critical entities defined by Eq. (9.41) when (a) $\Delta G_p = 0$ and (b) $\Delta G_p = 50$ MJ/m³. Assume that martensitic transformation starts at $T = 700$ K (427°C). (c) Explain why there is retained austenite after martensitic transformation and how martensite benefits mechanical properties. Given data set:

$\Delta G_V = 180$ MJ/m³	$S_{\gamma\alpha'} = 15 \times 10^{-3}$ J/m²	$K_\varepsilon = 2.5$ GPa

Consider this data set as average experimental values..

Solution

(a) If martensite homogeneous nucleation takes place without any distortion of the habit plane, then $\Delta G_p = 0$ and

$$r_c = \frac{4 K_\varepsilon S_{\gamma\alpha'}}{(\Delta G_p - \Delta G_V)^2} = \frac{4 \left(2.5 \times 10^9\right) \left(15 \times 10^{-3}\right)}{\left(-180 \times 10^6\right)^2}$$

$$r_c = 4.6296 \times 10^{-9} \text{ m} = 4.63 \text{ nm}$$

$$h_c = -\frac{2S_{\gamma\alpha'}}{\Delta G_p - \Delta G_V} = -\frac{2\left(15 \times 10^{-3}\right)}{-180 \times 10^6} = 1.67 \times 10^{-10} \text{ m} \simeq 0.17 \text{ nm}$$

$$\Delta G_{c,\text{hom}} = \frac{32\pi K_\varepsilon^2 S_{\gamma\alpha'}^3}{3\left(\Delta G_p - \Delta G_V\right)^4} = \frac{32\pi \left(2.5 \times 10^9\right)^2 \left(15 \times 10^{-3}\right)^3}{3\left(-180 \times 10^6\right)^4}$$

$$\Delta G_{c,\text{hom}} = 6.73 \times 10^{-19} J = 4.21 \text{ eV}$$

(b) For an elastic strain energy due to distortion of the habit plane during martensitic transformation, $\Delta G_p = 50 \text{ MJ/m}^3$ and

$$r_c = \frac{4K_\varepsilon S_{\gamma\alpha'}}{\left(\Delta G_p - \Delta G_V\right)^2} = \frac{4\left(2.5 \times 10^9\right)\left(15 \times 10^{-3}\right)}{\left(50 \times 10^6 - 180 \times 10^6\right)^2} = 8.88 \text{ nm}$$

$$h_c = -\frac{2S_{\gamma\alpha'}}{\Delta G_p - \Delta G_V} = -\frac{2\left(15 \times 10^{-3}\right)}{50 \times 10^6 - 180 \times 10^6} = 0.23 \text{ nm}$$

$$\Delta G_{c,\text{hom}} = \frac{32\pi K_\varepsilon^2 S_{\gamma\alpha'}^3}{3\left(\Delta G_p - \Delta G_V\right)^4} = \frac{32\pi \left(2.5 \times 10^9\right)^2 \left(15 \times 10^{-3}\right)^3}{3\left(50 \times 10^6 - 180 \times 10^6\right)^4}$$

$$\Delta G_{c,\text{hom}} = 2.47 \times 10^{-18} \text{ J} = 15.47 \text{ eV}$$

Additionally,

$$E_s = \frac{K_\varepsilon h_c}{r_c} = \frac{\left(2.5 \times 10^9\right)(8.88)}{0.23} = 96.52 \text{ GJ/m}^3$$

$$60k_B T = (60)\left(1.38 \times 10^{-23} \text{ J/K}\right)(700 \text{ K}) = 9.66 \times 10^{-21} \text{ J} = 3.60 \text{ eV}$$

Therefore, these results indicate that martensitic transformation must take place under a heterogeneous nucleation mechanism since $\Delta G_{c,\text{hom}} > 60k_B T = 3.60 \text{ eV}$ when $\Delta G_p = 0$ and $\Delta G_p = 50 \text{ MJ/m}^3$ at $T = 700$ K.

(c) It is acceptable nowadays that the martensitic transformation is mostly an incomplete transformation of the destabilized austenite in steels. Thus, the martensitic phase is embedded in the retained austenite matrix or contains traces of retained austenite. Initially, this metallurgical phenomenon is attributed to randomly localized nucleation of martensite nuclei in a displacive manner causing localized volume changes and lattice distortion, and consequently, some austenite FCC crystals are simply retained after quenching. At a macroscale, a martensitic steel is classified as a brittle material with very low fracture toughness and high yield strength. Therefore, martensite strengthens steel at the expense of ductility and toughness. Fundamentally, ductile retained austenite benefits fracture toughness to an extent. In spite of the fact that small quantities of retained austenite may be present in a martensitic microstructure, subsequent heat treatment, such as tempering, may cause thermal instability of retained austenite leading to embrittlement.

9.10 Solid Transformation Thermal Resistance

The goal in this section is to derive the dimensionless temperature (T_θ) equation for one-dimensional analysis of solid-phase transformation in a plate, cylinder, or sphere initially at a temperature T_i and subsequently cooled in a fluid at a temperature T_f.

For a plate, the governing heat transfer equation, the continuity of temperature, initial condition (IC), and boundary condition (BC) are

$$\frac{\partial T}{\partial t} = \alpha \frac{\partial^2 T}{\partial x^2} \tag{9.43a}$$

$$T(x, 0) = T_i \quad \text{(IC)} \tag{9.43b}$$

$$\frac{\partial T(0, t)}{\partial x} = 0 \quad \text{(BC)} \tag{9.43c}$$

$$k \frac{\partial T(x_c, t)}{\partial x} + h \left[T(x_c, t) - T_f \right] = 0 \quad \text{(Energy balance)} \tag{9.43d}$$

where x_c denotes half-thickness. Following Poirier–Geiger ([24, p. 294] and [25, p. 290]) with $\theta = T - T_f$,

$$\frac{\partial \theta}{\partial t} = \alpha \frac{\partial^2 \theta}{\partial x^2} \tag{9.44a}$$

$$\theta(x, 0) = T_i - T_f = \theta_i \quad \text{(IC)} \tag{9.44b}$$

$$\frac{\partial \theta(0, t)}{\partial x} = 0 \quad \text{(BC)} \tag{9.44c}$$

$$\frac{\partial \theta(x_c, t)}{\partial x} + \frac{h}{k} \theta(x_c, t) = 0 \quad \text{(Energy balance)} \tag{9.44d}$$

The method of separation of variables, $\theta(x, t) = F(x) \cdot G(t)$, yields the solution to Eq. (9.44a) as (see Appendix 9B)

$$\frac{\theta}{\theta_i} = \frac{T - T_f}{T_i - T_f} = 2 \sum_{n=1}^{\infty} \frac{\sin(\lambda_n x_c) \cos(\lambda_n x)}{\lambda_n x_c + \sin(\lambda_n x_c) \cos(\lambda_n x_c)} \exp\left(-\lambda_n^2 \alpha t\right) \tag{9.45a}$$

$$B_i = \lambda_n x_c \tan(\lambda_n x_c) \tag{9.45b}$$

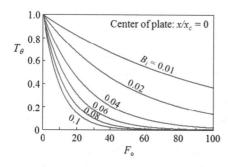 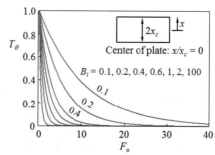

Fig. 9.20 Dimensionless temperature response of an infinite plate initially at T_i and subjected to cooling due to heat convection in a fluid at T_f

Let $F_o = \alpha t / x_c^2$ (Fourier number), $B_i = h x_c / k$ (Biot number) and manipulate Eq. (9.45) to get $T_\theta = f(\phi_n, B_i, F_o)$

$$T_\theta = \frac{T - T_f}{T_i - T_f} = 2 \sum_{n=1}^{\infty} \frac{\lambda_n x_c \sin(\lambda_n x_c) \cos[\lambda x_c (x/x_c)]}{(\lambda_n x_c)^2 + B_i \cos^2(\lambda_n x_c)} \exp\left(-(\lambda_n x_c)^2 F_o\right)$$

(9.46)

where $(n-1) \leq \phi_n = \lambda_n x_c \leq n\pi$ are the roots of Eq. (9.45b). If $n = 1$, $\phi = \lambda_1 x_c$ and $0 \leq x_\theta = x/x_c \leq 1$ in Eq. (9.46) yields the relative temperature equation for an isothermal solid-phase transformation as

$$T_\theta = \frac{T - T_f}{T_i - T_f} = \frac{2\phi \sin(\phi) \cos(\phi x_\theta)}{\phi^2 + B_i \cos^2(\phi)} \exp\left(-\phi^2 F_o\right)$$

(9.47a)

$$B_i = \phi \tan(\phi) = (h/k) x_c \quad \text{(plate)}$$

(9.47b)

Thus, plotting Eq. (9.47a) for a plate initially at T_i and subjected to heat convection in a fluid at T_f with $x_\theta = 0$ (center of a flat plate) yields Fig. 9.20.

Note that $T_\theta = f(F_o)$ is an exponential-dependent function for fixed values of ϕ, B_i, where ϕ is an eigenvalue determined using Eq. (9.47b). Applying Eq. (9.45) to cylinders and spheres with $x = r \leq r_c$ and $x_c = r_c$ yields

$$T_\theta = \frac{T - T_f}{T_i - T_f} = \frac{2 \sin(\lambda r_c) \cos[\lambda x_c (r/r_c)]}{\lambda r_c + (\lambda r_c / B_i) \sin(\lambda r_c)} \exp\left[-(\lambda r_c)^2 F_o\right]$$

(9.48a)

$$B_i = \lambda r_c \tan(\lambda r_c) = (h/k) r_c \quad \text{(cylinder or sphere)}$$

(9.48b)

Other plots can be generated for plates with $0 < x/x_c \leq 1$ and contoured geometries with $0 \leq r/r_c \leq 1$. These plots are referred to as Gurney–Lurie charts for small B_i and F_o or Heisler charts for $0.01 \leq B_i \leq \infty$ and $F_o \to \infty$.

9.10.1 Lumped System Analysis (LSA)

Transient heat transfer analysis of solid-state phase transformation can be used to study the quenching process to induce phase transition. This is a significant heat transfer phenomenon in the steelmaking industry since time is crucial for obtaining high quality heat treated components. Thus, quenching is a time-dependent technique related to a suitable cooling media and it can be characterized using the lump system analysis (LSA).

The lumped system analysis (LSA) is a transient heat transfer analysis with a particular criterion $Bi < 0.1$ and specific conditions, such as the temperature of a solid remains uniform within or throughout the body at all times and it changes with time only. Figure 9.20a shows some $T_\theta = (F_o)_{B_i}$ curves for $Bi < 0.1$. Hence, $T = T(t)$ and $\partial T(x, t)/\partial x = 0$ when $Bi < 0.1$.

Mathematically, Eqs. (6.12d) and (6.3a), respectively, are recast in this section for characterizing the lump system analysis. Thus,

$$B_i = \frac{R_{cond}}{R_{conv}} = \frac{V}{A_s}\frac{h}{k} = \frac{L_c h}{k} \tag{9.49a}$$

$$h = \frac{q_x}{A_s(T_s - T_o)} \tag{9.49b}$$

where i denotes conduction (cond) and convection (conv), V denotes the volume (m^3) of the component (part, sample, or specimen) being quenched, A_s denotes the surface area (m^2) of the component, $L_c = V/A_s$ denote the characteristic length (m) of the component, h denotes the heat transfer coefficient $(W/m^2 \cdot K)$ at the component–medium interface, k denotes the thermal conductivity $(W/m.K)$ of the component being quenched, and $q_x = dQ/dt$ denotes heat flux (W).

From Eq. (9.49a), the lump system analysis requires that $B_i \to 0$ as $k/L_c \to \infty$ for a small L_c or large k. This supports the LSA criterion which requires that $Bi < 0.1$.

The heat conduction by convection in a cooling medium is of interest because the Biot number (B_i) is enormously helpful in setting boundary conditions. The Biot number is defined as an external-to-internal conductance ratio or simply a thermal resistance (R_i) ratio.

Example 9A.6 Consider the isothermal solid-phase transformations indicated by the IT trajectories in the eutectoid TTT diagram given below.

(a) For a 5-cm thick 1080 steel plate initially at 750°C and immersed in a fluid at $T_f = 600°C$, calculate the time for complete pearlitic transformation $(\gamma \to p)$ when the temperature at center is $T = 650°C$. (b) Repeat the calculation for bainitic transformation $(\gamma \to B)$ in another 5-cm thick 1080 steel plate immersed in a similar fluid when $T_f = 400°C$ and $T = 450°C$. Compare the results with the read-off time from the TTT diagram. Thermophysical data set:

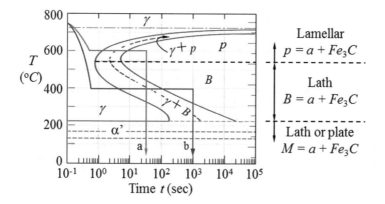

$$h = 50 \text{ BTu/} \left(\text{h.ft}^2.^{\circ}\text{F}\right) = (50)\,(5.6783) = 283.92 \text{ W/m}^2.\text{K}$$
$$\alpha = 0.46 \text{ ft}^2/h = (0.46)\left(2.5807 \times 10^{-5}\right) = 1.1871 \times 10^{-5} \text{ m}^2/\text{s}$$
$$k = 20 \text{ BTu/} (\text{h.ft.}^{\circ}\text{F}) = (20)\,(1.7307) = 34.614 \text{ W/m.K}$$

Solution First of all, the semi-thickness is $x_c = (5/2)\left(10^{-2}\right) = 0.025$ m. Subsequent calculations give the dimensionless temperature and the Biot number as

$$T_\theta = \frac{T - T_f}{T_i - T_f} = \frac{650 - 600}{750 - 600} = 0.33$$

$$B_i = hx_c/k = \frac{\left(283.92 \text{ W/m}^2.\text{K}\right)(0.025 \text{ m})}{34.614 \text{ W/m.K}} = 0.21$$

Notice that the Biot number is greater than 0.1 as required by the LSA criterion. This is attributed to non-uniform temperature within a sample.

(a) Pearlitic transformation $\gamma \rightarrow p$: Let $\phi = \lambda x_c$, $x_c = 0.025$ m and $x/x_c = 0$ so that Eq. (9.48b) with $0 \leq \phi \leq \pi$ gives

$$\lambda x_c \tan (\lambda x_c) = (h/k)\, x_c = B_i \quad \text{(plate)}$$
$$\phi \tan (\phi) = 0.21 \implies \phi = \lambda x_c = 0.44282$$

From Eq. (9.48a),

$$F_o = \frac{1}{(\lambda x_c)^2} \ln \left\{ \frac{1}{T_\theta} \frac{2 \sin (\lambda x_c) \cos [\lambda x_c\,(x/x_c)]}{\lambda x_c + (\lambda x_c/B_i) \sin (\lambda x_c)} \right\}$$

$$F_o = \frac{1}{(0.44282)^2} \ln \left(\frac{1}{0.33} \frac{2 \sin (0.44282) \cos (0)}{0.44282 + (0.44282/0.21) \sin (0.44282)} \right)$$

$$F_o = 3.35 = \alpha t / x_c^2$$

Then, the required time is

$$t = \frac{x_c^2 F_o}{\alpha} = \frac{(0.025)^2 (3.35)}{1.1871 \times 10^{-5}} = 176.38 \text{ s} = 2.94 \text{ min}$$

This result means that a $x = 2x_c = 5$-cm thick γ-phase specimen can be transformed to pearlite $\gamma \rightarrow p = \alpha + Fe_3C$ at $t = 176.38$ s. According to the the TTT diagram, the transformation time is $t \simeq 8$ s, which corresponds to a specimen with a thickness of

$$x_c = \sqrt{\alpha t / F_o} = \sqrt{(1.1871 \times 10^{-5})(8)/3.35} \simeq 0.5 \text{ cm}$$

(b) Similarly, the specimen being heat treated for the production of a bainitic microstructure requires that

$$T_\theta = \frac{T - T_f}{T_i - T_f} = \frac{450 - 400}{750 - 400} = 0.14$$

$$B_i = hx_c/k = \frac{(283.92 \text{ W/m}^2.K)(0.01 \text{ m})}{34.614 \text{ W/m.K}} = 0.21$$

$$\phi \tan(\phi) = 0.21 \implies \phi = 0.44282$$

and

$$F_o = \frac{1}{(0.44282)^2} \ln\left(\frac{1}{0.14} \frac{2 \sin(0.44282)\cos(0)}{0.44282 + (0.44282/0.21)\sin(0.44282)}\right)$$

$$F_o = 7.72 = \alpha t / x_c^2$$

$$t = \frac{x_c^2 F_o}{\alpha} = \frac{(0.025 \text{ m})^2 (7.72)}{1.1871 \times 10^{-5} \text{ m}^2/\text{s}} = 406.45 \text{ s} = 6.77 \text{ min}$$

This result means that a $x = 2x_c = 5$-cm thick γ-phase specimen can be transformed to bainite $\gamma \rightarrow B$ at $t = 406.45$ s. Accordingly, $t \simeq 97$ s the TTT diagram which corresponds to a specimen with a thickness of

$$x_c = \sqrt{\alpha t / F_o} = \sqrt{(1.1871 \times 10^{-5})(97)/3.35} \simeq 1.9 \text{ cm}$$

Therefore, the transformation time strongly depends on the specimen thickness and temperature of the cooling medium.

9.11 Strain-Induced Phase Transformation

It is evident that certain crystalline materials undergo strain-induced phase transformation during plastic deformation, referred to as martensitic transformation which results from the austenite habit plane via a displacive (dimensionless) atomic movement of atoms. This is a fast phenomenon that can occur in some metallic and nonmetallic materials and it is entirely different from some diffusion-controlled solid-state transformations. Nevertheless, the martensitic phase forms as plates or needles being embedded in the austenitic parent phase (matrix) along certain crystallographic planes, inducing high residual stresses, lattice distortion, and even cracks.

In addition, $\gamma \rightarrow \alpha'$ transformation is a dynamic strain-hardening (work-hardening) process since martensite is a hard phase compared to room temperature ferrite and pearlite. However, cold working can also induce martensitic transformation in some metallic materials. For instance, the typical stainless steel 304 austenite microstructure shown in Fig. 9.21a when subjected to tension loading partially transforms to martensite as depicted in Fig. 9.21b, in which the austenite (light) and martensite (dark) grains are elongated along the tension direction. Moreover, the γ-α' microstructure in Fig. 9.21b, revealed out of a fractured specimen tested in tension at approximately 10^{-4} s^{-1} slow strain rate, illustrates the effect of cold plastic deformation on the martensitic transformation.

The transformation kinetics at temperatures T, crystallography, and the morphology of the newly formed α' phase are important aspects in the characterization of the structure-property relation. Hence, one must consider analytical or numerical solutions to strain-induced and thermal-induced martensitic transformation problems for a complete assessment of a particular material's mechanical behavior in service.

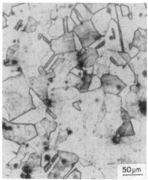 

(a) Before tension testing   (b) After tension testing

Fig. 9.21 IM 304 stainless steel microstructures. (**a**) Austenite before tension testing and (**b**) austenite (light)-martensite (dark) after tension testing at room temperature. Images taken at a 200x magnification

For the sake of clarity, strain-induced through mechanical deformation methods (tension, compression, torsion, bending, or a combination of these) and thermally-induced through heating/cooling in a particular environment (air, water, gas, etc.) may cause phase transformations. These techniques may or may not produce the same type of phase (such as martensite), but the resultant morphology of the microstructure is significantly influenced by the rate of transformation. Thus, the characterization of the transformation kinetics and related morphology of the newly formed phase can be accomplished using X-ray or electron diffraction and microscopy.

Martensitic transformation (MT) in ferrous (steel) and non-ferrous (Nitinol) alloys is nowadays an important technological process for specific applications. The reader should be aware of the fact that the former has been studied thoroughly over the years and the latter is more recent, acquiring the name shape-memory materials (SMM) or smart materials (SM). Nevertheless, no long-range diffusion is involved in the martensitic transformation process, which is fundamentally a massive transformation that causes lattice distortion. For comparison, a diffusion-controlled phase transformation is slow and allows precipitation of secondary particles, which harden the ductile or soft matrix.

As mentioned before, the habit plane (interface) between the matrix and the new phase corresponds to a specific crystallographic plane, where the macroscopic strain of the latter is adequately defined during the transformation.

In addition, the concept of stress-induced (σ-induced) phase transformation applies to those alloys exhibiting a reversible deformation without the aid of temperature. On the other hand, strain-induced (ϵ-induced) phase transformation may be considered an irreversible deformation at a fixed temperature, but the phase deformation may revert if the material is heated at a different temperature.

9.11.1 X-Ray Diffraction Patterns

This section introduces X-ray diffraction (XRD) data sets for determining or verifying the orientation relationships (OR's) theory characterized in a previous section. The emphasis here is to show XDR data for thermally-induced and strain-induced martensitic transformation. Thus, one can characterize the phase transition from γ-FCC austenite to α'-BCT martensite phase or to α'-BCT and ϵ-hexagonal martensitic phases.

For instance, Fig. 9.22a shows X-ray diffraction (XRD) patterns for the AISI 5160 steel (0.56%-0.64C, 0.7-0.9%Cr, 0.75-1.00Mn) being subjected to a high-speed quenching (HSQ) process in water at room temperature as per Martinez-Cazares et al. [26]. Notice that the strongest peaks clearly correspond to the $(111)_\gamma \parallel (110)_{\alpha'}$ orientation relationship (OR).

On the other hand, Fig. 9.22b depicts the same $(111)_\gamma \parallel (110)_{\alpha'}$ orientation relationship for a strain-induced martensitic transformation in a AISI 301 stainless steel (0.15%C max., 16%-18%Cr, 2%Mn max.) at variable strain ε values as per

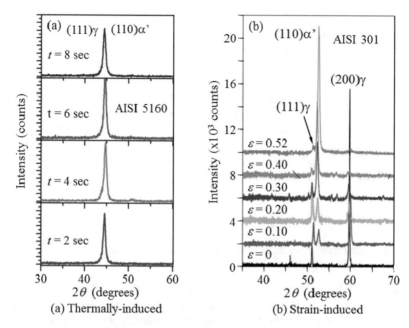

Fig. 9.22 (**a**) X-ray patterns for thermally-induced martensitic transformation in a AISI 5160 steel. The data shows the orientation relationships (OR's) of austenite and martensite peaks. After Martinez et al. [26]. (**b**) X-ray diffractograms for strain-induced martensitic transformation in AISI 301 stainless steel. After Celada-Casero et al. [27]

Celada-Casero et al. [27]. In this case, one can see in Fig. 9.22b that there is a visible misfit angle 2θ. Therefore, the above X-ray diffraction data confirm that there is an orientation relationship (OR) between γ and α' phase-crystallography. Obviously, the martensitic ε-phase is not revealed in this XDR patterns.

Moreover, some austenite γ peaks are revealed suggesting that a portion of this phase is retained during martensitic transformation. In this case, the metallic composite defines the microstructure of the material exhibiting a slight toughness.

The goal now is to introduce hybrid techniques for characterizing phase transformation and eventually, the mechanical behavior of a material. For instance, in general, in-situ X-ray diffraction coupled with an applied 1stress [28] and in-situ microscopy coupled with an applied temperature gradient [29] are interesting research approaches. The former case is for revealing the crystalline structure of phases and the latter case is for revealing the phase morphology during transformation.

Actually, the X-ray diffraction data set and microscopic work are important in thermodynamic and kinetic modeling of the microstructure evolution during phase transformation in the solid state. Thus, the rate of transformation (df_s/dT) and the microstructure morphology are reliably attached to observed mechanical behavior of the material in question.

It is clear now that one has to consider available experimental and even numerical techniques in order to validate the relevant measurements, such as electrical resistivity at temperatures T. Additionally, one important aspect of martensitic transformation is the determination of the kinetics of martensitic transformations in magnetic and non-magnetic fields under athermal and isothermal conditions [30].

Furthermore, martensitic transformation represents one of the most important technological processes for hardening steels and undoubtedly, X-ray diffraction data has its place in the assessment of crystallographic transformation. Hence, the solid-state phase transformation, $\gamma\,(FCC) \rightarrow \alpha'\,(BCT)$, is one of the relevant invariant reaction for producing martensitic steels with a variety of microstructures and related mechanical properties. Moreover, the quenching media play an important role in the steelmaking industry.

9.11.2 Volume Fraction of Martensite

In general, mechanically-induced and thermally-induced phase transformation methodologies provide experimental data for characterizing the phase transition from austenite to martensite. Neglecting the martensite morphology for the moment, it is now convenient to use the volume fraction of martensite as a dependent variable on the elastic strain for characterizing the martensitic transformation.

Among relevant experimental data sets available in the literature, Fig. 9.23a shows the martensite volume fraction ($f_{\alpha'}$) of initially deformed AISI 301 stainless steel specimens loaded in tension mode. This data set was published by Celada et al. [27] using techniques like light optical microscopy (LOM), magnetization measurements (SQUID = superconducting quantum interference device), and X-ray diffraction (XRD).

The kinetic curve fitting functions $f_{\alpha'} = f\,(\varepsilon)$ and $df_{\alpha'}/d\varepsilon$ for the data shown in Fig. 9.23a are defined as [31]

$$f_{\alpha'} = a\left[1 - \exp\left(-k_x\varepsilon^n\right)\right] \tag{9.50a}$$

$$\frac{df_{\alpha'}}{d\varepsilon} = ak_x n\varepsilon^{n-1}\exp\left(-k_x\varepsilon^n\right) \tag{9.50b}$$

$$\frac{df_{\alpha'}}{d\varepsilon} = ak_x n\varepsilon^{n-1}\left(1 - f_{\alpha'}/a\right) \tag{9.50c}$$

Here, a, k_x, n are curve fitting constants. Here, $a = 0.92, k_x = 13.4$, and $n = 2.2$ [27] are used in Eq. (9.50) and subsequently, curve fitting yields Fig. 9.23b, from which the functions $df_{\alpha'}/d\varepsilon = g\,(\varepsilon)$ and $df_{\alpha'}/d\varepsilon = g\,(f_{\alpha'})$ are evaluated as the rate of the α'-martensite transformation ($f_{\alpha'}$).

Note that these rate functions coincide at point "p," where

$$df_{\alpha'}/d\varepsilon = g\,(\varepsilon = 0.22) \simeq 2.8 \tag{9.51a}$$

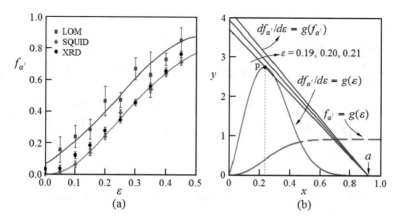

Fig. 9.23 (a) Room temperature strain-induced martensite volume fraction in AISI 301 steel [27] and (b) Rate of martensite volume fraction

$$df_{\alpha'}/d\varepsilon = g\,(f_{\alpha'} = 0.22) \simeq 2.8 \qquad\qquad (9.51b)$$

Therefore, $f_{\alpha'}$ as a function of applied stress is expected to exhibit a sigmoidal shape at a particular strain rate $(d\varepsilon/dt)$.

Example 9A.7 Show that martensitic transformation is driven by an undercooling $\Delta T_{\gamma\to\alpha'}$ and that it does not occur at $T > M_s$.

Solution At the equilibrium temperature T_o, the Gibbs energy change for martensitic transformation is $\Delta G_{\gamma\to\alpha'} = \Delta H_{\gamma\to\alpha'} - T_o\Delta S = 0$. Then, the entropy change of formation is $\Delta S = \Delta H_{\gamma\to\alpha'}/T_o$ at $T_o = M_s$. For martensitic transformation to occur, $T < M_s$ and

$$\Delta G_{\gamma\to\alpha'} = \Delta H_{\gamma\to\alpha'} - M_s\Delta S = \Delta H_{\gamma\to\alpha'} - \frac{M_s\,\Delta H_{\gamma\to\alpha'}}{T_o}$$

$$\Delta G_{\gamma\to\alpha'} = \Delta H_{\gamma\to\alpha'}\left(\frac{T_o - M_s}{T_o}\right) = \Delta H_{\gamma\to\alpha'}\left(\frac{\Delta T}{T_o}\right)_{\gamma\to\alpha'}$$

Therefore, $\Delta G_{\gamma\to\alpha'} \neq 0$ is driven by an undercooling $\Delta T_{\gamma\to\alpha'}$. This means that martensitic transformation is not possible at temperatures $T > M_s$.

9.11.3   Shape-Memory Effect

In effect, shape-memory alloys (SMA) and shape memory polymers (SMP) are smart solids that exhibit memory effects induced by temperature in their deformed state. The former memory effect is due to austenite-to-martensite-to-austenite crystallographic transformation, while the former's memory effect is induced by the glass transition temperature and by a magnetic field [32, 33]. Despite that these

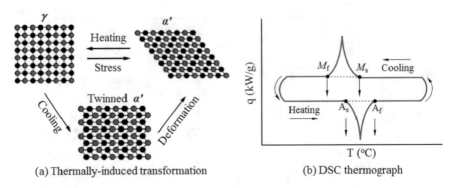

Fig. 9.24 Nitinol transformation cycle. (**a**) Thermally-induced and (**b**) schematic DSC thermograph

smart materials are different, they have potential applications in several engineering and science fields due to their unique capability to undergo reversible phase transformations when treated thermo-mechanically. Characteristics like lightweight, high strength, corrosion resistance, actuating capacity, and so forth make these materials potential candidates in broad fields of engineering and science as a result of their memory effects.

This section describes the shape-memory effect (SME) on crystalline materials due to a combination of the material's atomic or molecular structure and a suitable external stimuli, such as heat or strain. As a result, materials that have the ability to recover a predetermined crystallographic shape induced by external stimuli are classified as shape-memory materials (SMM) or shape-memory alloys (SMA).

If the external stimulus is heat, then a SMA, such as the equiatomic Ti-Ni alloy referred to as Nitinol, has a high-temperature phase called austenite (γ) and a low-temperature phase called martensite (α'). Thus, there must exist a critical or transformation temperature (T_c) so $\gamma \rightarrow \alpha'$ at $T > T_c$ and $\alpha' \rightarrow \gamma$ at $T < T_c$. This is mechanism is illustrated in Fig. 9.24. According to the order of the above invariant reactions, the γ-austenite is the predetermined structure that represents the shape-memory effect in some crystalline or polymeric materials.

Furthermore, Fig. 9.24a schematically shows thermally-induced and strain-induced (deformation) phase transformation cycles at a suitable temperature, while Fig. 9.24b schematically illustrates the counterclockwise differential scanning calorimetry (DSC) thermograph for determining the heat flow, $q = f(T)$, during the start "s" and finish "f" of austenite and martensite transformation. Note the downward arrows related the temperature range for transformation. Hence, thermally-induced phase transformation upon heat treatment leads to the reactions $\gamma \rightarrow \alpha'$ or $\alpha' \rightarrow \gamma$.

Note that $\gamma \rightarrow \alpha'$ and the reverse $\alpha' \rightarrow \gamma$ transformation do not take place at the same temperature. Therefore, T_c described above can be assigned to the peak temperature for either γ-phase or α'-phase and the shape-memory effect, on the other hand, is evident during the process of restoring the original shape of a plastically deformed structure by heating it. Moreover, when comparing shape-

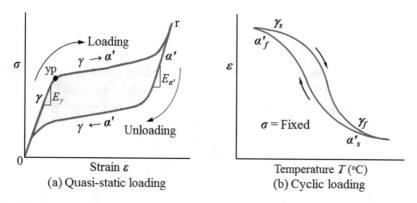

Fig. 9.25 SMA strain-induced transformation. (**a**) Quasi-static loading and (**b**) cyclic loading representing a hysteresis loop. After Hartl–Lagoudas [34, p. 63]

memory polymers (SMP's) and shape-memory alloys (SMA's) one must consider the maximum mechanical deformation, cost effective, thermal processing, and suitable applications.

In addition, Fig. 9.25 schematically shows the mechanical aspects of the strain-induced phase transformation [34, p. 63]. For instance, Fig. 9.25a depicts the stress–strain (σ-ε) cycle during loading and unloading a specimen at a temperature T. As a result, Nitinol undergoes strain-induced phase transformation. Among the mechanical properties obtained from a σ-ε curve, the modulus of elasticity (E_γ or $E_{\alpha'}$) of austenite (E_γ) or martensite ($E_{\alpha'}$) is determined using Hooke's law $E = \sigma/\varepsilon$, where ε is the elastic macrostrain which can have a maximum up to the yield point. One attractive engineering feature of Ni-Ti SMA's is their superelasticity (or possibly pseudoelasticity) response to mechanical deformation beyond the yield point as shown in Fig. 9.24a before fracture. In particular, SMA materials are nowadays useful in certain applications due to amazing superelasticity mechanical behavior upon unloading. For the sake of clarity, the schematic stress–strain curves shown in Fig. 9.25a indicate that the hypothetical SMA material undergoes a stress-induced phase transformation without temperature application due to the instability of phase atomic structures (austenite and martensite).

Thus, a solid-state phase deformation ($\gamma \rightarrow \alpha'$ or $\alpha' \rightarrow \gamma$) is induced by a mechanical plastic deformation. Moreover, Fig. 9.25b schematically shows a cyclic thermal loading (hysteresis loop) of an SMA specimen undergoing phase change at a constant applied stress. Additional details on SMA's and SMP's can be found elsewhere [35–39].

Furthermore, the area under the tension stress–strain (σ-ε) curve at a quasi-static strain rate ($d\sigma/d\varepsilon$) is known as the strain-energy density (W) required to deform the material mechanically. The total strain-energy density is defined by

$$W = \int_o^\varepsilon \sigma \, d\varepsilon = \int_o^{\varepsilon_{yp}} \sigma \, d\varepsilon + \int_{\varepsilon_{yp}}^\varepsilon \sigma \, d\varepsilon \qquad (9.52a)$$

$$W = \frac{\sigma_{ys}^2}{2E} + \int_{\varepsilon_{yp}}^{\varepsilon} \sigma \, d\varepsilon \qquad (9.52b)$$

where $\sigma = \varepsilon E$ (Hooke's law) is used to solve the first integral, σ denotes an elastic stress, $\varepsilon \leq \varepsilon_{yp}$ denotes the strain at the yield point (yp), σ_{ys} denotes the yield strength at ε_{yp}, E_γ denotes the modulus elasticity from zero to ε_{yp} for austenite, and $E_{\alpha'}$ denotes the modulus elasticity from point "r" for martensite. Nonetheless, once the plastic stress function $\sigma = f(\epsilon)$ is defined the second integral is solved.

Mathematically, this shear stress can be defined as $\tau_{max} = \sigma_{ys}/2$, where the yield strength (σ_{ys}) is the transition stress at the elastic–plastic boundary, located at the yield point (yp) in Fig. 9.25a.

In addition, the σ-ε curve at $\varepsilon > \varepsilon_{yp}$ represents the $\gamma \rightarrow \alpha'$ transformation upon loading. However, the reverse invariant reaction $\alpha' \rightarrow \gamma$ occurs upon unloading from point "r" and eventually, the shape of Nitinol is recovered at $\varepsilon = 0$.

Accordingly, one can determine the net strain-energy density (W_{net}) for transformation as the net area under the σ-ε curve (Fig. 9.25a). Mathematically,

$$W_{net} = W_{or} - W_{ro} = \left(\int_o^{\varepsilon_r} \sigma \, d\varepsilon \right)_\gamma - \left(\int_o^{\varepsilon_r} \sigma \, d\varepsilon \right)_{\alpha'} \qquad (9.53)$$

where the variables W_{or}, W_{ro} denote the areas under the σ-ε curves illustrated in Fig. 9.25a as "or" and "ro" trajectories.

According to Eq. (9.51), a loaded ductile specimen in tension undergoes elastic and plastic deformation before fracture. In this case, the normal strain in the loading direction induces a change in volume, while generated plastic stresses and shear strains cause lattice distortion in the form of dislocations.

Further, engineering materials are subjected to be characterized (analyzed) according to design failure theories for ductile and brittle response. In particular, the Maximum Shear Stress theory predicts the maximum shear stress that causes plastic deformation (due to formation of dislocation) at a macroscale. Analytical details on failure theories are out of the scope of this book.

9.12   Summary

In solid-phase transformation studies, X-ray diffraction is used as an essential tool for revealing the orientation relationship (OR) between the initial and final phases, and for determining the invariant reactions associated with this process. In fact, thermally-induced and strain-induced transformations have particular characteristics with respect to their phase-transformation crystallography and their interface coherency.

Of particular interest in this chapter is the dislocation theory that describes the elastic fields related to linear defects known as edge, screw, and mixed dislocations. This leads to modeling dislocations as cylindrical shells for developing the stress and strain, and strain-energy mathematical models.

Additionally, solid-state phase transformation is also characterized at a macroscale using TTT diagrams, which are mostly used for heat treating steels isothermally or continuously. Thus, isothermal transformation (IT) and continuous cooling transformation (CCT) diagrams can be superimposed on a TTT diagram for designing the most suitable heat treatment process in order to produce (obtain) a desire microstructure for a particular alloy.

9.13   Appendix 9A: TTT Diagrams

Below are the TTT diagrams for 1080 and 4340 steels adopted from [19] and Callister–Rethwisch [9, p. 378–379], respectively.

Figure 9A.1 shows the TTT diagram for hypoeutectoid 1050 steel [19, p. 15].

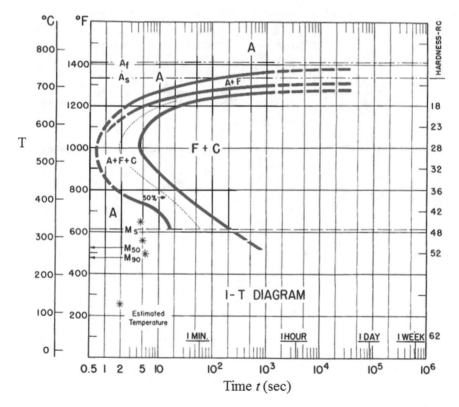

Fig. 9A.1 TTT diagram for a hypoeutectoid 1050 steel adapted from Callister and Rethwisch book [9, p. 378]. Legend: A = Austenite (γ), B = Bainite, F = Ferrite (α), M = Martensite (α'), and P = Pearlite

Figure 9A.2 depicts the TTT diagram for 1080 eutectoid steel [19, p. 16].

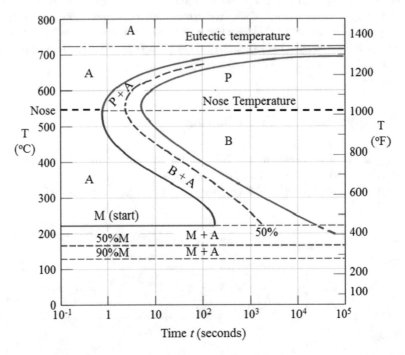

Fig. 9A.2 TTT diagram for a eutectoid 1080 steel adapted from references [9, p. 378] and [19, p. 16]

Figure 9A.3 exhibits the TTT diagram for 9440 alloy steel [19, p. 38].

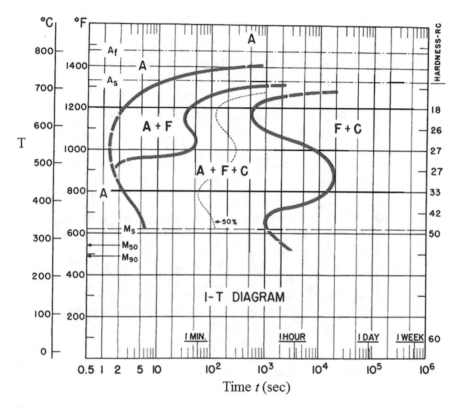

Fig. 9A.3 TTT diagram for 9440 alloy steel adapted from [19, p. 38]

Figure 9A.4 illustrates the TTT diagram for 4340 alloy steel [19, p. 35].

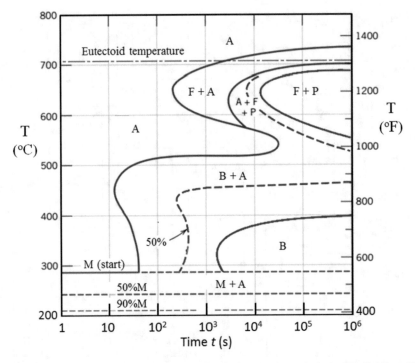

Fig. 9A.4 TTT diagram for a 4340 alloy steel adapted from Callister–Rethwisch book [19, p. 35]

9.14   Appendix 9B: Separation of Variables

This Appendix includes the analytical procedure for deriving Eq. (9.41). Again, following Poirier–Geiger ([24, p. 294] and [25, p. 290]), let $\theta = T - T_f$ so that

$$\frac{\partial \theta}{\partial t} = \alpha \frac{\partial^2 \theta}{\partial x^2} \tag{9B.1a}$$

$$\theta\,(x, 0) = T_i - T_f = \theta_i \quad \text{(IC)} \tag{9B.1b}$$

$$\frac{\partial \theta\,(0, t)}{\partial x} = 0 \qquad \text{(BC)} \tag{9B.1c}$$

$$\frac{\partial \theta\,(x_c, t)}{\partial x} + \frac{h}{k}\theta\,(x_c, t) = 0 \quad \text{(Energy balance)} \tag{9B.1d}$$

In order to solve Eq. (9B.1a) by the method of separation of variables, the product solution of the form $\theta\,(x,t) = F\,(x) \cdot G\,(t)$ is used for such a purpose. Thus, the derivatives of $\theta\,(x,t)$ are

$$\theta\,(x,t) = F\,(x) \cdot G\,(t) \tag{9B.2a}$$

$$\frac{\partial\theta\,(x,t)}{\partial x} = G\,(t)\,\frac{dF\,(x)}{dx} \tag{9B.2b}$$

$$\frac{\partial^2\theta\,(x,t)}{\partial x^2} = G\,(t)\,\frac{d^2F\,(x)}{dx^2} \tag{9B.2c}$$

and

$$\frac{\partial\theta\,(x,t)}{\partial t} = F\,(x)\,\frac{dG\,(t)}{dt} \tag{9B.3}$$

Inserting Eqs. (9B.2c) and (9B.3) into (9B.1a) with $\theta\,(x,t) = \theta$ yields the following expression

$$\frac{1}{F\,(x)}\,\frac{d^2F\,(x)}{dx^2} = \frac{1}{\alpha G\,(t)}\,\frac{dG\,(t)}{dt} = -\lambda^2 \tag{9B.4}$$

from which two homogeneous ordinary differential equations (ODE's) arise

$$\frac{d^2F\,(x)}{dx^2} + \lambda^2 F\,(x) = 0 \tag{9B.5a}$$

$$\frac{dG\,(t)}{dt} + \lambda^2 \alpha G\,(t) = 0 \tag{9B.5b}$$

The solutions of Eq. (9B.5) are

$$F\,(x) = a_1 \cos(\lambda x) + a_2 \sin(\lambda x) \tag{9B.6a}$$

$$G\,(t) = \exp\left(-\lambda^2 \alpha t\right) \tag{9B.6b}$$

where λ is the eigenvalue representing a characteristic root or scalar associated with an eigenstate during linear transformation of a vector space and it is a value corresponding to a quantity being measured. The eigenvalue λ may have a positive or negative value for a non-zero eigenvector and it must satisfy the characteristic equations.

Continuing with the current analytical procedure, note that Eq. (9B.6a) satisfies (9B.1c) if $a_2 = 0$; that is,

$$\frac{\partial F\,(x)}{dx} = -\lambda a_1 \sin(\lambda x) + \lambda a_2 \cos(\lambda x) \tag{9B.7a}$$

$$\left[\frac{\partial F(x)}{dx}\right]_{x=0} = \lambda a_1 \sin(0) - \lambda a_2 \cos(0) = 0 \tag{9B.7b}$$

$$a_2 = 0 \tag{9B.7c}$$

which means that the coefficient $a_2 = 0$ does not contribute to Eq. (9B.6a). Consequently, Eq. (9B.6a) simplifies to

$$F(x) = a_1 \cos(\lambda x) \tag{9B.8a}$$

$$\frac{dF(x)}{dx} = -\lambda a_1 \sin(\lambda x) \tag{9B.8b}$$

From Eq. (9B.1d),

$$\frac{dF(x_c, t)}{dx} + \frac{h}{k} F(x_c, t) = 0 \tag{9B.9a}$$

$$-a_1 \lambda \sin(\lambda x_c) + \frac{h}{k} a_1 \cos(\lambda x_c) = 0 \tag{9B.9b}$$

$$\frac{h}{k} \cos(\lambda x_c) = \lambda \sin(\lambda x_c) \tag{9B.9c}$$

$$\cot(\lambda x_c) = \frac{\lambda}{h/k} = \frac{\lambda x_c}{(h/k) x_c} = \frac{\lambda x_c}{B_i} \tag{9B.9d}$$

$$\lambda x_c \tan(\lambda x_c) = B_i \tag{9B.9e}$$

Substituting Eqs. (9B.6b) and (9B.8a) into (9B.2a) and generalizing the resultant expression for all values of λ_n satisfying (9B.9d) yields $\theta(x, t) = T - T_f$ and

$$\theta(x, t) = \sum_{n=1}^{\infty} F_n \cos(\lambda_n x) \exp\left(-\lambda_n^2 \alpha t\right) \tag{9B.10}$$

Using the initial condition (IC) defined by Eq. (9B.1b) gives $\theta_i = T_i - T_f$ and

$$\theta_i = \sum_{n=1}^{\infty} F_n \cos(\lambda_n x) \tag{9B.11}$$

In classical Fourier analysis, use of orthogonality integrals is a standard procedure for finding each of the Fourier coefficients (amplitudes) F_n (Appendix 8A). Multiply Eq. (9B.11) by a weight function, say, cosine function $\cos(\lambda_m x) dx$ and integrate the resultant expression with appropriate limits of integration illustrates the application of the orthogonality integral approach. Thus,

$$\theta_i \int_0^{x_c} \cos(\lambda_m x)\, dx = \int_0^{x_c} \sum_{n=1}^{\infty} F_n \cos(\lambda_n x) \cos(\lambda_m x)\, dx \qquad (9B.12)$$

and for a non-zero integral, $m = n$ and $\lambda_n x_c = n\pi/x_c$. Then, Eq. (9B.12) yields

$$\theta_i \left[\frac{1}{\lambda_n} \sin(\lambda_n x_c) \right] = F_n \left[\frac{x_c}{2} + \frac{1}{2\lambda_n} \sin(\lambda_n x_c) \cos(\lambda_n x_c) \right] \qquad (9B.13a)$$

$$F_n = \frac{2\theta_i \sin(\lambda_n x_c)}{\lambda_n x_c + \sin(\lambda_n x_c) \cos(\lambda_n x_c)} \qquad (9B.13b)$$

Substitute Eq. (9B.14b) into (9B.10) to get the sought equation

$$\frac{\theta}{\theta_i} = \frac{T - T_f}{T_i - T_f} = 2 \sum_{n=1}^{\infty} \frac{\sin(\lambda_n x) \cos(\lambda_n x)}{\lambda_n x_c + \sin(\lambda_n x_c) \cos(\lambda_n x_c)} \exp\left(-\lambda_n^2 \alpha t\right) \qquad (9B.14)$$

which is defined as Eq. (9.45a) in the text.

9.15 Problems

9.1 Given the grain high angles, calculate the surface energy ratio $\gamma_{\alpha\alpha}/\gamma_{\alpha\beta}$.

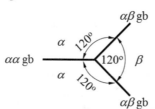

9.2 Calculate the displacement involved in martensitic transformation in a carbon steel plate containing 0.80% by weight.

9.3 If a 1080 steel bar undergoes martensitic transformation, then the lattice deformation occurs by contraction along the $[001]_\gamma$ crystallographic direction and the uniform expansion on the $(001)_\gamma$ plane. (a) Calculate the respective Bain strains, the maximum displacement and write down the results in matrix format, and (b) determine the orientation relationship between the austenite (parent) and martensite (product) lattices.

9.4 Consider $(111)_\gamma \rightarrow (110)_{\alpha'}$ during martensitic transformation in steel as shown in Fig. 9.13 and write down the Miller indices for the crystallographic directions corresponding to orientation relationships in the Bain model.

9.5 For a 1060 steel bar undergoing martensitic transformation, (a) calculate the respective Bain strains, the maximum displacement and write down the Bain matrix, and (b) determine the orientation relationship between the austenite (parent) and martensite (product) lattices.

9.6 For a 1030 steel bar undergoing martensitic transformation, (a) calculate the respective Bain strains, the maximum displacement and write down the Bain matrix, and (b) determine the orientation relationship between the austenite (parent) and martensite (product) lattices.

9.7 For a carbon steel plate being quenched in agitated water at 25°C, calculate the critical entities defined by Eq. (9.37) when (a) $\Delta G_p = 0$ and (b) $\Delta G_p = 50 \, \text{MJ/m}^3$ and assume a homogeneous nucleation of martensite without composition change. Assume that the martensitic transformation starts at $T = 700 \, \text{K} \, (427°C)$ and

$\Delta G_V = 180 \, \text{MJ/m}^3$	$S_{\gamma\alpha'} = 15 \times 10^{-3} \, \text{J/m}^2$	$K_\varepsilon = 2.5 \, \text{GPa}$

9.8 For a carbon steel plate being quenched in agitated water at 25°C, calculate the critical entities defined by Eq. (9.37) when (a) $\Delta G_p = 0$ and (b) $\Delta G_p = 60 \, \text{MJ/m}^3$. Assume that the martensitic transformation starts at $T = 700 \, \text{K} \, (427°C)$ with no composition change for homogeneous nucleation of martensite. Use the following data

$\Delta G_V = 180 \, \text{MJ/m}^3$	$S_{\gamma\alpha'} = 15 \times 10^{-3} \, \text{J/m}^2$	$K_\varepsilon = 2.5 \, \text{GPa}$

9.9 Consider two isothermal solid-phase transformations for two 6-cm thick 1080 steel plates as indicated by the IT trajectories in the eutectoid TTT diagram given in Example 9A.6. (a) If one plate is initially at 750°C and immersed in a fluid at $T_f = 600°C$, then calculate the time for complete pearlitic transformation ($\gamma \rightarrow p$) when the temperature at center is $T = 650°C$. (b) Repeat the calculation for bainitic transformation ($\gamma \rightarrow B$) in the other 6-cm thick 1080 steel plate immersed in a similar fluid when $T_f = 400°C$ and $T = 450°C$. Compare the results with the read-off time from the TTT diagram. Given data:

$$h = 283.92 \, \text{W/m}^2.\text{K}; \, \alpha = 1.1871 \times 10^{-5} \, \text{m}^2/\text{s}; \, k = 34.614 \, \text{W/m.K}$$

9.10 Plot (a) the Fourier number $F_o = f(\phi)$ at $0 \leq \phi = \lambda x_c \leq 0.8$, $B_i = 0.25$ and $x/x_c = 0, \, 0.3, \, 0.4, \, 0.5$, (b) the Biot number $B_i = f(\phi)$ at $0 \leq \phi \leq 2$ and $F_o = f(B_i)$ at $0 \leq B_i \leq 2$. Let $T_\theta = 0.33$. The resultant curves simply show the non-linear trends of the functions $F_o = f(\phi)$, $B_i = f(\phi)$, and $F_o = f(B_i)$. You are on your own to characterize the non-linearity of the curves.

9.11 For a quenched carbon steel plate in agitated water at 25°C, calculate the critical entities defined by Eq. (9.37) when (a) $\Delta G_p = 0$ and (b) $\Delta G_p = 60$ MJ/m^3. Use the following data $\Delta G_V = 182$ MJ/m^3, $S_{\gamma\alpha'} = 16 \times 10^{-3}$ J/m^2, $K_\varepsilon = 2.6$ GPa, and assume that the martensitic transformation starts at $T = 700$ K (427°C) with no composition change for homogeneous nucleation. [Solution: (a) $\Delta G_{c,\text{hom}} = 5.28$ eV and (b) $\Delta G_{c,\text{hom}} = 27.90$ eV].

9.12 Assume that two 4-cm thick 1080 steel plates are subjected to isothermal solid-phase transformation in a similar manner as illustrated in Example 9A.6. (a) For one plate initially at 750°C and immersed in a fluid at $T_f = 600$°C, calculate the time for complete pearlitic transformation ($\gamma \rightarrow p$) when the temperature at center is $T = 650$°C. (b) Repeat the calculation for bainitic transformation ($\gamma \rightarrow B$) when $T_f = 400$°C and $T = 450$°C. Compare the results with the read-off time from the TTT diagram. Given data: $h = 283.92$ W/m^2.K, $\alpha = 1.1871 \times 10^{-5}$ m^2/s and $k = 34.614$ W/m.K. [Solution: (a) $t = 124.34$ s and (b) $t = 314.72$ s].

9.13 Suppose that two 2-cm thick 1080 steel plates are initially at 750°C and are subjected to isothermal solid-phase transformation. Calculate the transformation time (a) for pearlitic transformation ($\gamma \rightarrow p$) when the fluid is at $T_f = 600$°C and the temperature of one plate at center is $T = 650$°C, and (b) for bainitic transformation ($\gamma \rightarrow B$) when the fluid is at $T_f = 400$°C and the center of the other plate is at $T = 450$°C. Compare the results with the read-off time from the TTT diagram. Given data: $h = 283.92$ W/m^2.K, $\alpha = 1.1871 \times 10^{-5}$ m^2/s and $k = 34.614$ W/m.K. [Solution: (a) $t = 32.43$ s and (b) $t = 64.61$ s].

References

1. D.A. Porter, K. Easterling, M.Y. Sherif, *Phase Transformations in Metals and Alloys*, 3rd edn. (CRC Press, Taylor & Francis Group, LLC, Boca Raton, FL, USA, 2009)
2. A.G. Guy, J.J. Hren, *Elements of Physical Metallurgy*, 3rd edn. (Addison-Wesley Publishing, Reading, MA, USA, 1974)
3. C.S. Smith, *Metal Interfaces* (ASM, Cleveland, OH, 1952), pp. 65–113
4. W.T. Read, Jr., *Dislocations in Crystals* (McGraw-Hill Book Company, New York, 1953)
5. W.T. Read, W. Shockley, in *Dislocation Models of Grain Boundaries*, ed. by W. Shockley, J.H. Hollomon, R. Maurer, F. Seitz. Imperfections in Nearly Perfect Crystals, Chapter 13. Pocono Manor (Pennsylvania) Symposium (October 12–14, 1950) (Wiley, New York, 1952)
6. W.T. Read, W. Shockley, Dislocation models of crystal grain boundaries. Phys. Rev. **78**(3), 275–289 (1950)
7. A.T. Blumenau, *The Modelling of Dislocations in Semiconductor Crystals*, Ph.D. Thesis (English), Department of Physics, Faculty of Science, University of Paderborn, Paderborn, Germany, 2002
8. D. Hull, D.J. Bacon, *Introduction to Dislocations*, 5th edn. (Butterworth-Heinemann, Elsevier Ltd., New York, USA, 2011)
9. W.D. Callister, Jr., D.G. Rethwisch, *Materials Science and Engineering: An Introduction*, 9th edn. (Wiley, New York, USA, 2014)

10. W.F. Hosford, *Mechanical Behavior of Materials*, 2nd edn. (Cambridge University Press, New York, USA, 2010)

11. D.R. Askeland, P.P. Fulay, W.J. Wright, *The Science and Engineering of Materials*, 6th edn. (Cengage Learning, New York, USA, 2010)

12. Z. Nishiyama, *Martensitic Transformation* (Academic Press, New York, 1978)

13. B.D. Cullity, S.R. Stock, *Elements of X-Ray Diffraction*, 3rd edn. (Pearson Education Limited, London, UK, 2014)

14. C. Simsir, C. Hakan Gur, in *Simulation of Quenching*, ed. by C. Hakan Gur, J. Pan. Handbook of Thermal Process Modeling of Steels, Chapter 9 (CRC Press, Taylor & Francis Group, LLC, Boca Raton, FL, USA, 2009)

15. M. Munirajulu, B.K. Dhindaw, A. Biswas, A. Roy, *Modelling of Eutectoid Transformation in Plain Carbon Steel*, vol. 34(4) (ISIJ International, 1994), pp. 355–358

16. M. Avrami, Kinetics of phase change. I general theory. J. Chem. Phys. **7**(12), 1103–1112 (1939)

17. I. Avramov, J. Sestak, *Generalized Kinetics of Overall Phase Transition in Terms of Logistic Equation* (2015). https://arxiv.org/abs/1510.02250v1

18. A. Kulawik, Modeling of thermomechanical phenomena of welding process of steel pipe/Modelowanie Zjawisk Termomechanicznych Procesu Spawania Rury Stalowej. Archives Metallurgy Mater. **57**(4), 1229–1238 (2012)

19. G.F. Vander Voort (ed.), *Atlas of Time-Temperature Diagrams for Irons and Steels* (ASM International, Metals Park, OH, USA, 1991)

20. L. Kaufman, M. Cohen, Thermodynamics and kinetics of martensitic transformations. Progress Metal Phys. **7**, 165–246 (1958)

21. G.B. Olson, M. Cohen, in *Dislocation Theory of Martensitic Transformations*, ed. by F.R.N. Nabarro. Dislocations in Solids, vol. 7 (North Holland Publishing, Amsterdam, 1986)

22. A.K. Sinha, *Physical Metallurgy Handbook* (The McGraw-Hill Companies, New York, USA, 2003)

23. J.D. Eshelby, *Elastic Inclusions and Inhomogeneities*. Prog. Solid Mech., vol. 2 (Amsterdam, 1961), pp. 89–140

24. G.H. Geiger, D.R. Poirier, *Transport Phenomena in Metallurgy* (Addison-Wesley Publishing, Reading, MA, USA, 1973)

25. D.R. Poirier, G.H. Geiger, *Transport Phenomena in Materials Processing* (The Minerals, Metals & Materials Society, A Publication of TMS, Springer, Switzerland, 2016)

26. G.M. Martınez-Cazares, D.E. Lozano, M.P. Guerrero-Mata, R. Colas, G.E. Totten, High-speed quenching of high carbon steel. Mater. Perform. Charact. **3**(4), 256–267 (2014)

27. C. Celada-Casero, H. Kooiker, M. Groen, J. Post, D. San-Martin, In-situ investigation of strain-induced martensitic transformation kinetics in an austenitic stainless steel by inductive measurements. Metals **7**(7), 271 (2017)

28. A.A. Klopotov, A.I. Potekaev, E.S. Marchenko, O.M. Loskutov, G.A. Baygonakova, V.E. Gunther, X-ray diffraction studies of martensitic transformations in situ of the parameters of a thin crystalline structure in TiNi under external loading, in *SMBIM Conference Proceedings Shape Memory Biomaterials and Implants in Medicine, KnE Materials Science*, vol. 2017, pp. 149–158 (2017)

29. X. Yu, S.S. Babu, J.C. Lippold, H. Terasaki, Y.I. Komizo, In-situ observations of martensitic transformation in blast-resistant steel. Metall. Mater. Trans. A **43**(5), 1538–1546 (2011)

30. T. Kakeshita, J.M. Nam, T. Fukuda, Kinetics of martensitic transformations in magnetic field or under hydrostatic pressure. Natl. Inst. Mater. Sci. Sci. Technol. Adv. Mater. **12**(1), 015004 (2011)

31. H.C. Shin, T.K. Ha, Y.W. Chang, Kinetics of deformation induced martensitic transformation in a 304 stainless steel. Scr. Mater. **45**(7), 823–829 (2001)

32. R.D. James, M. Wuttig, Magnetostriction of martensite. Philos. Mag. A **77**(5), 1273–1299 (1998)

33. R.C. O'Handley, Model for strain and magnetization in magnetic shape-memory alloys. J. Appl. Phys. **83**(6), 3263–3270 (1998)

34. D.J. Hartl, D.C. Lagoudas, in *Thermomechanical Characterization of Shape Memory Alloy Materials*, Chapter 2, ed. by D.C. Lagoudas. Shape Memory Alloys: Modeling and Engineering Application (Springer Science, New York, USA, 2008)
35. C. Cismasiu (ed.), *Shape Memory Alloys* (Published by Sciyo, Croatia, 2010)
36. D.C. Lagoudas (ed.), *Shape Memory Alloys: Modeling and Engineering Applications* (Springer Science + Business Media, LLC, New York, USA, 2008)
37. A. Lendlein (ed.), *Shape-Memory Polymers* (Springer, Berlin, Heidelberg, New York, USA, 2010)
38. J. Leng, S. Du (eds.), *Shape-Memory Polymers and Multifunctional Composites* (Taylor and Francis Group, LLC, Boca Raton, FL, USA, 2010)
39. J. Hu, *Shape Memory Polymers and Textiles* (CRC Press, Woodhead Publishing Limited, Boca Raton, FL, USA, 2007)

Chapter 10
Solidification Defects

10.1 Introduction

This chapter introduces the theory of solidification and cooling shrinkage during casting. The most convenient academic approach is to include a compilation of analytical solutions available in the literature for shrinkage problems in the production of casting devices with simple or complicated geometries. In fact, solidification defects due to excessive shrinkage is an undesirable problem, but it can be controlled in order to produce a sound casting device.

Producing defect-free and dimensional fit metal castings close to design specifications is a challenging process for foundries since most metals shrink to some degree during cooling. Moreover, solidification and cooling shrinkage is a contraction process related to phase transformation followed by solid-phase cooling. The solidification shrinkage process may cause casting defects and solid-phase cooling may cause internal stresses responsible for crack formation during solid-phase cooling. Thus, liquid shrinkage refers to the contraction of molten metal prior to solidification and solidification shrinkage denotes liquid-to-solid shrinkage, and cooling shrinkage refers to contraction in the solid state.

Obviously, the integrity of the casting geometry (device) may be compromised by shrinkage problems, such as porosity and cracking, and it can be avoided by feeding liquid metal to the casting device using a riser, which in turn, is the liquid metal feeder or simply the liquid metal reservoir within the mold cavity. Simply put, the casting feeder compensates for solidification shrinkage of the casting geometry and it is essentially a recyclable part of the casting cavity. Other parts such as the sprue and gating systems are also recyclable.

© Springer Nature Switzerland AG 2020
N. Perez, *Phase Transformation in Metals*,
https://doi.org/10.1007/978-3-030-49168-0_10

10.2   Riser Feeding Distance

The goal in this section is to introduce simple mathematical models to determine proper design variables related to casting and to avoid solidification defects. Among the simplest and yet, classical mathematical model for designing castings is the well-known Chvorinov's rule (introduced in Chap. 5) which predicts the solidification time as a function of volume-to-area ratio (V/A) of the entire or part of the casting, such as a riser. For instance, the essence is to design the feeder and the casting geometry with particular casting modulus; $M_f/M_c > 1$, where $M_f = (V/A)_f$ for the feeder and $M_c = (V/A)_c$ for the casting geometry. Additional details on this topic can be found elsewhere [1–5].

Assume that melt feeding is required to avoid casting defects, such as dispersed gas porosity and solidification shrinkage, during solidification. In such case, consider the side riser-runner-casting geometry and one metal chill plate shown in Fig. 10.1a, where the feeder is just a cylindrical riser used to prevent cavities due to shrinkage. The riser feeding distance (F_d) is the maximum dimension a riser can feed a casting section and it defines the edge of the riser safety zone shown as a circle in Fig. 10.1b.

On the other hand, Fig. 10.1c depicts a top riser and two metal chill plates and Fig. 10.1d illustrates the top view with the maximum riser feeding distance F_d. With respect to Fig. 10.1a, runner or gating design is as important as the riser design because both directly impact casting quality.

The general information on design is related to industrial castings of simple and complex geometries using sand molds with specific characteristics to ensure high

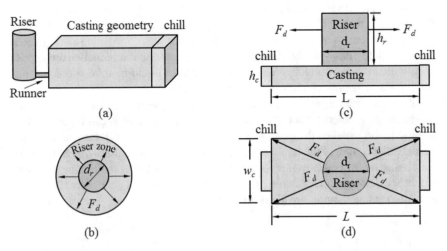

Fig. 10.1 (a) Schematic riser and bar casting ($w_c \le 3h_c$), (b) top view of the riser showing the feeding radial distance (zone) F_d and a chill, (c) plate front view ($w_c > 3h_c$) with a top riser and chills, and (d) plate top view and top riser showing the maximum feeding distance F_d

quality as per industry guidelines. Below are some variables that must be analyzed in order to produce high quality or defect-free castings.

| Moisture | Green strength | Permeability | pH | Mold hardness |

From this point forward, assume that sand molds have appropriate properties, including collapsibility, and that the melt is at a suitable pouring temperature T_p, where $T_p > T_f$ for pure metals and $T_p > T_f^*$ for alloys with a nominal composition C_o and a solidification temperature range $T_s' \leq T_f^* \leq T_l$.

10.2.1 Rectangular Casting

Additionally, a bar and plate width are defined as $w_c \leq 3h_c$ and $w_c > 3h_c$, respectively, and internal chill plates are used to promote solidification, in the above case, at the edges of the casting geometries. Moreover, recall that solidification starts from the mold walls towards the center of the casting geometry and conveniently, the feeding distance (F_d) is defined as the maximum length from the edge of a riser in the radial direction (Fig. 10.1c) to the edge of the casting geometry. For flat plates, F_d is the diagonal distance from the riser edge to the plate's corner (Fig. 10.1d).

Following Campbell's book on castings [5, p. 678], Fig. 10.2 illustrates Pellini's method for placing the risers at the optimal position and for defining the riser feeding distances in order to avoid solidification shrinkage porosity.

Importantly, these particular riser locations on top of plates are considered adequate for producing defect-free casting geometries, provided that other mold characteristics and melt pouring temperature are strictly controlled. However, the number of risers and their optimal locations depend on the riser and casting geometry dimensions, the type of metal or alloy, the metal casting designer's experience, and the industry guidelines.

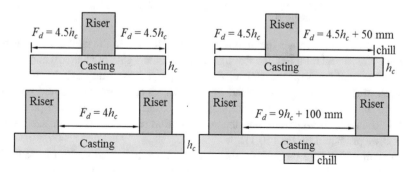

Fig. 10.2 Pellini's method for riser feeding distances since 1953

Since chills are metal plates with high heat transfer capabilities, they can promote (assist) directional solidification of the casting towards the feeder and subsequently, they provide a localized cooling effect on the casting. As a result, casting soundness can be achieved and the casting production may be cost effective.

Comparing the casting and the riser dimensional conditions one must assure that the riser solidifies last by controlling or designing the volume to surface area ratio. This calls for the Chvorinov's rule which states that the casting with large surface area and small volume $(V/A_s)_1$ cools faster than a casting with small surface area and a large volume $(V/A_s)_2$. Thus, $(V/A_s)_{riser} > (V/A_s)_{casting}$ and the definition of V/A_s is used by the so-called Chvorinov's rule found in Chap. 5 and in the next section.

Extraordinarily, the need for a proper riser designs relates to the shape and dimensions of the casting geometry. However, cost-effective casting procedures are in high demand by the foundry industry. Nowadays, industries, specifically the automobile and aerospace engineering fields, demand high quality castings with machinability and recyclability characteristics. In order to comply with such a requirement, proper design of riser and gating systems and knowledge of phase transformation are the foremost aspects of casting techniques.

10.2.2 Designing Feeders

The main concern now is how to produce a sound cost-effective casting geometry. This can be accomplished by designing, at least, one riser as the melt feeder to the solidifying casting in order to avoid shrinkage porosity and solid contraction problems. Specifically, porosity is the main casting defect that imparts cracking sites.

Among the riser design methodology and related feeding distance methods, the Niyama's criterion [1] for radiographically sound castings is based on the riser feeding distance parameter F_r defined by

$$F_r = \frac{dT/dr}{\sqrt{dT/dt}} = \frac{G_r}{\sqrt{dT/dt}} > 0.1\sqrt{K/s}/mm \qquad (10.1)$$

Here, $G_r = dT/dr$ denotes the melt temperature gradient in the melt casting during the riser directional melt feeding through a short runner (gate or channel) and dT/dt denotes the melt cooling rate. Hence, $F_r > 0.1\sqrt{K/s}/mm$ is a threshold value.

Interestingly, the riser acts as the melt feeder since it provides liquid mass of metal to the solidifying casting geometry, As a result, the solidification process is slightly delayed, to an extent, and the solidification shrinkage is suppressed when $F_r > 0$, as predicted by Eq. (10.1).

The solidification time for a casting that solidifies in a sand mold, for example, can directly be read from a temperature–time curve defined a function $T = f(t)$. This requires embedded high-temperature resistant thermocouples in the mold cavity. As a result, the thermocouples become part of the casting after solidification is complete. Obviously, this is a good technique for comparing and characterizing the solidification time at different parts of the mold geometry. Instead, assume that the Chvorinov's rule, Eq. (5.36b), applies to the casting ("c") and riser ("r") so that the solidification times in castings are

$$t_c = B_c M_c^2 = B_c \left(\frac{V_c}{A_c} \right)^2 \tag{10.2a}$$

$$t_r = B_r M_r^2 = B_r \left(\frac{V_r}{A_r} \right)^2 \tag{10.2b}$$

where, M_c, M_r are the casting and riser moduli, B_c, B_r are Chvorinov's rule constants (M_c and M_r), and V_r is the riser solidified volume. Henceforward, the solidification time can be generalized as function of the form $t = f(B_i, M_i)$, where $i = c, r$.

Fundamentally, Eq. (10.2) indirectly includes the effects of shape and size of the riser and casting geometry, and superheat through the mold constants B_c and B_r. In particular, the riser constant B_r takes into account any superheating. Nonetheless, the mold constants (see Chap. 5) are defined by

$$B_r = \frac{\pi \alpha_m}{4} \left[\frac{\rho_s \Delta H_f + \rho_s c_p \left(T_p - T_f \right)}{k_m \left(T_f - T_o \right)} \right]^2 \tag{10.3a}$$

$$B_c = \frac{\pi \alpha_m}{4} \left[\frac{\rho_s \Delta H_f}{k_m \left(T_f - T_o \right)} \right]^2 \tag{10.3b}$$

These constants strongly depend on the evolved ΔH_f.

Method A Consider a practical safe approach to avoid solidification shrinkage problems outside the riser-zone length shown in Fig. 10.2. This calls for a safety factor of $\lambda = 5/4 = 1.25$ which is kept fixed hereafter for practical purposes [2, p. 275]. Let

$$t_r = \lambda t_c = \lambda B_c \left(\frac{V_c}{A_c} \right)^2 \tag{10.4a}$$

$$B_r \left(\frac{V_m}{A_r} \right)^2 = \lambda B_c \left(\frac{V_c}{A_c} \right)^2 \tag{10.4b}$$

so that the riser volume becomes

$$V_m = V_c \left(\frac{A_r}{A_c} \right) \sqrt{\frac{\lambda B_c}{B_r}} \tag{10.5}$$

Inserting Eq. (10.3) into (10.5) gives

$$V_r = \sqrt{\lambda} V_c \left(\frac{A_r}{A_c}\right) \left[\frac{\Delta H_f}{\Delta H_f + c_p \left(T_p - T_f\right)}\right] = \sqrt{\lambda} V_c \left(\frac{A_r}{A_c}\right) \left(\frac{\Delta H_f}{\Delta H_s}\right) \qquad (10.6)$$

where ΔH_s is also defined by Eq. (5.13b) for pure metals and by (7.35b) for binary alloys. Thus,

$$\Delta H_s = \Delta H_f + c_l \left(T_p - T_f\right) \quad \text{(metals)} \qquad (10.7a)$$

$$\Delta H_s = H_{f,A} + C_o \left|H_{f,B} - H_{f,A}\right| + c_s \left(T_p - T_o\right) \quad \text{(alloys)} \qquad (10.7b)$$

Furthermore, the solidification shrinkage porosity parameter ω, the riser efficiency parameter ε and the partial amount of solidified material V_m in the feeder can be written as [3, pp. 317–319]

$$\omega = \frac{\rho_s - \rho_l}{\rho_s} \qquad (10.8a)$$

$$\varepsilon = \frac{V_r - V_m}{V_r} \qquad (10.8b)$$

$$V_m = V_r - \omega \left(V_c + V_r\right) \qquad (10.8c)$$

$$V_m = (1 - \varepsilon) V_r \qquad (10.8d)$$

Here, ρ_s, ρ_l denote densities of solid and liquid phases, respectively. Now, combining Eqs. (10.5) and (10.8c) yields the sought riser design volume in a meaningful mathematical form for a sand mold casting practice

$$V_r = \frac{V_c}{1 - \omega} \left[\omega + \left(\frac{A_r}{A_c}\right) \sqrt{\frac{\lambda B_c}{B_r}}\right] \quad \text{for } B_c/B_r \neq 1 \qquad (10.9a)$$

$$V_r = \frac{V_c}{1 - \omega} \left[\omega + \left(\frac{A_r}{A_c}\right) \sqrt{\lambda}\right] \quad \text{for } B_c/B_r = 1 \qquad (10.9b)$$

According to Fredriksson–Akerlind [3, p. 318], $B_c/B_r = 1$ in Eq. (10.9b) is a special case. Substituting Eq. (10.3) into (10.9a) gives

$$V_r = \frac{V_c}{1 - \omega} \left[\omega + \left(\frac{A_r}{A_c}\right) \sqrt{\frac{\lambda \Delta H_f}{\Delta H_f + c_p \left(T_p - T_f\right)}}\right] \qquad (10.10)$$

Method B The main goal now is to introduce an alternative method for determining the riser dimensions [2–5]. The fundamental aspect of this method is based on the assumption that the riser modulus is slightly greater than that of the casting. This seems reasonable since the riser should be the last component of the casting system to solidified in order to avoid casting defects, such as shrinkage porosity.

Again, following DeGarmo's idea found in Black–Kohser book [2, p. 275] that the riser takes 25% longer than the casting to solidify, the thought relationship is $t_r = 1.25t_c$. Accordingly, let the riser and casting moduli differ by a factor $\lambda = 5/4$ and manipulate the resultant equation to solve for the riser volume V_r. Thus,

$$M_r = \lambda M_c \tag{10.11a}$$

$$\frac{V_r}{A_r} = \lambda \frac{V_c}{A_c} \tag{10.11b}$$

$$V_r = \lambda V_c \left(\frac{A_r}{A_c}\right) \tag{10.11c}$$

From Eqs. (10.2b) and (10.11c), the riser solidification time becomes

$$t_r = B_r M_r^2 = B_r \left[\frac{\lambda V_c}{A_r}\left(\frac{A_r}{A_c}\right)\right]^2 = B_r \left(\frac{\lambda V_c}{A_r}\right)^2 \left(\frac{A_r}{A_c}\right)^2 \tag{10.12a}$$

$$t_r = \frac{\pi \alpha_m}{4} \left[\frac{\rho_s \Delta H_f + \rho_s c_p \left(T_p - T_f\right)}{k_m \left(T_f - T_o\right)}\right]^2 \left(\frac{\lambda V_c}{A_r}\right)^2 \left(\frac{A_r}{A_c}\right)^2 \tag{10.12b}$$

Additionally, Eqs. (10.9b) and (10.11c) in the form $V_r/V_c = f\left(A_r/A_c\right)$ yield linear trends as shown in Fig. 10.3a, while plotting Eq. (10.12b) as $t_r = f\left(A_r/A_c\right)$ gives a non-linear behavior (Fig. 10.3b).

The above analytical procedure is simple, but effective for designing proper risers to promote defect-free solidification. One can find different analytical approaches in the literature for such a purpose, but for practical and reliable purposes, consult the USA Navy Foundry and the American Foundry Society (AFS) recommended procedures.

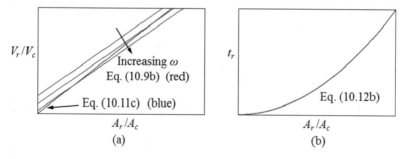

Fig. 10.3 (a) Volume ratio as a function or surface area ratio: $V_r/V_c = f\left(A_r/A_c\right)$ and (b) riser solidification time trend

Regardless of the foundry society one chooses to follow, the design of risers and the corresponding solidification time are two important aspects for producing the casting quality one desires. The goal is to produce defect-free casting components. For instance, gas and shrinkage porosity are defects caused by entrapment of gases and inappropriate riser design. However, hot tears and hot cracks are defects due to metal contraction that generates relatively a high triaxial state of stresses, which exceed the strength of the casting component while cooling to room temperature. Solidification defects are considered later.

Up to this point, the effect of filling velocity on the solidification process is ignored with respect to fluid laminar and turbulent flow and related heat transfer. Laminar flow is preferred in order to avoid mold erosion and associated casting problems. Hence, melt filling effect and riser solidification time must be controlled in casting defect-free components.

Example 10A.1 Consider the solidification of carbon steel. It is desired to cast a right-circular cylinder with a 10-cm diameter and 20-cm long in a green sand mold. Assume that the solidification shrinkage of the steel is 3.5%. Calculate the riser volume V_r with a riser efficiency of (a) $\varepsilon = 2\omega$ and (b) $\varepsilon = 4\omega$, and (c) plot $V_r = f(\varepsilon)$ profile for fixed $\omega = 3.5\%$.

Solution First of all, the volume of the casting is

$$V_c = \pi r^2 L = \pi \, (10 \text{ cm})^2 \, (20 \text{ cm}) = 6283.20 = 6283.20 \text{ cm}^3$$

Combining Eqs. (10.8c) and (10.8d) yields

$$V_r = \left(\frac{\omega}{\varepsilon - \omega} \right) V_c$$

(a) For $\varepsilon = 2\omega = 2 \, (3.5\%) = 7\%$,

$$V_r = \left(\frac{3.5}{7 - 3.5} \right) \left(6283.20 \text{ cm}^3 \right) = 6283.20 \text{ cm}^3$$

(b) For $\varepsilon = 4\omega = 4 \, (3.5\%) = 14\%$,

$$V_r = \left(\frac{3.5}{14 - 3.5} \right) \left(6283.20 \text{ cm}^3 \right) = 2094.40 \text{ cm}^3$$

Plotting the above equation yields a non-linear profile along with calculated V_r values

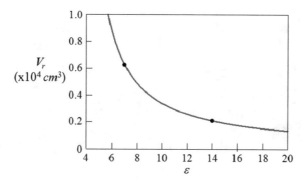

Therefore, the higher the riser efficiency the lower V_r. This leads to a reduction of casting weight and shrinkage porosity.

10.3 Morphology of Casting Defects

This section includes a few images of common casting defects and related techniques for detecting them. In effect, casting mechanical behavior depends on the final microstructure and associated defects.

Inappropriate casting design and pouring (filling) process convey to casting defects. For instance, Fig. 10.4a shows a machined bronze flange [5, p.74] exhibiting gas porosity due to the evolution of dissolved gases in the form of bubbles within the casting. This type of casting defect is attributed to the decrease in solubility in the solid metal. In contrast, Fig. 10.4b depicts shrinkage porosity and shrinkage pipe in a Cu-$10Sn$ bronze [5, p. 406] due to volume decrease, and density differences between liquid and solid during solidification. Nevertheless, these phenomena can occur simultaneously in a casting.

In addition, gas porosity is represented as an agglomeration of trapped bubbles (Fig. 10.4a) with smooth surfaces. On the other hand, Fig. 10.4b shows solidification shrinkage as a pipe and solidification porosity as a small elongated cavity. A proper riser design prevents the formation of shrinkage voids in the casting geometry upon solidification. Thus, a riser design is a dimensional-dependent thermal reservoir of liquid metal that provokes or simply induces shrinkage porosity outside the casting.

Figure 10.5 illustrates two radiographs showing solidification microporosity with irregular shape; Fig. 10.5a [5, p. 414] and b [5, p. 439].

This is evidence of shrinkage voids referred to as microporosity. This makes X-ray diffraction (XRD) a powerful nondestructive technique for characterizing the quality of castings, and determining the crystal structures of phases and related crystal size. In general, shrinkage porosity is an undesired casting defect since it affects mechanical properties, specifically fatigue behavior, or fatigue life of components subjected to either fluctuating or cyclic stresses in service.

Figure 10.6 shows hot tears in Al-$10Cu$ samples [6, 7]. The hot tear is revealed as an irregular linear defect that acts as a crack during contraction of the solidifying

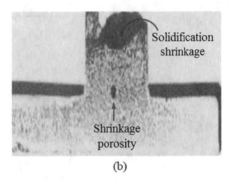

Fig. 10.4 (a) Gas porosity on top edge of a bronze casting flange after machining [5, p. 74] and (b) shrinkage solidification and shrinkage porosity in a bronze plate casting due to an inadequate riser (feeder) design [5, p. 406]

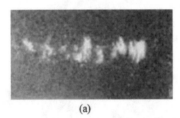
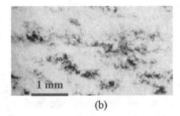

Fig. 10.5 (a) Radiography showing microporosity in a Ni-based alloy [5, p. 414] and (b) radiography showing interdendritic porosity in a carbon steel [5, p. 439]

liquid metal or alloy. Note that Fig. 10.6a exhibits a zigzag like hot tear and Fig. 10.6b depicts several hot tears at different orientations.

Unfortunately, pouring temperature and mold temperature are indirectly account-able for this type of casting defect. Consequently, the solidification hot tearing is defined henceforth as cracking process. Therefore, there must be a particular internal fracture stress σ_{ht} for evoking hot tearing [8].

Figure 10.7 compares two microscopy techniques for revealing casting defects, such as shrinkage porosity [9].

Figure 10.7a shows a typical microstructure of a hypoeutectic Al-Si alloy containing a solidification shrinkage pore. Figure 10.7b, on the other hand, depicts a scanning electron microscopic (SEM) image of the same alloy tested under fatigue loading. The presence of casting defects, such as pores and small cracks, in the final solidified products is especially detrimental to the material fatigue life (service life). Notice that the shrinkage pore is the fatigue initiation site and fatigue crack growth direction is indicated by the radial lines. Therefore, the production of high quality castings depends on the basic understanding of solidification since macroporosity is due to hot spots and microporosity is due to interdendritic shrinkage.

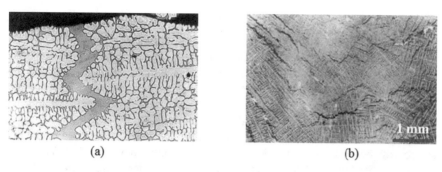

Fig. 10.6 (a) Optical and (b) radiography methods for revealing hot tear in Al-10Cu alloy [6, 7]

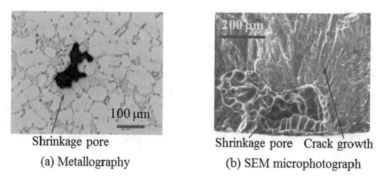

Fig. 10.7 (a) Metallographic specimen showing a solidification shrinkage pore and (b) a scanning electron microscopic (SEM) fatigue fractured surface exhibiting the solidification shrinkage pore as the crack initiation site and the fatigue crack growth direction. After ref. [9]

10.4   Mechanism of Porosity Formation

The source of gas porosity is mainly due to dissolved gases in the melt attributed to chemical reactions between the hot melt and the sand mold organic constituents. Chemical reactions with dissolved oxygen and nitrogen in the melt promote slag inclusions (metal oxides and metal nitrides), which may be trapped during the pouring process. Therefore, gas porosity and slag inclusions are significant solidification problems because of the degraded quality of the final casting product.

The mechanism of porosity formation during liquid-to-solid-phase transformation in casting and welding processes is attributed to low solubility of dissolved hydrogen, oxygen, nitrogen, and the like in the solidified melt at $T < T_f$ for pure metals and $T < T_f^*$ for alloys. The significant variable for gas porosity formation is the solubility of gases in the melt and it is much higher than that in the solid phases.

10.4.1   Bubble Absolute Pressure

The pressure (P) inside a hydrogen (H_2), nitrogen (N_2), or oxygen (O_2) bubble is controlled by the equilibrium condition between the liquid or solid phase and the gas phase. From general thermodynamics, this pressure can be derived assuming that a spherical bubble or pore is under axial surface tension forces around the circumference of the circle (equatorial circle) illustrated in Fig. 10.8a.

The gas is considered an incompressible fluid located at a depth h_c below the surface of the liquid or solid phase as schematically shown in Fig. 10.8b. The column shown in Fig. 10.8c is treated as an element under equilibrium forces and it is assumed to have uniform cross-sectional area A. Moreover, diffusion of gas within the melt attracts each other to form relatively small bubbles, which in turn, get trapped in the solid casting at a pressure P higher than the atmospheric pressure P_{atm}. This implies that pressure, temperature, and composition of the melt influence the solubility of a gas. Therefore, the excess gas in the molten metal forms the deleterious bubbles. Therefore, degassing provides a viable approach to significantly reduce the excess amount of gases via oxidation of certain metals.

According to Newton's second law of motion with zero acceleration implies that the net force acting on the column is conventionally represented as the free-body diagram (FBD) shown in Fig. 10.8c. Consequently, the sum all forces along the vertical direction (y-axis) takes the form

$$\sum F_y = AP_o - AP_{atm} - W - A\Delta P_b = 0 \tag{10.13a}$$

$$P_o = P_{atm} + \rho g h_c \tag{10.13b}$$

since the weight of the column (W) is defined by

$$W = mg = V\rho = Ah_c\rho g \tag{10.14}$$

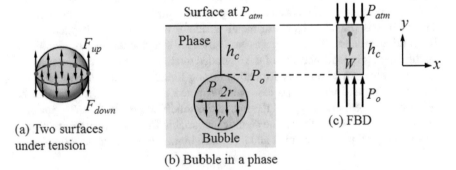

(a) Two surfaces under tension

(b) Bubble in a phase

(c) FBD

Fig. 10.8 (a) Bubble shown surface tension, (b) spherical bubble, and (c) free-body diagram (FBD)

where m denotes the mass of the column, $g = 9.81$ m/s^2 denotes the constant of acceleration, $P_{atm} = 1$ atm $= 101$ kPa denotes the atmospheric pressure, and ρ denotes the density of the liquid or solid phase.

Actually, for mechanical equilibrium, the weight (W) of the fluid balances the net force induced by the pressure differential. Thus, the relative magnitude and direction of all forces acting upon an element of the fluid are denoted as the free-body diagram (FBD) (Fig. 10.8c).

The net force $F_n = F_{up} - F_{down} = 0$ at equilibrium yields

$$F_{up} - F_{down} = 0 \tag{10.15a}$$

$$A_b (P - P_o) - C_b \gamma = 0 \tag{10.15b}$$

$$\left(\pi r^2 \right) (P - P_o) - (2\pi r) \gamma = 0 \tag{10.15c}$$

where $A_b = \pi r^2$ and $C_b = 2\pi r$ denote the area of the circle and the circumference of the bubble, respectively.

In addition, Eq. (10.15c) can be manipulated to yield the bubble internal (absolute) pressure written as

$$\Delta P_b = P - P_o = \frac{2\gamma}{r} \tag{10.16a}$$

$$P = P_o + \frac{2\gamma}{r} \tag{10.16b}$$

Here, ΔP_b denotes the difference in the pressure of the gas between the concave and convex side of the surface, r denotes the bubble radius of curvature (also known as the capillary radius), and γ denotes the surface or interfacial energy.

Moreover, the bubble surface remains at rest due to a balance between the internal and external pressures. Now, inserting Eq. (10.13b) into (10.16b) yields the bubble absolute pressure as

$$P = P_{atm} + \rho g h + \frac{2\gamma}{r} \tag{10.17}$$

where h denotes the depth of the bubble into the melt.

The bubble pressure P, as defined by Eq. (10.16b) or (10.17), is concerned with the mechanical equilibrium of the bubble surface, while the bubble is embedded in the solid phase and surrounded by the lattice of the solid on all sides. Therefore, finished casting geometries containing gas porosity is a major concern in the foundry engineering field.

Essentially, casting design problems can be related to inappropriate mold material, mold properties, gating and riser design systems, pouring temperature, and so forth. As a result, solid temperature gradients during the solidification

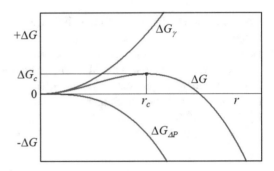

process associated with the heat transfer mechanisms can be a controlling factor
for obtaining near defect-free castings.

10.4.2   Work Done for Bubble Formation

Assume that random pockets of gas atoms undergo a homogeneous or heterogeneous nucleation process to form bubbles containing N_x number of atoms each. For
simplicity, the total work done $[\Delta G = -W_b$ as per Eq. (4.3b)] for the formation of
a spherical bubble homogeneously with radius r can be defined by [10, p. 178]

$$\Delta G = -\Delta G_{\Delta P} + \Delta G_\gamma = -\Delta P_b V_b + A_s \gamma \tag{10.18a}$$

$$\Delta G = -\left(\frac{4}{3}\pi r^3\right)\Delta P_b + \left(4\pi r^2\right)\gamma \tag{10.18b}$$

where $\Delta G_{\Delta P} = \Delta P_b V_b$ denotes the work due to a pressure change and $\Delta G_\gamma = A_s \gamma$ denotes the work for generating a surface tension at the liquid–gas interface.

In addition, plotting Eq. (10.18b) yields Fig. 10.9, which is similar to Fig. 4.4
for homogeneous solid nucleation during the solidification process. Thus, $\Delta G_c = \Delta G_{\max}$ constitute an energy barrier for further bubble growth.

Letting $[d\Delta G/dr]_{r=r_c} = 0$ or solving Eq. (10.16a) for $r = r_c$ yields the bubble
critical radius for homogeneous or heterogeneous nucleation

$$r_c = \frac{2\gamma}{\Delta P_b} \tag{10.19}$$

Substituting Eq. (10.19) into (10.18b) gives $\Delta G_c = f(r_c)$ which becomes the
critical Gibbs energy change or the bubble energy barrier at the critical bubble radius
r_c under the assumed homogeneous bubble nucleation

$$\Delta G_c = -\left(\frac{4}{3}\pi r_c^3\right)\Delta P_b + \left(4\pi r_c^2\right)\gamma = \frac{16\pi\gamma^3}{3\left(\Delta P_b\right)^2} \tag{10.20a}$$

$$\Delta G_c = \frac{16\pi \gamma^3}{3\,(\Delta P_b)^2} \tag{10.20b}$$

So far Eq. (10.20b) represents the driving force for bubble formation. This is a significant mathematical model in the casting defect analysis related to the detrimental or deleterious gas porosity.

Additionally, reliable castings are at the mercy of experienced foundry engineers, whom conduct numerical simulations prior to producing the casting geometries. Remarkably, nondestructive evaluation methods are used to assure the highest quality of castings at the expense of economy. Thus, imperfections, in general, are reduced to a great extent.

10.4.3 Hydrogen Gas

The main source of hydrogen content in conventional melts is the moisture or water in the surroundings, such as the furnace atmosphere. Amazingly, hydrogen-contaminated melts mainly arise from moisture in the air and incompletely dried furnace or ladle refractory. Obviously, the solubility of monoatomic hydrogen ($\underline{H}$) increases with increasing melt temperature. Therefore, the melt pouring temperature must not be too high in order to avoid easy hydrogen pick up (absorption) by the melt. However, once monoatomic hydrogen reaches its maximum solubility in the melt, the excess molecular hydrogen (H_2) has the tendency to form bubbles, specifically at the metal oxide–melt interface. Nonetheless, the solubility of ($\underline{H}$) in steels is strongly dependent on temperature, composition, and crystal structure, such as FCC austenite or BCC ferrite.

Inherently, released bubbles from the melt create pores on the casting surface and trapped bubbles generate solidification shrinkage porosity. For instance, the solubility of $\underline{H}$ in iron or steel depends on the temperature and it may be detrimental to components subjected to a mechanical loading due to a metallurgical phenomenon called *hydrogen embrittlement*. Accordingly, the Sievert's law implies that the amount of hydrogen in solution at a given temperature is proportional to the square root of the partial pressure of hydrogen.

In addition, degassing by injecting an inert gas into the melt removes the dissolved hydrogen and reduces the inherent hydrogen gas porosity. This is a practical technique to assure high quality castings. For a detailed assessment of gas porosity, one can consider the thermodynamics of most common $Fe\text{-}H$, $steel\text{-}H$, $Al\text{-}H$, and $Mg\text{-}H$ systems and related kinetics of hydrogen transport by diffusion.

Furthermore, diatomic hydrogen (H_2) in the steel melt can form bubbles during solidification and as a result, hydrogen porosity is builtup at oxide inclusion–steel interface. Moreover, molecular hydrogen (H_2) must be dissociated into the monoatomic form as per Eq. (10.21a), leading to the Sievert's law as defined by Eq. (10.22a).

10.4.4 Solubility of Gases in Metallic Solutions

The fundamental phenomena of gas bubble formation and subsequent bubble rise due to buoyancy can be assumed to contribute significantly to the hydrodynamics in gas–liquid melt. Logically, the initial size of bubbles is unstable and apparently, they grow until they reach an equilibrium size or simply reach a particular size and get trapped in the solidified melt.

In a general sense, the element forming the gas bubbles may have some solubility in the solid, but the excess gas randomly forms bubbles within a confined area. Some of these bubbles rise to the interface between the mold inner wall and the resultant solid. Obviously, they are trapped at this particular location within the solidified melt cavity.

Ideally, gas porosity is preventable in a vacuum environment, but certain situations are impractical due to lack of a cost-effective process or due to the casting application, where this type of imperfection is not accountable for casting quality. Additional details on porosity can be found in Campbell's book [10, chapter 6]. Moreover, the solubility of gases in molten metals or alloys (metallic liquids) involves dissociation of the gas molecules at a pressure P and temperature T. This process can be represented by the generalized chemical reaction of a diatomic gas "g_2" (typically, diatomic gases such as H_2, N_2, or O_2) into a dissolved gas "$\underline{X}$" ($\underline{H}$, $\underline{N}$, or $\underline{O}$)

$$X_2 \rightarrow 2\underline{X} \quad \text{at } P, T \tag{10.21a}$$

$$K_{sp} = \frac{[\underline{X}]^2}{[X_2]} \tag{10.21b}$$

for which the Sievert's law and the Arrhenius' equation are, respectively,

$$[\underline{X}] = \sqrt{K_{sp}\,[X_2]} = K_g\sqrt{[X_2]} = K_g\sqrt{\frac{P_g}{P_o}} \quad \text{(Sievert's law)} \tag{10.22a}$$

$$[\underline{X}] = A_g \exp\left(-\frac{\Delta G_c}{RT}\right) \quad \text{(Arrhenius equation)} \tag{10.22b}$$

Here, A_g denotes a constant known as the pre-exponent factor or frequency factor in this exponential decay law. It is convenient now to clarify that the melt pouring temperature is a few degrees above the freezing temperature and as a result, sufficient trapped air or moisture by the melt become the major source of gas bubbles, leading to gas porosity.

Example 10A.2 Assume that the solubility of atomic hydrogen, $[\underline{X}] = [H]$, in molten copper is $10^{-4}\%$ at $T_1 = 1083\,°C$ (1356 K) and $10^{-5}\%$ at $T_2 = 600\,°C$ (873 K). Calculate the concentration of the atomic hydrogen$[H]$ in percentage at $T_3 = 880\,°C$ (1153 K).

Solution From Eq. (10.22b) along with $R = 8.314\,\text{J/mol K}$,

$$\frac{[X]_{H,T_1}}{[X]_{H,T_2}} = \frac{10^{-4}\%}{10^{-5}\%} = 10 = \frac{A_g \exp[-\Delta G_c/RT_1]}{A_g \exp[-\Delta G_c/RT_2]} = \exp\left[\frac{\Delta G_c}{R}\left(\frac{1}{T_2} - \frac{1}{T_1}\right)\right]$$

$$\ln(10) = \frac{\Delta G_c}{R}\left(\frac{1}{T_2} - \frac{1}{T_1}\right) = \frac{\Delta G_c}{8.314}\left(\frac{1}{873} - \frac{1}{1356}\right)$$

$$\Delta G_c = 46{,}919\,\text{J/mol}$$

Then,

$$A_g = [\underline{g}]_{H,T_1} \exp[\Delta G_c/RT_1] = \left(10^{-4}\%\right)\exp\left[\frac{46{,}919}{(8.314)(1356)}\right]$$

$$A_g = 6.4184 \times 10^{-2}\,\%$$

Thus,

$$[\underline{g}]_{H,T_3} = A_g \exp[-\Delta G_c/RT_3]$$

$$[\underline{g}]_{H,T_3} = \left(6.4184 \times 10^{-2}\,\%\right)\exp\left(-\frac{46{,}919}{(8.314)(1153)}\right) = 4.81 \times 10^{-4}\,\%$$

Therefore, this is a small content of monoatomic hydrogen in solid solution (embedded) in the copper FCC lattice.

Example 10A.3 Following Fredriksson–Akerlind idea [3, p. 263] analytical procedure and Liu et al. [11] copper porous images, assume that dissolved monoatomic hydrogen lacks of equilibrium in the melt and consequently, it transforms to diatomic hydrogen as per the reversible chemical reaction $2H = H_2$, where $H = \underline{H}$ represents the dissolved monoatomic hydrogen in the melt. Initially, some molecules associate themselves to form spherical bubbles. Those dissolved hydrogen atoms having a non-equilibrium condition diffuse from the melt into the gas bubble and form hydrogen molecules. Hence, bubble growth at a pressure P_{gas}. This bubble formation and growth process stops near the melt surface as solidification ceases.

Undoubtedly, atomic hydrogen dissolved partially and molecular hydrogen forming bubbles are trapped during solidification of molten copper. These bubbles are revealed as pores on prepared metallographic samples. Moreover, a substantial amount of hydrogen bubbles are supposed to migrate to the melt surface and eventually to the local environment.

Below are given three images of lotus-type copper being cast as vertical right-circular cylinders (50-mm diameter and 170-mm height) from the given pouring temperature. These images, published by Liu et al. [11], represent the porous microstructure of the cylinders being cut 20 mm from the bottom. The mean average and number of pores of these images are denoted as "d" and "n_y," respectively, as given below.

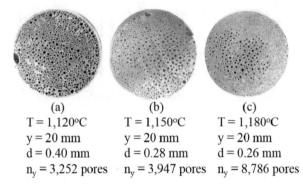

(a)	(b)	(c)
$T = 1{,}120°C$	$T = 1{,}150°C$	$T = 1{,}180°C$
$y = 20$ mm	$y = 20$ mm	$y = 20$ mm
$d = 0.40$ mm	$d = 0.28$ mm	$d = 0.26$ mm
$n_y = 3{,}252$ pores	$n_y = 3{,}947$ pores	$n_y = 8{,}786$ pores

Based on this information, use the non-equilibrium equation of state for gases and Fick's law of steady-state diffusion to derive an expression for the growth rate of spherical pores.

Solution The non-equilibrium equation of state for the number of molecular hydrogen along with Eq. (10.16b) takes the form

$$N_{H_2} = \frac{P_{gas} V_{sphere}}{R_{H_2} T} = \frac{P_{gas} V_s}{RT} \tag{E1}$$

$$N_{H_2} = \frac{(P_o + 2\gamma/r)\left(4\pi r^3/3\right)}{RT} = \frac{4\pi \left(P_o r^3 + 2\gamma r^2\right)}{3RT} \tag{E2}$$

$$\frac{dN_{H_2}}{dr} = \frac{4\pi \left(3 P_o r^2 + 4\gamma r\right)}{3RT} \tag{E3}$$

Notice that dN_{H_2}/dr resembles the hydrogen concentration gradient and depends on the bubble radius r and the temperature T.

For steady-state diffusion of monoatomic hydrogen, the diffusion flux is defined two mathematical forms. Thus,

$$J_H = \frac{dN_H}{A_{sphere} dt} = \frac{1}{A_s} \frac{dN_H}{dt} \tag{E4}$$

$$J_H = -D_H \frac{\partial C_H}{\partial r} = -D_H \frac{C - \overline{C}_H}{r} = -D_H \frac{\left(X_H - \overline{X}_H\right)}{r V_m} \tag{E5}$$

$$\frac{\partial C_H}{\partial r} = \frac{X_H - \overline{X}_H}{r V_m} \tag{E6}$$

where V_m is the molar volume. Equating Eqs. (E4) and (E5) yields

$$\frac{dN_H}{dt} = -\frac{A_s D_H \left(X_H - \overline{X}_H\right)}{V_m} \frac{1}{r} = \frac{4\pi r^2 D_H \left(\overline{X}_H - X_H\right)}{V_m} \frac{1}{r} \tag{E7}$$

$$\frac{dN_H}{dt} = \frac{4\pi r D_H \left(\overline{X}_H - X_H\right)}{V_m} \tag{E8}$$

Accordingly, the hydrogen concentration is $C_H = X_H/V_m$, where X_H denotes the mole fraction in the melt, $\overline{X}_H$ denotes an average mole fraction close to the pore surface and V_m denotes the molar volume so that C_H acquires the $mol/volume$ units.

Equating the number of hydrogen atoms is twice the number of hydrogen molecules, and using the Chain rule and Eq. (E3) and yields

$$\frac{dN_H}{dr} = 2\frac{dN_{H_2}}{dr} \tag{E9}$$

$$\frac{dN_H}{dt} = \frac{dN_H}{dr}\frac{dr}{dt} = 2\left(\frac{dN_{H_2}}{dr}\right)\frac{dr}{dt} \tag{E10}$$

Substituting Eq. (E3) into (E10) gives

$$\frac{dN_H}{dt} = \frac{8\pi \left(3P_o r^2 + 4\gamma r\right)}{3RT}\frac{dr}{dt} \tag{E11}$$

Combining Eqs. (E8) and (E11) yields the growth rate of spherical pores

$$\frac{dr}{dt} = \frac{3RT D_H \left(\overline{X}_H - X_H\right)}{2V_m \left(3P_o r + 4\gamma\right)} \tag{E12}$$

which is the theoretical or idealized growth rate of spherical pores. Moreover, the fraction of porosity is defined by [11]

$$P_r = 1 - \frac{m}{\rho V} \tag{E13}$$

Additionally, the rate of steady-state nucleation of pores per unit volume (I_p) can be described by Eq. (9.34a) or (9.34b). Thus, either expression is redefined or adopted here as

$$I_p = \frac{n_o k_B T}{h}\exp\left(-\frac{\Delta G_c}{k_B T}\right)\exp\left(-\frac{Q_d}{k_B T}\right) \tag{E14}$$

In summary, dr/dt and I_p are relevant variables that describe some characteristics of the pore formation mechanism. For instance, the former is inversely proportional to the pore radius ($dr/dt \propto 1/r$) and both are strongly dependent on the temperature. Moreover, the reader can use the data provided below the images to perform non-linear curve fitting.

10.5   Melt Heat Loss

10.5.1   Gas Concentration

During the pouring process of a particular melt, some moisture and oxygen from the local environment react with the hot liquid metal to form metal oxides called slag, and some hydrogen, oxygen, and nitrogen gases may be absorbed by the melt. Specifically, oxygen gas (O_2) in an steel melt causes slag formation and excess dissolved hydrogen $(\underline{H})$ or nitrogen $(\underline{N})$ is the source of porosity during solidification.

Fundamentally, hydrogen and nitrogen dissolve in steel up to their solubility limits during solidification. However, the excess gas being rejected causes casting porosity as already mentioned previously. Nitrogen, for example, may also form hard nitrides. Therefore, hydrogen and nitrogen must be removed using a technique called degassing. Oxygen, on the other hand, is removed by high-temperature reactions to form floating slag.

Adding the analysis of gas concentration to the solidification characterization, one can predict the concentration of a dissolved gas (g), such as hydrogen $(\underline{H})$, in the melt. Assuming a steady-state diffusion of a dissolved gas to prefer areas within the melt, the one-dimensional analysis of the Fick's first law of diffusion is written as an ordinary differential equation of the form [3, p. 260]

$$\frac{dm_g}{dt} = -D_g A_x \rho \frac{dC_g}{dy} = -D_g A_x \rho \left(\frac{C_g - C_{eq}}{\delta} \right) \qquad (10.23)$$

where m_g denotes the mass of the diffused hydrogen (kg), A_x denotes the effective area of the melt stream during pouring (m^2), D_g denotes the diffusion coefficient of the dissolved gas in the melt (m^2/s), $C_g = \rho W_g$ denotes the dissolved gas concentration, W_g denotes the weight fraction of the dissolved gas, ρ denotes the melt density, dC_g denotes the specific gas concentration (kg/m^3), dC_g/dy denotes the specific concentration gradient, C_{eq} denotes the solubility limit of hydrogen in the melt at a given temperature, and δ denotes the thickness of the boundary layer.

A convenient definition of the differential mass is $dm_g/dt = M_g dC_g/dt$, where M_g is the molecular weight. Then, Eq. (10.23) becomes

$$\frac{dC_g}{dt} = -\frac{D_g A_x \rho}{M_g} \left(\frac{C_g - C_{eq}}{\delta} \right) \qquad (10.24)$$

from which

$$\frac{dC_g}{C_g - C_{eq}} = -\frac{D_g A_x}{M_g \delta} dt \qquad (10.25)$$

Integrating Eq. (10.25) from an initial hydrogen concentration C_i to some value C_h at time $t > 0$ yields

$$\int_{C_i}^{C_g} \frac{dC_g}{C_g - C_{eq}} = -\frac{D_g A_x \rho}{M_g \delta} \int_0^t dt \qquad (10.26a)$$

$$\left[\ln\left(C_g - C_{eq}\right)\right]_{C_i}^{C_g} = -\frac{D_g A_x \rho}{M_g \delta} t \qquad (10.26b)$$

Evaluating the left-hand side of Eq. (10.26b) gives

$$\ln\left(C_g - C_{eq}\right) - \ln\left(C_i - C_{eq}\right) = -\frac{D_g A_x \rho}{M_g \delta} t \qquad (10.27)$$

Rearranging Eq. (10.27) yields

$$\ln\left(\frac{C_g - C_{eq}}{C_i - C_{eq}}\right) = -\frac{D_g A_x \rho}{M_g \delta} t \qquad (10.28)$$

Manipulating Eq. (10.28) provides the sought expression for C_g. Accordingly, the gas concentration becomes

$$C_g = C_{eq} + \left(C_i - C_{eq}\right) \exp\left(-\frac{D_g A_x \rho}{M_g \delta} t\right) \qquad (10.29)$$

When the concentration is very small, the common unit is "ppm = parts per million," but the most common unit is weight percent ($\%wt$ or $\%$). In this case, the conversion of units is 1 ppm = 10^{-4} wt% and the weight refers to mass percentage.

10.5.2 Coupled Convection and Radiation

The purpose of this section is to include a general quantitative assessment of the heat loss during typical large casting production, such an ingot. Recall that casting is introduced in Chap. 4, specifically see the relevant pictures in Fig. 4.4a and in Example 4.1.

In physical and mathematical modeling studies on solidification, the amount of heat loss due to natural convection and radiation on to a falling melt jet stream are conventionally given by

$$Q_{conv} = c_p \rho_s V (T - T_0) = c_p m (T - T_i) \qquad (10.30a)$$

$$Q_{rad} = A_x q_{rad} = A_x \epsilon \sigma \left(T^4 - T_o^4 \right) t \tag{10.30b}$$

where T is the melt temperature leaving the ladle, T_0 is the mold temperature, T_i is the melt temperature entering the mold, V is the volume of the melt, m is the total mass of the melt in the ladle, $\sigma = 5.67 \times 10^{-8}$ W/(K^4 m^2) is the Stefan–Boltzmann constant and t is the pouring time to fill the mold using the molten jet stream.

In order to simplify the mathematical modeling methodology, assume that $Q_{conv} = Q_{rad}$ so that the surface area of the jet stream becomes

$$A_x = \frac{c_p m \left(T - T_i \right)}{\epsilon \sigma \left(T^4 - T_o^4 \right) t} \tag{10.31}$$

Substituting Eq. (10.31) into (10.29) eliminates the time t and the resultant expression is

$$C_g = C_{eq} + \left(C_i - C_{eq} \right) \exp \left(-\frac{D_g A_x \rho_s}{\delta} \frac{c_p \left(T - T_0 \right)}{\epsilon \sigma \left(T^4 - T_i^4 \right)} \right) \tag{10.32}$$

which is the gas concentration due to the coupled heat loss methodology.

Example 10A.4   Assume that a 10-ton carbon steel melt at 1520 °C initially contains 4 ppm of hydrogen and subsequent hydrogen from local moisture in the air diffuses through a boundary layer of 10-μm thick, and that it takes 120 s to fill a large mold cavity with the melt at 1500 °C. Also assume that the heat loss is due to convection and radiation. From Fredriksson and Akerlind book [3, p. 260], $c_p = 0.76$ kJ/(kg K); $C_{eq} = 20$ ppm; $\rho_{Fe} = 7.80 \times 10^3$ kg/m^3. Let the hydrogen diffusion coefficient be $D_H = 10^{-8}$ m^2/s and $Q_{conv} = Q_{rad}$. Calculate (a) the total amount of hydrogen (C_H) in the liquid steel during the pouring process for filling a mold cavity, and (b) Q_{conv} and Q_{rad} in units of kJ. (c) Repeat the calculations if the steel melt is initially at 1510 °C. Is there any significant difference in C_H? Explain.

Solution

(a) **Steel melt at** 1520 °C = 1793 K: From Eq. (10.31) with $\sigma = 5.67 \times 10^{-8}$ (J/s) K^4/m^2, the surface area of the jet stream is

$$A_x = \frac{c_p m \left(T - T_0 \right)}{\epsilon \sigma \left(T^4 - T_i^4 \right) t} = \frac{\left(0.76 \times 10^3 \right) \left(10 \times 10^3 \right) (1793 - 1773)}{(0.6) \left(5.67 \times 10^{-8} \right) \left[(1793)^4 - (293)^4 \right] (120)}$$

$$A_x = 3.60510 \text{ m}^2$$

From Eq. (10.29) with $D_g = D_H$, $C_g = C_H$ and $\rho = \rho_{Fe}$,

$$\frac{D_g A_x \rho t}{m \delta} = \frac{\left(10^{-8} \text{ m}^2/\text{s} \right) \left(3.60510 \text{ m}^2 \right) \left(7.80 \times 10^3 \text{ kg/m}^3 \right) (120 \text{ s})}{\left(10 \times 10^3 \text{ kg} \right) \left(10 \times 10^{-6} \text{ m} \right)}$$

$$= 0.33744$$

$$C_H = C_{eq} + (C_i - C_{eq}) \exp \left(-\frac{D_g A_x \rho}{m\delta} t \right)$$

$$C_H = 20 + (4 - 20) \exp(-0.33744) \simeq 8.58 \text{ ppm} = 8.58 \times 10^{-4} \text{ wt\%}$$

(b) Subsequently, the heat loss terms are

$$Q_{conv} = c_p m (T - T_i)$$

$$= \left(0.76 \times 10^3 \frac{\text{kJ}}{\text{kg K}} \right) \left(10 \times 10^3 \text{ kg} \right) (1793 - 1773) \ K$$

$$= 152 \times 10^6 \text{ kJ}$$

$$Q_{rad} = A_x \epsilon \sigma \left(T^4 - T_o^4 \right) t$$

$$= (3.60510)(0.6) \left(5.67 \times 10^{-8} \right) \left[(1793)^4 - (293)^4 \right] (120)$$

$$= 152 \times 10^6 \text{ kJ}$$

Therefore, $C_H = 2.145 C_i$ and $Q_{conv} = Q_{rad}$ is verified.

(c) **Steel melt at $T = 1510\,°C = 1783$ K**: From Eq. (10.31),

$$A_x = \frac{c_p m (T - T_0)}{\epsilon \sigma (T^4 - T_i^4) t} = \frac{(0.76 \times 10^3)(10 \times 10^3)(1783 - 1773)}{(0.6)(5.67 \times 10^{-8}) \left[(1783)^4 - (293)^4 \right] (120)}$$

$$A_x = 1.84340 \text{ m}^2$$

From Eq. (10.29) with $D_g = D_H$, $C_g = C_H$ and $\rho = \rho_{Fe}$,

$$\frac{D_g A_x \rho t}{m\delta} = \frac{(10^{-8} \text{ m}^2/\text{s})(1.84340 \text{ m}^2)(7.80 \times 10^3 \text{ kg/m}^3)(120 \text{ s})}{(10 \times 10^3 \text{ kg})(10 \times 10^{-6} \text{ m})}$$

$$= 0.17254$$

$$C_H = C_{eq} + (C_i - C_{eq}) \exp \left(-\frac{D_g A_x \rho}{m\delta} t \right)$$

$$C_H = 20 + (4 - 20) \exp(-0.17254) = 6.54 \text{ ppm} = 6.54 \times 10^{-4} \text{ wt\%}$$

Therefore, $C_H = 1.635 C_i$ during the pouring process for filling the mold.

(d) Heat loss due to convective and radioactive heat transfer:

$$Q_{conv} = c_p m (T - T_i)$$

$$= \left(0.76 \times 10^3 \frac{\text{kJ}}{\text{kg K}}\right) \left(10 \times 10^3 \text{ kg}\right) (1783 - 1773) \text{ K}$$

$$= 76 \times 10^6 \text{ kJ}$$

$$Q_{rad} = A_x \epsilon \sigma \left(T^4 - T_o^4\right) t$$

$$= (1.84340)(0.6) \left(5.67 \times 10^{-8}\right) \left[(1783)^4 - (293)^4\right] (120)$$

$$= 76 \times 10^6 \text{ kJ}$$

Again, $Q_{conv} = Q_{rad}$ is verified. A $10\,°C$ difference between the steel melt initial temperatures makes $C_H^{(a)} > C_H^{(c)}$. Therefore, the effect of this temperature difference in the melt initial conditions can be determined through the percentage error based on an average hydrogen concentration. Thus,

$$Error = \frac{C_H^{(a)} - C_H^{(c)}}{\left(C_H^{(a)} + C_H^{(c)}\right)/2} = \frac{8.58 - 6.54}{(8.58 + 6.54)/2} \times 100 \simeq 27\%$$

For $C_H = f(T)$ in the closed interval $[1510, 1525]\,°C$ using increments of $5\,°C$ yields the following analytical and curve fitting equations, and the corresponding plot

$$C_H = 20 + (4 - 20) \exp\left(-\frac{7.1136 \times 10^5 T - 1.2613 \times 10^9}{4.0824 \times 10^{-6} T^4 - 30087}\right) \qquad \text{(analytical)}$$

$$C_H = -7072.70 + 7.7147 T - 0.0021 T^2 \qquad \text{(fitting)}$$

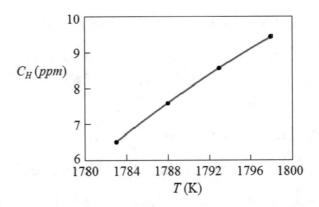

10.5.3 Heat Flow Through Composite Cylindrical Walls

Consider a cylindrical ladle or crucible (open vessel) composed of a steel hollow cylinder with an inner cylindrical lining made out of refractory materials. Principally, ladles are built according to industry specifications for holding and pouring requirements as per type of casting materials and quantity to be produced.

Following Gaskell's book [12, chapter 6], the ladle loaded with a steel melt (Fig. 10.10a) can be modeled as cylindrical composite walls or as multilayered composite cylinders. Figure 10.10b shows the top view of the cylindrical composite walls and Fig. 10.10c depicts the corresponding temperature profile through the composite thickness along with the electric circuit. Notice that this circuit includes the resistances R_i with $i = 1, 2, 3, 4$ connected in series [12, p. 253] and [13, p. 233].

Notice that the heat flux q_r (Fig. 10.10c) flows in the radial direction through the solid refractory lining and through the solid steel shell. Physically, q_r simply represents the rate of heat transfer per unit surface area that flows through the composite cylindrical walls as heat loss or extracted from the melt.

Moreover, the multilayered composite-vessel is just a fancy name for a big bucket lined with layers of a high-temperature resistant ceramic material. Large ladles are emptied (teemed) through hole in the bottom as shown in Fig. 10.10a. Thus, teeming melts from large ladles equipped with a bottom hole is commonly used in foundry practices.

The standard thermal-electrical analogy with Ohm's law provides some useful insights of the internal thermal resistance (R_i with $i = 1, 2, 3, 4$) to heat flow from the solidification system. Thus,

$$\sum R_i = \frac{T_1 - T_4}{q_r} = \frac{\text{Driving force}}{\text{Rate of heat loss}} \tag{10.33a}$$

$$R_1 + R_2 + R_3 + R_4 = \frac{T_1 - T_4}{q_r} \tag{10.33b}$$

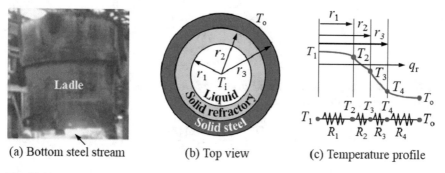

(a) Bottom steel stream   (b) Top view   (c) Temperature profile

Fig. 10.10 Multilayered cylindrical vessel. (**a**) Real ladle, (**b**) schematic top view of a composite cylindrical ladle, and (**c**) temperature profile and heat flow through the composite walls

$$q_r = \frac{\Delta T_i}{R_i} \tag{10.33c}$$

where $T_1 = T_i$ denotes the initial temperature of the liquid (melt) in the ladle, T_2 and T_3 denote intermediate temperatures, T_4 denote the outer surface temperature, and T_o denotes the air temperature.

Furthermore, the thermal resistances in Eq. (10.33b) due to heat transfer with convection for a hollow cylinder of length L are defined as

$$R_1 = \frac{1}{Ah_1} = \frac{1}{2\pi r_1 L h_1} \quad \text{(melt)} \tag{10.34a}$$

$$R_4 = \frac{1}{Ah_o} = \frac{1}{2\pi r_3 L h_4} \quad \text{(air)} \tag{10.34b}$$

Here, h_1, h_4 denote the heat transfer coefficients $[W/(m^2\,K)]$ of the melt and the surrounding air, respectively, and A denotes the effective surface area being exposed to convective heat transfer.

Essentially, heat flows from the liquid (inside) through the composite walls to the environment (outside). This implies that the thicker the walls (insulators), the lower the heat transfer rate. During this process, the heat encounters thermal resistances to convection heat transfer in the liquid and air, and to conduction in the solid phases. Therefore, there is a *coupled convection–conduction* mechanism taking place prior to pouring the melt into the mold cavity at the corresponding pouring or casting temperature T_p. Moreover, heat transfer in a steelmaking process, including the ladle refractory walls and mold material, must be evaluated thoroughly in order to optimize the heat loss during solidification.

For the rate of heat loss ($q_2 = dQ_2/dt$) due to radial heat transfer by conduction through the solid refractory material with k_2 as the thermal conductivity $[W/(m\,K]$,

$$q_2 = -A_r k_2 \frac{dT}{dr} = -2\pi r L k_2 \frac{dT}{dr} \tag{10.35a}$$

$$k_2 \int_{T_1}^{T_2} dT = -q_2 \int_{r_1}^{r_2} \frac{dr}{2\pi r L} \tag{10.35b}$$

$$q_2 = \frac{2\pi L k_2 (T_1 - T_2)}{\ln (r_2/r_1)} \tag{10.35c}$$

It is now convenient to rearrange Eq. (10.35c) and define the conduction thermal resistance R_2 as indicated below

$$R_2 = \frac{(T_1 - T_2)}{q_2} = \frac{\ln (r_2/r_1)}{2\pi L k_2} \tag{10.36}$$

Similarly, for the steel cylinder

$$q_3 = \frac{dQ_3}{dt} = \frac{2\pi L k_3 \, (T_2 - T_3)}{\ln (r_3/r_2)} \tag{10.37a}$$

$$R_3 = \frac{(T_2 - T_3)}{q_3} = \frac{\ln (r_3/r_2)}{2\pi L k_3} \tag{10.37b}$$

Substituting $R's$ into Eq. (10.33b) gives

$$R_x = \frac{T_1 - T_4}{q_r} = \frac{1}{2\pi r_1 L h_1} + \frac{\ln (r_2/r_1)}{2\pi L k_2} + \frac{\ln (r_3/r_2)}{2\pi L k_3} + \frac{1}{2\pi r_3 L h_4} \tag{10.38}$$

where R_x is the total thermal resistance of the solidification domain Γ, including the melt and ladle composite material.

Rearranging Eq. (10.38) yields the rate of heat loss through the multilayers shown in Fig. 10.10b

$$q_r = \frac{2\pi L \, (T_1 - T_4)}{1/(r_1 h_1) + \ln (r_2/r_1) /k_2 + \ln (r_3/r_2) /k_3 + 1/(r_3 h_4)} \tag{10.39}$$

This is the combined convective and radioactive heat transfer through the multilayered composite-wall ladle.

Pouring Time It is also essential to predict the total or partial pouring time to empty the ladle. From Example 4.1, the maximum stream velocity at the bottom nozzle is

$$v_2 = \frac{dL}{dt} = \sqrt{2gL} \tag{10.40}$$

and the time for emptying the ladle is

$$t = \left(\frac{d_1}{d_2}\right)^2 \frac{2}{\sqrt{2g}} \left(\sqrt{L_1} - \sqrt{L_2}\right) \tag{10.41}$$

where d_1, d_2 denote the ladle inner and the exit nozzle diameters, respectively. Also, L_1 denotes the initial height of the melt in the ladle and L_2 denotes the changing melt height at time $t > 0$.

Example 10A.5 A composite cylindrical refractory-steel ladle (Fig. 10.10) with dimensions $L = 3$ m, $r_1 = 1.40$ m, $r_2 = 1.45$ m, and $r_3 = 1.50$ m is used to cast 5 tons of steel at 1520 °C into sand molds at 25 °C. The thermal conductivity of the refractory material and the steel hollow cylinder are $k_2 = 0.04$ W/m K and $k_3 = 40$ W/m K, respectively, and the heat transfer coefficients of the melt and air are $h_1 = 145$ W/m K and $h_4 = 30$ W/m K, respectively. Calculate the corresponding temperature.

Solution From Eqs. (10.34), (10.36), and (10.37b), the thermal resistances are

$$R_1 = \frac{1}{2\pi r_1 L h_1} = \frac{1}{2\pi \, (1.40 \text{ m}) \, (3 \text{ m}) \, \left(145 \text{ W/m}^2 \text{ K}\right)} = 2.6134 \times 10^{-4} \text{ K/W}$$

$$R_2 = \frac{\ln \, (r_2/r_1)}{2\pi \, L k_2} = \frac{\ln \, (1.45/1.40)}{2\pi \, (3 \text{ m}) \, (0.04 \text{ W/m K})} = 4.6541 \times 10^{-2} \text{ K/W}$$

$$R_3 = \frac{\ln \, (r_3/r_2)}{2\pi \, L k_3} = \frac{\ln \, (1.50/1.45)}{2\pi \, (3 \text{ m}) \, (40 \text{ W/m K})} = 4.4963 \times 10^{-5} \text{ K/W}$$

$$R_4 = \frac{1}{2\pi r_3 L h_4} = \frac{1}{2\pi \, (1.50 \text{ m}) \, (3 \text{ m}) \, \left(30 \text{ W/m}^2 \text{ K}\right)} = 1.1789 \times 10^{-3} \text{ K/W}$$

Therefore, the refractory lining exhibits a significant thermal resistance since $R_2 > R_3$ and its thickness can be defined as $B = r_2 - r_1$.

From Eq. (10.33b), the total thermal resistance is

$$\frac{T_1 - T_4}{q_r} = R_1 + R_2 + R_3 + R_4 = R_x$$

$$R_x = 4.8026 \times 10^{-2} \text{ K/W (total)}$$

Then, the heat loss under the assumption of steady heat flow becomes

$$q_r = \frac{(1520 + 273) \text{ K} - (25 + 273) \text{ K}}{4.8026 \times 10^{-2} \text{ K/W}} = 31,129 \text{ W} = 31,129 \text{ J/s}$$

Using this quantity of heat loss the melt–refractory interface temperature is

$$\frac{T_1 - T_2}{q_r} = R_1$$

$$T_2 = T_1 - q_r R_1 = (1520 + 273) \text{ K} - (31,129 \text{ W}) \left(2.6134 \times 10^{-4} \text{ K/W}\right)$$

$$T_2 = 1784.90 \text{ K} = 1511.90 \,^{\circ}\text{C}$$

Similarly, the refractory–steel interface temperature based on the calculated heat loss is

$$\frac{T_2 - T_3}{q_r} = R_2$$

$$T_3 = T_2 - q_r R_2 = 1784.90 \text{ K} - (31,129 \text{ W}) \left(4.6541 \times 10^{-2} \text{ K/W}\right)$$

$$T_3 = 336.13 \text{ K} = 63.13 \,^{\circ}\text{C}$$

Finally, the temperature at the steel surface–air interface becomes

$$\frac{T_3 - T_4}{q_r} = R_3$$

$$T_4 = T_3 - q_r R_3 = 336.13 \text{ K} - (31{,}129 \text{ W}) \left(4.4963 \times 10^{-5} \text{ K/W}\right)$$

$$T_4 = 334.73 \text{ K} = 61.73\,^{\circ}\text{C}$$

Despite that the above calculations are based on a simple heat transfer methodology based on a thermal-electrical analogy, this example shows a simplified assessment of transfer of heat in a ladle thermal system, such as a composite cylindrical refractory-steel ladle. Recall that convective heat transfer involves the flow of heat in a liquid towards the ladle walls and air if the ladle is open, and conductive heat transfer through the refractory walls. Obviously, radioactive heat transfer is not included in this thermal-electrical analogy.

It is sufficiently clear now that the thermal-electrical analogy is fundamentally important in estimating the heat loss, which is directly linked to the desired thermal conditions within the loaded ladle containing a certain amount of melt.

Example 10A.6 Consider a small amount of the 5 metric tons of steel described in Example 10A.4 to cast a flat plate having design dimensional specifications ($h = 0.02$-m thick, $w = 0.08$-m wide, and $L = 0.15$-m long). Following Poirier–Geiger [13, p. 189], the plate is placed in a furnace at 1200 K for 30 min and the temperature is measured using appropriate thermocouples. In order to be in accordance with the polynomial introduced in Example 10A.4 as the resultant curve fitting equation, the temperature and gas concentration profiles can be determined using second-degree polynomials. Thus,

$$T(x) = a_1 + a_2 x + a_3 x^2$$

$$C(T) = b_1 + b_2 x + b_3 x^2$$

where x is the distance from the plate centerline, $T(x)_s = 1200$ K is the surface temperature, $T(0) = T_o = 400$ K is the centerline temperature, and b_i are the coefficients of $C(T)$ already given in Example 10A.4. In this example, only the polynomial, $T(x)$, is analyzed. Determine (a) the polynomial coefficients that contribute to the temperature trend and (b) an expression for the heat flux q_x in the x-direction.

Solution

(a) Polynomial coefficients: Initial condition

$$\left[\frac{\partial T}{\partial x}\right]_{x=0} = a_2 + 2a_3 x = 0$$

$$a_2 = 0$$

Then, the polynomial reduces to a power equation of the form

$$T(x) = a_1 + a_3 x^2$$

Boundary conditions:

$$T(0) = T_o = a_1 + a_3 (0)^2 = a_1$$
$$a_1 = T_o = 400 \text{ K}$$

Half-space across the thickness ($\delta = h/2 = 0.01$ m) due to symmetry:

$$T(\delta) = T_s = T_o + a_3 \left(\frac{\delta}{2}\right)^2$$

$$a_3 = \frac{4(T_s - T_o)}{\delta^2} = \frac{4(1200 - 400) \text{ K}}{(0.01 \text{ m})^2} = 3.20 \times 10^7 \text{ K/m}^2$$

Therefore, the only non-contributing coefficient is $a_2 = 0$.

(b) Temperature gradient:

$$\frac{\partial T}{\partial x} = 2a_3 x = \frac{8(T_s - T_o)x}{\delta^2} = \frac{8(1200 - 400) \text{ K}}{(0.01 \text{ m})^2} x$$

$$\frac{\partial T}{\partial x} = \left(6.40 \times 10^7 \text{ K/m}^2\right) x$$

Then, Fourier's law of conduction (with $k_3 = 40$ W/m K from Example 10A.5) in the half-space region yields the heat flux equation

$$q_x = -k_3 \frac{\partial T}{\partial x} = -\frac{8k_3(T_s - T_o)}{\delta^2} x$$

$$q_x = -\frac{8(40 \text{ W/m K})(1200 - 400) \text{ K}}{(0.01 \text{ m})^2} x$$

$$q_x = \left(-2.56 \times 10^9 \text{ W/m}^3\right) x = \left(-2560 \text{ MW/m}^3\right) x \quad \text{for } 0 < x \le \delta$$

which gives a linear trend with a negative slope for $0 < x \le \delta$. This result implies that the heat flux function is $q_x = f(x) < 0$ since it flows towards the plate surface. Moreover, the parameters a_i, where $i = 1, 2, 3$ in the polynomial $T(x)$, have physical definition since they have units, such that a_1 is in Kelvin degree (K), a_2 is in K/m, and a_3 is in K/m^2. However, $a_2 = 0$ is the only non-contributing parameter to melt temperature.

10.6 Melt Cleanliness

10.6.1 Degassing the Melt

This section deals with the metallurgical process known as degassing (gas flushing). Fundamentally, vacuum-degassing is used to remove dissolved gases (hydrogen, nitrogen, oxygen, and moisture) from their non-equilibrium state in the melt. Degassing can also be accomplished by injecting an inert gas through a lance (steel pipe) at a pressure $P > P_{atm} = 1$ atm $= 101$ kPa into the melt.

Ideally, degassing is the ultimate process to obtain a gas-free melt and to avoid the formation of gas porosity. Nonetheless, the melt cleanliness is a major issue in the casting industry due to the demand for high quality castings with design-property specifications.

10.6.2 Adjusting the Chemical Composition of the Melt

In order to adjust the chemical composition of the melt due to an excess content of a particular element, such as carbon (C) in the steel melt or moisture in the Al-alloy melt, injection of oxygen into the melt causes chemical reactions to occur at high temperatures. Thus, the ladle degassing is a practical technique for this purposes.

In steelmaking, excess carbon in the melt can be removed according to the chemical reaction

$$\underline{C} + \frac{1}{2}O_2 \rightarrow CO \quad \text{(gas)} \tag{10.42}$$

Thus, CO gas floats to the melt surface covered with an inevitable layer of slag (complex oxides), which must be removed manually or mechanically prior to filling the mold cavity.

On the other hand, an excess manganese Mn in a molten Mn-steel alloy and water from moist air in molten aluminum (Al) can be removed according to the following chemical reactions, respectively,

$$Mn + FeO \rightarrow MnO + Fe \tag{10.43a}$$

$$2Al + 3H_2O \rightarrow Al_3O_2 + 6H \tag{10.43b}$$

where magnesia MnO and alumina Al_3O_2 become part of the slag due to their low mass density. Moreover, Eqs. (10.42) and (10.43) represent two simple examples on how to adjust the chemical composition of a steel melt. These reactions, as written, occur if the standard Gibbs energy change is $\Delta G^o < 0$ at temperature T.

The slag formation is due to high-temperature reactions between molten metals, dissolved oxygen, and refractory materials. Despite that the refractory lining is used to control rapid heat loss from the melt into the surrounding environment, it may react with the metals in the melt. Therefore, one has to deal with molten metal-slag-refractory reactions at high temperatures prior to filling mold cavities with the ideal melt cleanliness. In other words, slag is an immiscible solution of oxides and sulfides in the molten state and it acts as a sink for impurities. Once a slag is floating on the melt, additional oxygen and hydrogen are prevented from dissolving in the melt and it also acts as thermal barrier to prevent heat transfer from the molten metal to the surrounding.

10.7   Thermodynamics of Metal Oxides

In the metalmaking industry, the chemistry and cleanliness of the melt must meet casting specifications so that any impurity is removed through oxidation reactions prior to pouring the melt into molds. Thus, the Gibbs energy change ΔG for a reaction is clearly defined.

10.7.1   Solubility Product

From the second law of thermodynamics, the change in free energy can be written as

$$dG = -SdT + VdP \tag{10.44}$$

For an isothermal (fixed $T =$) and isometric (fixed $V =$) system, Eq. (10.43) yields

$$\int_{\Delta G_j^o}^{\Delta G_j} dG = V \int_{P_o}^{P} dP \tag{10.45a}$$

$$\Delta G_j - \Delta G^o = V (P - P_o) \tag{10.45b}$$

Dividing this equation by RT and using the natural exponential function on the resultant expression gives

$$\exp \left[\frac{\Delta G - \Delta G^o}{RT} \right] = \exp \left[\frac{V (P - P_o)}{RT} \right] \tag{10.46}$$

where V denotes the substance volume, P denotes the actual pressure, P_o denotes the standard pressure (1 atm), $R = 8.314510$ J/(mol K) denotes the universal gas constant, and T denotes the absolute temperature (K).

Consider the hypothetical metal oxidation reaction at a relatively high temperature T corresponding to the formation of a metal oxide

$$xM + \frac{y}{2}O \rightarrow M_xO_y \tag{10.47}$$

In fact, M_xO_y represents a hypothetical ceramic or refractory compound (metal oxide). The solubility product K_{sp} for this reaction, Eq. (10.47), is defined in terms of the activity of each reaction component. Thus,

$$K_{sp} = \frac{\sum a_p}{\sum a_r} = \frac{\sum [\text{Product}]}{\sum [\text{Reactants}]} = \frac{[M_xO_y]}{[M]^x [O]^{y/2}} \tag{10.48}$$

Here, $a_p = [\text{product}]$ and $a_r = [\text{reactants}]$ denote the activities of the reaction components. Accordingly, $a_M = [M]$ for metal M, $a_O = [O]$ for oxygen and $a_{M_xO_y} = M_xO_y$ for metal oxide. By definition, the activity of a pure metal is defined as $[M] = 1$ and that for oxygen is $[O] = P_O/P_o$, where P denotes the reaction pressure and $P_o = 1$ atm denotes the standard pressure. Moreover, the activity of a gas is defined as $[gas] = P_{gas}/P_o$.

From Eqs. (10.46), the solubility product can be written as

$$K_{sp} = \exp\left(\frac{\Delta G - \Delta G^o}{RT}\right) \tag{10.49}$$

Solving Eq. (10.49) for the non-equilibrium Gibbs energy change ΔG gives

$$\Delta G = \Delta G^o + RT \ln\left(K_{sp}\right) \tag{10.50}$$

At standard conditions, $\Delta G = 0$ and Eq. (10.50) becomes

$$\Delta G^o = -RT \ln\left(K_{sp}\right) \tag{10.51}$$

Substituting Eq. (10.48) into (10.51) gives

$$\Delta G^o = -RT \ln\left(\frac{[M_xO_y]}{[M]^x [O]^{y/2}}\right) = -RT \ln\left(\frac{a_{M_xO_y}}{a_M^x a_O^{y/2}}\right) \tag{10.52}$$

If $\Delta G^o < 0$, then reaction proceeds as written in Eq. (10.47); otherwise, the reaction occurs in the opposite direction. Theoretically, the dimensionless activity of a component in the mixture can also be defined as

$$a_j = X_j \qquad \text{(ideal solution)} \tag{10.53a}$$

$$a_j = \gamma_j X_j = \gamma_j [\%j] \quad \text{(actual solution)} \tag{10.53b}$$

where $j = M_xO_y$, M or O in the melt, γ_j denotes the activity coefficient of species j due to deviations from ideal solutions, $[\%j]$ denotes the weight percent so that $[\%j] = [wt\%j]$, and X_j denotes the mole fraction of j.

For a dilute melt alloy, the activity coefficient can be calculated using the Wagner's equation [14]

$$\log(\gamma_i) = \frac{1}{2.3026} \ln(\gamma_i) = \sum e_i^j [\%i] \tag{10.54a}$$

$$\gamma_i = \exp\left(2.3026 \sum e_i^j [\%j]\right) \tag{10.54b}$$

In this particular case, e_j denotes the first-order interaction parameter that describes the influence of solute j on the activity coefficient of solute i.

10.7.2   Wagner Interaction Parameters

Following Dealy and Pehlke paper [15], the mathematical treatment developed by Wagner [14] for defining the activity coefficient of a solute element in a dilute alloy (low-solute concentration alloy) is based on the truncated Maclaurin's series about $X_i, X_j = 0$. Thus,

$$\ln(\gamma_i) = \ln(\gamma_i^o) + \left[X_i \left(\frac{\partial \ln \gamma_i}{\partial X_i}\right)_{X_i=0} + X_j \left(\frac{\partial \ln \gamma_i}{\partial X_j}\right)_{X_j=0} + \cdots \right]_{1st}$$

$$\tag{10.55}$$

$$+ \left[\frac{1}{2} X_i^2 \left(\frac{\partial^2 \ln \gamma_i}{\partial X_i^2}\right)_{X_i=X_j=0} + \frac{1}{2} X_i X_j \left(\frac{\partial^2 \ln \gamma_i}{\partial X_i \partial X_j}\right)_{X_j,X_j=0} + \cdots \right]_{2nd}$$

where γ_i^o denotes the activity coefficient of solute i at infinite dilution, X_i, X_j denote mole fractions, $[1st]$, $[2nd]$ denote the first- and second-order expansions in the Maclaurin's series. Moreover, in order to calculate the Wagner interaction parameters for each of the i components, one has the option to use as many terms as possible in the Maclaurin's series.

The solute mole fraction X_i and Wagner's first- (e_i^i, e_i^j) and second- (r_i^i, r_i^j) order interaction parameters along with the molecular weights M_i, M_j take the form, respectively,

$$X_i = \frac{[\%i] M_j}{100 M_i} \tag{10.56a}$$

$$e_i^i = \left(\frac{\partial \ln \gamma_i}{\partial X_i}\right)_{X_i=0} \quad \& \quad e_i^j = \left(\frac{\partial \ln \gamma_i}{\partial X_j}\right)_{X_i,X_j=0} \qquad (1st \; order) \tag{10.56b}$$

$$r_i^i = \left(\frac{\partial^2 \ln \gamma_i}{\partial X_i^2} \right)_{X_i, X_j = 0} \quad \& \quad r_i^j = \left(\frac{\partial^2 \ln \gamma_i}{\partial X_i \partial X_j} \right)_{X_i, X_j = 0} \quad \text{(2nd order)} \quad (10.56c)$$

Moreover, Eq. (10.55) can also be defined as

$$\ln(\gamma_i) = \ln(\gamma_i^o) + \sum_{j=2}^{N} e_i^j X_j + \sum_{j=2}^{N} r_i^j X_j^2 + \cdots \tag{10.57a}$$

$$\ln(\gamma_i) = \ln(\gamma_i^o) + X_i e_i^i + X_j e_i^j + \cdots + X_i^2 r_i^i + X_i X_j r_i^j + \cdots \tag{10.57b}$$

$$\ln(\gamma_i) \simeq \ln(\gamma_i^o) + \sum_{j=1}^{N} e_i^j X_j \tag{10.57c}$$

Here, e_i^j, r_i^j measure the effects of component j on the activity coefficient of component i in a dilute alloy. Obviously, additional coefficients in the Maclaurin's series convey to higher order interaction parameters and more accurate results.

If the temperature is taken into account, then the interaction parameter is related to the molar enthalpy [15]

$$\frac{\partial e_i^j}{\partial T} = -\frac{1}{RT^2} \left(\frac{\partial^2 H}{\partial X_i \partial X_j} \right)_{X_i, X_j = 0} \tag{10.58a}$$

$$\frac{\partial e_i^j}{\partial (1/T)} = -\frac{1}{R} \left(\frac{\partial^2 H}{\partial X_i \partial X_j} \right)_{X_i, X_j = 0} \tag{10.58b}$$

In addition, converting natural log to common log, and mole fraction to weight percent ($wt\%$) in Eq. (10.57c) yields

$$\log(\gamma_i) = \log(\gamma_i^o) + e_i^i [\%i] + e_i^j [\%j] + \cdots \tag{10.59}$$

with

$$e_i^j = \left(\frac{\partial \log \gamma_i}{\partial [\%j]} \right)_{[\%i],[\%j]=0} \tag{10.60a}$$

$$e_i^j = \left(\frac{M_i}{M_j} \right) e_j^i \tag{10.60b}$$

The reader should consult the literature for additional theoretical background on dilute alloys and related analytical compilations on interaction parameter formalisms published by Lehmann and Jung [16, chapter 5].

Example 10A.7 For a steel alloy with composition and interaction parameters

$$C = 3.98\%, \ Si = 1.60\%, \ Mn = 1.02\%, \ S = 0.04\% \text{ at } 1600\,^\circ C$$

$$e_S^S = -0.03, \ e_S^C = 0.24, \ e_S^{Si} = 0.07, \ e_S^{Mn} = -0.03$$

calculate the activity of sulfur.

Solution From Eq. (10.54a),

$$\sum e_S^j \,(\%i) = e_S^S \,(\%S) + e_S^C \,(\%C) + e_S^{Si} \,(\% \, Si) + e_S^{Mn} \,(\%Mn) = 1.0354$$

and from Eq. (10.54b),

$$\gamma_S = \exp\left[2.3026 \sum e_S^j \,(\%i)\right] = \gamma_S = \exp\left[(2.3026)\,(1.0354)\right] = 10.85$$

Then, Eq. (10.60a) yields the activity of sulfur (S) as

$$a_S = \gamma_S \,[\%S] = (10.85)\,(0.04) = 0.43$$

Therefore, $a_S = 0.43$ represents an approximation due to modified Wagner's equations available in the literature.

10.7.3   Aluminum-Oxygen Reaction in Molten Steel

The goal in this section is to characterize the effects of gas–solute interactions during the solidification of diluted steels. First of all, assume that an amount of pure aluminum (Al) reacts with dissolved oxygen in the molten steel and simultaneously another amount of Al reacts with the iron oxide (FeO) in the floating slag. Thus, the molten steel and the slag are fundamentally deoxidized by aluminum at the liquid steel temperature. Both processes are represented by the following high-temperature reactions

$$2Al + 3O \rightarrow Al_2O_3 \tag{10.61a}$$

$$3FeO + 2Al \rightarrow Al_2O_3 + 3Fe \tag{10.61b}$$

Next, consider the thermodynamics related to the $Al\text{-}O$ reaction described by Eq. (10.61a), as an example, to define the solubility product and the oxygen activity written as, respectively,

$$K_{sp} = \frac{a_{Al_2O_3}}{a_{Al}^2 a_O^3} = \exp\left(-\frac{\Delta G^o}{RT}\right) \tag{10.62a}$$

Table 10.1 Wagner interaction parameters for selected elements [18]

j	Al	C	Si	Mn	S	Cr	N
e_{Al}^i	0.045	0.091	0.006	0.070	0.030	0.025	−0.058
e_O^i	−1.17	−0.37	−0.14	−0.030	−0.104	−0.037	−0.123

$$a_O = \left(\frac{a_{Al_2O_3}}{a_{Al}^2 K_{sp}} \right)^{1/3} = \left[\frac{a_{Al_2O_3}}{a_{Al}^2 \exp\left(-\Delta G^o / RT\right)} \right]^{1/3} \tag{10.62b}$$

If ΔG^o is defined by [17, pp. 633–652]

$$\Delta G^o = -1.21 \times 10^6 + 386.71T \tag{10.63}$$

where ΔG^o is in J/mol and T is in absolute K units.

Then, the activity of oxygen becomes

$$a_O = \left[\frac{a_{Al_2O_3}}{a_{Al}^2 \exp\left[\left(1.21 \times 10^6 - 386.71T \right) / RT \right]} \right]^{1/3} \tag{10.64}$$

The activity coefficient γ_O for oxygen, written as $\log \gamma_O$, can be expanded as per Wagner's approach. Thus,

$$\log \gamma_O = \log \gamma_O^j + \sum_{j=2}^{n} e_O^j \, [\%j] + \sum_{j=2}^{n} \rho_O^j \, [\%j]^2 \simeq \log \gamma_O^j + \sum_{j=2}^{n} e_O^j \, [\%j] \tag{10.65}$$

Some values of e_i^j are listed in Table 10.1 as per Steneholm et al. paper [18].

Example 10A.8 Consider a molten steel containing (in weight percent)

0.5% Al	0.5% C	0.6% Si	2% Mn	0.002% S	0.4% Cr	0.0001% N

Determine if there is a deviation from the molten steel equilibrium state at high temperature (1550 °C) with respect to oxygen. Explain.

Solution Using Eq. (10.65) yields the oxygen activity coefficient as

$$\log \gamma_O \simeq \ln \gamma_O^0 - 0.92902 = 1 - 0.92902 = 0.07098$$

$$\gamma_O = 10^{0.07098} = 1.1776 \simeq 1.18$$

From Eq. (10.53b),

$$a_O = \gamma_O X_o = 1.18 X_O$$

Therefore, there is a deviation from the equilibrium state since $a_O > X_O$. Using Eq. (10.63) with $T = 1550\,°C = 1823$ K yields the Gibbs energy change

$$\Delta G^o = -\left(1.21 \times 10^6 \text{ J/mol}\right) + (386.71 \text{ J/mol K})\, T$$

$$\Delta G^o = -\left(1.21 \times 10^6 \text{ J/mol}\right) + (386.71 \text{ J/mol K})\,(1823 \text{ K})$$

$$\Delta G^o = -5.05 \times 10^5 \text{ J/mol} = 505 \text{ kJ/mol}$$

From Eq. (10.51), the solubility constant is

$$K_{sp} = \exp\left(-\frac{\Delta G^o}{RT}\right) = \exp\left[-\frac{(-5.05 \times 10^5 \text{ J/mol})}{(8.314 \text{ J/mol K})\,(1823 \text{ K})}\right]$$

$$K_{sp} = 2.95 \times 10^{14}$$

Therefore, solute interactions influence the activity coefficient of oxygen γ_O.

10.8 Energy Diagram for Metal Oxides

Thermodynamically, the driving force for the oxidation of a metal in gaseous environments is the Gibbs free energy of formation (ΔG) and consequently, the occurrence of an oxidation chemical reaction depends on the magnitude of ΔG. The Ellingham diagram shown in Fig. 10.11 is very valuable for determining the standard Gibbs energy change ΔG^o for metal oxide reductions under stable conditions so that $\Delta G^o < 0$. If $\Delta G^o = 0$ the metal oxides is at their equilibrium state and if $\Delta G^o > 0$ the metal oxides are unstable [19, p. 429].

Mathematically, the standard ΔG^o at a constant pressure is defined by

$$\Delta G^o = -RT \ln\left(K_{sp}\right) \tag{10.66a}$$

$$\Delta G^o = \Delta H^o - T\Delta S^o \tag{10.66b}$$

Actually, the Gibbs energy change (ΔG^o) is taken as the driving force for the chemical reactions cited in Fig. 10.11, it depends on the heat energy between the reaction and its surroundings called enthalpy change (ΔH^o).

The variations of both ΔH and ΔS with temperature are commonly written as

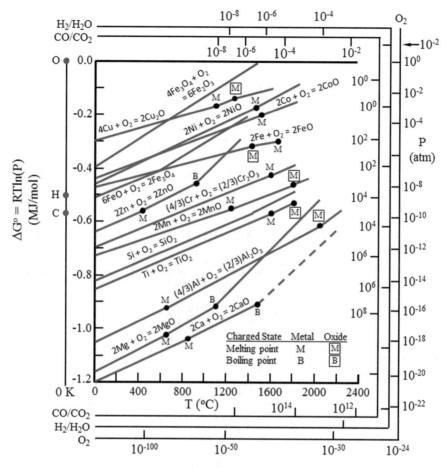

Fig. 10.11 Ellingham standard energy change diagram [19, p. 429]

$$\Delta H(T) = \Delta H^o(T_o) + \int_{T_o}^{T} C_p(T)dT \tag{10.67a}$$

$$\Delta S(T) = \Delta S^o(T_o) + \int_{T_o}^{T} \frac{C_p(T)}{T}dT \tag{10.67b}$$

where $\Delta G^o(T_o)$ denotes the standard Gibbs energy change, $\Delta H^o(T_o)$ denotes the standard enthalpy change which is a measure of the heat absorbed or released at constant gas pressure, $\Delta S^o(T_o)$ denotes the standard entropy change which is a measure of monoatomic disorder, $T_o = 298\,K = 25\,°C$ denotes the standard temperature, and C_p denotes the heat capacity at constant pressure P.

Consider some generalized high-temperature reactions for the formation of metal oxides having different chemical formulae. These reactions represent the metal–gas interactions being influenced by the gas pressure (oxygen O_2, carbon dioxide CO_2, and water H_2O) in the melt. Thus,

Reaction 1:

$$\frac{2x}{y}M + O_2 = \frac{2}{y}M_xO_y \tag{10.68a}$$

$$\Delta G^o = -RT \ln\left(\frac{[M_xO_y]^{2/y}}{[M]^{2x/y}[O_2]}\right) \tag{10.68b}$$

$$\Delta G^o = RT \ln\left(P_{O_2}\right) \tag{10.68c}$$

Reaction 2:

$$xM + yCO_2 = M_xO_y + yCO \tag{10.69a}$$

$$\Delta G^o = -RT \ln\left(\frac{[M_xO_y][CO]^y}{[M]^x[CO_2]^y}\right) \tag{10.69b}$$

$$\Delta G^o = -yRT \ln\left(\frac{P_{CO}}{P_{CO_2}}\right) \tag{10.69c}$$

Reaction 3:

$$xM + yH_2O = M_xO_y + yH_2 \tag{10.70a}$$

$$\Delta G^o = -RT \ln\left(\frac{[M_xO_y][H_2]^y}{[M]^x[H_2O]^y}\right) \tag{10.70b}$$

$$\Delta G^o = -yRT \ln\left(\frac{P_{H_2}}{P_{H_2O}}\right) \tag{10.70c}$$

where $[j] = a_j$ denotes the activity of species j, $[M] = 1$, $[M_xO_y] = 1$ and $P_{O_2}, P_{H_2}, P_{H_2O}$ denote the pressures (kPa). Here, $[j] = P_j/P_o$.

Gibbs Energy Criterion If $\Delta G^o < 0$, then it is a measure of negative deviation from equilibrium condition and the reaction proceeds from left to right as written in the above reactions. On the contrary, if $\Delta G^o > 0$, then a positive deviation from equilibrium implies that a reaction occurs in the reverse direction from right to left. However, if $\Delta G^o = 0$, the system is at equilibrium.

Example 10A.9 Assume that copper oxide (Cu_2O) scale forms at $887\,^\circ C$ as per reaction

$$4Cu + O_2 = 2Cu_2O$$

at 887 °C until they intercept the pertinent outer lines for the oxygen and pressure ratios. Determine (a) ΔG^o, (b) Which metals are oxidized and reduced based on ΔG^o. (c) How useful is the data in Fig. 10.11?

Solution

(a) Using a simplified version of Fig. 10.11 given below, draw straight lines (blue) from points "O," "H," and "C" through the Cu-O reaction line at $T = 887$ °C until these lines intercept the scale-lines for O_2, H_2/H_2O, and CO/CO_2 pressures.

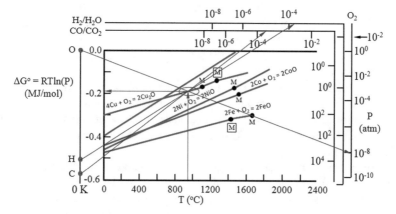

Then, read-off the corresponding pressure and Gibbs energy change values

$$P_{O_2} = 2.78 \times 10^{-9} \text{ atm}$$

$$P_{CO}/P_{CO_2} = 1.5 \times 10^{-4}$$

$$P_{H_2}/P_{H_2O} = 1.2 \times 10^{-4}$$

$$\Delta G^o = -190 \text{ kJ/mol of } O_2$$

(b) According to Fig. 10.11, pure metals having $\Delta G^o < -190$ kJ/mol (below) at 887 °C can be oxidized and those with $\Delta G^o > -190$ kJ/mol (above) at 887 °C can be reduced. However, Cu can be reduced by Ni because its line is below the $\Delta G^o < -190$ kJ/mol. A similar reasoning may be used at different Gibbs energy change and temperature values. Finally, metals with the Gibbs energy change lines above the $2C + O_2 = CO$ line can be reduced by carbon C as the affinity of metals for oxygen increases.

(c) Despite that Fig. 10.11 is normally used for reducing mineral compounds, such as oxides, it provides a basis for evaluating the possibility of chemical separation by oxidation. For example,

$$4Cu + O_2 \rightarrow 2Cu_2O \quad \text{at } \Delta G^o < 0 \quad \text{(reduction)}$$

$$4Cu + O_2 \leftarrow 2Cu_2O \quad \text{at } \Delta G^o > 0 \quad \text{(oxidation)}$$

These reactions are represented above with an equal sign instead.

10.9   Porosity-Induced Cracking

Generally, gas porosity (Figs. 10.4a and 10.12a), shrinkage porosity (Fig. 10.4b), and hot tear (Fig. 10.6) are inherent casting defects related to the solidification process. For instance, a hot tear can be treated as a preexisting crack, a shrinkage pore an irregular elongated shape may grow by acquiring a sharp tip and consequently, it becomes a large a large crack, and the circular edge of a gas pore may be the source of surface cracks as shown in Fig. 10.12b.

Porosity-induced crack initiation is detrimental to casting parts when subjected to external forces and exposed to a corrosive environment. For instance, Fig. 10.12a shows a uniform distribution of very small gas pores on a particular metallic surface and Fig. 10.12b reveals three (3) major surface cracks emanated from the circular edge of a single gas pore [20].

Cracks can be produced by high level of residual stresses during cooling of the solidified melt in the presence of high-temperature gradients. Therefore, a cast component having, at least, one crack being subjected to a quasi-static external force perpendicular to the crack plane may fracture, provided that the applied stress intensity factor (K_I) reaches the cast material inherent plane-strain fracture toughness (K_{IC}). Hence, fracture occurs if $K_I \geq K_{IC}$.

The initiation and growth of a crack from a pore are attributed to slip mechanism induced by the residual stresses. Once a crack develops in a structural component, the static or fatigue life becomes a major concern for engineering structural integrity. For instance, Fig. 10.13 shows pore-induced fatigue cracks; 40-μm (Fig. 10.13a) and 65-μm (Fig. 10.13b) long cracks [21, 22].

(a) Gas porosity   (b) Cracks around a gas pore

Fig. 10.12   (a) Gas porosity and (b) disclosed cracks emanated from a gas pore [20]

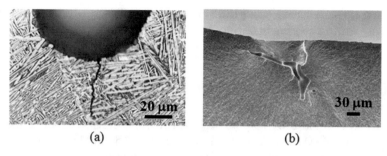

(a)   (b)

Fig. 10.13 Pore cracking. (**a**) A 45-μm long fatigue crack on a post-weld heat treatment Ti-6Al-4V (Grade 5) titanium alloy after 10^7 cycles [22] and (**b**) CMSX4 single-crystal turbine material [23]

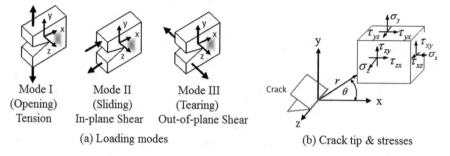

Mode I	Mode II	Mode III
(Opening)	(Sliding)	(Tearing)
Tension	In-plane Shear	Out-of-plane Shear

(a) Loading modes   (b) Crack tip & stresses

Fig. 10.14 (**a**) Stress loading modes for through-thickness cracks and (**b**) crack coordinate system with an element at a distance r from the crack tip at an angle θ

Furthermore, the above crack formation and crack growth cases require an in-depth analysis of the field of fracture mechanics (applied mechanics field) [23] and corrosion (mostly electrochemical in nature) [24].

10.10   Linear-Elastic Fracture Mechanics

This section introduces the concept of "fracture mechanics" as the discipline concerned with the behavior of materials containing cracks. Thus, the characterization of crack growth requires the determination of the fracture toughness of the material being subjected to quasi-static or dynamic (fatigue) loading.

Solid bodies containing cracks can be characterized by defining a state of stress near a crack tip. The theory of linear-elastic fracture mechanics (LEFM) is briefly integrated in this chapter using an analytical approach that will provide the reader with useful insights for characterizing the intensity of an applied stress within a small plastic zone ahead of the crack tip as schematically shown in Fig. 10.14. If this plastic zone is sufficiently small ($r << a$), the small-scale yielding (SSY) approach is used for characterizing brittle solids and for determining the stress–strain fields.

In contrast, a large-scale yielding (LSY) is for ductile solids, in which $r \geq a$. Hence, plasticity is also a concern in fracture mechanics.

10.10.1 Modes of Loading

A crack in a brittle body may be subjected to three different types of loading, which involve displacements of the crack surfaces. The mechanical behavior of a solid containing a crack of a specific geometry and size can be predicted by evaluating the stress intensity factors K_I, K_{II}, and K_{III} as per the stress modes or the modes of loading shown in Fig. 10.14. Although a combined loading can be encountered in structural components, but K_I for mode I is the most studied and evaluated experimentally for determining its critical value called plane-strain fracture toughness (K_{IC}), which is a material property.

Furthermore, K_{II} for shear mode II and K_{III} for tearing mode III are also significant in fracture mechanics studies. In reality, most cracks are under a mixed-mode interaction where K_I, K_{II}, K_{III} must be characterized in order to determine which mode of loading dominates fracture. Thus, structural components having flaws or cracks can be loaded to various levels of the applied stress intensity factor for a particular stress mode as shown in Fig. 10.14a.

For elastic or brittle materials, the crack driving force, known as the strain-energy release rate G_i, is related to the stress intensity factor K_i and to the modulus of elasticity E, where $i = I, II, III$ indicate the type of mechanical loading as illustrated in Fig. 10.14. Commonly, mode I prevails in most fracture mechanics studies.

For convenience, consider mode I only so that the relationship between G_I and K_I can be written as

$$G_I = \frac{K_I^2}{E'} \quad \text{(before fracture)} \tag{10.71a}$$

$$G_{IC} = \frac{K_{IC}^2}{E'} \quad \text{(at fracture)} \tag{10.71b}$$

These expression are well-known generalized equations in the field of linear fracture mechanics, where $E' = E$ for plane-stress condition (MPa), $E =$ Elastic modulus of elasticity (MPa), $E' = E/(1 - v^2)$ for plane-strain condition (MPa), and $v =$ Poisson's ratio. Moreover, the expression defined by Eq. (10.71a) is the fundamental mathematical model in the field of fracture mechanics which is used to emphasize that the applied K_I value drives the crack to stably grow in a self-similar manner; along the crack plane. Essentially, Eq. (10.71a) is denoted the crack driving force. On the other hand, (10.71b) represents the critical crack driving force for crack instability, which means that the crack propagates very rapidly. Thus, Eq. (10.71b) represents the plain-strain fracture toughness (property) of a material.

10.10.2   Stress Field for Mode I

For crack growth under tension mode I, the stress field equations in Cartesian coordinates are defined by [23, p. 134]

$$\sigma_x = \frac{K_I}{\sqrt{2\pi r}} \cos \frac{\theta}{2} \left(1 - \sin \frac{\theta}{2} \sin \frac{3\theta}{2} \right) \tag{10.72a}$$

$$\sigma_y = \frac{K_I}{\sqrt{2\pi r}} \cos \frac{\theta}{2} \left(1 + \sin \frac{\theta}{2} \sin \frac{3\theta}{2} \right) \tag{10.72b}$$

$$\tau_{xy} = \frac{K_I}{\sqrt{2\pi r}} \cos \frac{\theta}{2} \sin \frac{\theta}{2} \cos \frac{3\theta}{2} \tag{10.72c}$$

The stress σ_z in the z-direction is of particular interest because it defines plane conditions as elucidated below using stress criteria. The term $K_I/\sqrt{2\pi r}$ of Eq. (10.72) measures the magnitude of the stress and the function $f(\theta)$ determines the distribution of the stresses ahead of the crack tip.

Plane Stress This is a stress condition used for thin bodies (plates), in which the specimen thickness must be $B \ll w$, where w is the width. Thus, the negligible stresses under the plane-stress condition are the through-thickness and the shear stresses

$$\sigma_z = \tau_{yz} = \tau_{zx} = 0 \tag{10.73}$$

This stress condition is vital in studying local stress fields near a crack tip in a solid body under a quasi-static or dynamic loading. The former loading is the most common in monotonic and fracture mechanics testing. This stress can be defined as $\sigma_z = 0$ at the surface and $\sigma_z \simeq 0$ at the mid-thickness plane.

Plane Strain This particular condition is for thick bodies, which develop a triaxial state of local stress at the crack tip. Thus, the through-thickness stress in Cartesian coordinates is

$$\sigma_z \simeq v \left(\sigma_x + \sigma_y \right) \tag{10.74}$$

Inserting Eq. (10.71) into (10.73) yields

$$\sigma_z = \frac{2v K_I}{\sqrt{2\pi r}} \cos \frac{\theta}{2} \tag{10.75}$$

Letting $\theta = 0$ implies that the crack grows along its own crack plane (x-axis) under plane-strain conditions. Then, Eq. (10.72) yields

$$\sigma_x = \frac{K_I}{\sqrt{2\pi r}} \qquad \text{for } \theta = 0 \qquad\qquad (10.76a)$$

$$\sigma_y = \frac{K_I}{\sqrt{2\pi r}} \qquad \text{for } \theta = 0 \qquad\qquad (10.76b)$$

$$\tau_{xy} = 0 \qquad\qquad\ \text{for } \theta = 0 \qquad\qquad (10.76c)$$

$$\sigma_z = \frac{2v K_I}{\sqrt{2\pi r}} \qquad \text{for } \theta = 0 \qquad\qquad (10.76d)$$

Notice that $\sigma_x, \sigma_y \to \infty$ as $r \to 0$ is referred to as a singularity condition of order $r^{-1/2}$. The accuracy of the magnitude of the stress intensity factor (K_I) depends on the magnitude of the plastic zone size (r) in the range $0 < r < \infty$ and it measures the intensity of the stress ahead of the crack tip that causes crack growth. In principle, K_I measures the strength of the singularity so that $K_I = f(\sigma, r)$, where σ is the applied stress.

10.10.3　Finite Specimens

Consider two fracture mechanics cracked plates having different crack configurations. The plate in Fig. 10.15a is modeled as a through-thickness single-edge crack with a crack length "a." On the other hand, the plate in Fig. 10.15b is modeled as a part-through semi-elliptical central crack with major axis "$2c$" and minor axis "a" (crack depth).

A fracture analysis of cracked structures is based on a pre-crack size (a) in order to predict crack initiation and crack propagation in brittle materials. Crack initiation and crack propagation refer to stable and unstable crack growth, respectively. The latter condition leads to fracture at a high crack speed.

In order for fracture mechanics to be a practical discipline in engineering applications, $K_I = f(\sigma, a)$ or $K_I = f(\sigma, a_e)$, where $a_e = a+r$ denotes an effective crack length. This implies that K_I for pre-crack finite structural components is normally corrected using a geometry factor, say, $f(a, w)$. This is considered in the next section for simple crack configurations under mode I loading.

Fig. 10.15 Fracture mechanics specimens. (**a**) Through-thickness single-edge and (**b**) part-through central cracks

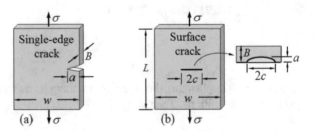

Notice that only the tension mode I has been introduced for brittle solids. Other loading modes can also be used to simulate in-service performance as an essential condition for characterizing the effect of porosity-induced cracking.

The stress intensity factor for a particular crack configuration and specimen geometry can be defined as a general function of the form $K_I = f(r, a, \sigma, T)$, provided that there exists a large enough plastic zone with size r. Usually, $r \ll a$ for most brittle materials so that $K_I = f(a, \sigma)$ at temperature T.

For mode I loading, the stress intensity factor K_I and the corresponding stress perpendicular to the crack line, where $y = 0$ and $\theta = 0$, take the form [23, p. 86]

$$K_I = \lim_{r \to 0} \left(\sigma_{yy} \sqrt{2\pi r} \right) f_I(\theta) = \sigma_y \sqrt{2\pi r} \quad at \ \sigma_y = \sigma_y \, (r, \theta = 0) \qquad (10.77a)$$

$$\sigma_y = \sigma \sqrt{\frac{a}{2r}} \qquad (10.77b)$$

Substituting Eq. (10.77b) into (10.77a) yields the K_I for a specimen containing a small crack

$$K_I = \sigma \sqrt{\pi a} \quad \text{(uncorrected)} \qquad (10.78a)$$

$$K_I = \alpha \sigma \sqrt{\pi a} \quad \text{(corrected)} \qquad (10.78b)$$

where α denotes a geometry correction factor. Now, assume that a specimen is loaded in tension to force crack growth to occur perpendicular to the loading direction. Commonly, the quasi-static and fatigue stress modes are applied to finite specimens under controlled testing conditions in order to determine the crack behavior and related fracture toughness.

Figure 10.16 schematically shows typical loading curves for quasi-static and fatigue testing schemes.

Figure 10.16a shows the point of failure for brittle solids under tension mode. This corresponds to a maximum load (mechanical force) at fracture. The most strict fracture mechanics testing method is found in the ASTM E399 Standard Test Method guidelines for calculating the plane-strain fracture toughness K_{IC}.

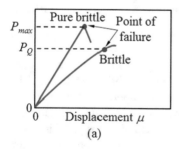
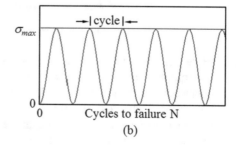

Fig. 10.16 Loading curves. (**a**) Quasi-static tension loading. (**b**) Fatigue loading

Moreover, a brittle material may show a slight degree of plasticity denoted with a small curvature in Fig 10.16a. In this case, the material is classified as an elastic-plastic solid with a small plastic zone $r << a$. However, for a specimen exhibiting significant plasticity ahead of the crack tip with $r < a$, the ASTM E1820 guidelines should be used.

Figure 10.16b schematically shows a cyclic fatigue loading history with a uniform stress range $\Delta\sigma = \sigma_{max} = \sigma_{min}$ as a function of cycles to failure N, where $\sigma_{min} = 0$ in this particular case.

Furthermore, comparing quasi-static and fatigue cases for specimens containing single-edge cracks (Fig. 10.15a), $\alpha = f(a/w)$ so that

$$K_I = f(a/w)\,\sigma\,\sqrt{\pi a} \quad \text{(single-edge crack)} \tag{10.79a}$$

$$\Delta K_I = f(a/w)\,\Delta\sigma\,\sqrt{\pi a} \tag{10.79b}$$

For specimens containing semi-elliptical cracks (Fig. 10.15b), $\alpha = f(a/c)$ and

$$K_I = f(a/c)\,\sigma\,\sqrt{\pi a} \quad \text{(surface crack)} \tag{10.80a}$$

$$\Delta K_I = f(a/c)\,\Delta\sigma\,\sqrt{\pi a} \tag{10.80b}$$

Additionally, K_I and ΔK_I depend on the crack morphology, specifically on longitudinal surface cracks with a groove appearance. In principle, fracture mechanics can be applied to any crack configuration and to a crack network such as crack branching. However, a cracked component may be subjected to a mixed-mode loading. In this particular case, failure occurs by a combination of the stress intensity factors.

10.10.4   Fracture Mechanics Criteria

This section aims at the fracture mechanics criteria for mode I loading at failure or fracture is defined as

$$K_I \rightarrow K_{IC} \text{ as } a \rightarrow a_c \text{ at } \sigma \quad \text{(static)} \tag{10.81a}$$

$$K_{max} \rightarrow K_{IC} \text{ as } a \rightarrow a_c \text{ at } \sigma_{max} \quad \text{(fatigue)} \tag{10.81b}$$

Here, K_{IC} is the material's property referred to as plane-strain fracture toughness, experimentally determined at the fracture stress for brittle materials. The mathematical statement $K_I \rightarrow K_{IC}$ implies that K_I approaches K_{IC} as the crack length reaches its critical value $a \rightarrow a_c$, which depends on the material's microstructural features and the magnitude of the remote applied stress. Similarly, fatigue failure occurs as $a \rightarrow a_c$ at a maximum stress σ_{max}, which opens the crack and causes crack growth.

For mode I loading, K_{IC} is used in designing schemes for controlling premature fracture by letting the design stress intensity factor be

$$K_{Id} = \frac{K_{IC}}{S_f} \qquad (10.82)$$

where S_f denotes the fracture mechanics safety factor. Moreover, K_{IIC} for shear mode II and K_{IIIC} for tearing mode III are also significant in fracture mechanics studies. In reality, most cracked structural components are under a mixed-mode interactions where K_I, K_{II}, K_{III} must be characterized together.

In general, structural components undergoing multisite crack nucleation due to the existence of deleterious defects and stress risers can be subjected to high enough stress levels in service. As a result, the design static or dynamic (fatigue) life is reduced. Therefore, monitoring microstructural features (dislocation density and grain morphology) and casting defects (gas porosity and microshrinkage) can indicate the level of casting soundness. Thus, in-service structural components subjected to dangerous loading must be monitored frequently.

Once a crack is detected on a component through a nondestructive evaluation (NDE) method, fracture mechanics dominates the engineering analysis in order to predict the component remaining life. This is purely an acceptable justification for considering fracture mechanics as an important engineering discipline. The case of multisite crack nucleation is even worse since several dominant cracks can evolve simultaneously. The final failure depends on the type of loading (axial, bending, torsion, or a combination of these) and the static or dynamic stress condition.

For comparison, the fracture stress σ_f from a conventional stress–strain tension test is higher than the one from a fracture mechanics mode I loading for the same material. In the latter case, the applied stress is potentially magnified at the crack tip. Consequently, the measured fracture or critical stress is $\sigma_c < \sigma_f$ at a macroscale.

Example 10A.10 Consider a large steel plate containing one 3 mm through-the-thickness single-edge crack being subjected to a tension load of 400 MPa. Calculate (a) the applied stress intensity factor K_I. Will fracture occur? (b) When would the plate fracture? Assume a brittle plate. (c) Repeat the calculations for an identical plate with a 3-mm deep surface crack, where $f(a/c) = 0.9621$. (d) Determine the strain-energy release rate G_I. Assume plane-strain conditions and let $K_{IC} = 55\ \text{MPa}\sqrt{\text{m}}$, $E = 200$ GPa and $v = 1/3$.

Solution

(a) For a large plate, $f(a/w) = 1.12$ since $w \gg a$ and [23, p. 89]

$$K_I = f(a/w)\,\sigma\sqrt{\pi a} = (1.12)\,(400\ \text{MPa})\sqrt{\pi\left(3 \times 10^{-3}\ \text{m}\right)}$$

$$K_I = 43.49\ \text{MPa}\sqrt{\text{m}}$$

Therefore, fracture will not occur since $K_I < K_{IC}$.

(b) Fracture would occur when

$$K_{IC} = f(a/w) \sigma \sqrt{\pi a_c}$$

$$a_c = \frac{1}{\pi} \left[\frac{K_{IC}}{f(a/w)\sigma} \right]^2 = \frac{1}{\pi} \left[\frac{55 \text{ MPa}\sqrt{m}}{(1.12)(400 \text{ MPa})} \right]^2 = 4.80 \text{ mm}$$

(c) For a surface crack with $f(a/c) = 0.9621$,

$$K_I = f(a/c)\sigma\sqrt{\pi a} = (0.9621)(400 \text{ MPa})\sqrt{\pi \left(3 \times 10^{-3} \text{ m}\right)}$$

$$K_I = 37.36 \text{ MPa}\sqrt{m}$$

fracture will not occur since $K_I < K_{IC}$ and

$$a_c = \frac{1}{\pi} \left[\frac{K_{IC}}{f(a/c)\sigma} \right]^2 = \frac{1}{\pi} \left[\frac{55 \text{ MPa}\sqrt{m}}{(0.9621)(400 \text{ MPa})} \right]^2 = 6.50 \text{ mm}$$

For constant crack velocity in each specimen, the surface cracked specimen allows a larger critical crack prior to crack propagation (fracture).

(d) Strain-energy release rate for the single-edge crack is

$$G_I = \frac{K_I^2}{E'} = \frac{(43.49 \text{ MPa}\sqrt{m})^2}{(200 \times 10^3 \text{ MPa})/(1 - 1/3)} = 6.30 \times 10^{-3} \text{ MPa m}$$

$$G_I = 6.30 \times 10^{-3} \text{ MJ/m}^2$$

and for the surface crack,

$$G_I = \frac{K_I^2}{E'} = \frac{(37.36 \text{ MPa}\sqrt{m})^2}{(200 \times 10^3 \text{ MPa})/(1 - 1/3)} = 4.65 \times 10^{-3} \text{ MPa m}$$

$$G_I = 4.65 \times 10^{-3} \text{ MJ/m}^2$$

Therefore, crack propagation or fracture does not occur because $K_I < K_{IC}$.

Example 10A.11 A large and thick plate containing a 4-mm long through-thickness center crack fractures when it is subjected to an external tensile stress of 7 MPa. Calculate the strain-energy release rate using (a) the Griffith's theory and (b) the LEFM approach. Should there be a significant difference between results? Explain. Data: $E = 62,000$ MPa and $v = 0.20$.

Solution

(a) For a total crack size of $2a = 4$ mm, Eq. (2.34) yields

$$\sigma_c = \sqrt{\frac{E'G_{IC}}{\pi a}}$$

$$G_{IC} = \frac{\pi a \sigma_f^2}{E'} = \frac{\pi \left(1 - v^2\right) a \sigma_f^2}{E}$$

$$G_{IC} = \frac{\pi \left(1 - 0.2^2\right) \left(2 \times 10^{-3} \text{ m}\right) (7 \text{ MPa})^2}{62,000 \text{ MPa}}$$

$$G_{IC} = 4.77 \times 10^{-6} \text{ MPa m} = 4.77 \text{ J/m}^2$$

(b) Using Eq. (3.29) along with $\alpha = \sqrt{\sec(\pi a/w)} = 1$ (see Table 3.1) and letting the stress intensity factor reach its critical value under plane-strain condition, $K_I = K_{IC}$, gives

$$K_{IC} = \alpha \sigma \sqrt{\pi a}$$

$$K_{IC} = (1)(7 \text{ MPa}) \sqrt{\pi \left(2 \times 10^{-3} \text{ m}\right)}$$

$$K_{IC} = 0.555 \text{ MPa}\sqrt{\text{m}}$$

This is a very small value, implying that the material is very brittle such as glass. From Eq. (2.34),

$$G_{IC} = \frac{K_{IC}^2}{E'} = \frac{\left(1 - v^2\right) K_{IC}^2}{E}$$

$$G_{IC} = \frac{\left(1 - 0.2^2\right) \left(0.555 \text{ MPa}\sqrt{\text{m}}\right)^2}{62,000 \text{ MPa}}$$

$$G_{IC} = 4.77 \times 10^{-6} \text{ MPa m} = 4.77 \text{ J/m}^2$$

These results indicate that there should not be any difference because either approach gives the same result.

10.11 Thermal Stress

This section aims at a particular thermo-mechanical problem related to hot tearing due to dimensional changes during cooling a solidified melt (solid) due to thermal gradient. As a result, thermal stresses and strains are developed.

Consider the simple models shown in Fig. 10.17 for thermal dimensional changes along the x-direction. For expansion, $\Delta L_x > 0$ (Fig. 10.17a) and for contraction, $\Delta L_x < 0$ (Fig. 10.17b).

Thermally, one-dimensional analysis yields the thermal deformation as

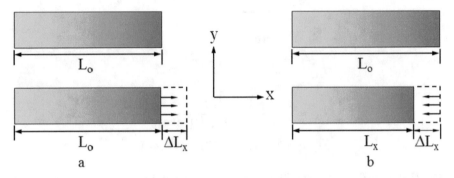

Fig. 10.17 Dimensional changes due to (**a**) thermal expansion and (**b**) thermal contraction

$$\Delta L_x = L_o \alpha_x \Delta T \tag{10.83a}$$

$$\epsilon_t = \frac{\Delta L_x}{L_o} = \alpha_x \Delta T > 0 \qquad \text{(expansion)} \tag{10.83b}$$

$$\epsilon_t = \frac{\Delta L_x}{L_o} = \alpha_x \Delta T < 0 \qquad \text{(contraction)} \tag{10.83c}$$

with

$$\Delta L_x = (L_x - L_o) > 0 \quad \text{(expansion)} \tag{10.84a}$$

$$\Delta L_x = (L_x - L_o) < 0 \quad \text{(contraction)} \tag{10.84b}$$

$$\Delta T = (T_f - T_i) > 0 \quad \text{(expansion)} \tag{10.84c}$$

$$\Delta T = (T_f - T_i) < 0 \quad \text{(contraction)} \tag{10.84d}$$

where $\delta_x = \Delta L_x$ denotes a dimensional change along the x-direction, ϵ_t denotes the thermal strain, α_x denotes the thermal expansion coefficient $(m/m\,^{\circ}C)$, and T_f, T_i denote the final and initial temperature, respectively.

Mechanically, one-dimensional analysis of a solid under elastic deformation along the x-direction gives the dimensional change ΔL_x and the Hooke's law defined, respectively, by

$$\Delta L_x = \frac{P_x L_o}{A_o E} = \frac{\sigma_x L_o}{E} \tag{10.85a}$$

$$\epsilon_x = \frac{\Delta L_x}{L_o} = \frac{\sigma_x}{E} \tag{10.85b}$$

$$\sigma_x = \epsilon_x E \tag{10.85c}$$

Letting $\epsilon_x = \epsilon_t$ yields the modified Hooke's law defining the thermal stress along the x-direction

$$\sigma_x = (\alpha_x \Delta T) E > 0 \quad \text{(expansion, tension)} \tag{10.86a}$$

$$\sigma_x = (\alpha_x \Delta T) E < 0 \quad \text{(contraction, compression)} \tag{10.86b}$$

Any temperature change between a specimen and the surrounding temperatures induces a coupled thermal stress–strain system under elastic deformation (thermoelastic). However, if the induced thermal stress during cooling a solidified melt causes hot cracking at a small scale, the resultant casting containing, at least, one crack which may undergo undetected due to lack of nondestructive evaluation (NDE). Hence, temperature gradients cause stress and strain to develop at hot spots on a solidified melt or cooled solid.

In order to make matters worse, assume that this cracked casting is put in-service under a mechanical loading condition, the crack may grow ($a \rightarrow a_c$) and reach a critical length (a_c), and consequently, the casting would simply fracture. This thought scenario is, indeed, not far from reality in certain cases. Additionally, if a crack is detected using NDE and the casting material has a known plane-strain fracture toughness K_{IC}, then the critical crack length under a quasi-stationary condition is predicted from Eq. (10.79a) or (10.80a)

$$a_c = \frac{1}{\pi} \left[\frac{K_{IC}}{f(a/w)\sigma} \right]^2 \tag{10.87a}$$

$$a_c = \frac{1}{\pi} \left[\frac{K_{IC}}{f(a/c)\sigma} \right]^2 \tag{10.87b}$$

On the other hand, if the cracked casting is under a fluctuating or cyclic stress condition, then the failure analysis using fracture mechanics is more complicated than the quasi-stationary case. In particular, fatigue in materials subjected to repeated cyclic loading can be defined as a progressive failure due to crack initiation (stage I), crack growth (stage II), and crack propagation (stage III) or instability stage.

Nonetheless, the fatigue crack growth rate can be characterized using Paris power law for stage II fatigue crack growth (FCG). Thus, the mathematical model is of the form [24]

$$\frac{da}{dN} = A(\Delta K)^n \tag{10.88}$$

where $\Delta K = K_{max} - K_{min} = (1 - R) K_{max}$ for $R \geq 0$; $\Delta K = K_{max}$ for $R \leq 0$; $A = \text{Constant } (m/cycles)$; $n = \text{Exponent}$; $R = \sigma_{min}/\sigma_{max} = K_{min}/K_{max}$ (stress ratio).

Continuing with solidification and deformation analysis, the initiation sites for hot tears in the casting are assumed to be located at stress risers. Recall that hot tears are cracks that initiate during solidification and they must be repaired by welding; otherwise, the casting must be scrapped. Accordingly, hot tearing leads to productivity and thermo-mechanical problems. In this case, the hot cracking process is considered a physical event related to lack of feeding molten metal to the casting geometry, causing thermal strains at specific hot spots. Consequently, thermo-mechanical strains may induce formation of non-linear cracks with intergranular crack growth. Obviously, some solidifying metals or alloys are more sensitive to hot cracking than others, and it appears that alloys are the most sensitive to this undesirable strain-induced solidification problem.

10.12   Summary

Understanding the concept of solidification defects requires knowledge of phase transformation and related heat transport. Interestingly, the Chvorinov's rule predicts the solidification time based on the casting modulus ($M = V/A$) and heat transfer factor B_c so that $t = B_c M^2$. Thus, controlling the solidification time of casting components using this approach allows the riser to solidify last by supplying hot liquid metal to the solidifying casting geometry.

It is well recognized that solidification defects are undesirable discontinuities in castings. Of particular interest is the solidification cracking related to hot cracks or cold cracks due to thermal stresses during contraction of the casting undergoing dimensional changes in the solid state. Other types of defects are gas-induced and shrinkage-induced porosity which are entirely detrimental to in-service performance of castings. It is evident that pores are potential sources of crack development.

Furthermore, the formation of metal oxides that float on the molten metal is characterized using the Ellington diagram, which is an important thermodynamic representation of a compilation of oxide reactions.

With respect to solidification cracking, linear-elastic fracture mechanics (LEFM) can be used to predict crack instability. The concept of stress intensity factor K_I and plane-strain fracture toughness K_{IC} for quasi-stationary cracks has been discussed as an important approach for assessing crack growth. Conversely, fatigue crack growth rate da/dN describes the dynamic state of cracks. Both fracture mechanics schemes are briefly introduced in this chapter for emphasizing the methodology used in dealing with existing cracks.

10.13   Problems

10.1 Determine the oxygen partial pressure stability domain for the Fe-O_2 system at $1000\,°C$. Assume that the following reaction occurs at this temperature.

$$2Fe + O_2 \rightarrow 2FeO \qquad \Delta G^o_{FeO} = -366 \text{ kJ/mol}$$

10.2 Determine (a) the oxygen partial pressure stability domain for the Fe-O_2 system at $1000\,°C$. Assume that the following reaction occurs at this temperature.

$$4Fe_3O_4 + O_2 \rightarrow 6Fe_2O_3 \qquad \Delta G^o_{FeO} = -120 \text{ kJ/mol}$$

(b) Calculate ΔS^o and ΔH^o.

10.3 Determine the partial pressures and the driving force (Fig. 10.11) for the reaction shown below at 1000 °C.

$$4Fe_3O_4 + O_2 = 6Fe_2O_3$$

10.4 Assume that the solubility of hydrogen ($[g] = [H]$) in molten copper is as given in the table below. Perform a linear curve fitting and calculate the hydrogen solubility at 1000 °C.

T (°C)	T (K)	$[g]_H$ ($\times 10^{-5}$ %)
1083	1356	10.05
900	1173	5.25
800	1073	3.35

10.5 Consider the solubility of hydrogen ($[g] = [H]$) in molten copper as given in the table below. Assume that the these data obey the Arrhenius equation. Determine (a) ΔG_c and A_x and (b) the hydrogen solubility at 1000 °C.

T (°C)	T (K)	$[g]_H$ ($\times 10^{-5}$ %)
1083	1356	10.05
900	1173	5.25

10.6 Assume that a 5-ton steel melt at 1520 °C initially contains 4 ppm of hydrogen and subsequent hydrogen from local moisture in the air diffuses through a boundary layer of 10 μm thick, and that it takes 140 s to fill a large mold cavity with the melt at 1500 °C. Also assume that the heat loss is due to convection and radiation. Consider that the hydrogen diffusion coefficient is $D_H = 10^{-8}$ m^2/s and that $Q_{conv} = Q_{rad}$. Calculate (a) the total amount of hydrogen (C_H) in the liquid steel during the pouring process for filling a mold cavity, and (b) Q_{conv} and Q_{rad} in units of GJ. Additional data:

$c_p = 0.76$ kJ/(kg K)	$C_{eq} = 20$ ppm	$\rho_{Fe} = 7.80 \times 10^3$ kg/m^3

10.7 A composite cylindrical refractory-steel ladle (Fig. 10.10) with dimensions $L = 2.5$ m, $r_1 = 1.40$ m, $r_2 = 1.45$ m, and $r_3 = 1.50$ m is used to cast 5 tons of steel at $1520\,°C$ into sand molds at $25\,°C$. The thermal conductivity of the refractory material and the steel hollow cylinder are $k_2 = 0.04$ W/m K and $k_3 = 40$ W/m K, respectively, and the heat transfer coefficients of the melt and air are $h_1 = 145$ W/m K and $h_4 = 30$ W/m K, respectively. Find T_4.

10.8 Consider a flat plate being solidified in a sand mold having design dimensional specifications $h = 0.02$-m thick, $w = 0.08$-m wide and $L = 0.15$-m long. Assume that the plate is heat treated in a furnace at 1200 K for 30 min and that the temperature profile is described by a polynomial of second order

$$T\,(x) = a_1 + a_2 x + a_3 x^2$$

where x is the distance from the plate centerline. If the temperature boundary conditions are $T\,(x)_s = 1200$ K and $T\,(0) = T_o = 400$ K, then calculate the heat flux $q_x = f\,(x)$ at the surface of the plate.

10.9 A large plate ($2a << w$ and $\alpha \rightarrow 1$) containing a through-thickness central crack 40-mm long is subjected to a tension stress. If the crack growth rate is 10 mm/month and fracture is expected at 10 months from now, calculate the fracture stress. Data: $K_{IC} = 30$ MPa$\sqrt{m}$.

10.10 A large and thick plate containing a 4-mm long through-thickness central crack fractures when it is subjected to an external tensile stress of 7 MPa. Calculate the strain-energy release rate. Should there be a significant difference between results? Explain. Data: $E = 62{,}000$ MPa and $v = 0.20$.

10.11 Calculate (a) the oxygen partial pressure stability domain for the $Fe\text{-}O_2$ system at $1200\,°C$. Assume that the following reaction occurs at this temperature.

$$4Fe_3O_4 + O_2 \rightarrow 6Fe_2O_3 \qquad \Delta G^o_{FeO} = -80 \text{ kJ/mol}$$

(b) Compute ΔS^o and ΔH^o for this reaction at the temperature range $200\text{--}1500\,°C$.

10.12 Consider a 4-ton steel melt at $1520\,°C$ initially contains 4 ppm of hydrogen. Assume that hydrogen from local moisture in the air diffuses through a boundary layer of 10 μm thick and it takes 140 s to fill a large mold cavity with the melt at $1500\,°C$. Therefore, heat loss is due to convection and radiation so that $Q_{conv} = Q_{rad}$ and the hydrogen diffusion coefficient is $D_H = 10^{-8}$ m^2/s. Calculate (a) the total amount of hydrogen (C_H) in the liquid steel during the pouring process for filling a mold cavity, and (b) Q_{conv} and Q_{rad} in units of GJ. Additional data:

$c_p = 0.76$ kJ/(kg K)	$C_{eq} = 20$ ppm	$\rho_{Fe} = 7.80 \times 10^3$ kg/m^3

10.13 Calculate the oxygen pressure (P_{O_2}) for the formation of solid MgO at $1000\,^{\circ}C$ and compare the result using the Ellingham diagram. This procedure is suitable for reducing oxygen from the Mg melt.

10.14 Calculate the Gibbs energy ΔG^o change for producing MgO solid oxide at $P_{O_2} = 2.70 \times 10^{-39}$ atm and $T = 1000\,^{\circ}C$. Compare the calculated result with the read-off value using the Ellingham diagram along with the pivotal point "O."

10.15 During the production of aluminum, the oxygen content in the melt is normally reduced through the reaction

$$\frac{4}{3}Al + O_2 \rightarrow \frac{2}{3}Al_2O_3$$

for which the standard and non-standard Gibbs energy changes are defined, respectively, by

$$\Delta G^o = -1.1273 \times 10^6 + 222.22T$$

$$\Delta G = \Delta G^o + RT \ln\left(K_p\right) = \Delta G^o + RT \ln\left(\frac{[Al_2O_3]^{2/3}}{[Al]^{4/3}[O_2]}\right)$$

where ΔG^o, ΔG have units of J/mol and T is Kelvin degree K. (a) For equilibrium conditions, calculate the temperature T and the corresponding ΔG^o value for the reaction to proceed as written at $P_{O_2} = 10^{-24}$ atm, and (b) determine T and ΔG^o using the Ellingham diagram. Compare results using both methods.

10.16 During the production of aluminum, the oxygen content in the melt is normally reduced through the reaction

$$Si + O_2 \rightarrow SiO_2$$

for which the standard Gibbs energy change is defined by

$$\Delta G^o = -8.6259 \times 10^5 + 172.41T$$

$$\Delta G = \Delta G^o + RT \ln\left(K_p\right) = \Delta G^o + RT \ln\left(\frac{[SiO_2]}{[Si][O_2]}\right)$$

where ΔG^o, ΔG have units of J/mol and T is Kelvin degree K. (a) For equilibrium conditions, calculate the temperature T and the corresponding ΔG^o value for the reaction to proceed as written at $P_{O_2} = 10^{-30}$ atm, and (b) determine T and ΔG^o using the Ellingham diagram. Compare results using both methods.

References

1. E. Niyama, T. Uchida, M. Morikawa, S. Saito, A method of shrinkage prediction and its application to steel castings practice. Am. Foundrymen's Soc. Int. Cast Met. J. **7**(3), 52–63 (1982)
2. J.T. Black, R.A. Kohser, *DeGarmo's Materials and Processes in Manufacturing*, 10th edn. (Wiley, New York, 2007)
3. H. Fredriksson, U. Akerlind, *Materials Processing During Casting* (Wiley, Chichester, 2006)
4. R.A. Flinn, *Fundamentals of Metal Casting* (Addison-Wesley, Reading, 1963)
5. J. Campbell, *Complete Casting Handbook: Metal Casting Processes, Metallurgy, Techniques and Design* (Butterworth-Heinemann, New York, 2011)
6. J.A. Spittle, A.A. Cushway, Influences of superheat and grain structure on hot-tearing susceptibilities of Al-Cu alloy castings. Metals Technol. **10**(1), 6–13 (1983)
7. R.A. Rosenberg, M.C. Flemings, H.F. Taylor, Nonferrous binary alloys hot tearing. AFS Trans. **69**, 518–528
8. D.J. Lahaie, M. Bouchard, Physical modeling of the deformation mechanisms of semisolid bodies and a mechanical criterion for hot tearing. Metall. Mater. Trans. B **32**(4), 697–705 (2001)
9. G. Nicoletto, G. Anzelotti, R. Konecna, X-ray computed tomography vs. metallography for pore sizing and fatigue of cast Al-alloys. Procedia Eng. **2**(1), 547–554 (2010)
10. J. Campbell, *Castings* (Butterworth-Heinemann, Elsevier Science, New York, 2003)
11. D.R. Gaskell, *An Introduction to Transport Phenomena in Materials Engineering*, 2nd edn. (Momentum Press, New York, 2013)
12. D.R. Poirier, G.H. Geiger, *Transport Phenomena in Materials Processing* (Springer, New York, 2016)
13. C. Wagner, *Thermodynamics of Alloys*, Translated by S. Mellgren and J.H. Westbrook (Addison-Wesley, Reading, 1952)
14. J.M. Dealy, R.D. Pehlke, Activities in dilute molten alloys. The University of Michigan, Industry Program of the College of Engineering, United States Atomic Energy Commission under contract No. AEC-AT (11-1)-979 (1961), pp. 1–22
15. J. Lehmann, I.-H. Jung, Thermodynamics, in *Treatise on Process Metallurgy: Process Phenomena*, vol. 2, Editor-in-Chief S. Seetharaman (Elsevier, Waltham, 2014), pp. 588–642
16. P.C. Hayes, *Process Principles in Minerals and Materials Production* (Hayes Publishing, Brisbane, 2003). ISBN 10: 0958919739 and ISBN 13: 9780958919739
17. K. Steneholm, M. Andersson, M. Nzotta, P. Jonsson, Effect of top slag composition on inclusion characteristics during vacuum degassing of tool steel. Process Metall. Steel Res. Int. **78**(7), 522–530 (2007)
18. D.R. Gaskell, *Introduction to the Thermodynamics of Materials*, 4th edn. (Taylor & Francis e-Library, New York, 2009)
19. Radiograph Interpretation - Castings - NDT Resource Center (Online link heading)
20. F. Fomin, N. Kashaev, Influence of porosity on the high cycle fatigue behaviour of laser beam welded Ti-6Al-4V butt joints, 3rd inter. symp. on fatigue design and materials defects, FDMD 2017, Lecco, Italy (19–22 September 2017). Procedia Struct. Integr. **7**, 415–422 (2017)
21. P.A.S. Reed, M.D. Miller, Comparison of low cycle (notch) fatigue behaviour at temperature in single crystal turbine blade materials, in *Superalloys 2008, Eleventh International Symposium on Superalloys* (The Minerals, Metals & Materials Society, Warrendale, 2008), pp. 527–533
22. N. Perez, *Fracture Mechanics*, 2nd edn. (Springer, New York, 2016)
23. N. Perez, *Electrochemistry and Corrosion Science*, 2nd edn. (Springer, New York, 2016)
24. P.C. Paris, F. Erdogan, A critical analysis of crack propagation laws. J. Basic Eng. Trans. ASME **85**, 528–534 (1963)

Appendix A
Metric Conversion Tables

Prefixes

Factor	Prefix	SI symbol
10^{18}	exa	E
10^{15}	peta	P
10^{12}	tera	T
10^9	giga	G
10^6	mega	M
10^3	kilo	k
10^{-3}	milli	m
10^{-6}	micro	μ
10^{-9}	nano	n
10^{-12}	pico	p
10^{-15}	femto	f
10^{-18}	atto	a

Greek alphabet

$A\ \alpha$	Alpha	$N\ \nu$	Nu
$B\ \beta$	Beta	$\Xi\ \xi$	Xi
$\Gamma\ \gamma$	Gamma	$O\ o$	Omicron
$\Delta\ \delta$	Delta	$\Pi\ \pi$	Pi
$E\ \epsilon$	Epsilon	$P\ \rho$	Rho
$Z\ \zeta$	Zeta	$\Sigma\ \sigma$	Sigma
$H\ \eta$	Eta	$T\ \tau$	Tau
$\Theta\ \theta$	Theta	$Y\ \upsilon$	Upsilon
$I\ i$	Iota	$\Phi\ \varphi$	Phi
$K\ \kappa$	Kappa	$X\ \chi$	Chi
$\Lambda\ \lambda$	Lambda	$\Psi\ \psi$	Psi
$M\ \mu$	Mu	$\Omega\ \omega$	Omega

Physical constants

Avogadro's number	$N_A = 6.023 \times 10^{23}$ atom/mol
Boltzmann's constant	$k = 1.38 \times 10^{-23}$ J/atom $^\circ$K
Gas constant	$R = 8.315$ J/$^\circ$K mol
Plank's constant	$h = 6.63 \times 10^{-34}$ J s

© Springer Nature Switzerland AG 2020
N. Perez, *Phase Transformation in Metals*,
https://doi.org/10.1007/978-3-030-49168-0

Length

$1\,m = 10^{10}\,\text{Å}$	$1\,\text{Å} = 10^{-10}\,m$
$1\,m = 10^{9}\,nm$	$1\,nm = 10^{-9}\,m$
$1\,m = 10^{6}\,\mu m$	$1\,\mu m = 10^{-6}\,m$
$1\,m = 10^{3}\,mm$	$1\,mm = 10^{-3}\,m$
$1\,m = 10^{2}\,cm$	$1\,cm = 10^{-2}\,m$
$1\,m = 3.28\,ft$	$1\,ft = 0.3049\,m$
$1\,m = 39.36\,in$	$1\,in = 0.0275\,m$
$1\,cm = 10\,mm$	$1\,mm = 0.10\,cm$
$1\,cm = 3.28 \times 10^{-2}\,ft$	$1\,ft = 30.48\,cm$
$1\,cm = 0.394\,in$	$1\,in = 2.54\,cm$
$1\,mm = 3.28 \times 10^{-3}\,ft$	$1\,ft = 304.8\,mm$
$1\,mm = 3.94 \times 10^{-2}\,in$	$1\,in = 25.4\,mm$

Area

$1\,m^2 = 10^{20}\,\text{Å}^2$	$1\,\text{Å}^2 = 10^{-20}\,m^2$
$1\,m^2 = 10^{18}\,nm^2$	$1\,nm^2 = 10^{-18}\,m^2$
$1\,m^2 = 10^{12}\,\mu m^2$	$1\,\mu m^2 = 10^{-12}\,m^2$
$1\,m^2 = 10^{6}\,mm^2$	$1\,mm^2 = 10^{-6}\,m^2$
$1\,m^2 = 10^{4}\,cm^2$	$1\,cm^2 = 10^{-4}\,m^2$
$1\,m^2 = 10.76\,ft^2$	$1\,ft^2 = 9.29 \times 10^{-2}\,m^2$
$1\,m^2 = 1.55 \times 10^{3}\,in^2$	$1\,in^2 = 6.45 \times 10^{-4}\,m^2$

Volume

$1\,m^3 = 10^{27}\,nm^3$	$1\,nm^3 = 10^{-27}\,m^3$
$1\,m^3 = 10^{18}\,\mu m^3$	$1\,\mu m^3 = 10^{-18}\,m^3$
$1\,m^3 = 10^{9}\,mm^3$	$1\,mm^3 = 10^{-9}\,m^3$
$1\,m^3 = 10^{6}\,cm^3$	$1\,cm^3 = 10^{-6}\,m^3$
$1\,m^3 = 35.20\,ft^3$	$1\,ft^3 = 2.83 \times 10^{-2}\,m^3$
$1\,m^3 = 6.10 \times 10^{4}\,in^3$	$1\,in^3 = 1.64 \times 10^{-5}\,m^3$
$1\,cm^3 = 3.53 \times 10^{-5}\,ft^3$	$1\,ft^3 = 2.83 \times 10^{4}\,cm^3$
$1\,cm^3 = 6.010 \times 10^{-2}\,in^3$	$1\,in^3 = 16.39\,cm^3$
$1\,cm^3 = 2.642 \times 10^{-4}\,gal\ (US)$	$1\,gal\ (US) = 3.79 \times 10^{3}\,cm^3$
$1\,liter\ (l) = 10^{3}\,cm^3$	$1\,cm^3 = 10^{-3}\,l$
$1\,liter\ (l) = 0.2642\,gal\ (US)$	$1\,gal\ (US) = 3.785\,l$

Mass

$1\,kg = 10^{3}\,g$	$1\,g = 10^{-3}\,kg$
$1\,kg = 2.205\,lb_m$	$1\,lb_m = 0.454\,kg$
$1\,g = 2.205 \times 10^{-3}\,lb_m$	$1\,lb_m = 454\,g$
$1\,g = 3.53 \times 10^{-2}\,oz$	$1\,oz = 28.35\,g$
$1\,lb_m = 16\,oz$	$1\,oz = 6.25 \times 10^{-2}\,lb_m$

Density	
$1 \, \text{kg/m}^3 = 10^{-3} \, \text{g/cm}^3$	$1 \, \text{g/cm}^3 = 10^3 \, \text{kg/m}^3$
$1 \, \text{kg/m}^3 = 0.0624 \, \text{lb}_m/\text{ft}^3$	$\text{lb}_m/\text{ft}^3 = 16.03 \, \text{kg/m}^3$
$1 \, \text{kg/m}^3 = 3.61 \times 10^{-5} \, \text{lb}_m/\text{in}^3$	$\text{lb}_m/\text{in}^3 = 2.77 \times 10^4 \, \text{kg/cm}^3$
$1 \, \text{g/cm}^3 = 0.0361 \, \text{lb}_m/\text{in}^3$	$\text{lb}_m/\text{in}^3 = 27.70 \, \text{g/cm}^3$

Force	
$1 \, \text{N} = 1 \, \text{kg m/s}^2$	$1 \, \text{dyne} = 1 \, \text{g cm/s}^2$
$1 \, \text{N} = 10^5 \, \text{dynes}$	$1 \, \text{dyne} = 10^{-5} \, \text{N}$
$1 \, \text{N} = 0.2248 \, \text{lb}_f$	$1 \, \text{lb}_f = 4.448 \, \text{N}$
$1 \, \text{dyne} = 2.248 \times 10^{-6} \, \text{lb}_f$	$1 \, \text{lb}_f = 4.448 \times 10^5 \, \text{dyne}$

Stress	
$1 \, \text{MPa} = 0.145 \, \text{ksi}$	$1 \, \text{ksi} = 6.895 \, \text{MPa}$
$1 \, \text{MPa} = 145 \, \text{psi}$	$1 \, \text{psi} = 6.90 \times 10^{-3} \, \text{MPa}$
$1 \, \text{MPa} = 0.1019 \, \text{kg}_f/\text{mm}^2$	$1 \, \text{kg}_f/\text{mm}^2 = 9.81 \, \text{MPa}$
$1 \, \text{MPa} = 7.25 \times 10^{-2} \, \text{Ton}_f/\text{in}^2$	$1 \, \text{Ton}_f/\text{in}^2 = 13.79 \, \text{MPa}$

Energy	
$1 \, \text{J} = 10^7 \, \text{ergs}$	$1 \, \text{erg} = 10^{-7} \, \text{J}$
$1 \, \text{J} = 6.24 \times 10^{18} \, \text{eV}$	$1 \, \text{eV} = 1.60 \times 10^{-19} \, \text{J}$
$1 \, \text{J} = 0.239 \, \text{cal}$	$1 \, \text{cal} = 4.184 \, \text{J}$
$1 \, \text{J} = 9.48 \times 10^{-4} \, \text{Btu}$	$1 \, \text{Btu} = 1054 \, \text{J}$
$1 \, \text{J} = 1.3558 \, \text{ft lb}_f$	$1 \, \text{ft lb}_f = 0.7376 \, \text{J}$
$1 \, \text{cal} = 3.97 \times 10^{-3} \, \text{Btu}$	$1 \, \text{Btu} = 252 \, \text{cal}$

Fracture toughness	
$1 \, \text{MPa}\sqrt{m} = 0.91 \, \text{ksi}\sqrt{in}$	$1 \, \text{ksi}\sqrt{in} = 1.10 \, \text{MPa}\sqrt{m}$
$1 \, \text{MPa}\sqrt{m} = 910 \, \text{psi}\sqrt{in}$	$1 \, \text{psi}\sqrt{in} = 1.10 \times 10^{-3} \, \text{MPa}\sqrt{m}$
$1 \, \text{ksi}\sqrt{in} = 10^3 \, \text{psi}\sqrt{in}$	$1 \, \text{psi}\sqrt{in} = 10^{-3} \, \text{ksi}\sqrt{in}$

Appendix B
Solution to Problems

B.1 Chapter 1

1.1 (b) $\lambda = 0.36603$
1.4 (b) $\det r_{ij} = 1/729$ nm^2
1.5 (b) $\theta = 30°$
1.6 (b) $\theta = 35.26°$
1.7 $d_{hkl} = a/\sqrt{h^2 + k^2 + l^2}$
1.8 $d_{111} = 0.166$ nm
1.9 $\|\mathbf{g}_{200}\|$, $n = 2$
1.10 (a) $p_1 = 2\pi/a$, $p_2 = 0$, $q_1 = -2\pi/(\sqrt{3}a)$, $q_2 = -4\pi/(\sqrt{3}a)$

   (b) $\theta = 30°$

1.11 (c) $d_{110} = 0.203$ nm
1.12 (c) $\|\mathbf{g}_{200}\|$, $n = 2$

B.2 Chapter 2

2.1 (a) $X (111) \left(\sqrt{13} \times \sqrt{13}\right) R13.9°$
2.2 (c) $W_N = X (100) \left(\sqrt{2} \times 2\sqrt{2}\right) R45°$
2.3 (b) $\left(\sqrt{5} \times \sqrt{5}\right)$
2.4 $W_N = X (hkl) \left(\sqrt{3} \times \sqrt{3}\right) R30°$
2.6 (a) $\theta = 26.6°$ and (c) $A = 3a^2$

© Springer Nature Switzerland AG 2020
N. Perez, *Phase Transformation in Metals*,
https://doi.org/10.1007/978-3-030-49168-0

B.3   Chapter 3

3.1 (a) $v_x = 42,200$ cm/s, $D_x = 2.11 \times 10^{-5}$ cm^2/s, $\tau = 1.18 \times 10^{-14}$ s

(b) $v_x = 157,890$ cm/s, $D_x = 7.89 \times 10^{-5}$ cm^2/s, $\tau = 3.17 \times 10^{-15}$ s

3.2 (a) $N = 5 \times 10^{17}$ atoms/m^2

(b) $x_j = 1.21$ µm

3.3 (b) $x = 1.07$ µm

3.5 (c) $t = 166.67$ s $= 2.78$ min

3.6 (a) $t = 15,721$ s $= 4.37$ h

(b) $t = 29,331$ s $= 8.15$ h

3.7 (a) $t = 13,304$ s $= 3.70$ h

(b) $t = 30,342$ s $= 8.43$ h

3.8 $t = 18,656$ s $= 5.18$ h

3.9 $C_b = 0.20\%$

3.10 $C_b = 0.20\%$

3.11 (a) $N = 5 \times 10^{17}$ atoms/m^2

(b) $x_j = 1.21$ µm

3.12 (a) $J_x = 7.42 \times 10_{-10}$ mol/(cm^2 s)

(b) $\partial^2 C / dx^2 = 1.93 \times 10^{-4}$ mol/cm^5

3.13 (a) $J_x = 2.77 \times 10_{-19}$ mol/(cm^2 s)

(b) $\partial^2 C / dx^2 = 9.6321 \times 10^{-5}$ mol/cm^5

$\partial C / dt = 1.93 \times 10^{-9}$ mol/(cm^3 s)

3.14 (a) $v_x = 48,728$ cm/s, $D_x = 2.44 \times 10^{-5}$ cm^2/s, $\tau = 1.02 \times 10^{-14}$ s

(b) $v_x = 182,320$ cm/s, $D_x = 9.12 \times 10^{-5}$ cm^2/s, $\tau = 2.74 \times 10^{-15}$ s

3.15 (a) $v_x = 59,679$ cm/s, $D_x = 2.98 \times 10^{-5}$ cm^2/s, $\tau = 8.39 \times 10^{-15}$ s

(b) $v_x = 223,30$ cm/s, $D_x = 1.12 \times 10^{-4}$ cm^2/s, $\tau = 2.23 \times 10^{-15}$ s

3.16 (a) $t = 15,214$ s $= 4.23$ h

(b) $t = 28,385$ s $= 7.88$ h

3.17 $C_b = 0.20\%$

3.18 (a) $J_x = 2.97 \times 10_{-9}$ mol/(cm^2 s)

(b) $\partial^2 C / dx^2 = 7.0516 \times 10^{-5}$ mol/cm^5

$\partial C / dt = 1.41 \times 10^{-9}$ mol/(cm^3 s)

3.19 $C_b = 0.30\%$

3.20 $C_b = 0.27\%$

B.4   Chapter 4

4.1 (a) $r_c = 0.67$ nm, $\Delta G_c = 1.32 \times 10^{-19}$ $J = 0.82$ eV

(b) $N_{cell} = 28$ cells and $N_{atom} = 112$ atoms

4.2 (a) $r_c = 1.01$ nm, $\Delta G_c = 2.03 \times 10^{-19}$ $J = 1.27$ eV

(b) $N_{cell} = 101$ cells and $N_{atom} = 404$ atoms

4.5 (b) $W = -\Delta G = 3$ kJ/mol

4.6 (a) $X_c = 1/2$ and $\Omega_c = 2$

(b) $\Delta G_c/RT = -0.19315$

4.7 (a) $n_s = 4 \times 10^{-15}$ clusters/cm^3

(b) $r_c = 1.06$ nm and $n_s \simeq 0$ clusters/cm^3

(c) $r_c = 0.665$ nm and $n_s = 10 \times 10^{10}$ clusters/cm^3

(d) $N_{cell} = 28$ cells and $N_{atom} = 112$ atoms

4.8 (a) $n_s = 3 \times 10^{-6}$ clusters/cm^3

(b) $r_c = 1.06$ nm and $n_s = 24$ clusters/cm^3

(c) $r_c = 1.01$ nm and $n_s = 1238$ clusters/cm^3

(d) $N_{cell} = 185$ cells and $N = 370$ atoms

4.9 (a) $n_s \simeq 0$ clusters/cm^3

(b) $r_c = 1.61$ nm and $n_s \simeq 0$ clusters/cm^3

(c) $N_{cell} = 78$ cells and $N = 312$ atoms

4.10 (a) $r_c = 0.665$ nm

(b) $\Delta G_c = 4.73 \times 10^{-19}$ J

(c) $N_{cell} = 28$ cells

(d) $N_{atom} = 112$ atoms

4.11 (a) $n_o = 8 \times 10^{28}$ atoms/m^3 and $I_o = 2.38 \times 10^{42}$ atoms/(m^3 s)

(b) $\Delta T = 252$ K and $T = 1358$ K

(c) $r_c = 1.17$ nm, $N_c = 572$ atoms
$\Delta G_{c,hom} = 1.02 \times 10^{-18}$ J and $\Delta G_{c,het} = 5.09 \times 10^{-19}$ J
$n_{s,hom} = 1$ clusters/m^3 and $n_{s,het} = 3 \times 10^{14}$ clusters/m^3
$I_{s,hom} = 1.53 \times 10^{13}$ clusters/m^3 and $I_{s,het} = 5 \times 10^{27}$ clusters/m^3

4.12 Process ab: $\Delta S_{ab} = 0.51$ J/K, $\Delta H_{ab} = 305.56$ J, $\Delta G_{ab} = 4.66$ J
Process bc: $\Delta S_{bc} = -8.02$ J/K, $\Delta H_{bc} = -4810$ J, $\Delta G_{bc} = 0$ J
Process cd: $\Delta S_{cd} = -0.49$ J/K, $\Delta H_{cd} = -293.61$ J, $\Delta G_{cd} = -4.51$ J
Process da: $\Delta S_{da} = -8$ J/K, $\Delta H_{da} = -4798.10$ J, $\Delta G_{da} = 0.15$ J
$S_{gen} = 0.15$ J/K

4.13 (a) $n_o = 9 \times 10^{28}$ atoms/m^3 and $I_o = 3 \times 10^{42}$ atoms/(m^3 s)

$\Delta H_f = 2.21 \times 10^5$ J/kg, $\Delta S_f = 122.03$ J/kg K, $\Gamma_{ls} = 2.12 \times 10^{-7}$ K m

(b) $\Delta T = 338\ K$ and $T = 1473\ K$
(c) $r_c = 1.25$ nm, $N_c = 765\ atoms$
  $\Delta G_{c,hom} = 1.35 \times 10^{-18}$ J and $\Delta G_{c,het} = 6.77 \times 10^{-19}$ J
  $n_{s,hom} = 1$ clusters/m^3 and $n_{s,het} = 3 \times 10^{14}$ clusters/m^3
  $I_{s,hom} = 2 \times 10^{13}$ clusters/m^3 s and $I_{s,het} = 6 \times 10^{27}$ clusters/m^3 s

4.14 $T_f = 1279.40\ K = 1006.40\,°C$
4.15 (b) $W = -\Delta G = 3.90$ kJ/mol
4.16 (a) $\Gamma_m = 7969.60$ J/mol

   (b) $Zn = 1.0491$
   (c) $a_{Zn} = 0.8393$ and $a_{Cd} = 0.4306$
   (d) $\Delta G_m = -2053$ J/mol, $\Delta H_m = 1286.20$ J/mol, $\Delta S_m = 4.1740$ J/mol K

4.17 (b) $\gamma_{Zn} = 0.14621$ and $a_{Zn} = 0.04386$
       $\gamma_{Cu} = 0.70247$ and $a_{Cu} = 0.49173$
4.18 (a) $\gamma_{Al} = 0.61199$ and $a_{Al} = 0.37$

   (b) $\Omega_m = -47,576$ J/mol, $\gamma_{Al} = 0.51332$, $a_{Al} = 0.31$
   (c) $\gamma_{Al} = 0.0010722$, $a_{Al} = 0.000054$

4.19 (a) $r_c = 0.71$ nm

   (b) $\Delta G_c = 5.38 \times 10^{-19}$ J
   (c) $N_{cell} = 34$ cells
   (d) $N_{atom} = 136$ atoms

4.21 (b) $\Omega_m = -18,543$ J/mol and $a_{Al} = 0.78$
4.24 $a_{Cu} = 0.40833$
4.25 (a) $ln(\gamma_{Ti}) = $ Area $= 0.51534$

   (b) $a_{Ti} = 0.78687$

4.26 (a) $ln(\gamma_{Ti}) = $ Area $= 0.78159$

   (b) $a_{Ti} = 1.3765$

B.5   Chapter 5

5.2 $t_{spheroid} = 4.43$ min and $t_{sphere} = 0.87$ min
5.3 (a) $t = 3.21$ min

   (b) $t = 1.33$ min

5.4 Yes, $n = 2 \pm 0.05$
5.5 $T_i = 1387\ K = 1114\,°C$ and $q_k = 36,125$ W/m^2
5.6 (a) $t = 1.43$ s

(b) $= 7 \times 10^3$ m/s

(c) $(\partial T/\partial x)_{x_s} = -34,859$ K/m and $q_{x_s} = 11.50$ MW/m^2

5.7 (a) $t = 15.97$ s and $dx_s/dt = 2253.60$ mm/h

(b) Plots: $T_m = 25\,°C + (1042.8\,°C)\,\mathrm{erfc}(-47.966x)$ (mold)
$T_s = 1067.80\,°C + (699.53°C)\,\mathrm{erf}(34.286x)$ (solid)

5.8 (a) $t = 60.62$ s and $dx_s/dt = 1188$ mm/h

(b) Plots: $T_m = 25\,°C + (1036.8°C)\,\mathrm{erfc}(-24.619x)$ (mold)
$T_s = 1067.80°C + (695.35\,°C)\,\mathrm{erf}(17.598x)$ (solid)

5.9 The cube solidifies faster because $t_{cube} < t_{cyl} < t_{sph}$
$t_{cube} = 41$ min, $t_{cyl} = 48$ min, $t_{sph} = 63$ min

5.10 (b) $t_{min} = 3.56$ min

5.11 $t_{cylinder} = 3.24$ min

5.12 $t_{cube} = 4.32$ min

5.13 (a) $t = 6.67$ s and $((dx_s)/(dt)) = 5400$ mm/h

(b) Plots: $T_m = 25\,°C + (654.40\,°C)\,\mathrm{erfc}(-74.22x)$ (mold)
$T_s = 679.40\,°C + (73.31\,°C)\,\mathrm{erf}(17.392x)$ (solid)

5.16 (a) $T_i = 482.80\,°C$, $t = 24.53$ s, $ds(t)/dt = 8.15 \times 10^{-4}$ m/s

(b) Plots: $T_m = 25\,°C + (457.80\,°C)\,\mathrm{erfc}(-19.523x)$ (mold)
$T_s = 482.80\,°C + (349.54\,°C)\,\mathrm{erf}(12.118x)$ (solid)
$T_l = 700\,°C + (118.98\,°C)\,\mathrm{erfc}(17.002x)$ (liquid)

5.17 $t = 3.31$ min

B.6 Chapter 6

6.1 $t_{plate} = 6.27$ s, $t_{cylinder} = 1.15$ s, $t_{sphere} = 0.47$ s

6.2 $t = 355.87$ s $= 5.93$ min, $ds(t)/dt = 114.12$ mm/h
$\partial T \partial r = -58,185$ K/m, $\partial T/\partial t = -63.17\,°C/s$

6.3 (a) $F_o = 0.27$ and $t = 2.82$ s, (b) $F_o = 0.39$ and $t = 4.07$ s

(c) $S_t = 0.32$

6.4 $F_o = 0.52$

6.5 $F_o = 0.23$

6.6 (a) Heat transfer:approach
Slab: $t = 47.39$ min and $ds(t)/dt = 12.67$ mm/h
Cylinder: $t = 7.67$ min and $ds(t)/dt = 46.80$ mm/h
Sphere: $t = 3.08$ min and $ds(t)/dt = 80.28$ mm/h

(b) The Chvorinov's rule

Slab: $t = 47.39\,\mathrm{min}$
Cylinder: $t = 11.85\,\mathrm{min}$
Sphere: $t = 5.26\,\mathrm{min}$

(c) $T_m = 25\,^\circ\mathrm{C} + (1057.50\,^\circ\mathrm{C})\,\mathrm{erfc}(-18.329x)$
$T_s = 1082.50\,^\circ\mathrm{C} + (26.322\,^\circ\mathrm{C})\,\mathrm{erfc}(0.8424x)$

6.7 (a) $v_p = 167.40\,\mathrm{mm/h}$ and $t = 17.92\,\mathrm{h}$

(b) $v_r = 275.04\,\mathrm{mm/h}$ and $t = 0.73\,\mathrm{h}$

6.8 (a) Heat transfer:approach
Slab: $t = 48.62\,\mathrm{min}$ and $ds(t)/dt = 12.35\,\mathrm{mm/h}$
Cylinder: $t = 7.83\,\mathrm{min}$ and $ds(t)/dt = 45.72\,\mathrm{mm/h}$
Sphere: $t = 3.14\,\mathrm{min}$ and $ds(t)/dt = 78.84\,\mathrm{mm/h}$

(b) The Chvorinov's rule
Slab: $t = 48.62\,\mathrm{min}$, $ds(t)/dt$—not computed
Cylinder: $t = 12.16\,\mathrm{min}$, $ds(t)/dt$—not computed
Sphere: $t = 5.40\,\mathrm{min}$, $ds(t)/dt$—not computed

(c) $T_m = 25\,^\circ\mathrm{C} + (1057.60\,^\circ\mathrm{C})\,\mathrm{erfc}(-18.094x)$

$T_s = 1082.50\,^\circ\mathrm{C} + (21.33\,^\circ\mathrm{C})\,\mathrm{erfc}(0.83161x)$

6.9 (a) $v_p = 167.40\,\mathrm{mm/h}$ and $t = 11.95\,\mathrm{h}$

(b) $v_r = 275.04\,\mathrm{mm/h}$ and $t = 0.73\,\mathrm{h}$

6.10 (a) $v_p = 167.40\,\mathrm{mm/h}$ and $t = 8.96\,\mathrm{h}$

(b) $v_r = 275.04\,\mathrm{mm/h}$ and $t = 0.73\,\mathrm{h}$

6.11 (a) $F_o = 0.70$, (b) $B_i = 6.67$

(c) $h = 22,296\,\mathrm{W/(m^2\,K)}$
(d) $q_{x=0} = 23.59\,\mathrm{MW/m^2}$

6.12 (a) $F_o = 0.92$, (b) $B_i = 5.71$

(c) $h = 19,492\,\mathrm{W/(m^2\,K)}$
(d) $q_{x=0} = 20.62\,\mathrm{MW/m^2}$

6.13 (a) $F_o = 1.16$

(b) $B_i = 5.14$
(c) $h = 17,335\,\mathrm{W/(m^2\,K)}$
(d) $q_{x=0} = 18.34\,\mathrm{MW/m^2}$

6.14 (a) $F_o = 1.29$

(b) $B_i = 4.90$
(c) $h = 16,450\,\mathrm{W/(m^2\,K)}$
(d) $q_{x=0} = 17.40\,\mathrm{MW/m^2}$

6.15 (a) $F_o = 1.43$

  (b) $B_i = 4.65$
  (c) $h = 15,620 \, \text{W/(m}^2\,\text{K)}$
  (d) $q_{x=0} = 16.53 \, \text{MW/m}^2$

6.16 (a) $f_s = 1$—Complete

  (b) $s(t) = 0.04 \, \text{m}$ and $t = 6.65 \, \text{s}$
  (c) $q_{x=0} = 16.53 \, \text{MW/m}^2$

6.17 (a) $F_o = 1.4348$

  (b) $B_i = 4.0634$
  (c) $h = 19,409 \, \text{W/(m}^2\,\text{K)}$
  (d) $q_{x=0} = 20.535 \, \text{MW/m}^2$

6.18 $F_o = 0.52$

6.19 (a) $v_p = 167.40 \, \text{mm/h}$ and $t = 16.73 \, \text{h}$

  (b) $v_r = 275.04 \, \text{mm/h}$ and $t = 0.73 \, \text{h}$

6.20 (a) $v_p = 6.55 \, \text{mm/h}$ and $t = 305.34 \, \text{h}$

  (b) $v_p = 8 \, \text{mm/h}$ and $t = 250 \, \text{h}$

6.21 (a) $\partial T_l / \partial z < 549 \, \text{K/m}$

  (b) $v_p = 6.34 \, \text{mm/h}$

6.22 (a) $\partial T_l / \partial z < 1098 \, \text{K/m}$

  (b) $v_p = 5.44 \, \text{mm/h}$

B.7   Chapter 7

7.1 $k_{o,b} = 0.23$, $k_{o,c} = 0.31$ and $k_{o,b} = 0.33$

7.2 (a) $[\partial T_l(x)/\partial x]_{c,x=0} = 10^6 \, \text{K/m}$

  (b) A dendritic microstructure because $10^4 \, \text{K/m} < 10^6 \, \text{K/m}$

7.3 (a) $[\partial T_l(x)/\partial x]_{c,x=0} = 1.63 \times 10^6 \, \text{K/m}$

  (b) A dendritic microstructure due to $10^4 \, K/m < 1.63 \times 10^6 \, K/m$

7.4 (a) $[\partial T_l(x)/\partial x]_{c,x=0} = 2.17 \times 10^6 \, \text{K/m}$

  (b) A dendritic microstructure due to $10^4 \, \text{K/m} < 2.17 \times 10^6 \, \text{K/m}$

7.5 $[T_l(x)/dx]_{c,x=0}^{(3Cu)} = 1.00 \times 10^6 \, ^\circ\text{C/m}$
  $[T_l(x)/dx]_{c,x=0}^{(5Cu)} = 1.74 \times 10^5 \, ^\circ\text{C/m}$

For stability criterion, $[T_l(x)/dx]_{a,x=0}^{(3Cu)} \geq 1.00 \times 10^6\ {}^\circ\text{C/m}$

$$[T_l(x)/dx]_{a,x=0}^{(5Cu)} \geq 1.74 \times 10^5\ {}^\circ\text{C/m}$$

7.6 $[T_l(x)/dx]_{c,x=0}^{(3Cu)} = 2.00 \times 10^6\ {}^\circ\text{C/m}$

$[T_l(x)/dx]_{c,x=0}^{(5Cu)} = 3.48 \times 10^5\ {}^\circ\text{C/m}$

For stability criterion, $[T_l(x)/dx]_{a,x=0}^{(3Cu)} \geq 2.00 \times 10^5\ {}^\circ\text{C/m}$

$$[T_l(x)/dx]_{a,x=0}^{(5Cu)} \geq 3.48 \times 10^5\ {}^\circ\text{C/m}$$

7.7 (a) $t_1 = 3.52 \times 10^{-5}$ s—Eq. (7.42)

$t_2 = 2.57 \times 10^{-5}$ s—Eq. (7.50)

$t_3 = 6.67 \times 10^{-5}$ s—Eqs. (7.52) and (7.55)

(b) $[ds(t)/dt]_1 = 567.84$ mm/s—Eq. (7.52)

$[ds(t)/dt]_2 = 626.71$ mm/s—Eq. (7.36e)

$[ds(t)/dt]_3 = 275.91$ mm/s—Eq. (7.36e)

(c) $[\partial T_s/\partial t]_1 = -4.78 \times 10^6\ {}^\circ\text{C/s}$—Eq. (7.39b)

$[\partial T_s/\partial t]_2 = -7.23 \times 10^6\ {}^\circ\text{C/s}$—Eq. (7.39b)

$[\partial T_s/\partial t]_3 = -2.88 \times 10^6\ {}^\circ\text{C/s}$—Eq. (7.39b)

(d) $q_{s,1} = 1.59$ GW/(m^2 s)—Eq. (7.37) at $T_i = 475.68\,{}^\circ\text{C}$

$q_{s,2} = 1.86$ GW/(m^2 s)—Eq. (7.37) at $T_i = 475.68\,{}^\circ\text{C}$

$q_{s,3} = 1.68$ GW/(m^2 s)—Eq. (7.37) at $T_i = 304.76\,{}^\circ\text{C}$

7.8 (a) $t_1 = 4.96$ s—Eq. (7.42)

$t_2 = 3.61$ s—Eq. (7.50)

$t_3 = 9.37$ s—Eqs. (7.52) and (7.55)

(b) $[ds(t)/dt]_1 = 1.51$ mm/s—Eq. (7.52)

$[ds(t)/dt]_2 = 1.67$ mm/s—Eq. (7.36e)

$[ds(t)/dt]_3 = 0.74$ mm/s—Eq. (7.36e)

(c) $[\partial T_s/\partial t]_1 = -49.40\,{}^\circ\text{C/s}$—Eq. (7.39b) at $T_i = 475.68\,{}^\circ\text{C}$

$[\partial T_s/\partial t]_2 = -51.48\,{}^\circ\text{C/s}$—Eq. (7.39b) at $T_i = 475.68\,{}^\circ\text{C}$

$[\partial T_s/\partial t]_3 = -8.730\,{}^\circ\text{C/s}$—Eq. (7.39b) at $T_i = 304.76\,{}^\circ\text{C}$

(d) $q_{s,1} = 4.23 \times 10^6$ W/(m^2 s)—Eq. (7.37) at $T_i = 475.68\,{}^\circ\text{C}$

$q_{s,2} = 4.96 \times 10^6$ W/(m^2 s)—Eq. (7.37) at $T_i = 475.68\,{}^\circ\text{C}$

$q_{s,3} = 4.48 \times 10^6$ W/(m^2 s)—Eq. (7.37) at $T_i = 304.76\,{}^\circ\text{C}$

7.9 (a) $t_1 = 2.20$ s—Eq. (7.42)

$t_2 = 1.61$ s—Eq. (7.50)

$t_3 = 4.17$ s—Eqs. (7.52) and (7.55)

(b) $[ds(t)/dt]_1 = 2.27$ mm/s—Eq. (7.52)

$[ds(t)/dt]_2 = 2.50$ mm/s—Eq. (7.36e)

$[ds(t)/dt]_3 = 2.59$ mm/s—Eq. (7.36e)

(c) $[\partial T_s/\partial t]_1 = -82.45\,{}^\circ\text{C/s}$—Eq. (7.39b) at $T_i = 475.68\,{}^\circ\text{C}$

$[\partial T_s/\partial t]_2 = -115.32\,{}^\circ\text{C/s}$—Eq. (7.39b) at $T_i = 475.68\,{}^\circ\text{C}$

$[\partial T_s/\partial t]_3 = -19.61\,{}^\circ\text{C/s}$—Eq. (7.39b) at $T_i = 304.76\,{}^\circ\text{C}$

(d) $q_{s,1} = 6.35 \times 10^6$ W/(m^2 s)—Eq. (7.37) at $T_i = 475.68\,^\circ$C
$\;q_{s,2} = 7.42 \times 10^6$ W/(m^2 s)—Eq. (7.37) at $T_i = 475.68\,^\circ$C
$\;q_{s,3} = 6.72 \times 10^6$ W/(m^2 s)—Eq. (7.37) at $T_i = 304.76\,^\circ$C

7.10 (a) $t_1 = 2.57$ s—Eq. (7.42)

(b) $[ds(t)/dt]_1 = 1.95$ mm/s—Eq. (7.52)
(c) $[\partial T_s/\partial t]_1 = -46.16\,^\circ$C/s—Eq. (7.39b) at $T_i = 571.90\,^\circ$C
(d) $q_{s,1} = 2.28 \times 10^6$ W/(m^2 s)—Eq. (7.37) at $T_i = 571.90\,^\circ$C

B.8   Chapter 8

8.1 (a) $\Delta H_s = \Delta H_f^* = 908.74$ kJ/kg

(b) $k_o^\alpha = 0.1310$ and $k_o^\beta = 0.1442$
(c) $m_{l\alpha} = -6.5873\,^\circ$C/% and $m_{l\beta} = 9.5767\,^\circ$C/%
(d) $K_C = 7.2675 \times 10^{12}$ K s/m^2 and $K_R = 8.5454 \times 10^{-8}$ K m
(e) $\lambda_c = 10.84 \times 10^{-3}$ μm at $\upsilon_z = 100$ μm/s
$\;\Delta T_c = 15.76$ K at $\upsilon_z = 100$ μm/s
(f) $\lambda_c^2 \upsilon_z = 1.1751 \times 10^{-2}$ μm^3/s at $\upsilon_z = 100$ μm/s
$\;\Delta T_c/\sqrt{\upsilon_z} = 1.5760$ K$(\mu$m/s$)^{-1/2}$ at $\upsilon_z = 100$ μm/s
(g) $\Delta T_c = (0.17085\,\text{K}\,\mu\text{m})/\lambda_c$
(h) $\Delta T = \left(7.2675\,\text{K s}/\mu\text{m}^2\right)\upsilon_z\lambda + \left(8.5454 \times 10^{-2}\,\text{K}\,\mu\text{m}\right)/\lambda$
(i) $\Delta G_\upsilon = \Delta T \Delta S_f = 0$, no phase transformation

8.2 (a) $\Delta H_s = \Delta H_f^* = 908.74$ kJ/kg

(b) $k_o^\alpha = 0.1310$ and $k_o^\beta = 0.1442$
(c) $m_{l\alpha} = -6.8033\,^\circ$C/% and $m_{l\beta} = 9.5330\,^\circ$C/%
(d) $K_C = 16.6010 \times 10^{12}$ K s/m^2 and $K_R = 0.69282 \times 10^{-8}$ K m
(e) $\lambda_c = 20.43 \times 10^{-3}$ μm at $\upsilon_z = 100$ μm/s
$\;\Delta T_c = 67.83$ K at $\upsilon_z = 100$ μm/s
(f) $\lambda_c^2 \upsilon_z = 4.1738 \times 10^{-2}$ μm^3/s at $\upsilon_z = 100$ μm/s
$\;\Delta T_c/\sqrt{\upsilon_z} = 6.7830$ K$(\mu$m/s$)^{-1/2}$ at $\upsilon_z = 100$ μm/s
(g) $\Delta T_c = (1.3884\,\text{K}\,\mu\text{m})/\lambda_c$
(h) $\Delta T = \left(16.6010\,\text{K s}/\mu\text{m}^2\right)\upsilon_z\lambda + (0.69282\,\text{K}\,\mu\text{m})/\lambda$
(i) $\Delta G_\upsilon = \Delta T \Delta S_f = 0$, no phase transformation

8.3 $\Delta T_c = (0.11605\,\text{K}\,\mu\text{m})/\lambda_c$
$\;\Delta T = \left(4.2810\,\text{K s}/\mu\text{m}^2\right)\upsilon_z\lambda + (0.058008\,\text{K}\,\mu\text{m})/\lambda$

8.4 (a) $k_o^\alpha = 0.17169$ and $k_o^\beta = 0.58385$

(b) $m_{l\alpha} = -3.37\,^\circ$C/% and $m_{s\alpha} = -19.65\,^\circ$C/%
$\;m_{l\beta} = -28.12\,^\circ$C/% and $m_{s\beta} = 278.40\,^\circ$C/%
(c) $\Delta H_s = 502.60$ kJ/kg
$\;\Delta S_f = 0.92$ kJ/kg K

(d) $\Delta T_c = 2.15$ K

(e) $G_{\lambda_c} = 3.07$ K/μm

8.5 (a) $F_o = -0.38225$ and $F_n = 6.9246 \times 10^{-2}$

(b) $C_l(x, z) = 0.2 - 4.8643 \times 10^{-2} exp(-153.77z) - 0.38225 exp(-34.014z)$

8.6 (a) $k_o^{\alpha} = 0.92$

(b) $m_{l\alpha} = -2.95\,°C/\%$ and $m_{s\alpha} = -36.05\,°C/\%$

(c) $\Delta H_s = \Delta H_f^* = 433.85$ kJ/kg

$\Delta S_f = 0.68$ kJ/kg K

(d) $\Delta T_c = 0.30$ K and $\lambda_c = 9.12$ μm

(e) $G_{\lambda_c} = 3.29 \times 10^{-2}$ K/μm

8.7 (a) $k_o^{\alpha} = 0.81$

(b) $m_{l\alpha} = -2.33\,°C/\%$ and $m_{s\alpha} = -7.87\,°C/\%$

(c) $\Delta H_s = \Delta H_f^* = 43.38$ kJ/kg

$\Delta S_f = 0.24$ kJ/kg K

(d) $\Delta T_c = 0.73$ K and $\lambda_c = 1.59$ μm

(e) $G_{\lambda_c} = 0.46 \times 10^{-2}$ K/μm

8.8 (a) $\Delta H_s = 185.18$ kJ/kg

(b) $m_{l\alpha} = -9.83\,°C/\%$ and $m_{l\beta} = 5.75\,°C/\%$

(c) $k_o^{\alpha} = 0.50$ and $k_o^{\beta} = 0.03$

(d) $K_C = 4.9150$ K s/μm^2 and $K_R = 0.21638$ K μm

(e) $\Delta T = (491.50$ K s/μm$^2)\lambda + (0.21638$K μm)/λ

(f) $\Delta T_c = 20.63$ K, $\lambda_c = 2.10 \times 10^{-2}$ μm

$[d\Delta T/d\lambda]_{\lambda=\lambda_c} = 982.38$ K/μm

B.9   Chapter 9

9.1 $\gamma_{\alpha\alpha}/\gamma_{\alpha\beta} = 1$

9.2 $\mu = 0.0704$ nm

9.3 $\epsilon_{[100]_\gamma} = \epsilon_{[010]_\gamma} = 0.13$ and $\epsilon_{[001]_\gamma} = ((a_\alpha - a_\gamma)/(a_\gamma)) = -0.20$

$\mu = 0.07$ nm

$$B = \begin{pmatrix} \epsilon_{[100]_\gamma} & 0 & 0 \\ 0 & \epsilon_{[010]_\gamma} & 0 \\ 0 & 0 & \epsilon_{[001]_\gamma} \end{pmatrix} = \begin{pmatrix} 0.13 & 0 & 0 \\ 0 & 0.13 & 0 \\ 0 & 0 & -0.20 \end{pmatrix}$$

(b) Orientation relationships

$$[001]_\gamma \parallel [001]_{\alpha'} \qquad \left[0\bar{1}1\right]_\gamma \parallel [100]_{\alpha'} \qquad [111]_\gamma \parallel [010]_{\alpha'}$$

9.4   Miller indices for the crystallographic directions

$$[101]_\gamma \rightarrow [111]_{\alpha'}, \quad [110]_\gamma \rightarrow [100]_{\alpha'} \quad \& \quad [112]_\gamma \rightarrow [011]_{\alpha'}$$

9.5 $\epsilon_{[100]_\gamma} = \epsilon_{[010]_\gamma} = 0.13$ and $\epsilon_{[001]_\gamma} = ((a_\alpha - a_\gamma)/(a_\gamma)) = -0.20$
$\mu = 0.072\,\text{nm}$

$$B = \begin{pmatrix} \epsilon_{[100]_\gamma} & 0 & 0 \\ 0 & \epsilon_{[010]_\gamma} & 0 \\ 0 & 0 & \epsilon_{[001]_\gamma} \end{pmatrix} = \begin{pmatrix} 0.13 & 0 & 0 \\ 0 & 0.13 & 0 \\ 0 & 0 & -0.20 \end{pmatrix}$$

(b) Orientation relationships

$$[001]_\gamma \parallel [001]_{\alpha'} \qquad \left[0\bar{1}1\right]_\gamma \parallel [100]_{\alpha'} \qquad [111]_\gamma \parallel [010]_{\alpha'}$$

9.6 $\epsilon_{[100]_\gamma} = \epsilon_{[010]_\gamma} = 0.14$ and $\epsilon_{[001]_\gamma} = ((a_\alpha - a_\gamma)/(a_\gamma)) = -0.20$
$\mu = 0.072\,\text{nm}$

$$B = \begin{pmatrix} \epsilon_{[100]_\gamma} & 0 & 0 \\ 0 & \epsilon_{[010]_\gamma} & 0 \\ 0 & 0 & \epsilon_{[001]_\gamma} \end{pmatrix} = \begin{pmatrix} 0.14 & 0 & 0 \\ 0 & 0.14 & 0 \\ 0 & 0 & -0.20 \end{pmatrix}$$

(b) Orientation relationships

$$[001]_\gamma \parallel [001]_{\alpha'} \qquad \left[0\bar{1}1\right]_\gamma \parallel [100]_{\alpha'} \qquad [111]_\gamma \parallel [010]_{\alpha'}$$

9.7   (a) $r_c = 4.63\,\text{nm}$, $h_c = 0.17\,\text{nm}$, $\Delta G_{c,hom} = 4.20\,\text{eV}$

   (b) $r_c = 8.88\,\text{nm}$, $h_c = 0.23\,\text{nm}$, $\Delta G_{c,hom} = 15.42\,\text{eV}$

9.8   (a) $r_c = 4.63\,\text{nm}$, $h_c = 0.17\,\text{nm}$, $\Delta G_{c,hom} = 4.21\,\text{eV}$

   (b) $r_c = 10.42\,\text{nm}$, $h_c = 0.25\,\text{nm}$, $\Delta G_{c,hom} = 21.29\,\text{eV}$

9.9   (a) $t = 4.80\,\text{min}$

   (b) $t = 9.75\,\text{min}$

9.10   At $\lambda x_c\,(x/x_c) = 0$,

   (a) $F_o = 5.0997 \ln\left[(6.060\,6 \sin\phi)\,(0.44282 + 1.9204 \sin\phi)^{-1}\right]$
   (b) $F_o = 5.0997 \ln\left[(2.7991)\,(0.442\,82 + 0.22173/B_i)^{-1}\right]$

9.11   (a) $r_c = 5.02\,\text{nm}$, $h_c = 0.18\,\text{nm}$, $\Delta G_{c,hom} = 5.28\,\text{eV}$

   (b) $r_c = 11.56\,\text{nm}$, $h_c = 0.27\,\text{nm}$, $\Delta G_{c,hom} = 27.90\,\text{eV}$

9.12 (a) $t = 2.07$ min

    (b) $t = 5.25$ min

9.13 (a) $t = 32.43$ min

    (b) $t = 1.07$ min

B.10   Chapter 10

10.1 $P_{O_2} = 9.58 \times 10^{-16}$ atm

10.2 (a) $P_{O_2} = 1.19 \times 10^{-5}$ atm

    (b) $\Delta S^o = -d\Delta G^o/dt) = -0.24615$ J/mol K
       $\Delta H^o = -369.23$ J/mol

10.3 $P_{O_2} \simeq 10^{-5}$ atm
     $P_{CO}/P_{CO_2} \simeq 10^{-4}$ atm
     $P_{H_2}/P_{H_2O} \simeq 5.5 \times 10^{-4}$ atm
     $\Delta G^o - 115$ kJ/mol of O_2

10.4 $[g]_H = (6.5016 \times 10^{-3}\%)exp(-5658.60/T)$
     $[g]_H = 0.00008\%$ at $T = 1000\,°C = 1273$ K

10.5 (a) $[g]_H = A_x exp[-\Delta G_c/(RT_1)]$
       $\Delta G_c = 46,924$ J/mol and $A_x = 6.4535 \times 10^{-3}\%$

    (b) $[g]_{H,T_3} = 0.00008\%$

10.6 (a) $C_H = 8.58$ ppm $= 8.58 \times 10^{-4}$ wt%

    (b) $Q_{conv} = 76$ GJ and $Q_{rad} = 76$ GJ

10.7 $T_4 = 334.73\,K = 61.73\,°C$

10.8 $q_x = -25.60$ MW/m^2

10.9 $\sigma_f = 64$ MPa

10.10 $G_{IC} = 4.77$ J/m^2

10.11 (a) $P_{O_2} = 1.46 \times 10^{-3}$ atm

    (b) $\Delta S^o = -0.24615$ J/mol K and $\Delta H^o = -369.23$ J/mol

10.12 (a) $C_H = 8.58$ ppm $= 8.58 \times 10^{-4}$ wt%

    (b) $Q_{conv} = 76$ GJ and $Q_{rad} = 76$ GJ

10.13 (a) $P_{O_2} = 2.68 \times 10^{-39}$ atm

10.14 $\Delta G^o = 0.94$ MJ/mol

10.15 (a) $T = 1565\,K = 1292\,°C$
       $\Delta G^o = 0.78$ MJ/mol

    (b) $T \simeq 1292\,°C$
       $\Delta G^o \simeq 0.78$ MJ/mol

10.16 (a) $T = 1155\,K = 882\,°C$
$\Delta G^o = 0.66\,\text{MJ/mol}$

(b) $T \simeq 882\,°C$
$\Delta G^o \simeq 0.66\,\text{MJ/mol}$

Index

© Springer Nature Switzerland AG 2020
N. Perez, *Phase Transformation in Metals*,
https://doi.org/10.1007/978-3-030-49168-0

Printed in the United States
by Baker & Taylor Publisher Services